Fracture of Structural Materials

WILEY SERIES ON
THE SCIENCE AND TECHNOLOGY OF MATERIALS
Advisory Editors:
J. H. Hollomon, J. E. Burke, B. Chalmers, R. L. Sproull, A. T. Tobolsky

---

FRACTURE OF STRUCTURAL MATERIALS
  A. S. Tetelman and A. J. McEvily, Jr.
ORGANIC SEMICONDUCTORS
  F. Gutmann and L. E. Lyons
INTERMETALLIC COMPOUNDS
  J. H. Westbrook, editor
THE PHYSICAL PRINCIPLES OF MAGNETISM
  Allan H. Morrish
FRICTON AND WEAR OF MATERIALS
  Ernest Rabinowicz
HANDBOOK OF ELECTRON BEAM WELDING
  R. Bakish and S. S. White
PHYSICS OF MAGNETISM
  Sōshin Chikazumi
PHYSICS OF III-V COMPOUNDS
  Otfried Madelung (translation by D. Meyerhofer)
PRINCIPLES OF SOLIDIFICATION
  Bruce Chalmers
APPLIED SUPERCONDUCTIVITY
  Vernon L. Newhouse
THE MECHANICAL PROPERTIES OF MATTER
  A. H. Cottrell
THE ART AND SCIENCE OF GROWING CRYSTALS
  J. J. Gilman, editor
SELECTED VALUES OF THERMODYNAMIC PROPERTIES OF METALS AND ALLOYS
  Ralph Hultgren, Raymond L. Orr, Philip D. Anderson and Kenneth K. Kelly
PROCESSES OF CREEP AND FATIGUE IN METALS
  A. J. Kennedy
COLUMBIUM AND TANTALUM
  Frank T. Sisco and Edward Epremian, editors
MECHANICAL PROPERTIES OF METALS
  D. McLean
THE METALLURGY OF WELDING
  D. Séférian (translation by E. Bishop)
THERMODYNAMICS OF SOLIDS
  Richard A. Swalin
TRANSMISSION ELECTRON MICROSCOPY OF METALS
  Gareth Thomas
PLASTICITY AND CREEP OF METALS
  J. D. Lubahn and R. P. Felgar
INTRODUCTION TO CERAMICS
  W. D. Kingery
PROPERTIES AND STRUCTURE OF POLYMERS
  Arthur V. Tobolsky
PHYSICAL METALLURGY
  Bruce Chalmers
FERRITES
  J. Smit and H. P. J. Wijn
ZONE MELTING, SECOND EDITION
  William G. Pfann
THE METALLURGY OF VANADIUM
  William Rostoker

Fracture of
Structural Materials

A. S. TETELMAN
Associate Professor of Materials Science
    and Engineering Mechanics
Stanford University

A. J. McEVILY, JR.
Scientific Laboratory
Ford Motor Company

JOHN WILEY & SONS, INC.   New York · London · Sydney

Preface

During the last decade the rapid growth in a number of related fields has increased our knowledge of the fracture process and has tended to make structural design more reliable and efficient. At the same time it appears that a large communications gap has developed between the engineers who must design with materials of a given toughness and the metallurgists and materials scientists who are trying to improve them. Generally, the engineers do not seem to have appreciated that an understanding of the various toughness parameters requires some insight into the physical aspects of the fracture process. The metallurgists and materials scientists, in turn, do not seem to appreciate that data obtained from simple tests on small laboratory-size specimens provide only partial insight into the fracture mechanisms in large structures. These differences in approach to the general problem of fracture are reflected in the different *dimensional ranges* of interest to the two groups. Engineers have been concerned in *macroscopic parameters* such as the plate width W or thickness t, the flaw depth $2c$ and tip radius ρ, the plastic zone size R, and the crack opening displacement (C.O.D.). Metallurgists have been concerned with *microscopic parameters* such as the grain size $2d$, the spacing of dispersed particles λ, and the Burgers vector b. Typical values of these parameters are illustrated.

We believe that the point has now been reached at which an understanding of the critical aspects of fracture has been achieved in each dimensional range. It is necessary, therefore, to synthesize these macroscopic and microscopic aspects if further advances in structural and materials design (e.g., laminates and composites) are to be made. Accordingly we have discussed the problem of fracture from both points of view and whenever possible have attempted to bridge the gap between them.

Consequently, this book has been divided into five parts. Part 1 deals with the engineering (macroscopic or continuum) aspects of fracture which are particularly relevant to persons interested in de-

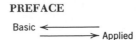

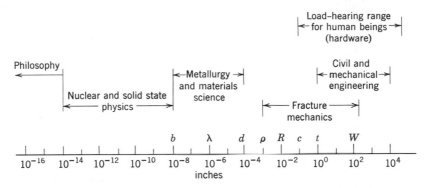

Typical dimensions of interest in structural engineering (inches).

signing against fracture in materials presently available. Part 2 deals with the scientific (microscopic or dislocation) aspects of fracture which are particularly relevant to persons interested in why materials have individual fracture characteristics and how alloy composition and microstructure determine them. Part 3 attempts to lay some semiquantitative and qualitative bridges between the two approaches. This is all that can be done at the present time; hopefully, additional synthesis will take place in the near future. Part 4 is concerned with time-dependent fracture under dynamic or static loading, another area in which quantitative understanding is still forthcoming. Part 5 deals with the microscopic aspects of fracture in individual materials and the guidelines that need to be followed in developing alloys with higher intrinsic fracture resistance.

We have tried to approach this extensive subject on an elementary level. A typical introductory course in physical metallurgy or materials science and one in elementary mechanics of materials provide sufficient background. Because a knowledge of the elementary concepts of dislocation theory is essential to an understanding of the microscopic aspects of fracture, we have included the pertinent aspects of this subject in Chapter 4. Hopefully, practicing engineers and research scientists will be able to understand the material covered without recourse to additional texts. Past experience has shown that the material in Parts 1 to 4 can be covered in a one-academic-quarter course of three lectures a week, and the entire text can be studied in a one-semester course. In both cases advanced undergraduate or first-year graduate students will have little difficulty in absorbing the significant concepts. Problem sets have been included for the benefit of both faculty and students.

Numerous colleagues contributed valuable suggestions that were incorporated in the book, and we are extremely grateful to them. We should particularly like to thank Dr. J. L. Low, Dr. F. A. McClintock, Dr. T. R. Wilshaw, Dr. T. L. Johnston, and Dr. J. B. Clark for their extensive review and discussion of the manuscript. Finally, we should like to thank Ellen Tetelman for her assistance during the writing and preparation of the manuscript and our wives Ellen and Jane for their patience and understanding.

A. S. TETELMAN

A. J. McEVILY, JR.

Stanford, California

Detroit, Michigan

December 1966

Contents

ix

PART ONE

Macroscopic Aspects
of Fracture

1

Conventional Design Concepts and Their Relation to the Occurrence of Fracture

The principal requirement of a properly designed structure is that it, or any part of it, be able to support the loads that are applied during its operating lifetime. These loads may include internal and external pressures, such as exist in pressure vessels or vacuum chambers, impact loads, such as applied to armor plate, and wind and earthquake loads that may be applied to structures such as buildings or oil field drilling rigs. In many cases the largest load may be the weight of the structure itself. The loads may be applied in a static, dynamic, or alternating fashion and can be axial, torsional, bending, or combinations of any of these. In statically determinate structures it is possible to determine the loads subjected to any member in the structure, but in indeterminate structures exact solutions are not always possible.

A component of load intensity (i.e., *stress*) that acts on a cross-sectional area A of a member is P/A, if P is a uniform load acting on the area A. The components of stress are normal (σ)—(tensile or compressive)—when they act perpendicular to the area that supports them and shear (τ) when they act in the plane of the supporting area. Depending on the choice of coordinates, both normal and shear stresses are simultaneously present throughout a loaded member. These stresses produce strains and hence deformations of the member, both of which increase with increasing applied stress. At very high stress levels the member can no longer support an increase in load and will fail if the load is increased. This may, in turn, lead to the failure of the entire structure (weakest-link principle). Alternatively, a structure designed such that the load-carrying members are in parallel may not fail when one member fails if the others can carry the increased load (fail-safe principle).

3

Failure can take place by several processes. *Elastic failure* (yielding) begins at a shear stress level above which large, permanent (plastic) deformations are produced by small increments of stress. Failure due to *mechanical instability* (e.g., buckling of a column loaded in uniaxial compression or necking of a bar under uniaxial tension) occurs when subsequent deformations can occur under decreasing applied loads. When the member separates into two or more parts, and its load-carrying capacity has dropped to zero, failure by *fracture* has taken place. Different criteria (e.g., a critical shear stress for yielding, a critical plastic strain for necking, a critical tensile stress at which cleavage fracture occurs) define the onset of the various failure modes. However, since stress, strain, and strain energy are related by the various laws of elastic and plastic deformation, it is possible to define each one of the failure event by a critical stress S_i (i.e., a *strength*)† at which the particular mode will occur.

Suppose that a particular failure event would impair the structural integrity of a member. The member must then be designed so that

$$\sigma = \frac{P}{A} < S_i$$

To support the same load in a lighter structure (smaller A), a material having a higher strength S_i would have to be utilized. Similarly, to increase load-carrying capacity either S_i or A or both would have to increase.‡

This simple illustration emphasizes that *both* applied stress and strength are related to design criterion. The former is a measure of the applied load; the latter is a measure of the maximum load that a material can withstand without failing. Both quantities must be known in order to predict whether failure will occur. Unfortunately, design engineers are usually concerned with the applied stress σ, and rely on handbooks to provide data on the strength S. In many instances this approach can prove dangerous because handbook data is applicable *only* to specific operating conditions (such as room temperature and absence of notches), which may not apply in practice. Similarly, metallurgists (or materials scientists as they are often

---

† The terms *stress* and *strength* are used by many workers to define the onset of failure. This can be misleading, unless the specific mode of failure is indicated. Thus "yield stress," "critical tensile (or compressive or shear) stress at which yielding begins," or "yield strength" are synonymous. Otherwise, stress refers to an applied load rather than to a material property.

‡ Elastic buckling of a long column is an exception since the buckling criterion is related to elastic modulus rather than to a yield or fracture stress.

called today) are primarily interested in producing materials having higher strengths S and have little knowledge of the stress states and stress levels under which they are required to operate. One of the aims of this book is to show that an understanding of both approaches toward materials design is necessary to avoid the problem of fracture in structural materials.

1.1   Stress-Strain Relations for Uniaxial Tensile Loading

Uniaxial tensile loading provides the simplest means to develop the stresses required to produce large-scale deflections, some kinds of mechanical instability, and fracture. Suppose that a given material is loaded in simple tension and that:

l_0 = initial length of specimen or gage length
A_0 = initial uniform cross-sectional area
P = applied tensile load
A = instantaneous uniform cross-sectional area after some deformation has occurred
l = instantaneous length after some deformation has occurred

The *engineering stress* is defined [1] as $\sigma_E = P/A_0$ and is based on the original cross-sectional area, whereas the *true tensile stress* $\sigma = P/A$ takes into account the fact that the load-bearing area decreases with increasing strain. The *engineering strain* ϵ_E is defined [1, 2, 3] as the change in length divided by the initial length

$$\epsilon_E = \frac{l - l_0}{l_0} = \frac{\Delta l}{l_0}$$

whereas the *true plastic strain* ϵ is the sum of all the instantaneous increments of an element of length l

$$\epsilon = \int_{l_0}^{l} \frac{dl}{l} = \ln\left(\frac{l}{l_0}\right) = \ln\left(\frac{l_0 + \Delta l}{l_0}\right) = \ln\left(1 + \epsilon_E\right)$$

Since $\ln(1 + x) \simeq x$ for $x < 0.10$, the engineering strain and the true plastic strain are the same for plastic strains less than 10%.

At low stresses and for short-time loading most materials exhibit *elastic behavior* and stress is proportional to strain. Hooke's law

$$\sigma = E\epsilon$$

then applies, where E is the elastic modulus. In the elastic range the strains are reversible with stress; that is, if the load is removed when $\sigma = C$ (Fig. 1.1a), the material returns to its original shape.

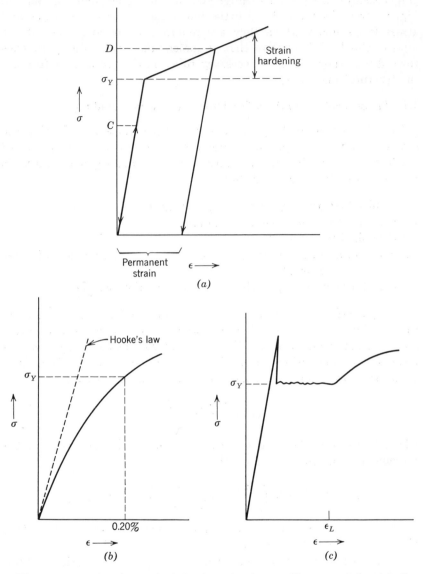

Fig 1.1. Typical stress-strain behavior of polycrystalline materials. (*a*) Generalized curve. (*b*) Curve for material that does not have a sharp yield point. (*c*) Curve for material that does have a sharp yield point and Luders strain ϵ_L.

If the material is loaded to a higher stress $\sigma = \sigma_Y = Y$, the tensile yield stress,† the deformation becomes permanent or *plastic* and Hooke's law no longer describes the relation between stress and strain. When the stress is removed from point $D > \sigma_Y$, for example, the unloading is elastic, but a permanent strain remains in the material. In general, the yield stress where plastic deformation begins is not so well defined (see Chapter 4) as shown in Fig. 1.1a. In many materials (e.g., copper and aluminum) there are deviations from Hooke's law almost from the first application of load, and the flow curve appears as shown in Fig. 1.1b. In this case the yield stress is taken to be the stress at a specified small value of strain, usually 0.2% offset. Some materials (e.g., mild steel) show a sharp yield point (Fig. 1.1c). The yield stress used in design considerations is then taken as the lower yield stress—the stress at which yielding begins (usually near a fillet) and propagates across the gage length as a "Luders band."

The shape of the stress-strain curve in the plastic region varies from one material to the next. In general, the stress required to produce additional strain after yielding begins increases with strain. This is referred to as *strain hardening* (i.e., the material becomes stronger as it deforms). The rate of strain hardening $d\sigma/d\epsilon$ is much less than the elastic modulus so that the stress-strain curve appears flatter in the plastic range (Fig. 1.1a).

Because plastic deformation occurs by a process of shear, there is essentially no change in the volume of the specimen during deformation. After the specimen has elongated to a length l,

$$Al = A_0 l_0$$

so that

$$\epsilon = \ln\left(\frac{l}{l_0}\right) = \ln\left(\frac{A_0}{A}\right) \tag{1.1}$$

As plastic deformation continues, the cross-sectional area decreases, but the load-carrying capacity increases because of strain hardening. Eventually an elongation is reached (point E, Fig. 1.2a) where the incremental increase in load-carrying capacity ($A\, d\sigma$) due to strain hardening becomes less than the incremental decrease in load-carrying capacity ($\sigma\, dA$) due to decreasing load-bearing area. At this point mechanical instability (necking) begins and further deformation takes place only in the necked region. The load is a maximum at point E

---

† σ_Y or Y are used to represent tensile yield stress, τ_Y or k represent shear yield stress.

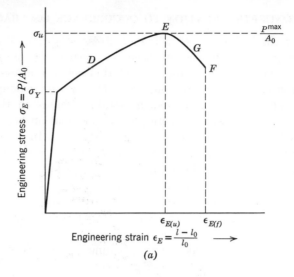

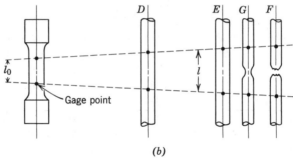

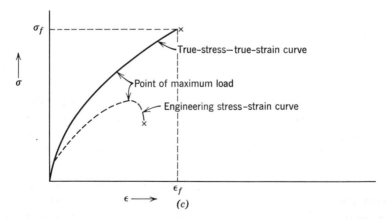

Fig. 1.2 (a) The engineering stress–engineering strain curve of a material that does not show a yield point. (b) The appearance of the gage length of a tensile specimen at the indicated stages of strain. (c) The relation between the engineering stress–engineering strain curve and the true-stress–true-strain curve.

and the *ultimate tensile stress* (UTS) is defined as $\sigma_u = P^{max}/A_0$. Taking the differential of the relation $P = A\sigma$ shows that $P = P^{max}$ when

$$dP = \sigma\, dA + A\, d\sigma = 0$$

or

$$\frac{d\sigma}{\sigma} = -\frac{dA}{A} = d\epsilon \qquad (1.2)$$

Further elongation in the necked region eventually leads to fracture at point F. The engineering strain at this point is $\epsilon_{E(f)}$. Figure 1.2b shows the shape of the tensile specimen at various stages of the deformation process.

It should be mentioned that the true-stress–true-strain curve for the material is quite different† (Fig. 1.2c) from the engineering stress-strain curve because $\sigma = \sigma_E(A_0/A)$ is greater than σ_E and because $\epsilon = \ln{(1 + \epsilon_E)}$ is less than ϵ_E, for $\epsilon > 10\%$. Furthermore, when necking begins, the strains and hence the true stresses in the necked region increase, even though the load and engineering stress decrease, because the cross-sectional area has decreased even more. Thus the true fracture strain‡ $\epsilon_f = \ln{(A_0/A_f)} = \ln{(1/\{1 - RA\})}$ is much greater than the measured engineering strain $\epsilon_{E(f)}$. Similarly, the fracture stress σ_f is usually much higher than the "ultimate" tensile stress, especially when necking precedes fracture. If fracture occurs without necking, then UTS $= \sigma_f(A_f/A_0)$.

For many low strength materials the equation

$$\sigma = \sigma_0\epsilon^n \qquad (1.3)$$

describes the approximate relation between true stress and plastic strain prior to necking, [2]; σ_0 is the value (or extrapolated value) of the flow stress at 100% plastic strain ($\epsilon = 1$), and n is called the strain-hardening exponent. Both are constants for a particular material under particular test conditions. Since

$$\frac{d\sigma}{d\epsilon} = n\sigma_0\epsilon^{n-1} = \frac{n\sigma}{\epsilon}$$

† This distinction is important when the terms *brittle* and *ductile* are applied to a material. To avoid any possible confusion, all schematic stress-strain curves in this book are *true*-stress–*true*-strain curves. Load-elongation curves are used to represent engineering stress and engineering strain.

‡ The per cent reduction in area at fracture, labeled % RA or simply RA is $(A_0 - A_f)/A_0 \times 100$.

then from (1.2)

$$\epsilon_u = n \tag{1.4}$$

is the "ideal" true tensile strain at necking. The UTS is therefore given by

$$\text{UTS} = \sigma_u = \sigma \left(\frac{A_u}{A_0}\right) = \sigma_0 \epsilon^n \left(\frac{A_u}{A_0}\right) = \sigma_0 \left(\frac{n}{e}\right)^n \tag{1.5}$$

where e is the base of natural logarithms; σ_u increases markedly with increasing strain-hardening capacity and with σ_0.

In conventional design practice the stresses that are of most interest are the yield stress and ultimate stress. The former defines the load at which large permanent deflections begin to take place (elastic failure)† and is important when close dimensional tolerances are required. The ultimate stress defines the maximum uniaxial load that the member can withstand without becoming mechanically unstable (buckling or necking). Owing to the possibility of defective material, errors in stress calculation, changes in material properties due to fabrication, and possible overloading (accidents), the working stresses of most structures are kept below the yield or ultimate stresses. The working stress σ_W is given by

$$\sigma_W = \frac{\sigma_Y}{N}$$

or   (1.6)

$$\sigma_W = \frac{\sigma_u}{N'}$$

where N and N' are called *factors of safety* [4, 5]. These factors are prescribed by various codes relating to the type of structure (buildings, pressure vessels, etc.) and to the construction materials. For low strength materials $N = 1.65 - 2.00$ and $N' \approx 4.0$. Because σ_Y is about $\frac{1}{2}\sigma_u$ for these materials, the two criteria can be roughly equivalent, with the one based on σ_u being a bit more conservative. They both predict that no large-scale yielding or elastic failure should occur during the operating lifetime of a structure.

1.2   Stress-Strain Relations for Combined Stress

Most structures are subjected to applied stresses that are more complicated than the simple tension described earlier, and it is necessary

---

† Except for long compression members which buckle before the yield stress is reached.

to develop criteria for the onset of elastic failure (yielding), plastic failure, and fracture.

Any system of stresses that acts on a body can be resolved [3, 6, 7], in Cartesian coordinates (x, y, z), into three normal components σ_{xx}, σ_{yy}, σ_{zz} which lie along orthogonal axes x, y, z, and three shear components $\tau_{xy} = \tau_{yx}$, $\tau_{xz} = \tau_{zx}$, $\tau_{yz} = \tau_{zy}$ which act on planes that are perpendicular to the axes x, y, z. The first subscript gives the direction of the shear component; the second is the axis perpendicular to the plane in which the shear stress lies. Once the applied stresses have been resolved in this manner, it is then possible to rotate the axes x, y, and z into another set of orthogonal axes (1, 2, 3) along which the shear stresses are zero. These axes are called *principal axes*, and the normal stresses resolved along them σ_1, σ_2, σ_3 are called *principal stresses*. By convention the axes are chosen so that $\sigma_1 > \sigma_2 > \sigma_3$. The problem of finding the principal stresses from the body stresses σ_{xx}, τ_{xy}, etc., is solved with the use of the Mohr's circle construction. This construction also allows calculation of the planes of maximum shear stress or *principal shear stress*. It turns out that these planes always bisect the right angle between two of the three principal axes, that is, they lie at 45° to two of the three principal axes and are parallel to the third (Fig. 1.3). The theory of elasticity shows that the three principal shear stresses are given by the relations

$$\tau_1 = \frac{\sigma_2 - \sigma_3}{2} \qquad \tau_2 = \frac{\sigma_1 - \sigma_3}{2} \qquad \tau_3 = \frac{\sigma_1 - \sigma_2}{2}$$

and since $\sigma_1 > \sigma_2 > \sigma_3$ by convention, the *maximum shear stress* is

$$\tau_{\max} = \frac{\sigma_1 - \sigma_3}{2} \tag{1.7}$$

For a two-dimensional state of stress it is sometimes more convenient to obtain the maximum shear stress directly from the components of stress by use of the relation

$$\tau_{\max} = \sqrt{\tfrac{1}{4}(\sigma_{xx} - \sigma_{yy})^2 + \tau_{xy}^2} \tag{1.8}$$

As in the case of the stresses, the elastic strains can be separated into a set of principal normal strains ϵ_1, ϵ_2, ϵ_3, and a set of shear strains γ_1, γ_2, γ_3. In isotropic materials the principal axes for stress and strain are coincident.

It is possible to determine experimentally criteria for elastic failure (yielding) and plastic failure for structural members subjected to

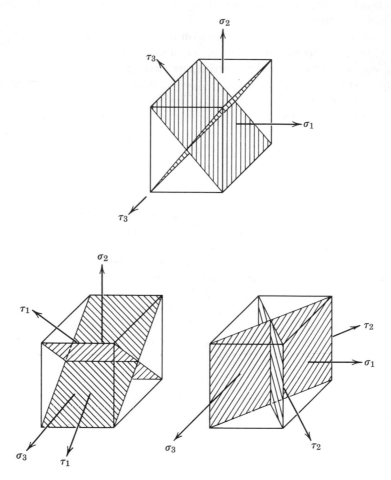

Fig. 1.3. Planes of principal shear stress for various combinations of principal stresses σ_1, σ_2, σ_3.

combined stresses. This, however, is an inconvenient and costly operation and it is simpler to develop failure criteria based on data obtained from simple tensile, torsion, or compressive tests. Three theories [7, 8, 9, 10] are used for this conversion.

The *maximum stress* or Rankine theory states that failure occurs when the largest principal stress σ_i $(i = 1, 2)$ is equal to the failure stress [elastic (σ_Y) or ultimate (σ_u)] of the same material tested in uniaxial tension. This theory does not agree with experimental test results when applied to yield and tensile strengths because these involve plastic deformation which occurs under the action of the *shear* component of the applied stress, and hence depend on the *difference* between principal stresses. It does, however, have considerable merit when failure occurs by tensile cleavage fracture, as will be shown in the next chapter.

The *maximum shear stress* or *Tresca theory* proposes that yielding occurs when the maximum shear stress acting on a member is equal to k, the yield stress in pure shear (torsional loading). From Eq. 1.7 the yield criterion can be written

$$\sigma_1 - \sigma_3 = 2k \tag{1.9}$$

The yield stress in simple tension (Y or σ_Y) is found by setting $\sigma_1 = Y$, $\sigma_2 = \sigma_3 = 0$ so that

$$\sigma_Y = Y = 2k \tag{1.10}$$

A similar approach can be made for analyzing plastic failure in terms of σ_u.

A third criterion, proposed by von Mises, states that yielding occurs when the shear strain energy of distortion per unit volume is greater than the shear strain energy per unit volume in a material strained to its uniaxial tensile (or compressive) yield stress. This criteria may be written, in terms of the principal stresses and the yield stress in pure shear, as

$$(\sigma_1 - \sigma_2)^2 + (\sigma_2 - \sigma_3)^2 + (\sigma_1 - \sigma_3)^2 = 6k^2 \tag{1.11}$$

so that the yield stress in uniaxial tension ($\sigma_Y = \sigma_1$, $\sigma_2 = \sigma_3 = 0$) is given by

$$\sigma_Y = \sqrt{3}\, k = 1.73k \tag{1.12}$$

The yield strength in shear k can be determined by means of torsion tests[2, 3] and compared with values of the tensile yield stress σ_Y. The experimental data[11] fall somewhere in between the two criteria represented by Eqs. 1.10 and 1.12, but are somewhat closer to the values predicted by the von Mises approach. Figure 1.4 shows how these theories can be represented graphically for the two-dimensional (plane-stress) case where $\sigma_3 = 0$. Elastic failure (yielding) occurs when the combined stresses reach the yield surface shown in Fig. 1.4a, and plastic failure occurs when the stresses reach the ultimate surface shown in Fig. 1.4b.

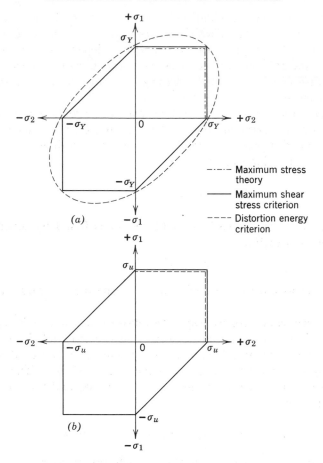

Fig. 1.4. Yield criteria ($\tau_{\max} = k$) according to various theories. (a) Elastic failure (plastic flow). (b) Plastic failure (instability).

1.3   Elastic-Plastic Structures

Elastic stress concentration

Suppose that a body of cross-sectional area A is loaded in uniaxial tension by a force P that is applied perpendicular to A. The stress $\sigma = P/A$ is the number of pounds carried by each square inch of area. The load P is, in effect, supported by lines of force, one line per square inch, with each line carrying P/A lb (Fig. 1.5a). Now suppose that a discontinuity, such as a hole or crack, is introduced into the body.

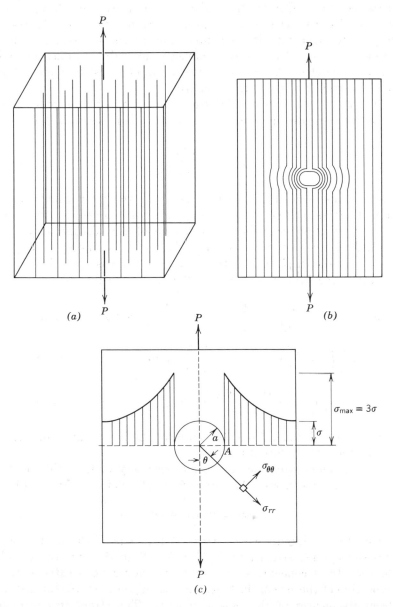

Fig. 1.5. The lines of force (stress lines) in a body subjected to uniaxial tension. (a) Homogeneous body. (b) Body containing a discontinuity. (c) The stress distribution in the vicinity of a hole in an elastic body under uniaxial loading [2].

15

Since the elastic modulus of a void is zero, stress cannot be transmitted across its faces. Consequently the lines of force must bend around the discontinuity, as shown (in projection) in Fig. 1.5b. In the vicinity of the discontinuity the lines of force (or stresses) are highly concentrated, but at distances removed from it the lines become more evenly distributed; the stress concentration effect is therefore a function of position relative to the discontinuity.

The concentration of elastic stress around the discontinuity is expressed analytically in terms of an *elastic stress concentration factor*

$$K_\sigma = \frac{\sigma \text{ local}}{\sigma \text{ nominal}} \tag{1.13}$$

which is the ratio of the local stress near the discontinuity to the nominal stress acting across the gross cross section. The elastic stress distribution depends on the shape of the discontinuity. The simplest case is that of the *circular hole* in an infinitely wide plate. It has been shown [5] that if the uniaxial stress σ is applied across the plate, then the stresses produced in the vicinity of the hole are

$$\sigma_{rr} = \frac{\sigma}{2}\left(1 - \frac{a^2}{r^2}\right) + \frac{\sigma}{2}\left[1 + \frac{3a^4}{r^4} - \frac{4a^2}{r^2}\right]\cos 2\theta \qquad (r > a)$$

$$\sigma_{\theta\theta} = \frac{\sigma}{2}\left(1 + \frac{a^2}{r^2}\right) - \frac{\sigma}{2}\left(1 + \frac{3a^4}{r^4}\right)\cos 2\theta \qquad (r > a) \tag{1.14}$$

$$\tau_{r\theta} = -\frac{\sigma}{2}\left(1 - \frac{3a^4}{r^4} + \frac{2a^2}{r^2}\right)\sin 2\theta \qquad (r > a)$$

where r is the distance from the center of the hole to any point in the plate and a is the radius of the hole.

The maximum stress occurs at the periphery of the hole and then decreases as r increases. At $r = a$, $\theta = \pi/2$ (Point A, Fig. 1.5c)

$$\sigma_{\max} = \sigma_{\theta\theta} = 3\sigma \tag{1.15}$$

so that the maximum tensile stress at the edge of the hole is three times the applied stress and $K_\sigma^{\max} = 3$. Note that for the circular hole the stress concentration factor can never be greater than 3, irrespective of the size of the hole, and that K_σ decreases as the distance r from the center of the hole increases. The stress concentration factor is also a function of the size of the plate. Figure 1.6a indicates that $K_\sigma^{\max}$ decreases from the value of 3 as the ratio of hole size to plate size increases.

The elastic stress concentration factors around discontinuities having more complicated shapes are difficult to determine analytically.   Many of the calculations that do exist have been performed by Neuber [12] and Savin [13] and solutions for a variety of problems are given in their books.   Experimental stress analysis has also been used [14, 15] to determine values of K_σ.   Stress concentrations exist when there is an abrupt change in the dimensions of a part [16].   Figure 1.6b indicates that when a plate has a step up in width from h to W, K_σ^{max} increases as the ratio W/h increases and as the radius of the shoulder decreases.   Similarly a step up in diameter (from d to D) of a shaft subject to torsional loading produces a concentration of stress (Fig. 1.6c) that increases with increasing D/d and decreasing fillet radius r.   (The important case of the elliptical hole (crack) is treated in Chapter 2.)

The fully plastic condition

All structural materials contain stress concentrations of some type (e.g., fillets and sharp dimensional changes).   Generally these have little effect on the load at which elastic and plastic failure of the entire member takes place, although they can induce premature fracture in brittle materials.   Suppose, for simplicity, that a small hole exists in a wide, thin plate† of a ductile material which has a well-defined tensile yield stress and which is *non strain hardening*.   The stress-strain curve of the material is shown in Fig. 1.7a.   Since $K_\sigma^{max} = 3$ for the hole, local yielding can occur in its vicinity at a load $P_Y/3$, where P_Y is the yield load for a plate having the same gross cross section but with the hole absent. $P_Y/3$ is called the *elastic limit load* and is that load which causes initial plastic deformation of the most highly stressed volume element in the plate.   Figure 1.7b shows the stress distribution that exists in the plate at the elastic limit load.   When the load is increased to $P_1 > P_Y/3$ the plastic zone spreads further from the hole (Fig. 1.7c), but the maximum tensile stress that exists in the plate is still σ_Y since the material is nonstrain hardening [1] and plane-stress conditions exist.   The strains at the tip of the hole are increased from A to B as shown on the figure.   A further increase in load from P_1 to P_{GY} causes the plastic zones to spread completely across the plate (Fig. 1.7d) so that it becomes completely plastic. P_{GY} is called the *fully plastic load or general yield load*.

When the applied load is much less than the fully plastic load in a ductile material (Fig. 1.7c), the outer regions of the plate behave elastically and obey Hooke's law.   Since the elastic and plastic regions

† Plane-stress loading condition.

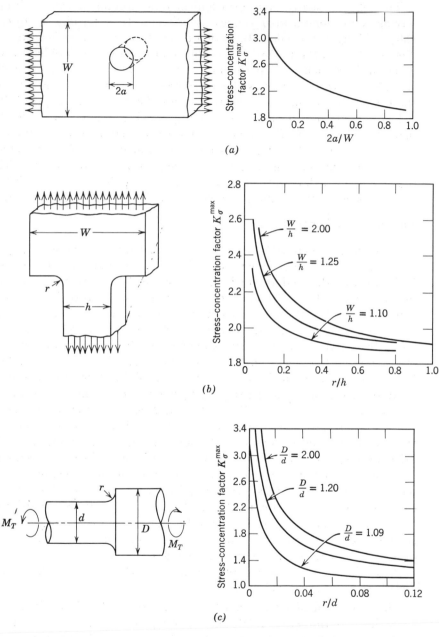

Fig. 1.6. The elastic stress concentration factor K_σ^{max} for some structural shapes under the indicated load [14].

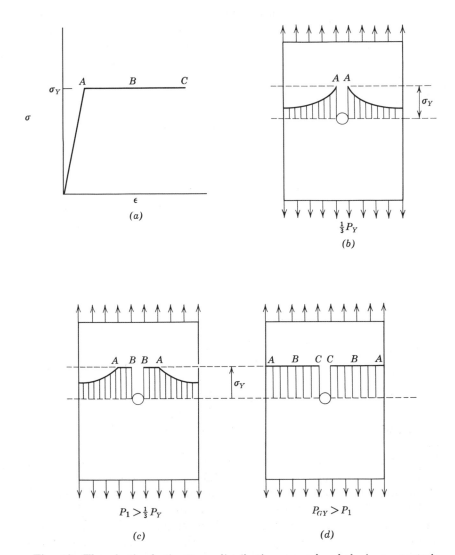

Fig. 1.7. The elastic-plastic stress distribution around a hole in a nonstrain hardening material under plane-stress tensile loading. (a) The stress-strain curve. (b) The onset of yielding at the periphery of the hole. The load P is $P_Y/3$, where P_Y is the yield load for a body of similar shape in the absence of a hole. (c) The stress distribution at a load $P_1 > P_Y/3$. (d) The stress distribution when $P = P_{GY}$, the fully plastic load.

are "attached" to one another, the overall deflection of the plate $\Delta l/l$ remains small and the plate as a whole behaves essentially in an elastic manner. Thus the localized plastic deformation does not prevent the plate from satisfactorily fulfilling its function as a load-supporting member in a structure, provided that $P < P_{GY}$. The plastic deformation becomes large compared with the elastic deformation when the plastic zones have spread over sufficiently large distances to cause a significant deflection of the member or structure.

The "elastic constraint" action described above prevents large-scale deformation of ductile members that have complicated shapes and contain stress concentrators such as fillets or sharp corners. It can also be utilized to achieve higher load-carrying capacity without elastic failure in members of simple shape that are *stressed nonuniformly*. Consider a beam of thickness $2t$ that is loaded uniformly. When the beam is entirely elastic, the bending stresses are

$$ \sigma = \frac{Mx}{I} \qquad -t < x < t $$

where M is the bending moment and I is the moment of inertia. At the center line of the beam ($x = 0$) the stresses are zero; this line is called the neutral axis. At the inner and outer elements the stresses are $-Mt/I$ (compressive) and $+Mt/I$ (tensile). When $\sigma = \sigma_Y$, the outer elements deform plastically so that $M = M_Y = \sigma_Y I/t$ is the elastic limit moment. At this point the inner core is still completely elastic because the elastic limit has not yet been reached there. When the applied load (bending moment) increases from M_Y to M_1, additional regions of the beam become plastically deformed (Fig. 1.8). Calculations show [1] that the *beam as a whole* does not become completely deformed until $M = M_p = 1.5M_Y$. Because large-scale deflections of the beam cannot take place until $M = M_p$, it is possible and more efficient to base design criteria for uniformly loaded beams on the fully plastic moment rather than on the elastic limit moment. Similar considerations apply to frameworks where one or more members of the frame may have yielded plastically but where the frame *as a whole* remains elastic because the stress in some of the members is still below the elastic limit. Elastic failure of the frame does not take place until *all* members of the frame have become plastically deformed. Analytical procedures known as *limit design* can be used to predict the fully plastic conditions for various structures and are used in engineering design.

The constraint action provided by the nondeforming elements of a

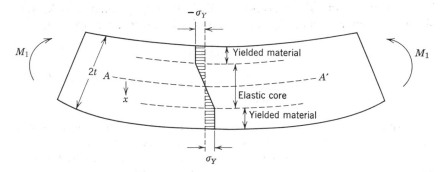

Fig. 1.8. The elastic-plastic stress distribution in a bent beam when $M_p > M_1 > M_Y$. M_1 is the applied bending moment, M_Y is the elastic limit moment at which yielding occurs at the inner and outer fibers of the beam, and M_p is the moment at which the beam becomes fully plastic [1].

body on the deforming elements is important in plasticity problems under *plane strain*† when triaxial stresses are present [17, 18]. The direction of maximum shear stress lies in the (1, 2) plane and the plane of maximum shear stress is parallel to the (3) direction. Since both yield criteria (Eqs. 1.10 and 1.12) are based on the shear stresses and hence on the *differences* between principal stresses, the equal addition or subtraction of a hydrostatic stress p has no effect on the yielding process. Consequently the principal stresses σ_1, σ_2, and σ_3 can be separated (Fig. 1.9) into a hydrostatic tension $(-p)$ super-

† In *plane-strain* plasticity theory the strain rate is proportional to stress, $\dot{\epsilon}^3 = 0$, $\sigma_3 = (\sigma_1 + \sigma_2)/2$. Plane stress implies that one of the principal stresses $\sigma_3 = 0$ but $\dot{\epsilon}_3 \neq 0$. In general, plane-strain conditions exist in thick bodies and plane-stress conditions exist in thin ones. Specific examples with reference to fracture behavior are discussed in the following chapters.

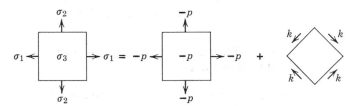

Fig. 1.9. The separation of a set of principal tensile stresses σ_1, σ_2, σ_3 into a hydrostatic tension $(-p)$ and a pure shear stress τ.

imposed on a pure shear stress τ. Yielding occurs when $\tau = k$, and since the theory assumes that the material is nonstrain hardening

$$\sigma_1 = -p + k$$

$$\sigma_2 = -p - k \qquad\qquad (1.16)$$

$$\sigma_3 = -p$$

According to the Tresca criterion, yielding begins when $\sigma_1 - \sigma_2 = 2k$. Much larger *hydrostatic* (tensile or compressive) stresses can therefore be present before general yielding begins under triaxial tension, for example, than in simple tension where $\sigma_1 = 2k$, $\sigma_2 = \sigma_3 = 0$.

The indentation of a semi-infinite solid by a flat punch [19, 20] under a compressive stress $-\sigma_2$ may be used to illustrate this principle. Yielding begins at the corners of the punch (Fig. 1.10a) when $\sigma_2 \ll \sigma_Y$ because of the stress concentration that exists there. However, in order to allow the punch to sink in and cause general yield, all of the material beneath the punch must be deformed and this in turn requires that the unpunched sections $ABCD$ and $EFGH$ be deformed. Because there is no nominal stress acting directly on $ABCD$ and $EFGH$, these unyielding masses act as compression blocks, which cause hydrostatic compressive stresses p to be set up at points such as x below the punch.

The *slip-line field theory* of plasticity [17, 18, 19, 20] shows that these hydrostatic stresses will be overcome and general yield can take place when

$$-\sigma_2 = (2 + \pi)k = 5.14k$$

$$= 2.57\sigma_Y \text{ (Tresca criterion)}$$

$$= 2.82\sigma_Y \text{ (von Mises criterion)} \qquad (1.17)$$

Thus the presence of triaxial constraint causes the resistance of a semi-infinite solid to indentation by a flat punch to be almost three times its resistance to deformation by uniaxial loading.

1.4 The Occurrence of Fracture and the Inadequacies of Conventional Design Concepts

The conventional design criteria described above are based on one or more of three principles: (1) keeping the loads which act on a member below the elastic limit (yield) load so that no large-scale deflection of the member takes place, (2) keeping the applied loads

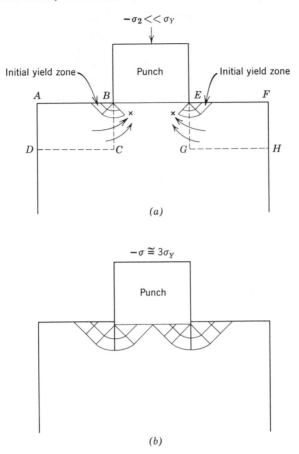

Fig. 1.10. The slip lines (lines of maximum shear stress) in the vinicity of punch that is being forced into a thick plate. (a) Initial deformation at the corners of the punch. (b) The slip line field at general yield when the punch sinks into the plate [17].

below the ultimate load so that mechanical instability (buckling or necking) does not occur, and (3) allowing for the inevitability of local yielding around discontinuities but relying on elastic or triaxial constraint to preserve the structural integrity of a member or a framework *as a whole*. Safety factors are used to account for any differences between calculated stresses and handbook strengths.

None of these standard design approaches is concerned with the problem of *fracture*. They all assume that the fracture strength is greater than the yield strength and equal to or greater than the ultimate strength (Fig. 1.11a); they rely on proper design involving past experience (feedback) and safety factors to keep the working loads below the yield and ultimate loads and hence below the fracture load. This procedure is acceptable for many engineering structures as attested to by the *relatively* few accidental fractures that have actually occurred in service.

The "accidental" failures that have occurred have often been quite spectacular [21]. Twelve persons died and 40 others were injured

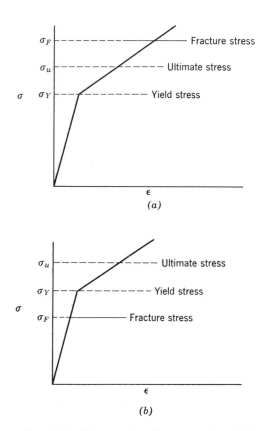

Fig. 1.11. The relation between the yield stress σ_Y, ultimate stress σ_u and fracture stress σ_F. (a) Ductile material—fracture occurs after extensive plastic deformation. (b) Brittle material—fracture occurs before the elastic limit is reached across the entire load-bearing cross section.

when a molasses tank suddenly fractured in Boston, Mass., in 1919. When a methane storage tank fractured in Cleveland, Ohio, in 1944, 128 persons were killed and property damage totaled almost seven million dollars. Subsequent investigations showed that the working loads in the molasses tank had been twice those allowed for in the building codes and that the methane tank had been constructed out of steel plate that had not been properly heat treated.

Accidental fractures such as these will continue to occur as long as structures are built, but it is reasonable to expect that improved design and inspection techniques will lower the frequency of their occurrence. It would, however, be desirable if some *warning* could be given *before* the fracture occurred so that operating loads might be reduced and personnel have time to leave the scene and escape injury. Furthermore, it would be desirable for the fracture to take place *slowly* because, if it occurs rapidly, considerable strain energy will be suddenly released, particularly if the structure is large. This can cause the structure to shatter, or even pulverize, and in many cases it is this shattering action which actually causes the most serious damage.

During World War II fractures were observed in 25% of the all-welded Liberty ships that were constructed in the United States [21]. Of the 4694 ships that were constructed, 1289 casualties (a structural failure) were reported, 233 being so serious (catastrophic) that the ships were lost or considered to be unsafe. Figure 1.12 shows one of these fractures which occurred in the T-2 tanker *Schenectady*, which failed in her fitting-out pier, without warning, in calm seas and mild weather. The ship broke in half in a matter of seconds. Subsequent investigation revealed that the maximum bending moments existing at the time of the fracture were only one-half of the bending moments allowed for in design.

This situation illustrates a major inadequacy of conventional design procedure: *it does not anticipate that unstable fractures can occur at stress levels that are below the design (elastic) limit*, as schematized in Fig. 1.11b. Low-stress fractures such as these can occur near defective welds or when design notches or cracks are present in the structures. As shown in Fig. 1.13 the fracture load decreases as the size of the defect increases; very small defects can produce unstable fracture, particularly in materials that have high yield strengths (in the absence of defects). The mechanics of the fracture process will be presented in the following chapter. It is only necessary to point out here that when conditions are right for an unstable fracture (e.g., presence of a notch, low service temperature), handbook values of the

Fig. 1.12. Photograph of a T-2 tanker that failed at pier. *Courtesy of E. Parker* [21].

"ultimate" tensile strength or yield strength are inadequate to handle the problem. Design criteria should then be based on *reliability against brittle fractures*.

Local yield is another factor that must be considered when fracture is a potential problem. Plastic design criteria allow for the presence of some local yielding around discontinuities because it is assumed that the unyielded elastic core will cause structural integrity to be maintained. Figure 1.14 shows [23] that a notched bar can fracture in a catastrophic manner when local yield has occurred around the notch; the local yield has, in fact, triggered the unstable fracture which could not be stopped in the elastic portion of the member. These unstable fractures were (and to some degree still are) particularly prevalent in large structures (ships, pipelines). In some cases conventional laboratory tests on small specimens indicate that a material is suitable for a particular application but a large plate of the material subsequently fractures in service. Thus there is a *size effect* that must be considered when designing with potentially brittle materials [21].

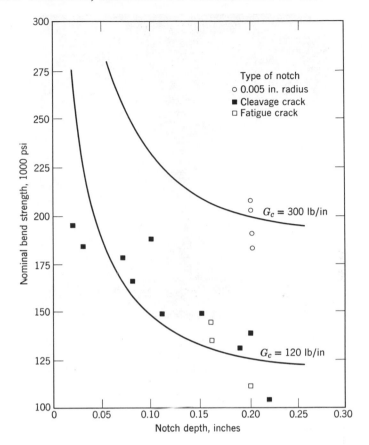

Fig. 1.13. The effect of notch depth and sharpness on the bend strength of a 1 in. square bar of a heat-treated alloy steel. *After Yukawa and McMullin* [22].

The use of conventional design criteria when *combined stresses* are present introduces a fourth possible problem. These design criteria are based on elastic or plastic instability and therefore are concerned with *shear stresses*. Unstable fracture, on the other hand, often occurs when a critical *tensile* stress is reached [24]. Suppose (Fig. 1.15) that fracture occurs at a tensile stress $\sigma = \sigma_f$. Under uniaxial tensile loading $\sigma_f > \sigma_Y$ and the ductility is ϵ_1. However, if triaxial tension is present [23] such that $\sigma_2 = \sigma_3$, then the yield stress is raised from $\sigma_1 = \sigma_Y$ to $\sigma_1 = \sigma_Y + \sigma_2 = \sigma'_Y$, according to the Tresca criterion, and the ductility is reduced to $\epsilon_2 < \epsilon_1$. In addition, biaxial tension can

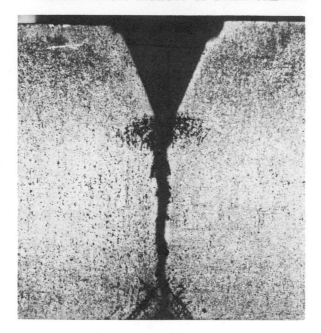

Fig. 1.14. Fractured halves of a notched bar that broke before general yielding occurred. Evidence of the local yielding that triggered off the fracture is revealed by etching near the root of the notch. 6×. *Courtesy J. Knott* [6] *and J. Iron Steel Institute.*

lead to a reduced ductility in anisotropic materials such as hexagonal metals (e.g., beryllium and titanium) or heavily rolled plate or sheet.

Probably the most serious drawback to the use of the handbook to obtain tensile strength (yield or ultimate) values for design against fracture is that these values are usually obtained by testing at room temperature (70°F), in atmospheric environment (air) and at moderate loading rates. Materials are often required to operate under conditions that may be quite different than these (e.g., alloy steel turbine blades at high temperatures and in corrosive (steam) environments, pressure vessels containing liquid fuels at cryogenic temperature), and it must be appreciated that the fracture behavior is extremely dependent on *environment*.

Service temperature is the most important environmental variable [21] that affects the probability of catastrophic fracture. Figure 1.16 shows the effect of test temperature on the load at which general

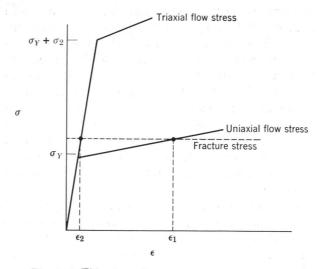

Fig. 1.15. The effect of stress state on ductility when fracture is determined by a maximum tensile stress criterion.

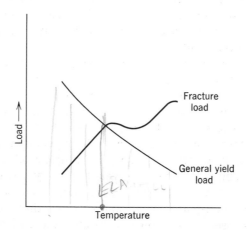

Fig. 1.16. The effect of test temperature on the general yield load and fracture load of notched bend specimens of mild steel.

yield or fracture occurs in notched specimens of mild steel that were loaded in four-point bending. At moderate temperature the fracture load is greater than the yield load and the steel exhibits some ductility, but at low temperature the fracture load is much less than the yield load (measured with the notch in compression) and the steel is extremely brittle. This fact could not be predicted from standard tensile data obtained at room temperature (70°F). Steels and many other materials as well show a *ductile-brittle transition temperature region,* below which they are unsafe and above which they may be safely used. Many of the catastrophic fractures that occurred in the Liberty ships are partially attributable to *low operating temperature* that existed at the time of the failures.

Alternatively materials that are loaded at *high operating temperature* can fail prematurely by creep rupture. Creep is the plastic deformation that occurs over a period of time in a material subjected to a constant stress that is typically below the elastic limit. Figure 1.17a shows a schematic creep curve and indicates that there are three stages to the creep process. During the first stage the strain rate (the slope of the strain-time curve) decreases; it remains constant during the second or steady-state stage and then increases rapidly until rupture occurs in stage 3. Increasing stress and increasing temperature cause the creep rate to be increased in all stages. Figure 1.17b shows the importance of engineering stress and temperature on the rupture life of an austenitic stainless steel.

In some applications the fracture strength of a material decreases during service lifetime because of the presence of *particular chemical environments.* Many high strength steels are susceptible to *hydrogen embrittlement* and fail over a period of time that varies with the applied tensile stress (Fig. 1.18) and amount of hydrogen dissolved in the steel. Numerous fractures have been reported in oil well casings (caused by dissociation of H_2S gas) after periods of time ranging from days to months. In no case was there any warning of the impending fracture. A few years ago a large number of fractures occurred in cadmium-plated steel landing gears that were used on jet aircraft. These were also shown to result from hydrogen embrittlement.

Stress corrosion cracking is another form of delayed fracture which occurs under static loading in materials that are normally quite ductile. These fractures also occur without warning, after a period of time that increases with decreasing applied tensile stress. In some cases there is a lower limiting stress below which fracture does not occur, but in other cases the limit is practically nonexistent (Fig. 1.19)

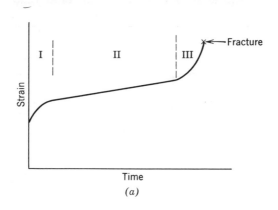

(a)

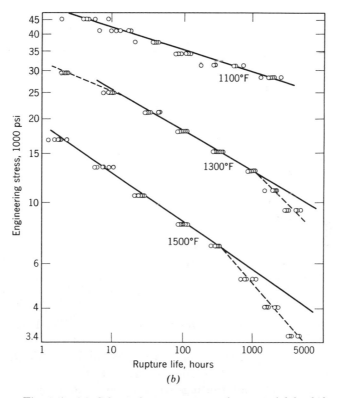

(b)

Fig. 1.17. (a) Schematic creep curve of a material loaded at high operating temperature under a constant stress. (b) Effect of initial applied stress and temperature on the rupture life of austenitic stainless steel. *After Garofalo et al.* [26].

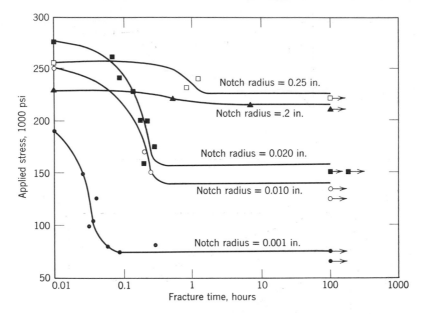

Fig. 1.18. The effect of applied (static) tensile stress and notch sharpness on the time for failure for hydrogen-embrittled high strength steel. *After Johnson et al. [27].*

and fracture can occur at stresses much below the yield stress. These fractures occur when particular materials are exposed to particular environments and a material that fractures in one environment will not necessarily fracture in another.

The rapid expansion in the use of nuclear power plants has brought about a new set of fracture problems. Fuel elements swell, creep, and perhaps fracture after operating for various periods of time. Of particular interest to the design engineer is the problem of *fast neutron embrittlement* of reactor pressure vessel steels after long operating lifetime.

Many structures such as crankshafts, axles, and aircraft wings are subjected to alternating tensile and compressive stresses or alternating tensile stresses. The alternating stresses can lead to *fatigue fracture* [29] after a period of time (number of stress cycles N) that varies inversely with the range of applied stress ($+\sigma$ to $-\sigma$) (Fig. 1.20) or strain amplitude. Certain materials such as steel show a *fatigue limit* or limiting stress below which fracture does not occur. Most materials, however, do not show such a limit and their fatigue strength

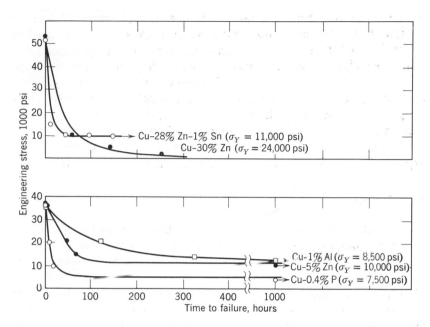

Fig. 1.19. The effect of applied stress on the time for failure of copper base alloys in an ammonia environment. *After Robertson and Tetelman* [28].

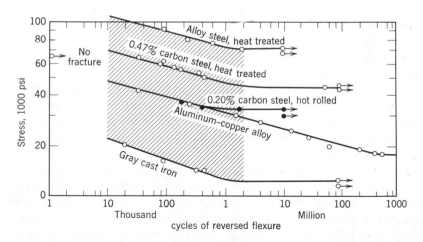

Fig. 1.20. The effect of applied stress on the number of cycles of loading before fatigue failure occurs in some engineering alloy. *American Society for Metals, Metals Handbook, 1948, p. 118.*

is the stress required to cause fracture after 10^7 cycles of loading. Fatigue is a particularly insidious problem because it is influenced by all of the factors (notches, corrosive environment, temperature, etc.) that affect static fractures plus the complicating effect of stress range, mean stress (when stresses are cycled about some fixed stress different from zero), combined stress, and surface condition. In addition to fatigue caused by externally applied stresses, *thermal fatigue* can take place when a material is alternately heated and cooled, because of the formation of thermal stresses due to anisotropic (nonuniform) thermal expansion of various parts of the material and thermal gradients.

1.5 Summary

1. An external load operating on a member produces normal and shear components of stress. Failure occurs when the sum of external and any internal stresses reaches a critical value known as the strength.

2. Elastic failure (yielding) occurs when the maximum shear stress or shear strain energy of distortion reaches a critical value. Since the shear stresses are related to the difference between the principal components of stress, a larger principal stress exists when yielding begins under triaxial tension or compression than under uniaxial tension or compression.

3. In members containing discontinuities which produce a concentration of stress, local yielding can occur in the vicinity of the discontinuity while regions removed from the discontinuity and the member as a whole remain elastic. The member is then in an elastic-plastic state of deformation.

4. The failure criterion for mechanical instability under tensile loading (necking) is determined solely by a material's capacity for strain hardening. In ductile materials the ultimate strength is a measure of load-carrying capacity; the stress required to cause fracture is much greater than the ultimate strength.

5. Conventional design criteria based on yield or ultimate strengths are inadequate for predicting the behavior of brittle materials where fracture can occur after small plastic strains, either on a local level (around a notch) or on a general level (over the entire member). Many materials that are ductile under normal operating conditions can become brittle when they are exposed to extremes of temperature, loading rates, chemical environment, alternating stress, or hydrostatic stress.

References

The asterisk indicates that published work is recommended for extensive or broad treatment.

*[1] F. B. Seely and J. O. Smith, *Resistance of Materials*, 4th ed., Wiley, New York (1957).

*[2] G. E. Dieter, Jr., *Mechanical Metallurgy*, McGraw-Hill, New York (1961).

*[3] F. A. McClintock and A. A. Argon, *Introduction to Mechanical Behavior of Materials*, Addison Wesley, Boston (1966).

 [4] J. Marin, *ASME Handbook; Metals Engineering Design*, McGraw-Hill, New York (1953), pp. 317, 328.

 [5] J. Marin, *Machine Design* (November 1941).

*[6] S. Timoshenko and J. N. Goodier, *Theory of Elasticity*, 2nd ed., McGraw-Hill, New York (1951).

*[7] A. Nadai, *Theory of Flow and Fracture of Solids*, 2nd ed., Vol. I, McGraw-Hill, New York (1950).

*[8] A. Phillips, *Introduction to Plasticity*, Ronald, New York (1956).

 [9] O. Hoffman and G. Sachs, *Introduction to the Theory of Plasticity for Engineers*, McGraw-Hill, New York (1953).

[10] W. Prager and P. G. Hodge, *Theory of Perfectly Plastic Solids*, Wiley, New York (1951).

[11] G. I. Taylor and H. Quinney, *Proc. Roy. Soc.* **230A**, 323 (1931).

[12] H. Neuber, *Theory of Notch Stresses* (English transl.), Edwards, Ann Arbor, Mich. (1946).

[13] G. N. Savin, *Stress Concentration Around Holes*, Pergamon, New York (1961).

[14] M. Hetenyi, *Handbook of Experimental Stress Analysis*, Wiley, New York (1950).

[15] M. M. Frocht, *Photoelasticity*, Wiley, New York (1941).

[16] G. Neugebauer, *Product Engineering*, **14**, 82 (1943).

[17] A. H. Cottrell, *Mechanical Properties of Matter*, Wiley, New York (1964).

*[18] R. Hill, *Mathematical Theory of Plasticity*, Oxford, New York (1950).

[19] L. Prandtl, *Nachr. Ges. Wiss. Gottingen*, **74**, (1920).

[20] D. Tabor, *The Hardness of Metals*, Oxford, New York (1951).

[21] E. R. Parker, *Brittle Behavior of Engineering Structures*, Wiley, New York (1957).

[22] S. Yukawa and J. G. McMullin, *Trans. ASME*, **83**, 541 (1961).

[23] J. F. Knott and A. H. Cottrell, *J. Iron Steel Inst.*, **201**, 249 (1963).

[24] P. Ludwick, *Elemente der Technologischen Mechanik*, Springer, Berlin (1909).

[25] E. Orowan, *Repts. Prog. Physics*, **12**, 185 (1948).

[26] F. Garofalo *et al.*, *Trans. AIME*, **221**, 310 (1961).

[27] H. H. Johnson, J. G. Morlet and A. R. Troiano, *Trans. AIME*, **212**, 526 (1958).

[28] W. D. Robertson and A. S. Tetelman, *Strengthening Mechanisms in Solids*, ASM, Cleveland (1962), p. 217.

[29] A. J. Kennedy, *Processes of Creep and Fatigue in Metals*, Wiley, New York (1963).

Problems

1. Derive a criterion for necking of a tensile specimen in terms of the true stress and engineering strain.

2. (a) Derive an expression for the true strain at necking and the UTS in terms of the parameters B, n, and σ_0 for a material whose flow curve is given by the relation $\sigma = \sigma_0(B + \epsilon)^n$, where B is a constant that defines the yield stress ($\epsilon = 0$).

(b) Assuming that $N = 2$ and $N' = 4$ in Eq. 1.6, what is the value of B such that the two criteria given by that equation are similar if $n = \frac{1}{2}$.

$$(1.86)$$

3. In certain BCC metals necking occurs immediately after yielding or during the Luder's extension. The instability condition is $\epsilon_L \gg n$, where ϵ_L is the Luder's strain. The friction stress σ_i is equal to the flow stress at very small microstrains and satisfies both Hooke's law and the relationship $\sigma = \sigma_0\epsilon^n$ simultaneously, as shown on the dia-

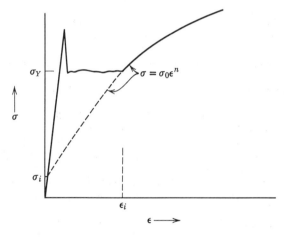

Fig. P. 1.3

gram. ϵ_L and σ_Y also define a point on the homogeneous strain-hardening curve which describes the plastic flow after the Luder's extension has ceased. The yield stress is given by the familiar relation $\sigma_Y = \sigma_i + k_y d^{-\frac{1}{2}}$, where $2d$ is the grain size.

Determine the conditions for mechanical instability in terms of d, σ_i, k_y, and the modulus E only. Should there be a transition temperature for this type of instability at a fixed grain size and should it be raised or lowered by dispersion hardening?

4. Suppose that the true stress–true strain curve (in tension) of a ductile material is given by the relation $\sigma = 200\epsilon^{1/2}$ in units of ksi. What is the increase in uniform strain that can be obtained if a round bar of the material is pulled in the presence of a hydrostatic pressure of 10 ksi applied uniformly perpendicular to the tensile axis?

(0.04)

5. Suppose that the flow curve of a material is given by the relation $\sigma = \sigma_0 (B + \epsilon)^n$ where $\sigma_0 = 100$ kpsi and $n = 0.5$ at all temperatures. Above $100°K$ the yield strength is directly proportional to $T^{-1/2}$, where T is the absolute temperature in degrees Kelvin, and is equal to 50 ksi at $400°K$. At this temperature, cleavage fracture occurs after 24% true strain. Assuming that the cleavage fracture strength is temperature-independent, what is the cleavage fracture strain at $300°K$?

(16%)

2

The Mechanics of Fracture

Fracture is an inhomogeneous process of deformation that causes regions of material to separate and load-carrying capacity to decrease to zero. It can be viewed on many levels, depending on the size of the fractured region that is of interest. At the *atomistic level* fracture occurs over regions whose dimensions are of the order of the atomic spacing (10^{-8} in.); at the *microscopic level* fracture occurs over regions whose dimensions are of the order of the grain size (about 5×10^{-4} in.); and at the *macroscopic level* fracture occurs over dimensions that are of the order of the size of flaws or notches (10^{-1} in. or greater).

At each level there are one or more criteria that describe the conditions under which fracture can occur. For example, at the *atomistic level* fracture occurs when bonds between atoms are broken across a fracture plane and new crack surface is created. This can occur by breaking bonds perpendicular to the fracture plane (Fig. 2.1*a*), a process called *cleavage,* or by shearing bonds across the fracture plane (Fig. 2.1*b*), a process called *shear.* At this level the fracture criteria are simple; fracture occurs when the local stresses build up either to the theoretical cohesive strength $\sigma_c \approx E/10$ or to the theoretical shear strength $\tau_c \approx G/10$, where E and G are the respective elastic and shear moduli.
moduli.

The high stresses required to break atomic bonds are concentrated at the edges of inhomogeneities that are called microcracks† or macrocracks (flaws, notches, cracks). At the *microscopic* and *macroscopic levels fracture results from the passage of a crack through a region of material.* The type of fracture that occurs is characterized by the type of crack responsible for the fracture. The criteria for fracture

† Microcracks are cracks that are one or two grain diameters in length. Macrocracks (usually just called cracks) are defects that are larger than this.

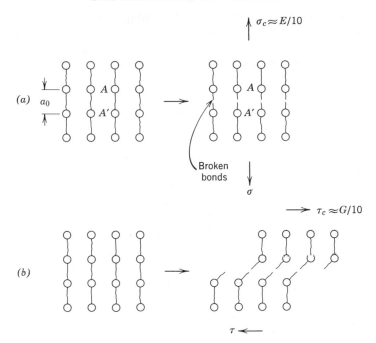

Fig. 2.1. Fracture viewed at the atomistic level in terms of the breaking of atomic bonds. (a) Cleavage. (b) Shear.

are complicated at these levels; it will require the remainder of this chapter to discuss the macroscopic criteria for fracture and Chapters 5 and 6 to discuss the microscopic criteria for fracture.

A description of the various types of fracture is presented in Section 2.1. In Section 2.2 the macroscopic fracture criterion for completely elastic materials is discussed. We show that fracture occurs when the product of the applied (nominal) stress and the stress concentration factor of a flaw reaches the cohesive stress, σ_c.

Few structural materials are completely elastic; localized plastic strain usually precedes fracture, even when the gross fracture strength is less than the gross yield strength. As we show in Section 2.3, fracture in these instances is initiated when a critical amount of local plastic strain or plastic work occurs at the tip of a flaw. From the principles of fracture mechanics it is possible to determine macroscopic fracture criteria in terms of the nominal fracture strength, the flaw length, and the critical amount of plastic work required to initiate unstable fracture (the fracture toughness). In Section 2.4 we discuss the physical significance of the fracture toughness parameters for the various types of fracture. Once we have established the boundary conditions under which fracture takes place, it is possible to consider procedures for testing for and designing against it; this is the subject of Chapter 3.

2.1 Types of Fracture That Occur Under Uniaxial Tensile Loading

Cleavage fractures occur when a cleavage crack spreads through a solid under a tensile component of the externally applied stress† (Fig. 2.2a). The material fractures because the concentrated tensile stresses at the crack tip are able to break atomic bonds, as shown in Fig. 2.1a. In many crystalline materials certain crystallographic planes of atoms are most easily separated by this process and these are called *cleavage planes*. Table 2.1 lists the preferred cleavage planes in the more important crystal structures.

Under uniaxial tensile loading the crack tends to propagate perpendicularly to the tensile axis. When viewed in profile, cleavage fractures appear "flat" or "square," and these terms are used to describe them [1, 2]. Most structural materials are polycrystalline. The orientation of the cleavage plane(s) in each grain of the aggregate is usually not perpendicular to the applied stress so that, on a microscopic scale, the fractures are not completely flat over distances larger than the grain size. In very brittle materials cleavage fractures can propagate continuously from one grain to the next. However, in materials such as mild steel the macroscopic cleavage fracture is actually discontinuous on a microscopic level; most of the grains fracture by cleavage but some of them fail in shear, causing the cleaved grains to link together by tearing.

Shear fracture, which occurs by the shearing of atomic bonds, is actually a process of extremely localized (inhomogeneous) plastic deformation. In crystalline solids, plastic deformation tends to be confined to crystallographic planes of atoms which have a low resistance to shear. These planes are called slip planes and are listed in Table 2.1 for various materials. Shear fracture in pure single crystals occurs when the two halves of the crystal slip apart on the crystallographic glide planes that have the largest amount of shear stress resolved across them [3]. When the shear occurs on only one set of parallel planes, a *slant fracture* is formed (Fig. 2.2b); *chisel-point fracture* occurs when the shear takes place in two directions.

---

† Internal (residual) stresses are only important in specific problems and will not be considered here.

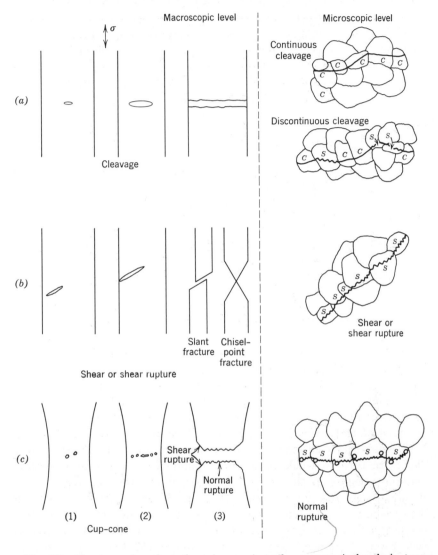

Fig. 2.2. Fracture viewed at the microscopic and macroscopic levels in terms of the passage of various types of cracks. (a) Cleavage: on microscopic level, cleavage can occur continuously, in which case each grain cleaves, or discontinuously, in which some grains cleave C; others fail in shear S. (b) Shear or shear rupture. (c) Cup-cone fracture in tensile specimen resulting from normal and shear rupture: on microscopic level, normal rupture results from formation of voids ° and shearing of material between them.

Table 2.1

Cleavage and Shear Planes for Various Crystal Structures and Materials

Crystal Structure	Example	Cleavage Plane	Primary Shear Planes
BCC	Li, Na, K, Fe, most steels, V, Cr, Mn, Cb, Mo, W, Ta	{100}	{112}, {110}
FCC	Cu, Ag, Au, Al, Ni, brass, 300 series stainless steels	None	{111}
HCP	Be, Mg, Zn, Sn, Ti, U, Cd, graphite	{1000}	{1122}, {1010}, {1000}
Diamond	diamond, Si, Ge	{111}	{111}
Rock salt	NaCl, LiF, MgO, AgCl	{100}	{110}
Zinc blend	ZnS, BeO	{110}	{111}
Fluorite	CaF_2, UO_2, ThO_2	{111}	{100}, {110}

In polycrystalline materials the advancing shear crack tends to follow the path of maximum resolved shear stress. This path is determined by both the applied stress system and the presence of internal stress concentrators such as voids, which are formed at the interface between impurity particles (e.g., nonmetallic inclusions) and the matrix material. Crack growth takes place by the formation of voids [4–8] and their subsequent coalescence by localized plastic strains. Shear fracture in thick plates and round tensile bars of structural materials begins in the center of the structure (necked region) and spreads outwards (Fig. 2.2c). The macroscopic fracture path is perpendicular to the tensile axis. On a microscopic scale the fracture is quite jagged, since the crack advances by shear failure (void coalescence) on alternating planes inclined at 30–45° to the tensile axis. This form of fracture is commonly labeled *normal rupture* (since the fracture path is normal to the tensile axis) or *fibrous fracture* (since the jagged fracture surface has a fibrous or silky appearance). Normal rupture forms the central (flat) region of the familiar cup-cone pattern. The structure finally fails by *shear rupture* (shear lip formation) on planes inclined at 45° to the tensile axis. This form of fracture is less jagged, appears smoother, and occurs more rapidly than the normal rupture which precedes it. Similarly noncleavage fracture in thin sheets of engineering materials occurs exclusively by shear rupture and the fracture profile appears similar to the slant fracture shown in Fig. 2.2b.

Under certain conditions the boundary between adjacent grains in the polycrystalline aggregate is weaker than the fracture planes in the grains themselves. Fracture then occurs *intergranularly*, by one of the processes mentioned above, rather than through the grains (*transgranular fracture*). Thus there are six possible modes of fracture: transgranular cleavage, transgranular shear rupture, transgranular normal rupture, and intergranular cleavage, intergranular shear rupture, intergranular normal rupture.

Fracture takes place by that mode which requires the least amount of *local strain* at the tip of the advancing crack. Both the environmental conditions and the state of *applied* (nominal) stress and strain determine the type of fracture which occurs, and only a few materials and structures fracture exclusively by one particular mode over a large variety of operating conditions (i.e., service temperature, corrosive environment, and so on). Furthermore, under any given condition more than one mode of fracture can cause failure of a structural member and the fracture is described as "mixed." This implies that the relative ease of one type of crack propagation can change, with respect to another type, as the overall fracture process takes place. For example, normal rupture, cleavage, and shear rupture are all observed on the fracture surfaces of notched mild steel specimens broken in impact at room temperature. (The characteristic features of the various types of fracture, as observed by optical and electron microscopy, will be described in Chapter 3.)

In order to analyze the fracture process under various types of stress systems, it is necessary to establish a coordinate system with respect to both the fracture plane, the direction of crack propagation, and the applied stress system. One of the difficulties encountered by engineers and scientists who are interested in a particular aspect of the fracture problem is the large mass of notation and coordinate systems used by other workers who have investigated similar problems. At the present time there is no standard and it seems most convenient to use a modification of the system employed by the ASTM Committee on Fracture Testing of High Strength Sheet Materials [9]. The system is illustrated in Fig. 2.3a. The crack whose length $= 2c$, height $= 2h$, and tip radius $= \rho$, lies in the xz plane of a plate of thickness t and width W. The propagation of the crack is treated as a

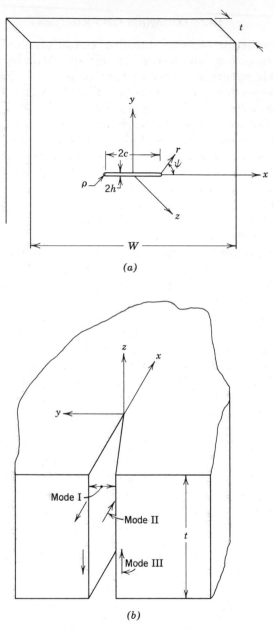

Fig. 2.3 (a) Coordinate system used in evaluating stress field at a point (r, Ψ) near a crack of length $2c$, tip radius ρ, and height $2h$ in a plate of width W and thickness t. (b) Modes of deformation applied to a cracked plate to produce the indicated modes of crack opening.

two-dimensional process since this introduces a relatively small error in computation and considerably simplifies the treatment of the problem. The direction of crack propagation is along the x axis and $x = 0$ at the center of the crack. The leading edge of the crack is parallel to the z axis. At the center of the plate $z = 0$ and equals $\pm t/2$ at the outer surfaces of the plate; (r, ψ) are the coordinates of a point in the xy plane near the crack tip. Three distinct modes of separation at the crack tip can occur, as shown in Fig. 2.3b.

Mode I. The tensile component of stress is applied in the y direction, normal to the faces of the crack, either under plane-strain (thick plate, t large) or plane-stress (thin plate, t small) conditions.

Mode II. The shear component of stress is applied normal to the leading edge of the crack either under plane-strain or plane-stress conditions.

Mode III. The shear component of stress is applied parallel to the leading edge of the crack (antiplane strain).

All stresses are computed in terms of the *gross section area Wt* unless stated otherwise, in which case the suffix "net" will be used (i.e., σ_{net}).

2.2 The Mechanics of (Elastic) Crack Propagation

Theoretical cohesive stress

At an atomistic level the fracture strength of a material will depend on the strength of its atomic bonds, such as that between atoms A and A' in Fig. 2.1a. To estimate this bond strength [2, 10] let a_0 be the equilibrium spacing between atomic planes in the absence of applied stress. The stress σ required to separate the planes to a distance $a > a_0$ increases until the theoretical strength σ_c is reached (Fig. 2.4) and the bonds are broken. Further displacement of the atoms can then occur under a decreasing applied stress. This stress-displacement curve can be approximated by a sine curve having wavelength λ, as shown on the figure. Thus

$$\sigma = \sigma_c \sin\left(\frac{2\pi x}{\lambda}\right) \tag{2.1}$$

where $x = (a - a_0)$ is the displacement from equilibrium. At small displacements the small angle approximation ($\sin x \cong x$) holds so that

$$\sigma \cong \sigma_c \frac{2\pi x}{\lambda}$$

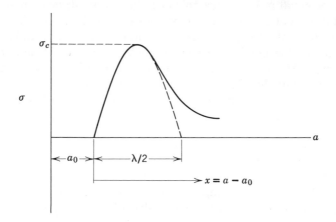

Fig. 2.4. The tensile stress required to separate atomic planes to a distance $a > a_0$; a_0 is the equilibrium separation at $\sigma = 0$. Fracture occurs when $\sigma = \sigma_c$.

If we assume that these small displacements also obey Hooke's law

$$\sigma = E\epsilon = \frac{Ex}{a_0} \tag{2.2}$$

then

$$\sigma_c = \frac{\lambda}{2\pi} \frac{E}{a_0}$$

For purposes of describing the energy relations during fracture, we now define a quantity called the *true surface energy* γ_s as the work done in creating new surface area by the breaking of atomic bonds. From Fig. 2.4 this is simply one-half the area under the stress-displacement curve since two new surfaces are created each time a bond is broken. Thus

$$2\gamma_s = \int_0^{\lambda/2} \sigma_c \sin\left(\frac{2\pi x}{\lambda}\right) dx = \frac{\lambda \sigma_c}{\pi} \tag{2.3}$$

so that from **2.2**

$$\sigma_c = \sqrt{\frac{E\gamma_s}{a_0}} \tag{2.4}$$

If we take typical values of $E = 10^{12}$ dynes per cm² (14.5 × 10⁶ psi), $a_0 = 3 \times 10^{-8}$ cm, and $\gamma_s = 10^3$ erg per cm² (5.8 × 10⁻³ in. lb per in.²) than $\sigma_c = 1.73 \times 10^{11}$ dynes per cm² = (2 × 10⁶ psi) $\approx E/7$.

Other calculations of the theoretical strength have been obtained [11] by using more precise force-separation laws and these give values of σ_c ranging from $E/4$ to $E/13$. In general,

$$\sigma_c = \frac{E}{10} \tag{2.5}$$

and

$$\gamma_s = \frac{Ea_0}{20}$$

are good measures of the theoretical strength and surface energy for most inorganic materials. Table 2.2 indicates that strengths of this

Table 2.2
$\sigma_f = $ **Maximum**
Observed Strength [11]

Material	(psi $\times 10^{-6}$)	E(psi $\times 10^{-6}$)	E/σ_f
Music wire	0.4	20	72
Silica fibers	3.5	14	4
Silica rods	1.9	14	7
Iron whiskers	1.9	43	23
Al_2O_3 whiskers	2.2	72	33
NaCl whiskers	0.16	6.3	40
BeO whiskers	2.8	49	17
Silicon whiskers	0.94	24	26
Silicon (bulk)	0.75	24	32
TiC (bulk)	0.80	70	87
Boron	0.35	51	145
Ausformed steel	0.45	29	64

order of magnitude have been achieved in a variety of materials when laboratory tests were performed on very thin specimens (fibers or whiskers) that were carefully prepared and carefully handled prior to testing.

For most metals the theoretical strength varies between 1,000,000 and 3,000,000 psi. However, structures that are commercially produced for engineering applications commonly fracture at *applied* stress levels 10 to 100 times *below* these values (i.e., at 10,000 to 300,000 psi), and the theoretical strength is rarely obtained in engineering practice. There are three reasons for this discrepancy. First, because *stress concentrators* such as cracks or notches exist (or are formed by plastic flow) in materials and these raise the applied stress up to the theoretical strength, at the crack tip. Closely allied to this reason is the *thermodynamic factor,* which shows that elastic cracks spread under applied stress when their growth lowers the *total energy* of the system (solid under stress). A second reason that cleavage fractures occur at applied stresses far below the theoretical value is the presence of "planes of weakness" where atomic bonds have been weakened by impurity atoms (e.g., grain boundary segregation) or particular chemical environments (e.g., stress corrosion cracking). Finally, we can see that the theoretical cohesive (cleavage) stress will not be reached if some other fracture process (e.g., shear or rupture), which occurs by plastic deformation, intervenes at a lower level of applied stress.

The concentration of stress around an elastic crack

Suppose that a tensile stress σ is applied across a thin plate (plane-stress case) of elastically isotropic material containing an elliptical hole whose major axis $2c$ is perpendicular to σ. This hole can be thought of as a crack having a length $= 2c$ and a height $= 2h$ (Fig. 2.3a). If the length of the crack is small compared to the width of the plate in which it lies, then the stresses in the vicinity of the crack can be directly obtained by the methods of elasticity theory [12, 13, 14]. The maximum tensile stress σ^{max} occurs at the end of the crack and is given by the expression†

$$\sigma^{max} = \sigma \left(1 + \frac{2c}{h} \right)$$

This stress is usually expressed in terms of the radius of curvature ρ at the end of the crack. Since $\rho = h^2/c$ for an ellipse

$$\sigma^{max} = \sigma \left[1 + 2 \sqrt{\frac{c}{\rho}} \right]$$

$$\cong 2\sigma \sqrt{\frac{c}{\rho}} \qquad \text{for } c \gg \rho \qquad (2.6)$$

so that the stress concentration factor for the "infinitely" sharp crack

$$K_\sigma = 2 \sqrt{\frac{c}{\rho}} = \frac{\sigma^{max}}{\sigma}$$

† Note that if $c = h$ the crack would be circular and the maximum stress concentration would be 3, as given by Eq. 1.15.

The crack half-height is related to the applied stress and crack length as [13]

$$h = \frac{2\sigma c}{E} \qquad (2.7)$$

At points (r, ψ) removed from the crack tip (Fig. 2.3) the elastic stresses are given by the relations [13]

$$\sigma_{xx} = \sigma \left(\frac{c}{2r}\right)^{\frac{1}{2}} \left[\frac{3}{4} \cos \frac{1\psi}{2} + \frac{1}{4} \cos \frac{5\psi}{2}\right] \qquad (2.8a)$$

$$\sigma_{yy} = \sigma \left(\frac{c}{2r}\right)^{\frac{1}{2}} \left[\frac{5}{4} \cos \frac{1\psi}{2} - \frac{1}{4} \cos \frac{5\psi}{2}\right] \qquad (2.8b)$$

$$\tau_{xy} = \sigma \left(\frac{c}{2r}\right)^{\frac{1}{2}} \left[\sin \frac{\psi}{2} \cos \frac{4\psi}{2} \cos \frac{3\psi}{2}\right] \qquad (2.8c)$$

which are valid for $c \gg r \gg \rho$. Directly ahead of the crack ($\psi = 0$) $\sigma_{xx} = \sigma_{yy} = \sigma(c/2r)^{\frac{1}{2}}$ and $\tau_{xy} = 0$. From Eq. 1.8 it is noted that there is no shear stress directly ahead of the crack in the xy plane.

At relative distances that are closer to the tip of a notch or crack (i.e., as $r \to \rho$) a useful *approximation* for the tensile stress acting across the yz plane ($\psi = 0$) is [15]

$$\sigma_{yy} = \sigma K_\sigma \sqrt{\frac{\rho}{\rho + 4r}} \qquad (2.9)$$

where K_σ is the elastic stress concentration factor for the particular notch at its tip ($r = 0$) and σ is the applied tensile stress. For the sharp crack $K_\sigma = 2\sqrt{c/\rho}$ and Eq. 2.9 reduces to Eq. 2.8b at $r = \rho$.

Equations 2.8 indicate that the local stresses near a flaw depend on the product of the nominal stress σ and the square root of the flaw depth $2c$. The importance of this fact was first appreciated by Irwin [16] who coined the term *stress intensity factor* to emphasize this fundamental relation. For an infinitely sharp elastic crack, in an infinitely wide plate, the stress intensity factor K is defined as

$$K = \sigma \sqrt{\pi c} \qquad (2.10a)$$

The local stresses near the crack are then written

$$\sigma_{xx} = \frac{K}{\sqrt{2\pi r}} F(\theta) \qquad \sigma_{yy} = \frac{K}{\sqrt{2\pi r}} F'(\theta) \qquad \tau_{xy} = \frac{K}{\sqrt{2\pi r}} F''(\theta)$$

$$(2.10b)$$

where the angular functions are similar to those given by Eq. 2.8.

A comparison of Eqs. 2.6 and 2.10 indicates that K is proportional to the limiting value of the elastic stress concentration factor as the root radius approaches zero [45] (i.e., $K = \lim\limits_{\rho \to 0} (\sigma^{max}/2) \sqrt{\pi\rho}$). K values can thus be determined for variously shaped flaws, in structures having *finite dimensions*, if values of the elastic stress concentration factors are known. Many of these can be obtained from standard texts [29, 30]; more complicated shapes can be analyzed by various mathematical techniques which have recently been reviewed extensively [20].

In all cases the stress intensity factor has the form

$$K = \sigma \sqrt{\alpha\pi c} \tag{2.11}$$

where α is a parameter depending on specimen and crack geometry. For example, for the important case of a center-cracked plate under tensile loading [9, 20]

$$K_I = \sigma \sqrt{\pi c} \left\{ \frac{W}{\pi c} \tan\frac{\pi c}{W} \right\}^{\frac{1}{2}}$$

where the term inside the brackets (α) varies from 1 to 1.20 as c/W varies from 0.037 to 0.29 [20]. Alternatively,

$$K_I = \sigma \sqrt{W} \left\{ \tan\frac{\pi c}{W} \right\}^{\frac{1}{2}} \tag{2.12}$$

which reduces to 2.10 for $c \ll W$.

The Griffith and Irwin relations for elastic crack propagation

A necessary condition for the propagation of the elliptical, elastic crack is that the maximum tensile stress level at its tip reach the theoretical cohesive stress σ_c. From Eqs. 2.4 and 2.6 we observe that the nominal stress $\sigma = \sigma_F$ required to satisfy this condition is

$$2\sigma_F \sqrt{\frac{c}{\rho}} = \sqrt{\frac{E\gamma_s}{a_0}}$$

or

$$\sigma_F = \sqrt{\frac{2E\gamma_s}{\pi c}\left(\frac{\pi\rho}{8a_0}\right)}$$

$$\sigma_F \cong \sqrt{\frac{2E\gamma_s}{\pi c}\left(\frac{\rho}{3a_0}\right)} \tag{2.13}$$

It is important to note that this relation *only applies* for completely brittle and completely elastic solids where (1) the elastic limit or yield strength σ_Y is much greater than the cohesive stress (approximately $E/10$), and (2) there is also no microplastic strain occurring near the crack tip.

Once the crack has started to spread, its length $2c$ increases. If the tip radius ρ remains constant or decreases as the crack extends (the crack gets flatter or "sharper") or if c/ρ remains constant (the crack shape remains the same), then the crack will spread across the material and cause complete fracture of the part, provided that σ is maintained constant during crack propagation. *Therefore, σ_F is the applied stress required to cause complete fracture (i.e., the fracture stress).* This is shown schematically in Fig. 2.5 where c_F is the crack length that begins to spread at $\sigma = \sigma_F$ and σ_M is the stress required to keep it in motion (curve AB).

An expression of this form was originally derived by Griffith [17] on the basis of thermodynamic considerations. He reasoned that the unstable propagation of a crack must result in a decrease in free energy of the system (cracked plate under stress with fixed grips) and proposed that a crack would advance when the incremental release of stored elastic strain energy $d(W_E)$ in a body became greater than the incremental increase of surface energy $d(W_S)$ as new crack surface was created. For the two-dimensional case in plane stress

$$W_E = \frac{\pi \sigma^2 c^2}{E} \tag{2.14}$$

and

$$W_S = 4c\gamma_s$$

since the internal crack spreads symmetrically about its center and two surfaces are created, each of length $2c$. The Griffith criterion can then be written as

$$\frac{\partial}{\partial c}\left(\frac{\pi \sigma^2 c^2}{E}\right) \geqq \frac{\partial}{\partial c}(4c\gamma_s)$$

so that

$$\sigma = \sigma_F = \sqrt{\frac{2E\gamma}{\pi c}} \tag{2.15}$$

Equation 2.15 is a necessary condition for elastic crack propagation. Comparison of this equation with 2.13 indicates that $\rho = 3a_0$ is a lower limit of the "effective" radius of an elastic crack. Cottrell [18] has pointed out that this results from the fact that, irrespective of crack sharpness, some surface energy must be created at the tip of the advancing crack so that σ_F cannot approach zero as ρ approaches 0. Thus, when $\rho < 3a_0$, the stress for unstable crack propagation is given by 2.15. When $\rho > 3a_0$ the fracture stress is given by Eq. 2.13 because it is then the more difficult of the two requirements. *Both Eqs. 2.13 and 2.15 must be satisfied for unstable fracture to occur*, in order that the local conditions at the crack tip permit the breaking of atomic bonds and that the overall free energy of the system be reduced. This point is very important when plastic deformation accompanies cleavage crack propagation, as we shall see shortly.

The Irwin analysis [5, 16] of fracture proposes that crack propagation occurs at $\sigma = \sigma_F$ when a parameter defined as the *crack extension force* $G = K^2/E$ (in plane stress) is equal to a critical value G_c, the critical strain-energy release rate for unstable crack extension, also called the *toughness* or the *crack-resistance force* of the material.

When $G = G_c$, $K = K_c$, where K_c is called the *fracture toughness*. Thus

$$\frac{K_c^2}{E} = G_c \tag{2.16}$$

and, for the elastic crack in an infinitely wide plate,

$$\frac{\sigma_F^2 \pi c}{E} = G_c$$

so that

$$\sigma_F = \sqrt{\frac{EG_c}{\pi c}} \tag{2.17}$$

A comparison of this equation with Eq. 2.15 indicates that

$$G_c = 2\gamma_s \tag{2.18}$$

so that the two approaches lead to the same result although their symbols and terminology are different.

Similar calculations have been performed for elastic crack propagation under plane stress [13], for the three-dimensional penny-shaped crack [19], for cracks exposed to the external surface and for cleavage under combined stress [2]. Table 2.3 lists some of these results[†] in terms of Griffith expressions and Irwin K_c values. It is significant that all of these relations are of the same form and that there are only

[†] A complete listing of expressions for K_I, K_{II}, and K_{III} for various types of deformation applied to cracks of more complicated shapes, in structures of finite width, can be found in the excellent review paper by Paris and Sih [20].

Table 2.3

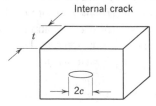

Internal crack

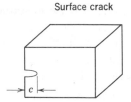

Surface crack

Fracture stress σ_F and fracture toughness K_c for plate containing an internal crack of length $2c$ or surface crack of length c. The width of the plate is assumed to be infinite (or semi-infinite for surface crack). In plate of finite thickness these relations apply for $\sigma < 0.6\sigma_Y$ or $\tau < 0.6k$.

$G_c = 2\gamma_P^*$ = work done in initiating unstable fracture at tip of crack when $\sigma = \sigma_F$.

E = elastic modulus $G = \dfrac{E}{2(1 + \nu)}$ = shear modulus ν = Poisson's ratio

---

Mode I deformation

$$\sigma_F = \sqrt{\frac{2E\gamma_P^*}{\pi c}} \quad \text{plane stress (t small)} \qquad K_{Ic} = \sigma_F \sqrt{\pi c} = \sqrt{G_c E}$$

$$\sigma_F = \sqrt{\frac{2E\gamma_P^*}{\pi c(1 - \nu^2)}} \quad \text{plane strain (t large)} \qquad K_{Ic} = \sigma_F \sqrt{\pi c} = \sqrt{\frac{G_c E}{(1 - \nu^2)}}$$

$$\sigma_F = \sqrt{\frac{\pi E\gamma_P^*}{2c(1 - \nu^2)}} \quad \begin{array}{l}\text{plane strain (penny-}\\ \text{shaped crack)}\end{array} \qquad K_{Ic} = \frac{2\sigma_F}{\pi} \sqrt{\pi c} = \sqrt{\frac{G_c E}{(1 - \nu^2)}}$$

Mode II deformation

$$\tau_F = \sqrt{\frac{2G\gamma_P^*}{\pi c(1 - \nu)}} \quad \text{plane strain} \qquad K_{IIc} = \tau_F \sqrt{\pi c} = \sqrt{\frac{G G_c}{(1 - \nu)}}$$

Mode III deformation

$$\tau_F = \sqrt{\frac{2G\gamma_P^*}{\pi c}} \quad \text{antiplane strain} \qquad K_{IIIc} = \tau_F \sqrt{\pi c} = \sqrt{G G_c}$$

---

small differences between them. It is convenient to have one form of these relations that can be used in a general sense when effects pertinent to all relations are considered. In this book we shall use the plane-stress tension equation (2.17) to describe the tensile fracture stress unless otherwise stated.

Velocity of cleavage

Once the crack has started to spread, its tip will be moving at some velocity $v_c = dc/dt$. As c increases, the semiminor axis h of the crack will also increase, according to Eq. 2.7. In order for h to increase, a (two-dimensional) volume element of material $(dx\ dy)$ near the sides of the crack must be displaced perpendicular to the crack plane; the rate at which this material is moved limits [21] the speed at which the crack tip can advance. This sideways movement of material can be considered as a kinetic effect, so that the spreading crack will have a kinetic energy K_E in addition to a strain energy W_E and surface energy W_S. Because the total energy of the system remains constant under constant stress σ, we can write that

$$K_E = W_E - W_S = W_E \left[1 - \frac{W_S}{W_E} \right] \tag{2.19}$$

On dimensional grounds K_E must be written as [21]

$$K_E = \frac{\kappa\, \Delta v_c{}^2}{2E} \frac{\sigma^2 c^2}{E} = \frac{\kappa\, \Delta v_c{}^2}{2E\pi} W_E \tag{2.20}$$

where Δ is the density of the material and κ is a numerical constant. Therefore

$$v_c{}^2 = \frac{2E\pi}{\kappa\Delta} \left[1 - \frac{W_S}{W_E} \right]$$

$$v_c = \sqrt{\frac{2\pi}{\kappa}} \sqrt{\frac{E}{\Delta}} \left[1 - \frac{W_S}{W_E} \right]^{\frac{1}{2}}$$

$$v_c = 0.38v_0 \left[1 - \frac{W_S}{W_E} \right]^{\frac{1}{2}} = 0.38v_0 \left[1 - \frac{4E\gamma_s}{\pi\sigma_c{}^2} \right]^{\frac{1}{2}} \tag{2.21}$$

where $v_0 = \sqrt{E/\Delta}$ is the velocity of sound and the constant $\sqrt{2\pi/\kappa} \cong 0.38$. As c becomes large so that the limiting crack velocity should be about 40% of the velocity of sound, W_S/W_E approaches zero. This has been observed experimentally [22] in a number of materials. Since $v_0 = 16,000$ fps in steel, for example, it is easy to see how large structures can fracture in very short times when cleavage cracks begin to run through them.

Crack propagation accompanied by plastic deformation

Equations 2.13 and 2.15 define the conditions for the propagation of an elastic crack in a completely brittle solid—one which does not yield (deform plastically) at the crack tip. Engineering materials do not contain cracks such as these, nor do they fracture in a completely elastic manner. If they did, they would be so brittle that they would be unreliable for use as structural members. In most materials, localized plastic deformation is produced by the high stresses near crack tips and this deformation gives the materials some toughness, or resistance to crack propagation.

The nature and importance of the localized plastic deformation that accompanies crack propagation were first appreciated by Orowan and his coworkers [2, 23, 24] and by Irwin [16]. These workers recognized that when plastic deformation occurs near the crack tip, a certain amount of *plastic work* γ_P is expended during crack propagation, in addition to the elastic work γ_s that is required to create the two fracture surfaces. The mechanics of the fracture process, when plastic relaxation occurs at the crack tip, depends on the magnitude of γ_P. This, in turn, depends on crack velocity and temperature, as well as on the nature of the material (Chapter 5).

When the crack velocity is high, the temperature is low, and the material is intrinsically brittle, γ_P *is relatively low* (i.e., $\gamma_P < 10\gamma_s$). Low values of γ_P, in turn, imply: (1) that the plastic strains in the vicinity of the crack are also small, of the order of the microstrains that occur at stresses below the yield stress (Chapter 4); (2) that the yield strength σ_Y of regions adjacent to the tip is very high, of the order of $E/10$ or greater. *Consequently crack propagation is still continuous and elastic.* However, the total work done in expanding the crack, per unit area of fracture surface, is increased from γ_s to $(\gamma_P + \gamma_s)$. The Griffith equation is modified to read (in plane stress)

$$\sigma_M = \sqrt{\frac{2E}{\pi c}(\gamma_s + \gamma_P)}$$

$$= \sqrt{\frac{2E\gamma_s}{\pi c}\left(1 + \frac{\gamma_P}{\gamma_s}\right)} \tag{2.22}$$

when $\gamma_P/\gamma_s > 1$,

$$\sigma_M = \sqrt{\frac{2E\gamma_s}{\pi c}\left(\frac{\gamma_P}{\gamma_s}\right)} = \sqrt{\frac{2E}{\pi c}\gamma_P} \tag{2.23}$$

Neglecting the factor 1 in Eq. 2.22, we can use γ_P as the *total* work done when a crack spreads through a solid. *The stress required to keep the crack in motion is* σ_M; the subscript $_M$ emphasizes the fact that we are considering the work done around a *moving* crack. Similar relations, of course, can be written for all of the fracture criteria presented in Table 2.3.

The plastic deformation occurring near the crack tip "blunts" the crack [25] (Chapter 5) and thereby relaxes some of its concentrated stresses by increasing ρ. This is apparent from a comparison of Eqs. 2.13 and 2.23 since

$$\sqrt{\frac{2E\gamma_s}{\pi c}\left(\frac{\rho}{3a_0}\right)} = \sqrt{\frac{2E\gamma_s}{\pi c}\left(\frac{\gamma_P}{\gamma_s}\right)}$$

so that

$$\frac{\rho}{3a_0} = \frac{\gamma_P}{\gamma_s} \qquad (\gamma_P > \gamma_s) \qquad (2.24)$$

Note that although the additive term γ_S in Eq. **2.22** has been neglected as being small compared with γ_P, γ_S is still very much a part of the fracture process because it is a *multiplying factor* in Eq. **2.24** for determining γ_P at a particular value of ρ.

Suppose that the crack which started propagating elastically at $c = c_F$, $\sigma = \sigma_F$ (Fig. 2.5) becomes blunted by some plastic deformation when $c = c_x$ so that γ_P becomes greater than γ_S. The stress σ_M is then represented by the curve DE. If σ is maintained constant during propagation, the crack will still be able to continue spreading at $c = c_x$ because $\sigma > \sigma_M$. However, if the plastic work γ_P is sufficiently extensive that σ_M is described by the curve FG, then the Griffith criterion is no longer satisfied at c (because $\sigma < \sigma_M$) and the crack will decelerate and stop. *Localized plastic flow can therefore increase the toughness of a material.*

2.3   The Mechanics of Plastically Induced Fracture

General considerations

It is important to remember that the Griffith equations **2.13** and **2.23** *only apply to continuous, elastic fractures*, that is, under conditions (low temperature, high crack velocity, intrinsically brittle material) where the yield strength σ_Y of regions adjacent to the tip is raised above the cohesive stress $E/10$. This allows the tensile stresses at the crack tip σ_{yy} to build up to $E/10$ and cause atomic bonds to be broken there.

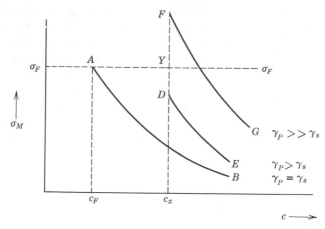

Fig. 2.5. The stress σ_M required to continue unstable propagation of a crack of length $2c$ as a function of the total work γ_P required for crack propagation. σ_F is the stress at which fracture initiates near the tip of a flaw of length $2c_F$.

Alternatively, when the crack is stopped, when the temperature is higher, and when the material is intrinsically tougher, the yield strength of regions near the tip will be less than $E/10$. In the absence of appreciable strain hardening, σ_{yy} cannot be raised much above σ_Y, as shown in Chapter 1 for the case where local yielding occurs around a circular hole (Fig. 1.7). Since $\sigma_{yy} \approx \sigma_Y < E/10$ there is insufficient tensile stress to break atomic bonds at the crack tip. *Consequently, continuous, elastic crack propagation is not possible and the Griffith relation (Eqs. 2.13 and 2.23) cannot be used as fracture criteria.* (We shall develop criteria for these cases in this section.)

Low values of σ_Y relative to $E/10$ imply: (1) relatively large plastic zones near the crack tip, (2) relatively large values of γ_P (i.e., $\gamma_P \approx 10^3\gamma_s$), and (3) relatively large plastic strains in regions adjacent to the crack tip. In fact, it is these localized plastic strains which cause the crack to grow; *consequently, fracture is plastically induced.*

All types of *crack propagation*, with the exception of continuous cleavage (Fig. 2.2a), fall into this category. Normal rupture and shear rupture, which occur by void formation (at small particles) and void coalescence require plastic deformation to cause the voids to join together.   Discontinuous cleavage, which is actually a form of normal rupture [26, 27, 28] since it is a series of individual, microcleavage crack initiations (hole formation) followed by shearing of the un-cleaved gains (hole coalescence), also requires plastic deformation for the same reason.

In structural materials the initiation of unstable fracture usually occurs at or near the tip of a preinduced crack or notch.   These de-fects are introduced into the structure by such factors as fatigue, in-complete welding, and corrosion.   *In all structural metals and most structural nonmetals as well, fracture initiation is plastically induced* since $\sigma_Y < E/10$ near the tip of a slowly moving or stopped crack.

Suppose, for example, that a flaw of half-length $c = c_0$ exists in a plate that is loaded in tension (mode I).   Basically there are three ways by which this flaw can develop into an unstable fracture.

1. Go–No Go Situation Under Increasing Applied Stress (as in a Tensile Test)

At a given applied stress $\sigma = \sigma_F$, a dynamic microcleavage crack is formed, by plastic deformation [31–35] in the plastic zone near the pre-induced flaw.†   If this microcrack can grow unstably (either con-tinuously or discontinuously) without any increase in the applied stress, complete fracture of the plate occurs.   Thus σ_F is the nominal fracture stress; $c \to \infty$ at $\sigma = \sigma_F$ (Fig. 2.6a).   This type of behavior is noted in materials tested below their ductile-brittle transition tem-perature.

2. Slow Go Situation Under Increasing Applied Stress (as in a Tensile Test)

When the material is more plastic than in situation 1, the first micro-crack or void which forms ahead of the crack cannot propagate un-stably.   A whole series of new microcracks or voids must be formed and joined together (by plastic tearing) to allow the overall crack front to advance.   Initially, crack growth proceeds slowly, in a stable manner [36], with increasing stresses above σ_1 (Fig. 2.6b) required for crack growth.   This can be shown as a dashed curve AB, but usually in structural materials it is a series of bursts of crack growth which can be represented by the solid steps in the figure.   Eventually the crack becomes long enough, and the applied stress high enough, to allow unstable propagation; $c \to \infty$ at $\sigma = \sigma_F$.   The degree of slow crack growth that occurs before instability is one measure of a

---

† The mechanisms of cleavage crack formation at stress levels σ_Y that are less than $E/10$ on a *microscopic scale* will be considered in Chapter 6.

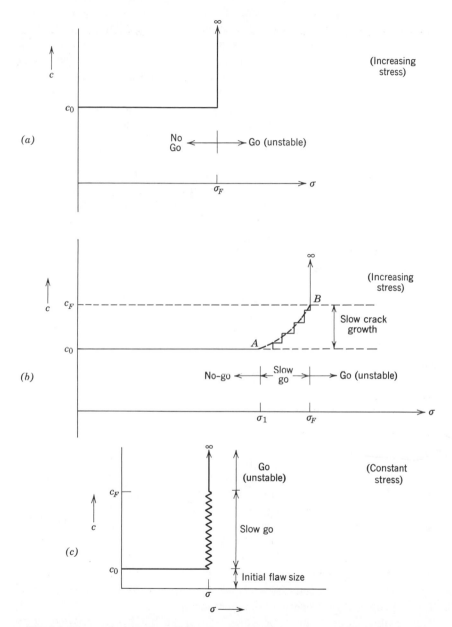

Fig. 2.6. The variation of crack length with applied stress. (a) Unstable fracture initiated at a stress $\sigma = \sigma_F$ without prior slow growth of a crack of length $c_0 = c_F$. (b) Slow crack growth from c_0 to c_F occurs before unstable fracture at $\sigma = \sigma_F$, (c) Slow growth at constant stress from c_0 to c_F.

59

material's toughness. For example, in low strength BCC metals such as mild steel, stable fibrous (normal) rupture can develop into unstable, discontinuous cleavage at the lower end of the brittle-ductile transition temperature region. Above the ductile-brittle transition the toughness is so high that for all practical purposes crack growth is always stable and occurs under increasing stress. Alternatively in thick plates of high strength material, such as 7075-T6 aluminum alloy, the normal rupture itself absorbs little plastic deformation and only small amounts of microscopic slow crack growth precede instability, or "pop in" ($\sigma_F \approx \sigma_1$).

Slow crack growth also occurs when more than one *macroscopic mode* of fracture is operating simultaneously [9] (i.e., normal rupture in the center and shear rupture at the outer edges of a thin plate of a high strength alloy). In this case the combined crack front starts to grow slowly when $\sigma = \sigma_1$, where σ_1 now represents (approximately) the fracture stress of a thick plate of the same alloy. This macroscopic form of slow crack growth will be discussed in more detail in the following chapter.

3. Slow Crack Growth Under Constant Applied Stress (Static Test)

Slow crack growth can also occur under *constant stress*, over a *period of time*, when certain materials are subjected to particular environments (e.g., corrosive) or alternating loading (which leads to fatigue). Under these conditions the crack extends slowly until it becomes long enough to propagate unstably at the given level of applied stress. $c \to \infty$ at $\sigma = \sigma_F$ (Fig. 2.6c), after a period of time that decreases as the applied stress increases (see Figs. 1.18 to 1.20).

The criterion for crack growth

Since each of the fracture processes discussed above is plastically induced, the fracture criterion must be related to the amount of plastic strain that occurs in a minute region near the crack tip. To a first approximation suppose that the material ahead of the crack tip contains a series of "hypothetical, miniature tensile specimens," having a gage length l and width w (Fig. 2.7a). Crack growth occurs when the specimen adjacent to the crack is fractured [28]. When failure of the first adjacent specimen immediately causes the next adjacent specimen to fail, then the overall fracture process is unstable and crack propagation occurs under decreasing stress. When it does not, the applied stress must be increased if stable crack growth is to continue.

Consider, for example, a cracked plate under tensile (mode I) load-

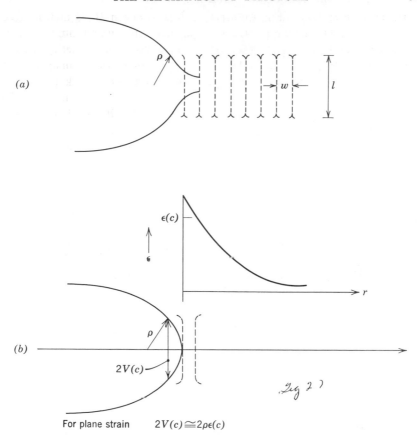

For plane strain $2V(c) \cong 2\rho\epsilon(c)$

Fig 2.7. (a) The region ahead of a tensile crack represented as a series of "hypothetical" tensile specimens of gage length l and width w. The crack advances when the specimen adjacent to its tip has been fractured. (b) The relation between crack tip displacement $2V(c)$, root radius ρ, and strain $\epsilon(c)$ in specimen adjacent to crack tip in plane strain. The strain distribution ahead of the crack is also shown schematically.

ing. Let the plate be sufficiently thick that *plane-strain conditions prevail*. In plane strain the plastic deformation at the crack tip is confined to narrow bands whose thickness (the gage length of the specimen l) is of the order of the diameter 2ρ of the crack tip (Fig. 2.7b). If $\epsilon(c)$ is the tensile strain in the specimen adjacent to the tip, then the displacement of the crack faces, at the crack tip, $2V(c)$ is

$$2V(c) = 2\rho\epsilon(c) \qquad \text{(plane strain)} \qquad (2.25a)$$

Unstable fracture occurs when the strains in the specimen adjacent to the tip build up to the tensile ductility† of the specimens (i.e., when $\epsilon(c) = \epsilon_f(c)$). *At the point of instability, $V(c)$ has reached a critical value $V^*(c)$,*

$$V(c) = V^*(c) = \rho\epsilon_f(c) \tag{2.25b}$$

Similar criteria apply for fractures that are initiated in plane-stress tensile loading [38] and pure-shear (Mode III) loading [36, 37]. In all cases unstable fracture occurs when [28]

$$V(c) = V^*(c) \tag{2.26}$$

It is now necessary to relate $V(c)$ to nominal stress and crack length so that the fracture strength of a cracked plate can be determined if $V^*(c)$ is known. This is the subject of the following section.

The formation of plastic zones and tip displacements near a stopped crack under tensile loading

Equations 2.8 indicate that the elastic stress in the vicinity of a crack can be very large at $r \ll c$. Localized plastic deformation occurs when the appropriate yield criterion is satisfied in the vicinity of the crack and a *plastic zone* is created near the crack tip. In non-strain hardening materials the shear stresses inside the plastic zone are equal to k, the yield stress in pure shear, but outside the plastic zone the shear stresses are less than k. The size and shape of the plastic zone depend on the mode of deformation that acts on the crack and on the criterion for yielding (Tresca or von Mises) that is applicable. The simplest method of determining the plastic zone size is to treat the problem as one of *plane stress*‡ and to assume that yielding occurs in those regions where the stresses σ_{yy} at the crack tip are greater than the tensile yield stress σ_Y. The material is assumed to be nonstrain hardening unless otherwise stated.

The yield zones spread out on 45° planes that bisect the angle between the tensile axis and the leading edge of the crack [38, 39] and are parallel to the direction of crack propagation (Fig. 2.8a). In the plane of the crack ($\psi = 0$) the zones will extend a distance

---

†For reasons that will be discussed in Chapter 7 the ductility of the miniature tensile specimens is proportional to, but does not always equal, the ductility as measured in a conventional tensile test under the same conditions.

‡ Elastic-plastic deformation under plane strain conditions will be discussed in Chapter 7.

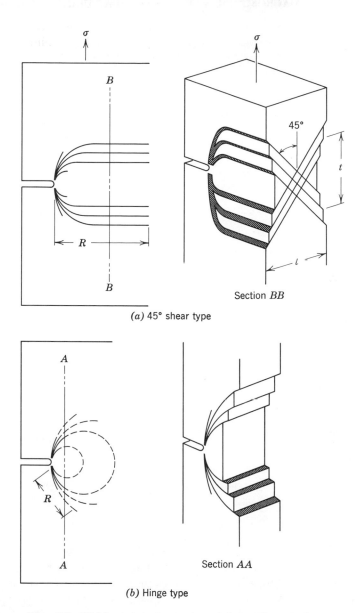

(a) 45° shear type

(b) Hinge type

Fig. 2.8. Yield zones observed on the surface and cross section of a cracked sheet under uniaxial tensile loading in plane stress *(a)* and plane strain *(b)*. *After Hahn and Rosenfield* [38].

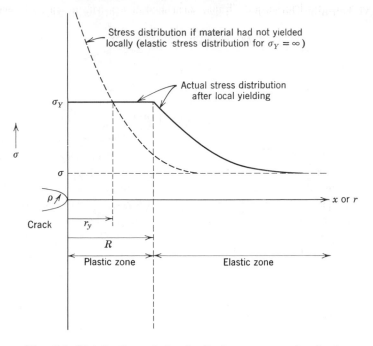

Fig. 2.9 Distribution of longitudinal stress σ_{yy} ahead of a crack in a thin plate under plane-stress tensile loading. Plastic deformation has occurred over a distance R ahead of the tip.

$r = r_y$, that can be approximated [5] from Eq. **2.8**b by setting $r = r_y$, $\sigma_{yy} = \sigma_Y$, so that†

$$\sigma_Y = \sigma \left(\frac{c}{2r_y} \right)^{\!\frac{1}{2}}$$

$$r_y = \left(\frac{\sigma}{\sigma_Y} \right)^{\!2} \frac{c}{2}$$

Within the plastic zone the maximum tensile stress is reduced from the elastic value, shown by dotted lines in Fig. 2.9, to the tensile yield stress σ_Y. Actually, in order to take up the load that was carried by the elastic material in the region $0 < r < r_y$, the plastic zone must extend to a distance $R = \alpha r_y$ so that the area under the stress-distance

† We shall not consider the shape of the plastic zone here since it is the zone size, as measured in the crack plane ($y = 0$), that is of primary interest to fracture initiation.

curve remains the same. Thus, to a second approximation [9, 40, 41]

$$R\sigma_y \cong \beta r_y \sigma_Y \cong \int_0^{r_y} \sigma_{yy} \, dr \qquad (2.27)$$

so

$$\beta = 2$$

and

$$R = 2r_y \cong \left(\frac{\sigma}{\sigma_Y}\right)^2 c \qquad (2.28)$$

is the plastic zone radius. The stress distribution around the relaxed crack is approximately the same outside the plastic zone $(r > R)$ as the elastic stress distribution around an elastic crack of half-length $(c + r_y)$ [9]. As far as the K values of stress intensity around the crack are concerned, Irwin has postulated that the local stress relaxation by plastic strain in the region $r = r_y = R/2$ is essentially the same as that due to an additional length of "elastic" crack equal to

$$r_Y = \frac{K^2}{2\pi\sigma_Y^2} \qquad (2.29a)$$

so that 2.12 becomes [9]

$$K^2 = \sigma^2 W \tan\left[\left(\frac{\pi c}{W}\right) + \frac{K^2}{2W\sigma_Y^2}\right]. \qquad (2.29b)$$

A more precise method for determining the plastic zone size for the sharp tensile crack under plane stress in a nonstrain-hardening material has been given by Dugdale [42]. Suppose that $a = R + c$ is the distance from the center of the crack to the elastic-plastic zone boundary and that the plastic zone is narrow. The combined crack and plastic zone may be regarded as a flattened ellipse. To determine R, it is assumed that the crack of length $2c$ has spread elastically to a distance $2a$, but that the crack has been closed up in the plastic zone by an internal tensile stress which acts across the crack faces in the region $|a| > |x| > |c|$ (Fig. 2.10). Since internally applied forces are in static equilibrium, this internal tensile stress must equal σ_Y because σ_Y is the tensile stress existing in the plastic zone. When this internal stress field is superimposed on the elastic stress field of the crack in the presence of an externally applied tensile stress σ, and the restriction is imposed that the stress at the end of the plastic zone $(x = a)$ be finite, then the plastic zone size is determined by the relation

$$\frac{c}{a} = \cos\left(\frac{\pi}{2}\frac{\sigma}{\sigma_Y}\right) \qquad (2.30)$$

or

$$\frac{R}{c} = \frac{a - c}{c} = \sec\left(\frac{\pi}{2}\frac{\sigma}{\sigma_Y}\right) - 1 \qquad \text{(plane stress)} \qquad (2.31)$$

For $\sigma \ll \sigma_Y$ this relation reduces to

$$\frac{R}{c} = \frac{\pi^2}{8}\left(\frac{\sigma}{\sigma_Y}\right)^2 \qquad (2.32)$$

which is essentially the same as Eq. 2.28.

As the plastic zones spread from the tip of the crack, the crack opening displacements $2V(c)$ produced at the tip will increase. These

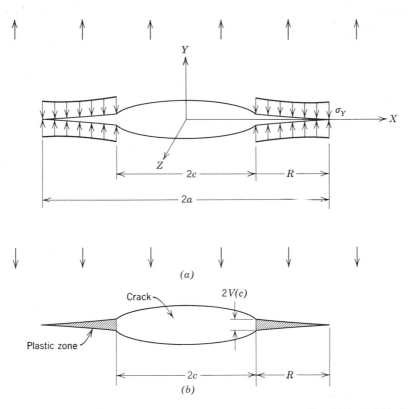

Fig. 2.10. (a) Internal stress distribution used in the Dugdale model of elastic-plastic deformation near a crack of length 2c under plane-stress tensile loading. (b) Displacements 2V associated with crack opening. *After Hahn and Rosenfield* [38].

displacements are related to the plastic zone size R as [43]

$$V(c) = \frac{4\sigma_Y c}{\pi E} \ln\left(\frac{a}{c}\right) = \frac{4\sigma_Y c}{\pi E} \ln\left(\frac{R + c}{c}\right) \qquad (2.33)$$

and consequently are a function of the applied stress σ

$$V(c) = \frac{4\sigma_Y c}{\pi E} \ln\left[\sec\frac{\pi\sigma}{2\sigma_Y}\right] \qquad (2.34)$$

Since $\ln x \cong x - 1$ for x small, Eq. 2.33 can be written, for $\sigma \ll \sigma_Y$, as

$$V(c) = \frac{4\sigma_Y c}{\pi E}\left(\frac{a - c}{c}\right) = \frac{4\sigma_Y c}{\pi E}\left(\frac{R}{c}\right) \qquad (\sigma \ll \sigma_Y) \qquad (2.35)$$

As $\sigma_Y/E = \epsilon_Y$, the yield strain, Eq. 2.35 indicates that the tip displacement is approximately equal to the total plastic displacement $\epsilon_Y R$ inside the plastic accommodation zone.

The 45° type of shear zone (Fig. 2.8a) characteristic of plane-stress deformation begins to develop [38] when the yield zones have spread a distance $R \cong \pm t/2$ in a sheet of thickness t. When $R = 4t$ the deformation is entirely confined to a narrow, tapered zone, as predicted by the Dugdale plane-stress model. Recent experimental investigations have shown [38, 44] that measured values of R are in good agreement with Eq. 2.33 when $R \geq 4t$.

Under *plane-stress tensile loading* the strains at the crack tip are distributed over a "tensile specimen" whose gage length l (Fig. 2.7a) is of the order of the sheet thickness (Fig. 2.8a). Consequently the tensile displacement at the crack tip is [38]

$$2V(c) = \epsilon(c)t \qquad \text{(plane stress)} \qquad (2.36a)$$

and the critical displacement for fracture is

$$2V^*(c) = \epsilon_f(c)t \qquad (2.36b)$$

At low stress levels or in thicker plates *plane-strain conditions* exist, at least near the center of the plate, and the plastic deformation is confined to the xy plane. The plastic zone has either an elongated shape [53], similar to that shown in Fig. 2.10b but confined to the xy plane, or the form of hinges, similar to those shown in Fig. 2.8b.

At present there is no analytical method for determining the variation of plastic zone size R and root displacement $V(c)$ as a function of applied stress, notch depth, and root radius for the case of plane-strain tensile loading. Some approximations can be made and these will be discussed in Chapter 7. It is only necessary to point out two significant facts at this point. First, the plastic zone size is smaller, at a constant value of stress intensification, in plane-strain deformation than in plane-stress deformation (Fig. 2.8c) [39, 54]. Second, the degree of plane strain depends on the ratio of the (plane-strain) plastic zone size to plate thickness. When R/t is less than about $\frac{1}{2}$, the deformation is predominantly plane strain [38]. When R/t is greater than about 2, the deformation becomes predominantly plane stress [9, 55].

At the present time there is also no exact method for determining the local root strains in a strain-hardening material or at applied stress levels that are greater than the yield stress. In these cases the Neuber *approximation* [46] is useful

$$K_{\epsilon(p)} K_{\sigma(p)} = K_\sigma{}^2 \qquad (2.37)$$

where $K_{\epsilon\,(p)}$ and $K_{\sigma\,(p)}$ are the plastic strain and stress concentration factors a distance r from the crack tip, and K_σ is the elastic stress concentration of the particular notch at r, as given by Eq. 2.9. For a material whose tensile flow curve is given [15] by the relation $\sigma = \sigma_0 \epsilon^n$,

$$K_{\epsilon(p)} = K_\sigma^{\left(\frac{2}{1+n}\right)} \qquad (2.38)$$

The macroscopic criteria for unstable crack propagation

Tensile Loading

In the preceding section it was shown that as the applied stress increases, plastic-yield zones spread over increasing distances R from the tip of a notch or crack. These zones are accompanied by tensile displacements at the crack tip which are, in turn, functions of the applied stress and crack length according to Eqs. 2.34 and 2.35. For $\sigma < 0.6\sigma_Y$, Eqs. 2.32 and 2.35 can be combined to give

$$V(c) = \frac{\pi c \sigma^2}{2E\sigma_Y} \qquad (\sigma < 0.6\sigma_Y) \qquad (2.39)$$

According to Eq. 2.26 the criterion for crack extension is that $V(c)$ be equal to a critical value $V^*(c)$. The stress required for unstable extension is then

$$\sigma = \left\{ \frac{2E}{\pi c} \sigma_Y V^*(c) \right\}^{\frac{1}{2}} \qquad (\sigma < 0.6\sigma_Y) \qquad (2.40)$$

This expression is of the same form as the Griffith relations (2.13 and 2.23) that were derived for continuous, elastic crack propagation

$$\sigma_M = \sqrt{\frac{2E\gamma_P}{\pi c}} \qquad (\sigma_M < 0.6\sigma_Y) \qquad (2.41)$$

with

$$\gamma_P = \sigma_Y V^*(c) \tag{2.42}$$

as the plastic work required to keep a fast-moving crack in motion.

Similarly these relations apply for the *initiation of unstable fracture* near the tip of a slowly moving or stopped crack.

$$\sigma_F = \sqrt{\frac{2E\gamma_P{}^*}{\pi c}} = \sqrt{\frac{EG_c}{\pi c}} \qquad (\sigma_F < 0.6\sigma_Y) \tag{2.43}$$

where

$$\gamma_P{}^* = \frac{G_c}{2} = \sigma_Y V^*(c) \tag{2.44}$$

is the critical amount of work required to *initiate* an unstable fracture. In strain-rate insensitive materials $\gamma_P{}^* \cong \gamma_P$. However, when the critical opening displacement $V''(c)$ and the yield stress σ_Y are strain-rate dependent, then $\gamma^*_P \neq \gamma_P$ since regions near the tip of a fast-moving crack will be strained more rapidly than those near a stopped crack.

At higher stress levels the simplifying approximations (2.32 and 2.35) cannot be used and the fracture stress is obtained from Eq. 2.34,

$$\sigma_F = \frac{2\sigma_Y}{\pi} \cos^{-1}\left(\exp - \frac{\pi EG_c}{8\sigma_Y{}^2 c}\right) \tag{2.45}$$

In the notation of fracture mechanics the fracture criterion is always given by Eq. 2.16

$$K_c{}^2 = EG_c \tag{2.16}$$

At low stress levels K is given by Eq. 2.10 so that Eqs. 2.16 and 2.43 are identical. At higher stress levels K is given by Eq. 2.29b; this predicts a value of σ_F that is approximately the same as the one given by Eq. 2.45. Similarly Eq. 2.16 can be used to determine the stress at which plastically induced fracture occurs near any defect of arbitrary shape, provided that the stress intensification factor K can be calculated and G_c is known.

However, it must be noted that none of these criteria applies when fracture occurs after general yield ($\sigma_F > \sigma_Y$). At the present time there are no adequate criteria to handle these problems.

When unstable crack extension is entirely flat (all cleavage or normal rupture) then

$$G_c = G_{Ic} \qquad \text{(plane strain)} \tag{2.46}$$

where the subscript I signifies that all crack extension is taking place under the opening (tensile) mode of deformation. In general, mode I extension occurs under plane-strain conditions, and consequently G_{Ic} *is known as the plane-strain toughness.*

The plane-strain fracture toughness K_{Ic} is

$$K_{Ic} = \left[\frac{EG_{Ic}}{(1 - \nu^2)}\right]^{1/2} \tag{2.47}$$

Alternatively when fracture occurs entirely by shear rupture, on planes inclined at 45° to the tensile axis, then

$$G_c = G_c(45°) \qquad \text{(plane stress)} \tag{2.48}$$

where $G_c(45°)$ is the plane-stress toughness for shear rupture.

In general, both plane-strain and plane-stress modes of fracture operate in parallel (plane strain at the center, plane stress at the outer edges) during fracture of a plate; the value of G_c then lies somewhere in-between the two extremes given by Eqs. 2.46 and 2.48. This is discussed in Chapter 3 (cf., Fig. 3.36).

Antiplane Strain (Mode III) Loading

The relations between applied stress, crack length, plastic zone size, and tip displacement are similar [47] for antiplane-strain loading and tensile loading. The elastic shear stresses, a distance r from the tip of an infinitely sharp crack, are given by

$$\tau_{yz} = \tau \left(\frac{c}{2r}\right)^{1/2} \tag{2.49}$$

which is analogous to Eq. 2.8b, with τ the applied (nominal) shear stress. Once yielding begins, $\tau_{yz} = k$ inside the plastic zone. At low stress levels the plastic zone size is given by [36, 37]

$$R = \left(\frac{\tau}{k}\right)^2 c \tag{2.50}$$

At intermediate stress levels

$$R = \left(\frac{\tau}{k}\right)^2 c \left[1 + \frac{(4/\pi)(\tau/k)^2}{1 - (\tau/k)^2}\right] \tag{2.51}$$

and the effective crack length is increased, over that given by Eq. 2.50, by the term inside the brackets. This is similar to the Irwin correction given by Eq. 2.29 for tensile cracks. At high stress levels

$$R = \frac{4c}{\pi[1 - (\tau/k)^2]} \tag{2.52}$$

For a nonstrain-hardening material the plastic strains vary with distance r from the crack tip as

$$\epsilon_{yz} = \left(\frac{k}{G}\right)\frac{R}{r} \qquad (2.53)$$

so that the shear strains decrease with increasing distance r from the crack tip until the plastic zone boundary is reached ($r = R$) and $\epsilon_{yz} = \epsilon_Y$, the shear yield strain.

Under antiplane strain loading (pure torsion), unstable shear ruptures develop when a critical strain is achieved over a critical distance r from the crack tip, producing a critical shear displacement [36, 37]

$$\Phi^*(c) = \frac{k}{G} R^* \qquad (2\ 54)$$

where R^* is the critical plastic zone size required for fracture. At low stress levels, fracture occurs at a stress $\tau = \tau_F$, that is, from Eq. 2.50

$$\tau_F = \sqrt{\frac{Gk\Phi^*(c)}{\pi c}} = \sqrt{\frac{GG_c}{\pi c}} \qquad \left(\frac{\tau}{k} \rightarrow 0\right) \qquad (2.55)$$

with

$$G_c = k\Phi^*(c) = G_{IIIc} \qquad (2.56)$$

as the critical work required for initiating the unstable antiplane-strain fracture. At higher stress τ_F is determined from combinations of Eqs. 2.54 with 2.51 or 2.52.

2.4 The Physical Significance of Fracture Toughness

The work done per unit area at the crack tip during crack extension is [48]

$$2\gamma_P = 2\int_0^\infty \sigma_{yy}\, dV \qquad (2.57)$$

where σ_{yy} is the maximum stress in the miniature tensile specimen adjacent to the crack tip (Fig. 2.7a) and $2dV$ is the displacement occurring when the crack advances by dc. This relation is a generalized form of Eq. 2.3 which was derived for elastic fracture.

Because σ_{yy} drops to zero when $V > V^*(c)$ and the "specimen" breaks,

$$2\gamma_P \approx 2\sigma_{yy}^* V^*(c) \qquad (2.58)$$

where σ_{yy}^* is the stress that exists in the volume element when the crack advances by one of the particular modes. The work done in

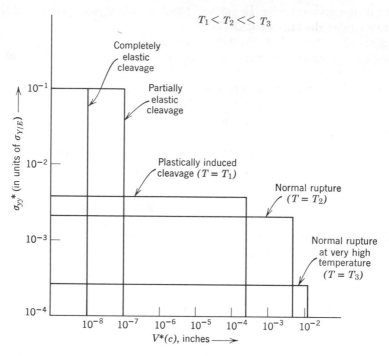

Fig. 2.11. The longitudinal tensile stresses σ_{yy}^* at the tip of an advancing crack, as a function of crack-opening displacement $2V^*(c)$ required for propagation. The toughness is proportional to the area inside the indicated rectangles.

crack extension may then be represented (Fig. 2.11) as the area under the curve of a plot of σ_{yy}^* versus $V^*(c)$.

Similarly, for the *initiation of fracture*,

$$2\gamma_P^* = G_c \approx 2\sigma_{yy}^* V^*(c) \tag{2.59}$$

so that the fracture toughness can be estimated from the areas under the curves in Fig. 2.11. Consider the following examples.

Plane-strain fracture $(G_c = G_{Ic})$

Case 1. Completely Elastic Cleavage Fracture. (*Example: mica or glass.*) In this case $V^*(c) \approx a_0$, the displacement required to break the atomic bonds between the cleavage planes, and σ_{yy}^* is the theoretical cohesive strength $E/10$ so that

$$G_{Ic} = \frac{E}{10} a_0 = 2\gamma_s \approx (10^{-2} \text{ in. lb per in.}^2) \tag{2.60}$$

as given previously by Eq. 2.5. Fracture is initiated directly at the crack tip, as shown in Fig. 2.12a.

Case 2. Partially Elastic Cleavage Feature. (Example: ionic crystals at low temperature.) In this case $\rho > \alpha_0$ and $V^*(c) = \rho$; σ_{yy}^* is still equal to $E/10$, so that

$$G_{Ic} = 2\gamma_P^* = \frac{Ea_0}{10}\frac{\rho}{a_0} = 2\gamma_s\frac{\rho_0}{a_0} \tag{2.61}$$

which is essentially the same as Eq. 2.24. As shown in Fig. 2.11, G_{Ic} has been increased, as compared with Case 1. Typically, $G_{1c} \approx 20\gamma_s$ ($\approx 10^{-1}$ in^2 lb per in.^{2}) for these types of fracture, which still initiate directly at the crack tip as in Fig. 2.12a.

Case 3. Plastically Induced Cleavage Fracture. (Example: mild steel at $-150°C$.) In this case σ_{yy}^* is of the order of σ_Y and G_{Ic} is given by Eq. 2.44 (i.e., $G_{Ic} \cong 2\sigma_Y V^*(c)$). According to Eq. 2.25 $V^*(c)$ is, in turn, equal to $\rho\epsilon_f(c)$. Consequently

$$G_{Ic} \cong 2\sigma_Y\rho\epsilon_f(c) \tag{2.62}$$

where $\epsilon_f(c)$ is the ductility of the specimen at the crack tip. Thus G_{Ic} is proportional to the area under the stress-strain curve of the nonstrain-hardening specimen. (When strain hardening does occur, a better estimate of G_{Ic} would be based on the entire area under the curve, not just the product of σ_Y and ϵ_f.)

The effective radius of the crack is ρ. The word "effective" is used because the plastic deformation processes which initiate cleavage require some minimum volume in which to operate [28]. Thus $G_{Ic} \neq 0$ as $\rho \to 0$. Instead there is a lower limiting value of ρ, ρ_{min}, below which further decreases in root radius do not produce a decrease in toughness. Typically this dimension is of the order of the grain size or twin-band spacing when cleavage fracture is initiated by plastic deformation† (i.e., $\rho_{min} \approx 5 \times 10^{-3}$ in. $\approx 10\ \mu \approx 40 \times 10^4a_0$).

Taking $\epsilon_f(c) \approx 0.05$ and $\sigma_Y = E/200$ gives

$$G_{Ic} \cong 2\frac{E}{200}(40 \times 10^4a_0 \times 0.05) \cong 200\gamma_s \approx (20 \text{ in. lb per in.}^2)$$

As shown in Fig. 2.11 the toughness is considerably greater than in Case 2. Fracture is now initiated *ahead of the crack tip*, as shown

---

† When unstable normal rupture occurs in high strength materials, Eq. 2.62 can be used to determine G_{Ic}. As in the case of cleavage fracture ρ has a lower limiting value which varies from one material to another [50, 51].

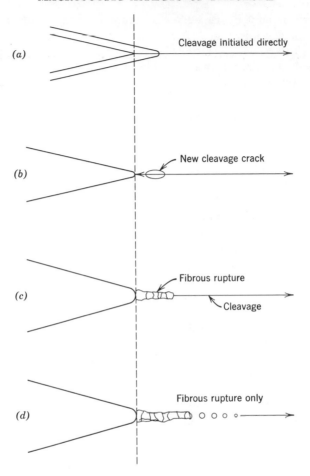

Fig. 2.12. The various processes by which fracture is initiated at or near the tip of a stopped crack or notch.

in Fig. **2.12***b*.   The microcrack both joins the crack tip and propagates forward (from left to right), causing complete failure.

Case 4. Plastically Induced Normal Rupture. (Example: mild steel at about 0°C.)   Equation **2.62** is still applicable. For fibrous fractures in mild steel ρ_{min} is somewhat larger than in Case 3, of the order of 10^{-2} in. $\sigma_Y \approx E/500$ and $\epsilon_f(c) \approx 0.75$, the tensile ductility;

$$G_{Ic} \cong 2\left(\frac{E}{500}\right)(80 \times 10^4 a_0 \times 0.75) \cong 2400\gamma_s \approx 240 \text{ in. lb per in.}^2$$

Fracture initiates by slow fibrous cracking, which can develop into a fast-running cleavage crack below the ductile-brittle transition temperature (Fig. 2.12c). Above the transition-temperature region, fracture occurs by fibrous rupture only (Fig. 2.12d).

These examples indicate that for reasons to be explained in Chapter 7 a *small decrease in* σ_Y *(due to increasing temperature*, for example) *produces a large increase in* $V^*(c)$ *and hence a large increase in* G_{Ic}. This argument is not restricted to the cleavage fractures that were used as examples here; G_{Ic} also decreases with increasing σ_Y in high strength materials (Chapter 3) where fracture occurs exclusively by normal rupture. Consequently the toughness of a material can be increased by causing *microscopically weaker* (smaller σ_{yy}^*) but *macroscopically stronger* (larger G_{Ic}) modes of fracture initiation to operate [48].

However, if the yield strength is extremely low (e.g., near the melting point) G_{Ic} will again be low even though $V^*(c)$ is large, as shown in Fig. 2.11. Thus, there is a certain balance between strength and ductility at which the toughness is a maximum. Alternatively materials that have low elastic moduli (e.g., rubber) are tough [49] because notches become severely blunted by elastic deformation.

It is important to realize that when a cleavage fracture has been initiated at (or below) the notch root and begins to spread in a continuous or discontinuous manner, the work expended in keeping the crack in motion, γ_P, is less than the work that was done in initiating the fracture, γ_P^*, because of the high strain rate at the tip of the moving crack. Consequently the measured fracture stress σ_F will be greater than the stress to keep the crack in motion σ_M and the stress required for crack propagation will follow the curve ABC (Fig. 2.13). Suppose that a stress σ is applied across a plate of material containing a crack of length c_F and that $\sigma < \sigma_F$ so the unstable fracture should *not* initiate. However, if the material is overloaded *locally* (e.g., owing to residual stresses around welds) so that in the vicinity Δc of the tip the stresses are $\sigma + \Delta\sigma = \sigma_F$, then an unstable fracture will initiate and continue to spread unstably if $\sigma > \sigma_M$ (at $c + \Delta c$) as shown on the figure. Similar behavior would result if σ_F *were lowered locally* (e.g., by corrosion) so that σ_F dropped to σ. This illustration indicates that operating at stress levels below those which are required to *initiate* cleavage fracture does not guarantee that cleavage fracture will not occur. The only way that it can be guaranteed is to operate a structure above its *crack-arrest temperature*, the temperature at which moving cleavage cracks will be stopped for a given level of applied stress.

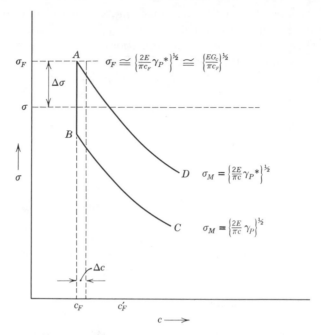

Fig. 2.13. The stress required to continue propagation of a crack that becomes stable when $c = c_F$, $\sigma = \sigma_F$. Curve AD represents a case where work done in propagation γ_P is the same as work done in crack initiation $\gamma_P{}^*$. Curve ABC represents a case where the total work done in propagation γ_P is less than $\gamma_P{}^*$.

Alternatively when unstable tear fractures develop in relatively strain-rate insensitive materials such as high strength aluminum alloys, the work done in keeping the crack in motion, γ_P, is about the same as $\gamma_P{}^*$, the work done in initiating the instability. Consequently the fracture stress-crack length relation follows along the curve AD in Fig. 2.13, and overloading over a distance Δc will not induce catastrophic fracture in a structure containing a crack of length $2c_F$, at a nominal stress σ. Unstable fracture can only develop if the crack extends (e.g., by fatigue and corrosion) from c_F to c'_F. Thus if all other factors are held constant, strain-rate insensitive materials tend to be more reliable than strain-rate sensitive ones.

Plane-stress fractures $(G_c = G_c(45°))$

When a tensile stress is applied across a *thin sheet* containing a notch, the large plastic strains near the tip can (if cleavage does not

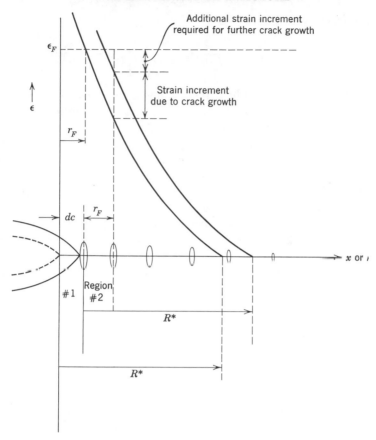

Fig. 2.14. Redistribution of local strain at crack tip when crack advances by incremental distance dc. R^* is critical plastic zone size for unstable fracture.

intervene) cause the formation and subsequent coalescence of voids, as shown in Fig. 2.14. Since the material between the coalescing voids behaves the same, on a miniature scale, as a tensile specimen [28, 36, 38] which fractures by necking, the maximum tensile strain near the crack tip is approximately the same as the tensile ductility ϵ_f, as measured in an unnotched tensile test under the same environmental conditions. Consequently $V^*(c) \cong \frac{1}{2}\epsilon_f t$ from Eq. 2.36 [38], and hence

$$\gamma_P{}^* = \frac{G_c}{2} \cong \frac{\sigma_Y}{2}\,\epsilon_f t \qquad \text{(plane-stress, shear rupture)} \qquad (2.63)$$

and the fracture toughness

$$K_c = \{EG_c\}^{1/2} \cong \{E\sigma_Y\epsilon_f t\}^{1/2} \qquad (2.64)$$

Under uniaxial tensile loading, these plane-stress shear ruptures occur on planes inclined at 45° to the tensile axis. Equation 2.58 indicates that the toughness decreases with decreasing sheet thickness. Consequently *very thin* sheets of normally "ductile" materials (large ϵ_f) are susceptible to unstable fracture at stresses below the yield stress, when deep cracks are present.

On a *microscopic scale slow (stable) crack extension* often precedes the development of an unstable shear rupture (or normal rupture). This effect results from the redistribution of strain at the tip of an advancing crack [36]. Suppose (Fig. 2.14) that region 1, between the crack tip and the void immediately adjacent to the tip, can neck down (coalesce) when the crack half length $c = c_1$, at an applied stress σ_1 that is less than σ_F (Fig. 2.6b). This coalescence allows the crack to advance an incremental length dc. The next coalescing region (2) is then moved, relatively, into the position adjacent to the tip. Because the strains in the vicinity of the tip increase with decreasing distance r from the tip (e.g., Eq. 2.53), as well as with increasing plastic zone size, the strains in region 2 are thereby increased. At first the strain increment produced by the growth dc is *not* sufficient to cause coalescence because ϵ is less than a critical value (about ϵ_f) across the entire coalescing region. Hence the applied stress must increase [to increase R, hence $V(c)$ and $\epsilon(c)$] to cause the crack to extend. This process is repeated until $c = c_F$, $\sigma = \sigma_F$ (Fig. 2.6b) and no further increase in applied stress is required; the growth is then unstable.

The relation between toughness and ductility

On the basis of the discussion presented above it is apparent that tough materials are those with a high capacity for deforming locally in the vicinity of the crack tip (high ϵ_f), whereas ϵ_f is small in brittle materials. A good rule of thumb is that a tough material is one whose G_c value is about 10^8 erg per cm² (50 ft lb per in.²) and brittle materials are those whose G_c value is less than this. For example, for a tough steel $E = 30 \times 10^6$ psi so that $\sigma_F \cong 75,000$ psi for a structure containing a 1 in. crack. Alternatively a crack would have to be about 2 in. long before it could propagate unstably at a stress level of 50,000 psi. Cracks of this length would certainly be detected by the crudest of inspection techniques. However, if $G_c = 10^6$ ergs per cm^2 (0.5 ft-lb per in.^{2}), a crack of only 0.02 in. ($\frac{1}{2}$ mm) could be spread at 50,000 psi and this crack would be more difficult to detect.

Cleavage fractures are always brittle; shear fractures are almost always tough, except in: (1) *very thin sheets,* (2) *ultra-high strength materials* in the form of thick plates where ϵ_f can be extremely small, (3) *near the melting point* of a material where the yield stress σ_Y is so small that the area $\sigma_Y V^*(c)$ (Fig. 2.11) is small despite the high value of ϵ_f, and (4) *chemically reactive environments.*

Ductility is a term that can be used to describe either the overall plastic deformation of a material under stress or the way in which the material fractures. If a deep notch or crack is present such that unstable fracture will occur at a stress that is less than or equal to the yield stress σ_Y (Fig. 2.15a), then the material is brittle by either definition.

However, in the absence of preinduced flaws (e.g., unnotched tensile test) these two definitions of ductility do not always agree. In some *engineering terminology, ductile materials* are those which undergo large uniform plastic deformations (uniform elongation in a uniaxial tensile test), whereas brittle materials either break or deform inhomogeneously (neck) after small deformations. In materials terminology *brittle materials* are those in which the inhomogeneous strains accompanying fracture ϵ_f are small whereas ϵ_f is large in a ductile material. The differences between the two definitions are illustrated (Fig. 2.15b and c) by load-elongation curves for two unnotched materials. Figure 2.15b shows a schematic curve for a material which fractures by cleavage after 20% elongation, and 2.15c is the curve for a material that begins to neck after 5% elongation and finally fractures by drawing down to a chisel point ($\approx 100\%$ reduction in area). In terms of uniform extension (b) is more "ductile" than (c), but in terms of resistance to catastrophic fracture (b) is more brittle because

$$\epsilon_f = \ln\left(\frac{A_0}{A_F}\right)$$

is smaller. One simple measure of toughness is the area under the true-stress–true-strain curve for the volume element of material undergoing fracture. It is quite apparent from Fig. 2.15d that (c) is the tougher of the two illustrative materials.

The low uniform elongation of material (c) results from its low strain hardening ability and this has no *direct* relation to fracture

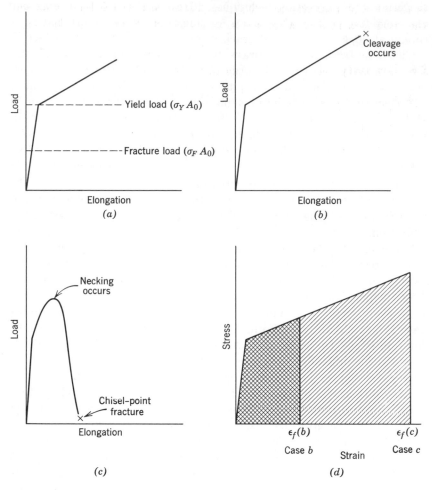

Fig. 2.15. Tensile load-elongation curves for: (a) Brittle material. (b) Partially brittle material. (c) Ductile material. The true-stress–true-strain curves for cases b and c are shown in (d).

toughness, although a low strain hardening capacity does make void coalescence somewhat easier [37, 52]. If A_0 is the initial cross-sectional area, A_u is the uniform area when necking begins and A_F is the cross-sectional area at fracture, then the total reduction in area at fracture

$$\frac{A_0 - A_F}{A_0} = \frac{A_0 - A_u}{A_0} + \frac{A_u - A_F}{A_0} \qquad (2.65)$$

is the sum of that due to reductions during uniform deformation and the reduction in area after mechanical instability (necking) begins.

2.5 Summary

1. Fracture is an inhomogeneous form of deformation which can be viewed on different levels. On an *atomistic level* it occurs by the breaking of atomic bonds, either perpendicular to a plane (cleavage) or across a plane (shear). On a *microscopic level* cleavage occurs by the formation and propagation of microcracks and the separation of grains along cleavage planes. Shear fracture (rupture) usually occurs by the formation of voids within grains and the separation of material between the voids by intense shear. On a *macroscopic level* cleavage occurs when a cleavage crack spreads essentially perpendicular to the axis of maximum tensile stress. Shear fracture occurs when a fibrous crack advances essentially perpendicular to the axis of maximum tensile stress (normal rupture) or along a plane of maximum shear stress (shear rupture). Fracture is said to be transgranular when microcrack propagation and void coalescence occur through the grains and intergranular when they occur along grain boundaries. More than one mode of crack propagation can contribute to the fracture of a structure. In general, cleavage fracture is favored by low temperatures.

2. Fracture occurs in a perfectly elastic solid when the stress level at the tip of a preinduced flaw reaches the theoretical cohesive stress $E/10$ and a sufficient amount of work γ_s is done to break atomic bonds and create free surface. Since the stress concentration factor of an elastic crack is proportional to $c^{1/2}$, the nominal fracture stress is proportional to $\sqrt{\gamma_s/c}$ as predicted by the Griffith equation.

3. When the yield stress of a material is less than $E/10$ (i.e., in a ductile material or in the vicinity of a stopped crack in a partially brittle one), plastic flow occurs near the crack tip and the stress level in the plastic zone is less than $E/10$. Consequently the crack cannot advance directly as an elastic Griffith crack. Fracture is then plastically induced and occurs when a critical plastic strain ϵ_f is produced over a critical distance ahead of the tip. For the cases of tensile or bend loading in thick structures the crack-tip displacement $2V(c) \cong 2\rho\epsilon(c)$ where ρ is the root radius. In thin sheets of thickness $t, 2V(c) = \epsilon(c)t$. The macroscopic criterion for fracture is that $\epsilon(c) = \epsilon_f$ or that $V(c) = V^*(c)$.

4. The crack-tip displacement $2V(c)$ increases with increasing plastic zone size R in an elastic-plastic solid ($\sigma < \sigma_Y$). The plastic zone size increases with increasing nominal stress σ and notch depth $2c$. Consequently the critical displacement criterion for unstable fracture before general yield implies a critical plastic zone size for unstable fracture and hence that the nominal fracture strength decreases as the notch depth increases. At low stress levels $R \propto \sigma^2 c$, so that $\sigma_F \propto \sqrt{G_c/c}$. No satisfactory relations between σ_F and c have been derived for the case of fracture after general yield (i.e., when $\sigma_F > \sigma_Y$).

5. Physically $G_c \cong 2\sigma_Y V^*(c)$ is the work done at the crack tip at the onset of unstable propagation. For reasons which will be discussed in Chapter 7, $V^*(c)$ increases rapidly as σ_Y decreases so that fracture toughness increases with decreasing yield strength (i.e., with increasing test temperature and decreasing applied strain rate). For plane-strain fractures G_{Ic} decreases with decreasing root radius ρ; for plane-stress fractures G_c decreases with decreasing sheet thickness t.

6. Toughness is a measure of the work for fracture and hence increases with the ductility which is a measure of the strain to fracture. Uniform elongation as measured in a tensile test is simply a measure of a material's capacity for strain hardening and has no direct relation to fracture toughness.

References

The asterisk indicates that published work is recommended for extensive or broad treatment.

*[1] J. E. Srawley and W. F. Brown, *Fracture Toughness Testing*, ASTM, Philadelphia, STP No. 381 (1965), p. 133.

[2] E. Orowan, *Repts. Prog. Phys.*, **XII**, 185 (1948).

[3] C. J. Beevers and R. W. K. Honeycombe, in *Fracture*, B. L. Averbach et al., eds., M.I.T., Wiley, New York (1959), p. 474.

[4] K. E. Puttick, *Phil. Mag.*, **4**, 964 (1959).

[5] C. D. Beachem, *Trans. ASM*, **56**, 318 (1963).

*[6] H. C. Rogers, *Trans. AIME*, **218**, 498 (1960).

[7] J. Gurland and J. Plateau, *Trans. ASM*, **56**, 318 (1963).

[8] F. A. McClintock, to be published.

*[9] *Fracture Testing of High Strength Sheet Materials*, ASTM Bull. (January 1960).

[10] J. J. Gilman, in *Fracture*, B. L. Averbach et al., eds., M.I.T., Wiley, New York (1959), p. 193.

[11] J. J. Gilman, *Strength of Ceramic Crystals*, Am. Ceram. Soc. Conf., New York (April 1962).

[12] C. Inglis, *Trans. Inst. Naval Archit.*, London, **55**, 219 (1913).

[13] I. N. Sneddon, *Proc. Roy. Soc.*, **A187**, 229 (1946).

[14] N. I. Muskhelishvilli, *Some Basic Problems of the Mathematical Theory of Elasticity,* Noordhoff, Groningen (1953).

[15] V. Weiss, *ASME Preprint,* 62-WA-270 (November 1962).

[16] G. R. Irwin, *Encyclopedia of Physics,* Vol. VI, Springer, Heidelberg (1958).

[17] A. A. Griffith, *Proc. Int. Congr. Appl. Mech.,* **55** (1924).

[18] A. H. Cottrell, *Mechanical Properties of Matter,* Wiley, New York (1964), p. 345.

[19] R. A. Sack, *Proc. Phys. Soc.,* **58**, 729 (1946).

*[20] P. C. Paris and G. C. M. Sih, *Fracture Toughness Testing,* source cited [1], p. 30.

[21] N. F. Mott, *Engineering,* **16**, 2 (1948).

[22] J. J. Gilman, C. Knudsen and W. P. Walsh, *J. Appl. Phys.,* **29**, 601 (1958).

[23] D. K. Felbeck and E. Orowan, *Welding J. Res. Suppl.* (1955), p. 570s.

[24] E. Orowan, *Trans. Inst. Eng. Shipbuild., Scotland,* **89**, 165 (1945).

[25] J. Friedel, in *Fracture,* B. L. Averbach et al., eds., M.I.T., Wiley, New York (1959), p. 498.

[26] C. F. Tipper, *J. Iron. Steel Inst.,* **185**, 4 (1957).

[27] P. L. Pratt and T. A. C. Stock, *Proc. Roy. Soc.,* **285**, 73 (1965).

[28] A. H. Cottrell, *Proc. Roy. Soc.,* **285**, 10 (1965).

[29] H. Neuber, *Theory of Notch Stresses,* Edwards, Ann Arbor, Mich. (1946).

[30] G. N. Savin, *Stress Concentration Around Holes,* Pergamon, New York (1961).

[31] H. G. Tattersall and F. J. P. Clarke, *Phil. Mag.,* **7**, 1977 (1962).

[32] R. J. Stokes and C. H. Li, *Fracture of Solids,* Interscience, New York, (1963) p. 289.

[33] J. F. Knott and A. H. Cottrell, *J. Iron Steel Inst.,* **201**, 249 (1963).

[34] A. S. Tetelman, *Acta Met.,* **12**, 993 (1964).

[35] A. S. Tetelman and T. L. J. Johnston, *Phil. Mag.,* **11**, 389 (1965).

*[36] F. A. McClintock, *J. Appl. Mech.,* **25**, 282 (1958).

[37] B. A. Bilby, A. H. Cottrell and K. H. Swinden, *Proc. Roy. Soc.,* **A272**, 304 (1963).

[38] G. T. Hahn and A. R. Rosenfield, *Acta Met.,* **13**, 293 (1965).

[39] F. A. McClintock and G. R. Irwin, *Fracture Toughness Testing,* source cited [1], p. 84.

[40] G. R. Irwin, *Metals Engineering Quart.* **3**, 24 (1963).

[41] A. A. Wells, *Brit. Weld. J.,* 855 (1963).

[42] D. S. Dugdale, *J. Mech. Phys. Solids,* **8**, 100 (1960).

[43] J. N. Goodier and F. A. Field, *Fracture of Solids,* Interscience, New York (1963), p. 103.

[44] A. R. Rosenfield, P. K. Dai and G. T. Hahn, *Int. Conf. on Fracture,* Sendai, Japan, A-179 (1965).

[45] G. R. Irwin, *Proc. First Symposium on Naval Structural Mechanics,* Pergamon, New York (1960).

[46] H. Neuber, *J. Appl. Mech.,* **28**, 544 (1961).

[47] J. A. H. Hult and F. A. McClintock, *Ninth Int. Congr. Appl. Mech.,* **8**, 51 (1957).

[48] A. H. Cottrell, *Proc. Roy. Soc.,* **282**, 2 (1964).

[49] A. H. Cottrell, *Proc. Roy. Soc.,* **A276**, 1 (1963).

[50] ASTM Committee, *Materials Research and Standards,* (1961), p. 877.

[51] J. H. Mulhern, D. F. Armiento and H. Markus, *ASME preprint* 63-WA-306 (1963).

[52] J. M. Krafft, *Appl. Materials Res.* **3**, 88 (1963).

[53] R. F. Koshelov and G. V. Ushich, 1*Zn. Akad Nauk SSSR Otn Mekhanika, I Mash.,* **1**, 111 (1959).

[54] V. Weiss and S. Yukawa, *Fracture Toughness Testing,* source cited in [1], p. 1.

[55] G. R. Irwin, *Materials For Missiles and Spacecraft,* McGraw-Hill, New York (1963) p. 204.

Problems

1. The Muskhelishvilli solution for the stresses a distance r from the tip of an elastic crack of half-length c are

$$\sigma_{yy} = \sigma \coth \alpha$$

where $\alpha = \cosh^{-1} x/a$, $x = c + r$. For small distances ahead of the crack where $x/c \to 1$, show that $\sigma_{yy} \cong \sigma\sqrt{c/2r}$. Use small angle approximations where necessary.

2. A two-dimensional Griffith crack of length $2c$ and height $2h$ in an elastic medium can be represented as an ellipse in rectangular coordinates

$$\left(\frac{x}{c}\right)^2 + \left(\frac{y}{h}\right)^2 = 1$$

Derive an expression for the elastic energy W_E that was stored inside the crack under a tensile stress σ.

3. The yield strength in shear of a material is 50,000 psi. What is the strain concentration factor at a point 0.1 in. in front of a 1 in. crack in this material when a stress of 10,000 psi is applied to the crack in antiplane-strain loading?   (2.0)

4. (a) The true stress–true strain curve of a ductile material can be represented by the relation $\sigma = 32{,}000\ \epsilon^{\frac{1}{2}}$ in units of psi. If the fracture strain in uniaxial tension is 0.8, what is the value of G_{Ic} when a 10 in. long notch, having a tip radius of 0.1 in., is present?

(b) If the modulus is 10^7 psi and $\nu = 0.3$, what is the (approximate) maximum tensile load that can be applied to the cracked material without causing unstable fracture?

                         (a) 306 in. lb per in.²; (b) 10 ksi

5. A sheet of aluminum skin 0.02 in. thick can support a load of 50,000 lb without failing by shear rupture in the presence of a **2** in.

notch. How much load can be supported by two skins, each 0.01 in. thick, if they each contain 2 in. notches and are loaded in parallel by tensile loading?   (35,500 lb)

6. The yield strength of a high strength aluminum alloy decreases linearly with absolute temperature T as $\sigma_Y = A - BT$. The tensile ductility varies with temperature as $\epsilon_f = CT^2$. For specimens containing cracks having a fixed length and root radius ρ, what is temperature at which the notch toughness is a maximum and what is the value of this maximum toughness?

7. Using a force-displacement (P, l) diagram, show that the net energy W_E tending to propagate an elastic crack is $\frac{1}{2} P \, dl$, whether the crack propagates under fixed grip conditions as described in the text or under dead loading.

3

Designing and Testing
for Fracture Resistance

In the preceding chapter it was shown that brittle (unstable) fracture initiates near the tip of a slowly moving or stopped flaw when

$$\sigma = \sigma_F = \sqrt{\frac{EG_c}{\alpha \pi c}} \qquad (\sigma_F < \sigma_Y) \qquad (3.1a)$$

or when

$$K^2 = EG_c \qquad (3.1b)$$

where

$$K = \sigma \sqrt{\alpha \pi c} \qquad (3.1c)$$

At the present time there are no exact analytic expressions that can be used for obtaining the fracture stress σ_F when fracture occurs after general yield ($\sigma_F > \sigma_Y$). It is reasonable to assume, however, that in these instances σ_F increases as the toughness G_c increases and the flaw depth $2c$ decreases.

Basically there are two approaches to fracture-safe design:

1. One approach is to choose a low-yield strength material with such a high toughness G_c that $\sigma_F > \sigma_Y$ unless unusually large and readily observable (to the naked eye) flaws are present.

2. The second approach is to choose a high-yield strength (brittle) material with a low value of G_c and maintain $\sigma < \sigma_F$ at all points in the member. σ_F can be determined from Eqs. 3.1 if the values of G_c, the flaw-shape factor α, and the size of the largest flaw that can go undetected are all known.

In the following section these two approaches to fracture-safe design are discussed. It is shown that (1) is a valid approach for low strength materials, where unstable fracture (low G_c) occurs by cleavage† at low operating temperature; at higher temperatures failure

† Except in the special cases of very thin sheets and at temperatures approaching the melting point.

occurs by ductile rupture and G_c is high. The transition from the former mode of failure to the latter is a *brittle-ductile transition*. Consequently these materials are characterized by a transition temperature region above which they may be safely used. Alternatively unstable fracture in high strength materials occurs predominantly by low energy rupture (low G_c) and is independent of temperature. Consequently there is no particular temperature above which these materials may be safely used. Fracture-safe design requires that $\sigma < \sigma_F$, as given by approach (2).

Terms such as "low strength" and "high strength" are relative. For the purpose of subsequent discussion, we define materials in terms of their *ambient temperature yield strengths* and choose the following values which are consistent with available fracture data:

	Approximate Relations	Steels	Aluminum Alloys	Titanium Alloys
Low strength	$\sigma_Y < E/300$	$\sigma_Y < 90$	$\sigma_Y < 30$	$\sigma_Y < 55$
Medium strength	$E/300 < \sigma_Y$ $\sigma_Y < E/150$	$90 < \sigma_Y < 180$	$30 < \sigma_Y < 60$	$55 < \sigma_Y < 110$
High strength	$E/150 < \sigma_Y$	$180 < \sigma_Y$	$60 < \sigma_Y$	$110 < \sigma_Y$

E is the elastic modulus.   All stresses are in units of ksi.

A complete analysis of fracture requires a detailed knowledge of the various modes of fracture that have taken place. In Section 3.2 the visual characteristics of the fracture modes, as determined by optical and electron microscopy, are discussed. We then consider the most useful tests for fracture resistance; Section 3.3 deals with comparative tests (e.g., is material A better than B) and Section 3.4 describes the specific tests that can be used to obtain data useful for design purposes (safe operating temperature, G_c value). Section 3.5 briefly describes some tests that may be used when fracture occurs after long periods of time. We then show how the principles described here can be used in analyzing failures that have occurred in service.

3.1 Principles of Fracture-Safe Design

Low strength materials

The Fracture Analysis Diagram. The fracture analysis diagram, developed by Pellini and Puzak [1, 2] at the United States Naval

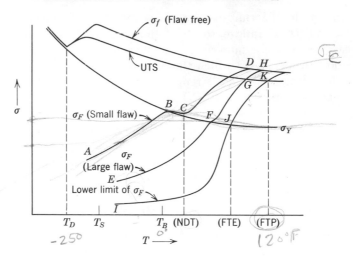

FIG. 3.1. Effect of temperature T on the yield strength σ_Y, UTS, and fracture strength σ_f of flaw-free materials. Also shown is the fracture strength σ_F of materials containing flaws of varying lengths. The reference temperatures are defined in the text.

Research Laboratory, is extremely useful for designing against unstable cleavage fractures in low strength materials. Although the diagram has only been quantitatively used to obtain fracture-safe design criteria for mild steel, it can, in theory, be extended to analyze fracture in any structure where *the unstable mode of fracture occurs by cleavage only* (i.e., the shear fractures which occur are so tough that they occur above the ultimate tensile stress and pose no practical problem. This is the case in mild steel).

At a given rate of loading,† the *operating temperature* is the most important variable that determines the brittleness of mild steel. This is so because both G_c and γ_P are strongly temperature-dependent, so that σ_F and σ_M will also be temperature-dependent. In addition, the mechanical properties of flaw-free materials (yield stress σ_Y, tensile strength UTS, and fracture strength σ_f) are also dependent on temperature. These variations are shown schematically in Fig. 3.1 for a structure of fixed dimensions.

† An increase in loading rate is equivalent to a decrease in service temperature. For mild steel, the relation between strain rate and temperature is semilogarithmic; a 10-fold increase in loading rate is equivalent to about a 60°F decrease in temperature, a 100-fold increase in loading rate is equivalent to a 120°F decrease in temperature.

For the *unnotched* (*flaw-free*) *structure* or test bar, σ_Y increases strongly with decreasing service temperature. At temperatures below the temperature T_D, cleavage fracture occurs on yielding ($\sigma_f = $ UTS $= \sigma_Y$) and the structure exhibits zero macroscopic ductility. Above T_D some yielding precedes cleavage fracture so that the UTS and σ_f are both greater than σ_Y. σ_f and, to a certain extent, the UTS increase with increasing temperature between T_D and T_S. At temperatures above T_S fracture initiates as a normal rupture; between T_S and T_B this rupture can change over into a cleavage fracture which in turn spreads radially to the outer surface of the structure until the remaining portion fractures by shear lip formation. At temperatures above T_B cleavage fracture no longer develops and the fracture is a mixture of normal and shear rupture (Fig. 3.7). Both the UTS and σ_f decrease with increasing temperatures above T_S, as shown in Fig. 3.1. Typically $T_D \simeq -250°F$ and $T_B \approx 0°F$ for a mild steel [3].

Suppose that a small (less than 0.25 in.) flaw or crack is introduced into a wide plate. For $W \gg c$, the gross-section yield stress of the plate σ_Y will be unaffected by the presence of the flaw. The fracture stress of the structure containing the flaw σ_F will now be lower than the fracture stress of the flaw-free structure σ_f, and its variation with temperature is shown by the curve $ABCD$. In the region AB, $\sigma_F < \sigma_Y$ and the structure breaks before yielding occurs *across the entire plate*. Of course, some *local yielding* has occurred in the plastic zone near the notch root to trigger off the fracture [4]. The increase in σ_F from A to B results from the fact that G_c increases with increasing temperature (see Chapter 7). $\sigma_F \cong \sigma_Y$ between B and C. At temperatures higher than that which corresponds to point C, $\sigma_F > \sigma_Y$ and σ_F increases sharply with increasing temperature until point D is reached. At D, $\sigma_F \cong$ UTS of the flaw-free material. The temperature corresponding to point C, the highest temperature at which the yield stress and fracture stress are contiguous, is called [1, 2] the *nil-ductility temperature* (NDT). At the NDT fracture occurs entirely by cleavage and crack propagation is brittle. At point D fracture initiates slowly and occurs by a mixture of cleavage and shear after large plastic strains. The temperature range between C and D is a *transition region* within which the ductility and toughness increase sharply.

If a larger flaw is introduced into the structure, the fracture stress is lowered even more and is given by the curve $EFGH$. At any temperature (fixed G_c) below the NDT σ_F varies† as the inverse square root of the flaw size (Eq. 3.1) when $\sigma_F < 0.6\sigma_Y$. Figure 3.1 shows

† Curves for σ_M follow essentially the same pattern.

that an increased flaw size also causes the transition temperature region to be shifted to higher temperatures. This is important for design considerations because it shows that laboratory tests on small specimens, that can only contain small flaws, may not reflect the degree of brittleness that can occur in service when large (say 2 in.) flaws are present.

It has been observed experimentally that at temperatures in the vicinity of the NDT there is a lower limiting stress [3] (σ_F or σ_M) that is required for unstable cleavage fracture, regardless of crack length (i.e., as $c \rightarrow \infty$). This limiting stress for all practical purposes is about 5,000 to 8,000 psi in steels. Above the NDT the stress required for unstable propagation of the "infinitely" long flaw (curve IJK) rises sharply with increasing temperature. This curve is called the *crack-arrest temperature curve* (CAT) and it defines the highest temperature T_R at which unstable propagation can occur at any stress level. At all points to the right of this curve unstable cleavage fractures cannot take place, regardless of stress level or temperature.

The *fracture-transition elastic* (FTE) is defined by the temperature where the CAT curve crosses the yield curve (point J). At point K the CAT curve intersects the ultimate tensile strength curve so that, from the point of view of load-carrying capacity, the material behaves as though it were flaw free. Point K is called the *fracture-transition plastic* (FTP). In mild steel the temperature range NDT–FTP, between which the small crack can still spread unstably at the yield stress and the largest crack cannot spread unstably at the ultimate stress, is about 120°F [1, 2].

The fracture analysis diagram is useful in structural design where cleavage may be a problem. Suppose that a particular design requires that a steel plate withstand a certain stress σ' at a minimum operating temperature T'. Safe operation could be achieved if the structure is designed so that fracture does not initiate under these conditions. This would be achieved by selecting a certain steel, experimentally determining G_c at the temperature and plate dimensions in question, and, from Eq. 3.1, calculating the maximum length of flaw that could exist without initiating fracture at σ'. Suppose that for this steel (steel A, Fig. 3.2) the critical flaw length is 1 in. and that T' is below the NDT. Then, in order to insure reliability, it is necessary that inspection techniques be able to detect the presence of any flaws of this length. It is also necessary that smaller flaws not exist in regions of high residual stress $\Delta\sigma$ (e.g., around welds) because these could initiate fracture if Eqs. 3.1 were satisfied.

A second procedure for insuring reliability is to choose steel B whose CAT is less than the temperature T', for a nominal applied

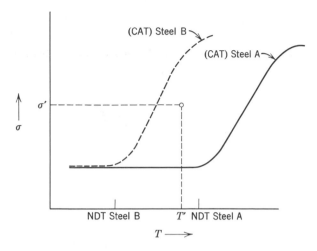

Fig. 3.2. Relative behavior of steel A (brittle) and steel B (tough) subjected to a tensile stress σ' at temperature T'.

stress σ'. Then, even if a large flaw is formed, it would be unable to propagate over large distances in a low strength material. This procedure is obviously simpler and more reliable than the design against initiation described above, for it is not necessary to know the stress at all points in the structure. In many cases, however, this is an expensive procedure because steel B, being tougher, costs more (per pound) than steel A. Furthermore, this is a very conservative design procedure because it is based on arresting a fracture that may never initiate at all.

Design Criteria for Mild Steel

Numerous tests have been conducted and evaluated by Pellini and his coworkers to establish safe and simple design criteria that may be used by engineers who employ structural steel in particular applications. If the structure does not contain any flaws at all, and if flaws will not develop during operating lifetime, then there is no need to utilize the fracture analysis diagram described above.

Most structures, especially large and welded ones, do contain flaws and in these cases the fracture analysis diagram, as shown in simplified form (Fig. 3.3), is valuable for design purposes. There are four primary reference temperatures [1, 2], relative to the NDT,† which establish the conditions under which unstable fracture will *not occur in mild steel* at a given level of applied (nominal) stress σ, relative to the yield stress σ_Y.

† For most mild steels the NDT varies between -60 and $+60°$F.

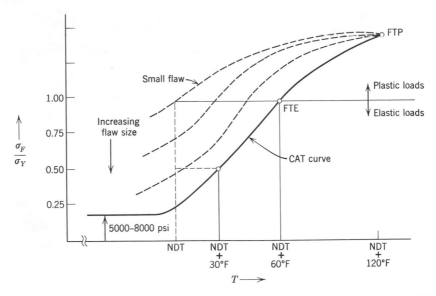

Fig. 3.3. Reference temperatures used for various design criteria for mild steel. *After W. Pellini and P. Puzak* [1].

(a) *NDT design criterion.* No guarantee of fracture arrest at $T \leq \text{NDT}$ unless $\sigma \lesssim 5,000\text{–}8,000$ psi. Fracture can only be controlled by preventing its initiation if applied stress is greater then 8,000 psi.

(b) *NDT + 30°F design criterion.* When $T = \text{NDT} + 30°\text{F}$, unstable cleavage crack propagation cannot occur if σ is maintained less than about $\frac{1}{2}\sigma_Y$.

(c) *NDT + 60°F design criterion.* When $T = \text{NDT} + 60°\text{F}$, unstable cleavage crack propagation cannot occur if σ is maintained less than about σ_Y.

(d) *NDT + 120°F design criterion.* Catastrophic fracture does not occur before tensile strength is reached, irrespective of the applied stress level. Material behaves as if all flaws were absent.

The choice of which of these design criteria to use depends on many factors, such as prior experience, cost, and degree of reliability required. If a structure is to be operated at nominal stress levels that are one-half the yield stress, as is often the case, then (b) is the most efficient criterion. If a more efficient design is required (e.g., the structure is operated at stress levels closer to the yield stress), then (c) must be used to guarantee that unstable fracture will not occur. If there exists the possibility of plastic overloading because of accidents or explosive attack (military applications), then (d) would probably be best.

Suppose, for example, that engineering considerations indicate the use of design criterion (c) and the lowest anticipated service temperature to be 20°F. Then the most inexpensive steel, whose NDT temperature is 60°F below the operating temperature (i.e., −40°F), would be chosen for the particular application. (In Section 3.4 we will discuss the various means of evaluating the NDT temperature.)

Criterion (a) could be used only if there was an excellent chance of preventing the initiation of fracture, since unstable propagation occurs easily below the NDT. This would require that $\sigma < \sigma_F$, as determined by Eq. 3.1. Some of the difficulties associated with this approach are discussed below in connection with high strength materials.

Design Criteria for Other Low Strength Materials

It is important to remember that design criteria (b)–(d) are based on the facts that: (1) G_c for cleavage fracture increases with increasing temperature so that there are crack-arrest temperatures, at various values of σ/σ_Y, above which unstable fracture does not occur, and (2) in low strength materials the normal and/or shear ruptures which then take place are stable unless unusually large cracks are present.

In theory, every material has a CAT curve that can be determined experimentally at various values of σ/σ_Y. For example, cleavage fracture *never occurs* in materials having an FCC or HCP crystal structure (except for zinc and beryllium) so that all operating temperatures T and relative stress levels σ/σ_Y are above the CAT for these materials. Other low strength structural materials, such as the BCC refractory metals, nonmetallic crystals, and many composites do fracture by cleavage at low temperatures. Except in a few isolated cases, design criteria similar to those developed for mild steel have not yet been reported. It is unlikely that the criteria developed for mild steel can be safely used for these materials and more experimentation is needed to obtain exact correlation between the NDT, FTE, and FTP temperatures for these materials.

High-strength materials and medium-strength materials

For reasons that are discussed in Chapter 7, G_c for normal and/or shear rupture decreases as the yield strength σ_Y increases. In high strength materials the value of G_c can be so low and σ_Y can be sufficiently high that catastrophic fracture can initiate at stress levels σ_F

that are less than σ_Y, especially when the initiating flaw is large (Fig. 3.4). These noncleavage modes of unstable fracture are called "low-energy tear" [6]. *Unlike in the case of mild steel and other low strength materials, the occurrence of noncleavage modes of crack propagation is no longer a guarantee that fracture has occurred or would occur in a stable manner.*

Tear fractures in high strength materials, whether stable or unstable, are only slightly influenced by temperature because their value of G_{Ic} is essentially temperature-independent. Consequently there is no "transition temperature" above which they do not occur. If unstable fracture is possible at one temperature it will be possible at all temperatures and one cannot design against it on the basis of a crack-arrest temperature.

Medium strength materials do undergo a ductile-to-brittle transition as the operating temperature is decreased. The mode of fracture changes from essentially stable normal rupture to predominantly low-energy tear. The reason that this ductile-brittle transition occurs is that σ_Y increases as the temperature decreases (Fig. 3.1). Consequently at low temperatures the yield strength equals or exceeds $E/150$ and the materials *behave as though they were high strength materials which operate at ambient temperature.* Thus G_c decreases gradually with decreasing temperature and, at low temperatures, $\sigma_F < \sigma_Y$.

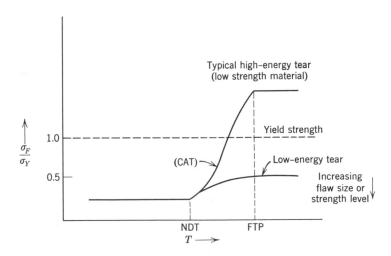

Fig. 3.4. Effect of temperature on the mechanical behavior of materials showing high-energy and low-energy tear fracture.

The only safe design that can be successfully used when low-energy tear is a potential problem is one based on preventing the initiation of fracture. The four parameters σ, α, c, and G_c that comprise Eqs. 3.1 determine the fracture criterion. Three of them are independent variables and the fourth is a dependent variable. Theoretically it should be possible to obtain a value of the maximum allowable design stress σ_F that a structure can withstand without breaking if G_c, α, and c are known. In practice there are certain difficulties that make an exact evaluation of all parameters extremely difficult. Some of these difficulties are as follows.

The value of α depends on the planar shape of the flaw, its root radius, the dimensions of the structure within which the flaw resides, and, at high stress levels, the plastic zone size. $\alpha = 1$ for unidimensional propagation of an infinitely sharp crack in an infinitely wide plate. However, most often flaws propagate in two dimensions, have complicated shapes, have finite root radii, and lie in structures having finite dimensions. In these instances $\alpha \neq 1$ and the methods of analytical fracture mechanics must be used to calculate α (or K) for a given stress σ and semimajor axis c of the flaw (see Eq. 2.29b). This can be a complicated procedure and some approximations (e.g., worst possible case) may have to be made to insure safe design.

The value of c is difficult to determine since fatigue and/or corrosion can cause preinduced flaws (because of welding, machining, or sharp corners) to grow slowly during service. Inspection techniques may not be able to determine the dimensions of flaws introduced before the structure is put into service, let alone any changes that occur during operation. A "worst case" situation can be assumed for design purposes in which the value of c, prior to operation, is set equal to the largest flaw size that can go undetected. Changes in c during operation can be estimated on the bases of known rates of fatigue and/or corrosion crack growth.

The value of the toughness G_c is rarely a constant for a given material. It is strongly dependent on the mode of fracture initiation and this, in turn, depends on size of the structure (particularly thickness), environmental conditions, strength level, and prior heat treatment. Except near the melting point, G_c decreases as the yield strength σ_Y of the material increases. It is also possible to lower G_c by certain corrosive environments and heat treatments (e.g., those which produce temper brittleness in heat-treated steels). Consequently safe design requires that G_c be estimated on the basis of the worst possible conditions that a material will be exposed to during service.

The value of the stress σ can also be difficult to determine, especially in structures having complicated shapes and subjected to complex loading. The presence of residual stresses (e.g., around welds) may raise the stress level, on a local level up to σ_F. As shown in the preceding chapter (see Fig. 2.13), this increase can lead to unstable fracture in strain-rate sensitive materials where the amount of work required to continue fast propagation is less than the work required to initiate the fracture.

Despite these difficulties the fracture mechanics approach is extremely useful for producing fracture-safe designs in *high strength materials at all temperatures and in low and medium strength materials at low temperatures*. However, it must be emphasized that the approach is only as good as: (1) *the nondestructive testing procedure* [63, 64, 65] that is used to detect the flaws and their slow growth; (2) *the stress analysis procedures* [66, 67] that are used to determine the value of σ at all critical points in the structure; (3) *the analytical aspects of fracture mechanics* [68] which must be used to determine α (or K) values for complicated flaws in structures having complicated shapes; (4) *the techniques for determining G_c*.

The first three topics are outside the scope of this text; the fourth will be discussed in Section 3.4.

3.2   Visual Analysis of Fractures

General considerations

Most fractures occur in service when a tensile component (mode I) of deformation is present.† When a plate is loaded in uniaxial tension, for example, mode I is strongest across planes that are perpendicular to the stress axis, whereas modes II and III are strongest on planes inclined at 45° to the tensile axis (Fig. 3.5). Similar considerations apply to cylindrical plates that form the walls of pressure vessels. For these $\sigma = pD/2t$ is the maximum tensile stress in the plane of the sheet, where D is the diameter of the vessel, p is the internal pressure, and t is the wall thickness. In both cases the shear stresses along the 45° planes are $\sigma/2$. Mode I fractures (cleavage and normal rupture) tend to occur at the center of plates where plane-strain conditions of deformation exist. When these fractures spread outward (Fig. 3.6a), the thickness of the unfractured volume of material decreases and these regions are then loaded in plane stress. Unless the material is extremely resistant to plastic deformation, the

---

† The failure of a riveted joint in pure shear is a notable exception.

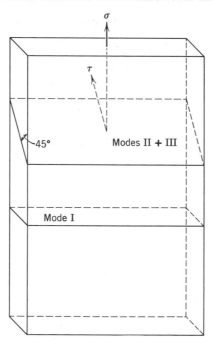

Fig. 3.5. Planes along which the various modes of crack propagation occur during fracture of either a plate or sheet under uniaxial tensile loading or a cylinder under internal pressure.

deformation can spread to the free surface in an unconstrained fashion (Fig. 3.6b) along 45° planes, giving rise to the formation of a *shear lip* [7] around the central fracture. Fractures that occur perpendicular to the stress axis (square fractures) are commonly called *plane-strain* fractures because they tend to occur under conditions of plane-strain deformation; 45° slant fractures are referred to as *plane-stress* fractures because they always occur under plane-stress loading.†

† This terminology sometimes causes confusion because the terms plane-strain and plane-stress refer strictly to the mode of plastic deformation around a crack which develops under the particular loading mode. For low energy tear there is really no problem since plane-strain fractures occur when the loading is plane strain, and vice versa. For cleavage propagation there may be confusion because this plane-strain mode of propagation can exist under plane-stress loading conditions at very low temperatures.

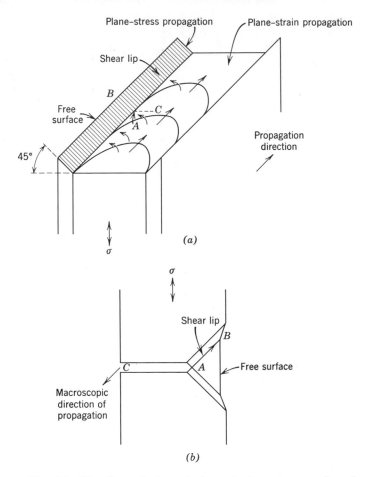

Fig. 3.6. Direction of plane-strain and plane-stress modes of crack propagation in a plate or sheet. (a) Top view. (b) Front view.

Each of the specific modes of fracture produces certain characteristic markings on the fracture surfaces of broken parts. Visual observation and interpretation of these markings provide a complete history of the fracture process and, hopefully, the reason for the occurrence of the fracture. In some cases the observation may be made with the naked eye or with a low power (20×) microscope, but in more complicated situations, particularly when observing the fracture pattern in a high strength material, optical (100–1,500×) or electron (2000–20,000×) microscopy is required. In the following paragraphs the characteristic features associated with each of the modes of fracture will be described, as well as cases where more than one mode has operated. Space limitation prohibits a description of the techniques used in obtaining the micrographs. This information can be obtained from the indicated references [8, 15, 16] at the end of the chapter.

Macroscopic observation of fracture surfaces

The smooth (unnotched) round tensile test [9, 10] provides a simple means of obtaining characteristic fracture surfaces and relating their macroscopic appearance to strength level and test temperature. Three fracture zones can be observed [22, 23] with the naked eye (Fig. 3.7a): (1) *The initiation or fibrous zone* within which failure occurs slowly, by stable normal rupture. This zone contains a series of markings that are perpendicular to the macroscopic direction of crack propagation. In the smooth, round, tensile test, fracture initiates at the specimen center and hence the markings in this zone are concentric with the tensile axis. Crack propagation tends to occur in a plane that is perpendicular to the tensile axis, but the fracture surface is jagged, as schematized in Fig. 2.2c. (2) *The radial zone*, within which fracture is rapid and unstable. The markings here are parallel to the direction of crack propagation and hence assume a spoke- or star-shaped appearance in the smooth, round tensile test. Radial markings result from either cleavage or low-energy tear-type propagation. Propagation occurs in a plane that is perpendicular to the tensile axis and, in general, *the finer (i.e., smaller) the markings, the more brittle (lower γ_P) is the propagation.* (3) *The shear lip zone*, within which fracture is rapid and occurs by shear rupture. The fracture surface is extremely smooth and tends to be inclined at 45° to the tensile axis.

The relative amount and degree of roughness of each of these three zones is primarily a function of the strength level of the material being tested and the test temperature. *In single-phase, low-strength materials* (e.g., pure iron, tungsten, magnesium oxide) fracture occurs either by cleavage† or shear rupture. Consequently the initiation zone of slow normal rupture is not present and the fracture surface is either all radial cleavage or all shear lip (Fig. 3.7b). The cleavage fractures are characterized by their extreme flatness and high reflectivity or "crystallinity," which causes laymen to attribute them (incorrectly) to "crystallization of the metal." The transition from cleavage to shear occurs abruptly, over only a few degrees of

† Except in all FCC and most HCP metals which do not fracture by cleavage.

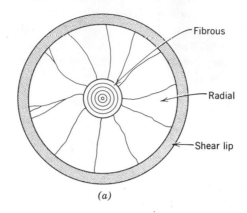

(a)

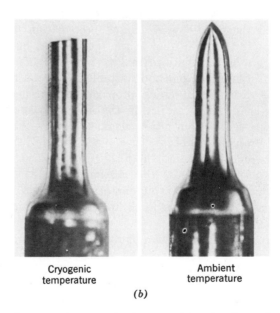

Cryogenic
temperature

Ambient
temperature

(b)

temperature. In single crystals the transition is also a function of the orientation of cleavage planes relative to the tensile axis [69].

Most low strength commercial materials are *polyphase*, containing distributions of hard particles (e.g., carbides in mild steel and oxides in tough pitch copper). These cause fracture initiation to occur by normal rupture in the initiation zone, except at very low temperatures ($T < T_S$, Fig. 3.1). At temperatures between T_S and T_B the slow

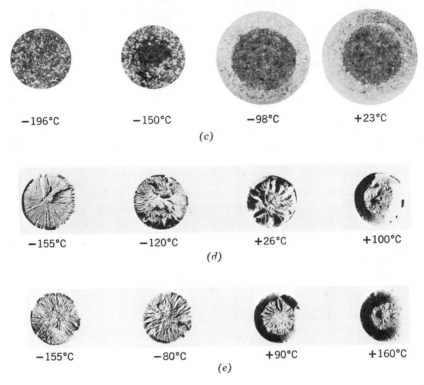

-196°C   -150°C   -98°C   +23°C

(c)

-155°C   -120°C   +26°C   +100°C

(d)

-155°C   -80°C   +90°C   +160°C

(e)

Fig. 3.7. Effect of temperature on the fracture appearance of unnotched smooth tensile specimens. (a) Schematic diagram showing various fracture zones. (b) Low strength, single-phase material (e.g., pure iron crystal ($\sigma_Y = 5$ ksi) in which ductile-to-brittle transition was obtained by varying orientation with respect to stress axis. Similar effect produced by slight changes in temperature for crystal of given orientation. *Courtesy B. Hopkins* [69] *and Institute of Metals.* (c) Low strength, polyphase material (mild steel, $\sigma_Y = 23$ ksi). *Courtesy C. A. Rau.* (d) Medium strength material (heat-treated 4340 steel, $\sigma_Y = 133$ ksi). *Courtesy F. Larson* [22] *and American Society for metals.* (e) High strength material (heat-treated 4340 steel, $\sigma_Y = 216$ ksi). *Courtesy F. Larson* [22] *and American Society for Metals.*

rupture can transform into the radial cleavage mode; above T_B cleavage does not occur and only zones 1 and 3 are observed (Fig. 3.7c). The transition from all radial cleavage to cup-cone (zones 1 and 3) occurs over a relatively broad temperature range.

Figures 3.7d and 3.7e indicate that a similar transition in fracture appearance occurs in *medium* and *high strength materials* as the temperature is varied. In these cases the radial zone results predominantly from *low-energy tear fracture rather than from cleavage.* Decreasing the test temperature or increasing the strength level† increases the relative amount of the radial zone and causes the height of the radial (star-shaped) markings to decrease [22, 23]. The relative amount of radial fracture can also be increased in a given alloy, tested at a given temperature, if the size of the specimen increases [24].

The radial markings are important in service failure analysis. It is usually possible to trace them back to their convergence point, which is the site of the fracture initiation, and thereby determine the cause of the failure. These sites are often a defective weld, the heat-affected zone near a weld, a corrosion pit (Fig. 3.8*a*), or a weld bead (Fig. 3.8*b*). The radial markings result from the fact that the crack front branches and propagates on different levels; this effect is most pronounced as the leading edge of the crack approaches a free surface. In wide plates the radial markings often bend outward, toward the free surface, to produce the chevron or herringbone pattern shown in Fig. 3.8*b*.

Microscopic observation of fracture surfaces

Cleavage Fracture. On a microscopic level it is observed that even within a single grain of a polycrystal, or in a large single crystal (Fig. 3.9), *transgranular cleavage fractures* propagate on more than one level. This gives rise to the formation of *river patterns* which converge at the site of initiation of fracture in the particular grain under observation [12, 13]. In polycrystalline materials adjacent grains are tilted and twisted with respect to one another. Large cleavage steps (differences in level of propagation) form when the crack crosses a grain boundary (Fig. 3.10) because the (crystallographic) cleavage planes in adjacent grains usually do not have a common line of intersection [14]. In BCC metals, deformation twins are usually formed near the tip of an advancing crack and these cause the crack to deviate off its plane; this produces [12] the block pattern shown in Fig. 3.11.

In fine-grained polycrystals it is necessary to use electron microscopy techniques [8, 15, 16] to resolve the various features on the fracture surfaces. Figure 3.12*a* indicates the presence of fine tear lines (rivers) within individual cleavage facets which typically have

---

† For example, by changing the tempering conditions applied to a heat-treatable steel.

Fig. 3.8. (*a*) Corrosion pit as fracture origin in pressure vessel. 8×. (*b*) Weld bead as fracture origin in H11 steel motor case. *Courtesy F. Larson* [9] *and American Society for Metals.*

the dimensions of the grain size of the matrix. The tear lines may be used to determine the direction of crack propagation on a local level. The microscopic direction of crack propagation is not always regular and, in some cases, the microscopic crack front may be moving in an opposite direction to the macroscopic direction of crack propagation. If the cleavage planes in a particular grain are poorly oriented for cleavage (the cleavage planes are far from perpendicular to the stress axis), the crack front may circumvent the grain and the grain

Fig. 3.9. River pattern formation associated with cleavage fracture of a single crystal of sodium chloride. 50×. *Courtesy R. Stokes* [11] *and American Society for Metals.*

Fig. 3.10. River pattern formation associated with cleavage crack propagation across a grain boundary in polycrystalline iron-3% silicon. 150×. *Courtesy J. R. Low* [13].

Fig. 3.11. Appearance of cleavage fracture surface of iron single crystal containing numerous twin traces. 75×. *Courtesy R. Honda* [12].

eventually fails by tearing.   This tearing, both within and outside of individual grains and grain boundaries, constitutes a large fraction of the plastic work γ_P expended during cleavage crack propagation [13, 14, 15].

The flat facets observed on the fracture surfaces of quenched and tempered steels that are broken at low temperature are similar to those shown in Fig. 3.12a.   However, the size of the facet is much larger than the size of the ferrite (martensite or bainite) grain and is, in fact, of the order of the size of the prior austenite grain.   Furthermore, the orientation of these facets is not precisely {100}, the normal cleavage plane of the ferrite.   These types of fracture facets are therefore referred to as "quasi-cleavage" to distinguish them from the true cleavage facets observed in single-phase metals [16] (Fig. 3.12a).   The characteristic features of quasi-cleavage are the high density of short, curved tear lines (Fig. 3.12b), which result from the formation of ridges when small cracks grow together within the facet.

On a macroscopic and microscopic basis it is difficult to distinguish between transgranular and intergranular cleavage.   This distinction is important because intergranular cleavage is often the result of improper heat treatment (Chapters 6 and 10).   Fortunately the electron microscope is able to distinguish between the two modes of cleavage.

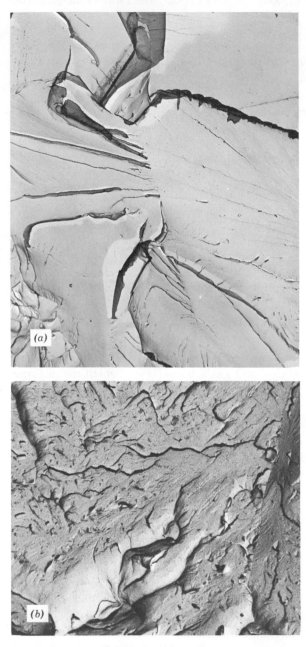

Fig. 3.12. Transgranular cleavage fracture surfaces as observed by replica electron microscopy. (a) Poly-crystalline iron. 3,400×. *Courtesy of J. R. Low* [13]. (b) Ridges associated with quasi-cleavage in high strength steel 10,050×. *Courtesy C. D. Beacham* [16] *and American Society for Metals.*

Fig. 3.13. Smooth fracture surface typical of intergranular fracture in iron. Electron fractograph. *Courtesy J. R. Low* [13].

Figure 3.13 shows, for example, that intergranular cleavage fracture surfaces do not contain river markings. This is so because these fractures closely follow the plane of a grain boundary which has been embrittled by second-phase particles or trace amounts of impurity which cannot even be detected by the electron microscope [13]. In these cases the fracture surfaces are extremely smooth. Another way of distinguishing between inter- and transgranular fracture is to observe the specimen profile near to the fracture surface. Nonpropagating microcracks are occasionally found there and the two possible paths are then easily delineated [14] (Fig. 3.14).

Pure Shear Fractures

Pure shear fractures occur in single-phase materials that do not fracture by cleavage ($T > $ FTP). They are rarely observed in structural materials because the latter are usually composed of two or more phases; these materials rupture rather than shear. Pure shear fractures tend to occur along the 45° planes of maximum-resolved

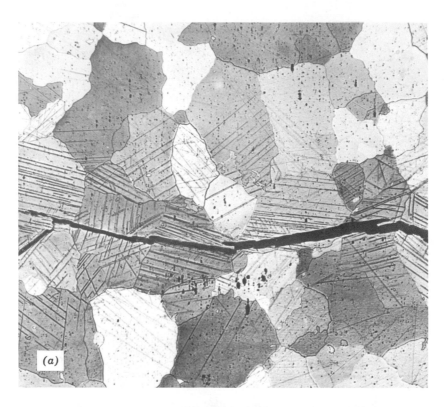

Fig. 3.14. Profiles of: (a) Transgranular and (b) intergranular fracture in iron. 15×. *Courtesy J. R. Low* [14] *and American Society for Metals.*

shear stress under uniaxial or biaxial tensile loading. Since shear rupture also occurs along these planes (see below) it is necessary to distinguish between these two modes on a microscopic basis.

In large single crystals, shear fractures occur when the two halves of the crystal slip apart along crystallographic glide planes. This process, called "glide plane decohesion" [17], produces straight scratches, on the slant or chisel-point fracture surfaces, which are visible to the naked eye. In polycrystals the deformation of any grain of the aggregate is partially constrained by the grains that surround it so that glide occurs on many acts of intersecting slip planes. This leads to the observation (Fig. 3.15) of "serpentine glide" markings on the fracture surfaces. In heavily cold-worked metals the glide markings that occur during fracture may be unresolvable, even under the electron microscope, and the fracture surface may appear featureless in certain areas. Fracture is then said [17] to have occurred by a process called "stretching."

Normal and Shear Rupture

In low strength materials the normal rupture surface is so rough that it can easily be distinguished, by the naked eye or low power microscope, from the flat cleavage mode (e.g., Fig. 3.7c). However, in high strength materials with low G_c values the normal rupture surface may be so smooth (Figs. 3.7d and 3.7e) that electron microscopy is required to make the distinction between the two modes [1]. The

Fig. 3.15. Fractograph of a pure shear fracture in copper.   3000×.
Courtesy C. D. Beacham [17].

presence of "equiaxed dimples" (Fig. 3.16) on the fracture surface is the characteristic feature of a normal rupture [18, 19]. As discussed in Chapter 2, these result from the formation of a void and its subsequent growth and coalescence (with another void) by stretching. The voids form at places where the nondeforming, hard particles have separated from the matrix, along the particle-matrix interface, or where the particle has itself fractured. In general, the density of particles, seen on most of the dimples, is higher than the average particle density in the material. The size of the voids is a measure of the fracture toughness. Shallow voids are indicative of low toughness whereas deep voids are indicative of high toughness [16, 20, 21].

If the voids grow and coalesce when a shear mode of deformation is present (e.g., along the 45° planes), then the dimples will be stretched [19] out in the shearing direction (Fig. 3.17) and appear parabolic. Fracture is then said to occur by *shear rupture*. Most of the fractures that form the shear lip outside a region of cleavage or normal rupture are of this type, unless the material is very pure. Both normal and shear ruptures can follow either transgranular or intergranular paths, depending on the distribution of hard particles in the material.

In cases where the loads are applied cyclically (e.g., the structure is repeatedly loaded and unloaded), fatigue cracks can form (Chapter

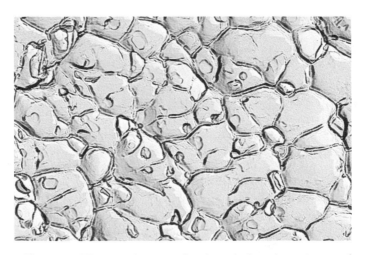

Fig. 3.16. Electron fractograph of typical region of normal rupture in a high strength aluminum alloy. 30,000×. *Courtesy C. D. Beacham* [17].

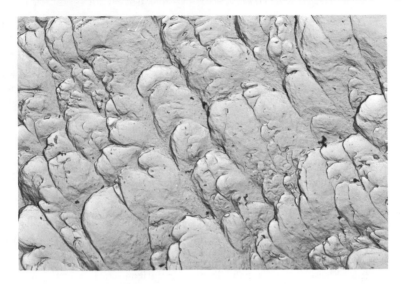

Fig. 3.17. Electron fractograph of typical region of shear rupture in a stainless steel. 30,000×. *Courtesy C. D. Beacham* [17].

8) and these, in turn, can trigger off the unstable fracture. Macroscopically fatigue fractures are usually characterized by a thumbnail pattern near the outer surface of the part (Fig. 3.40), similar to the corrosion pit shown in Fig. 3.8*b*. As observed by the electron microscope, fatigue crack growth is characterized by a high density of closely spaced striae [17, 25]. These are similar to those observed as serpentine glide markings in pure materials except that they do not cross over one another and are usually more pronounced. Figure 3.18 illustrates the fracture surface of an aluminum alloy where normal rupture, shown at the left of the micrograph, was triggered off by a fatigue crack moving from right to left.

3.3   Comparative Testing Procedures for Materials Evaluation

There are essentially five reasons why mechanical tests are carried out to evaluate a material's resistance to fracture:

1. to determine whether there is a strong probability that catastrophic fracture will occur in a particular material, loaded under a given set of conditions;

2. for use in choosing between potential materials for a particular application (i.e., is material A more "brittle" than material B?);

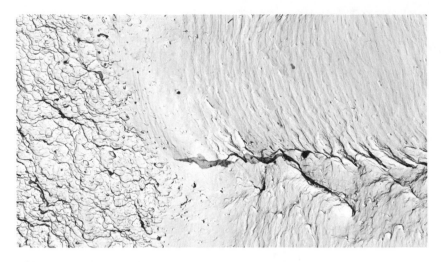

Fig. 3.18. Electron fractograph illustrating fatigue crack growth prior to normal rupture in an aluminum alloy. *Courtesy C. D. Beacham* [17].

3. for quality control tests by the materials producer;

4. for analysis of failures that have occurred in service;

5. to obtain data that can be directly used for design purposes (i.e., maximum working stress, minimum operating temperature, service lifetime).

During the past fifty years numerous tests have been developed [26, 27] for these purposes and it probably would take one or two textbooks to review them. Because of space limitations we shall only describe those tests which have, in our opinion, the greatest degree of usefulness to design and materials engineers. These recommendations are summarized in Table 3.1.

Basically, fracture testing can be divided into tests that are *comparative,* for use in connection with 1–3 above, and tests that are *specific,* for use in connection with reasons 4 and 5. Comparative tests should be reproducible, simple to operate, and inexpensive. Specific tests, used to obtain information that may be directly incorporated in structural design, tend to be expensive and consequently should be carried out only if the simpler test (or a service failure) has indicated that specific data are required for reliability. Thus, if a simple test indicates that catastrophic failure is unlikely in a particular structure made from a particular material, there is no need

Table 3.1

Potential Modes of Brittle Fracture in the Absence of Cyclic Loading or Reactive Environments† and Recommended Tests

Test Used For	Low Strength Materials, $\sigma_Y < E/300$	Medium Strength Materials, $E/300 < \sigma_Y < E/150$	High Strength Materials, $E/150 < \sigma_Y$
Brittle fracture modes	(1) Cleavage (in all materials except FCC metals) (2) Shear rupture (in very thin sheets)	(1) Cleavage (in all materials except FCC metals) (2) Normal rupture (at low temperature) (3) Shear rupture (in very thin sheets)	(1) Normal rupture (2) Shear rupture (in very thin sheets)
	Test	Test	Test
(1) Determining whether there is a high probability of catastrophic failure under service conditions	Charpy V appearance / Notch tensile (very thin sheet)	Charpy V appearance / DWTT energy / Notch tensile (very thin sheet)	DWTT energy value / K_{Ic} / K_c (very thin sheet)
(2) Choosing materials on a comparative basis	NDT / Charpy V appearance / DWTT / Notch tensile (very thin sheet)	Charpy V appearance and energy / DWTT / Notch tensile (very thin sheet)	K_{Ic} / K_c (very thin sheet)
(3) Quality control test by materials producer	NDT / Charpy V energy / Notch tensile (very thin sheet)	DWTT / Charpy V energy / Notch tensile (very thin sheet)	DWTT / K_{Ic} / K_c (very thin sheet)
(4) Service failure analysis	Visual analysis / NDT (for steel) / DWTT / Charpy V appearance and energy / K_c (at low temperature or in thin sheet)	Visual analysis / DWTT / Charpy V appearance and energy / K_c (at low temperature or in thin sheet)	Visual analysis / K_{Ic} / K_c (very thin sheet)
(5) Establishing design data (a) Safe working stress at given temperature and plate thickness	K_c and K_{Ic}	K_c and K_{Ic}	K_c and K_{Ic}
(b) Safe working temperature at given level of applied stress	NDT (for steel) / DWTT	DWTT and Charpy V appearance / K_c and K_{Ic}	Concept not applicable

† Tests to evaluate time-dependent fracture characteristics are discussed in Section 3.4.

to use an expensive and complicated test to evaluate the material's G_c or G_{Ic} value for use in a design criterion [21].

In the absence of embrittling environments the three types of catastrophic failure which can occur are (1) cleavage, (2) low-energy normal rupture in ultra-high strength materials, and (3) shear rupture in very thin sheets. The tendency for cleavage and low-energy tear fracture can best be evaluated, on a comparative basis, by means of the standard Charpy impact test, and the newly developed *drop weight tear test*. The notch tensile test can also be used for these purposes. These tests are discussed below. In Section 3.4 testing procedures are described which can be used to evaluate specific parameters, such as the NDT or G_c value of a material.

The Charpy V notch impact test

The Charpy V notch impact test measures the energy required to break a test bar of given dimensions. These dimensions are given in Fig. 3.19 for the standard bar containing a 2 mm deep notch whose root radius is 0.25 mm. The bar, cooled (or heated) to the desired test temperature, is placed in a holder which supports it at its ends, and is then struck with a hammer. The hammer is mounted on the end of a pendulum which is adjusted such that its kinetic energy, at

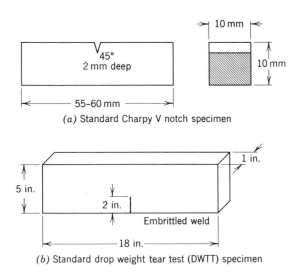

Fig. 3.19. Dimensions of specimen used in conducting (*a*) Charpy V notch impact test and (*b*) drop weight tear test.

the point where it strikes the bar, is 240 ft-lb (in the standard test). After breaking (or deforming) the test bar, the hammer swings through to a height which decreases as the amount of energy absorbed in fracture increases. This energy is recorded directly on a dial gage. It is important to remember that the Charpy test is *not* a specific test† because the energy values that are measured cannot be put into a fracture equation (such as 3.1) that can be directly used for design purposes. What can be measured, however, are *relative changes* of impact energy with test temperature and, in some cases, the tendency of a material to show a low-energy tear fracture.

The Charpy test is a very severe test that was developed for use on mild steel. It favors the plane-strain modes of fracture because the notch root is relatively sharp, the strain rate at the notch root is high, and the bar is thick enough to insure a large degree of plane-strain loading and triaxiality across almost all the notched cross section. This is fortunate for engineering evaluation because a material that can withstand a Charpy test at a given temperature without showing much cleavage fracture has an excellent chance of withstanding cleavage fracture when operated in service at that temperature. Similarly when a large amount of impact energy is required to produce a tear fracture, the material has an excellent chance of withstanding a brittle tear fracture during service.

Low Strength Materials

Figure 3.20 shows a typical Charpy curve obtained from a plate of mild steel that was broken at various temperatures [28]. Also included are curves showing the amount of crystalline or radial fracture observed at each of the test temperatures and the amount of root contraction that occurs at various temperatures. The latter is a measure of the strains required to initiate fracture and hence is a measure of the fracture toughness. Also included are some representative macrographs. As in the case of tensile specimens there are three possible macroscopic modes of fracture: (1) The flat fibrous mode, which is a stable form of normal rupture and occurs parallel to the plane of the notch, (2) the flat radial or crystalline mode, which is essentially a complete cleavage fracture in low strength materials although some grains in this zone fail in shear, and (3) the shear lip zone which occurs at 45° to the notch plane. In this region, fracture occurs by shear rupture.

---

† However, the instrumented Charpy test (Chapter 7) does allow determination of the fracture load at the high-strain rates, and hence provides a measure of G_{Ic}.

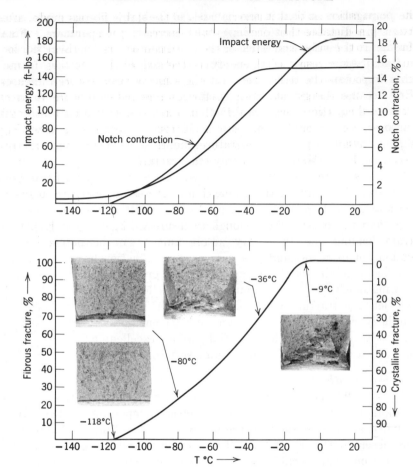

Fig. 3.20. Effect of temperature on the impact energy, % fibrous fracture, and root contraction for Charpy V notch specimens of a 3½% Ni, 0.1% carbon, mild steel. Some representative macrographs are also shown. *Courtesy R. Wullaert* [28].

At low temperature, fracture occurs entirely by cleavage (zone 2) and the impact energy increases slowly with increasing temperature. At −115°C (in this particular steel) a shear lip begins to form and the impact energy (ft-lb) begins to increase more rapidly with temperature. At −90°C (25 ft-lb) fracture is no longer initiated by cleavage but by fibrous tearing at the notch root. The tear mode is able to change over to the radial mode after running a short distance. The radial mode, in turn, is moving into a region that is under compression and is close to a free surface. These effects tend to retard its propagation so that it reverts back to the stable fibrous mode, after running a distance that decreases with increasing temperature. Final failure at the sides and back of the specimen occurs by shear lip formation. As a result of these effects the size of the "picture frame" that surrounds the radial zone increases as the temperature increases. Because the energy absorbed in fibrous and shear rupture is much greater than the energy absorbed in cleavage, the impact strength increases as the proportion of tear (fibrous) fracture increases (Fig. 3.20). Eventually a temperature is reached at which the fracture becomes 100% fibrous rupture (0% crystalline); at temperatures above this point the impact energy is essentially independent of temperature. It is known as the maximum Charpy energy, (c_V max) or the Charpy shelf energy.

Numerous criteria for evaluating materials have been based on curves such as these. They include (with values obtained for the curves shown in Fig. 3.20):

1. The average energy criterion (i.e., the temperature corresponding to an energy that is one-half the difference between the maximum and minimum energies).   −55°C
2. The temperature corresponding to an arbitrary low value of impact energy such as 15 ft-lb (commonly called the ductility-transition temperature).   −105°C
3. The highest temperature at which cleavage appears on the fracture surface (commonly called the propagation-transition temperature).   −10°C
4. The 50% crystalline temperature.   −50°C
5. The 70% crystalline temperature.   −75°C
6. The highest temperature at which the fracture is 100% crystalline.   −118°C
7. The lowest temperature where fracture initiates by normal rupture (commonly called the initiation-transition temperature).   −90°C

Often the "transition temperature" of a steel is reported in the literature or in specifications. It is necessary to know which transition is referred to, since it could be any one of the seven listed above.

The probability of cleavage crack propagation decreases at temperatures where the energy approaches its maximum value and, in some cases, this propagation-transition temperature coincides with the FTP temperature. Below the initiation-transition temperature the fractures are more than 80% crystalline and there is a strong possibility that a structure operated below this temperature would fracture in a brittle manner (if a large crack developed in it). Conversely, the best correlation between Charpy impact tests and service failures seems to be that *less than 70% crystalline appearance on the Charpy bar, when broken at a particular temperature, indicates a high probability that cleavage will not occur in service* [29, 30] *at or above the particular temperature*,† if the working stress is about $\frac{1}{2}\sigma_Y$.

In steels used in Liberty ship plates during World War II service fractures were never observed at temperatures at which the Charpy value of test pieces was 15 ft-lb or greater. This led to the concept that 15 ft-lb was a magical value, so to speak, and that all steels (and other materials as well!) could be operated at temperatures above the 15 ft-lb Charpy V value. *Detailed investigations* [1, 2] *have shown that this concept is fallacious.* In the Liberty ship steels the 15 ft-lb temperature happens to correspond to the 70% crystallinity transition temperature because so little energy is absorbed in fracture initiation. In the newer and stronger mild steels the Charpy curves are sharper, for reasons discussed in Chapter 7, and the curves are shifted to lower temperature. Because the work required to initiate fracture in these steels is so high, at 15 ft-lb the fractures do not even show a shear lip, and the NDT corresponds to the temperature at which 40 ft-lb are absorbed [2]. Thus an arbitrary criterion based on impact energy has no general relation to the NDT temperature and consequently has no physical meaning; *criteria based on fracture appearance are more realistic and should be used whenever possible in the absence of established NDT criteria.*

The Charpy test can be used to compare two materials whose Charpy curves do not cross. The material whose curve rises at the lower temperature would have the lower probability of exhibiting cleavage fracture when operated in service. It can also be used for quality control purposes [26] by observing whether samples from all batches or heats of material satisfy certain specifications (for example, less than 20% crystalline fracture in an impact test conducted at ambient temperature).

Medium and High Strength Materials

The Charpy test can also be used to evaluate medium and high strength materials in which the radial mode of brittle fracture occurs by low-energy tear as well as by cleavage [22, 23]. As shown in Fig. 3.21, the amount of radial fracture at a given temperature increases,

† Unfortunately there is no exact correlation between the NDT and a specific amount of noncrystalline fracture observed in a Charpy test. In general, the NDT correlates with about 85% crystalline fracture and the FTE with about 55% crystalline fracture. This indicates that (generally) the 70% crystalline appearance occurs at about NDT + 30°F, so that if the design stresses are about $\frac{1}{2}\sigma_Y$, unstable fracture should not occur, according to the criterion (p. 92).

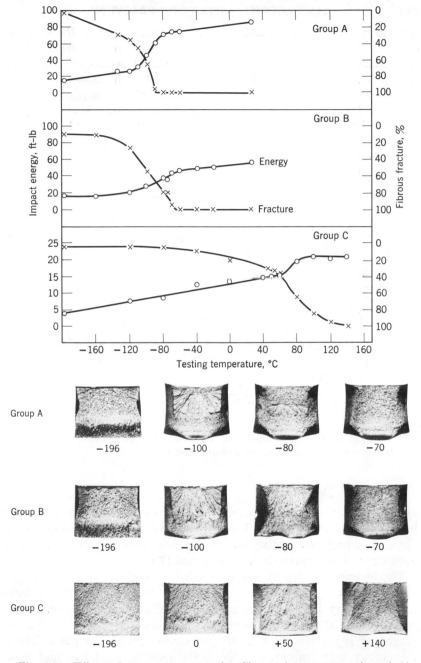

Fig. 3.21. Effect of temperature on the Charpy impact strength and % fibrous fracture for heat-treated 4340 steel. *Group A:* $\sigma_Y = 133$ ksi *and Group B:* $\sigma_Y = 166$ ksi are medium strength steels. *Group C:* $\sigma_Y = 216$ ksi is a high strength steel. (b) Characteristic appearance of the broken Charpy specimens. *Courtesy F. Larson* [9] *and A.S.T.M.*

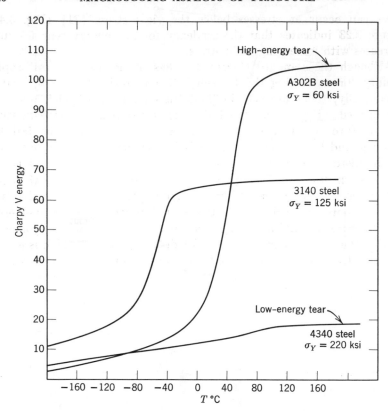

Fig. 3.22. Effect of yield strength level on the shape of Charpy V notch impact curves for some typical commercial steels.

and the impact energy decreases, with decreasing test temperature or increasing strength level. In heat treated medium strength steels there is a high probability of service failure below the 50% radial appearance temperature in a Charpy test [75].

The *maximum impact energies* (shelf energies) measured in the Charpy test C_V (max) are extremely useful in predicting the tendency for low-energy tear fracture in steels. When the maximum impact energy (at 100% tear) is greater than about 80 ft-lb, the fracture toughness is sufficiently high to guarantee against the possibility of low-energy tear fractures. This is the case in most low and many medium strength steels. In ultra-high-strength steels C_V (max) can be less than 30 ft-lb (Fig. 3.22), and unstable, low-energy tear frac-

tures can occur at stresses below the yield stress [21] (Fig. 3.4).
Figure 3.23 indicates that the tendency for low-energy tear fracture
increases with increasing yield stress.

Although the standard Charpy test has the greatest overall applicability for measuring the tendency of a material to fracture catastrophically, there are certain limitations on its use which must be
appreciated. First of all the test is not convenient for investigating
the fracture behavior of FCC metals since these do not fail by
cleavage and hence their toughness is independent of the temperature.
Furthermore, in FCC metals the value of C_V (max) is so low (≈ 25
ft-lb), even when low-energy tear fracture does not occur, that the
test cannot be easily used to discriminate between materials that are
and are not susceptible to brittle tear fracture [21]. Low-energy
tear fracture in FCC and HCP materials is best evaluated by means
of the *drop weight tear test* (DWTT) described below, which is essentially a Charpy test conducted on larger samples (1 in. thick or
greater). Second, in thick plates ($t > 1.0$ in.) the toughness of the

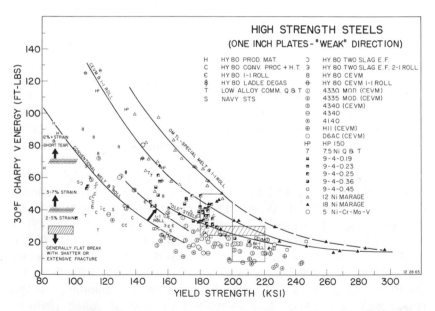

Fig. 3.23. Effect of yield strength level and processing history on the Charpy
V notch impact energy absorbed in a full tear fracture C_V (max) for medium
and high strength steels. The numbers at the left refer to the nominal
plastic strains at which fracture occurs in a thick plate containing a 2-in. flaw
that is subjected to an explosion tear test, ETT. *After W. Pellini* et al [21].

material may vary across the plate thickness, since variations in cooling rate after processing can produce a range of microstructures across the thickness. Consequently Charpy specimens cut from the center of the plate, or at $\frac{1}{4}$ thickness, can show higher transition temperatures and lower C_V (max) values than specimens cut out near the surface of the plate. Strictly speaking, the thick plate is a *composite structure* whose overall toughness in service is related to the properties of both interior and exterior sections and should be evaluated as such. This evaluation of the thick plate can best be carried out by the drop weight tear test. Finally, since the Charpy test occurs at high loading rates, it cannot be used to measure the tendency of a material to fracture over long periods of time (e.g., by creep at high temperature, hydrogen embrittlement, stress corrosion cracking). These types of fracture are best investigated by means of the notch tensile test carried out at low rates of loading, or by static loading tests where the time to fail is measured as a function of nominal stress (Section 3.4). Similarly the Charpy test cannot be used to measure a material's resistance to fatigue crack propagation under alternating stress (see Chapter 8).

The drop weight tear test (DWTT)

The drop weight tear test was developed for use in thick structures where toughness varies across the thickness, and for evaluation of low-energy tear fracture (i.e., $T >$ FTP) in high strength materials [21, 31, 32].† It is essentially a Charpy-type impact test carried out on large plates ($1 \times 5 \times 18$ in., Fig. 3.19b) containing a 2-in. long by 1 in. deep embrittled weld which serves as a starter crack or notch. The specimens are broken by means of a pendulum hammer in a 5000 ft-lb capacity impact machine. Since the plates used in the DWTT are considerably larger (3×1 in. load bearing cross-sectional area as compared with 8×10 mm) than in the Charpy test, more energy will be required to break them. For high strength steels, about 40–50 ft-lb are required to produce a full tear fracture in the DWTT per 1 ft-lb of Charpy shelf-energy. Figure 3.24 indicates that the DWTT energy, like the value of C_V (max) and the plane-strain fracture toughness K_{Ic}, decreases with increasing yield strength. Similar data have been obtained in high strength titanium and aluminum alloys (Figs. 3.25 and 3.26). The fracture toughness of rolled plate is *anisotropic*. It is lower parallel to the rolling direction (weak direc-

---

† Eventually it will be useful in determining CAT curves for cleavage fracture. This is discussed in the next section.

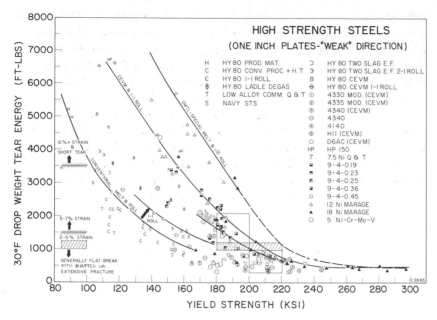

Fig. 3.24. Effect of yield strength level and processing history on the drop weight tear test energy absorbed in fracturing medium and high strength steels. The numbers at the left refer to ETT data. *After W. Pellini* et al [21].

tion) than perpendicular to it (strong direction). This fact must be kept in mind in comparing materials on the basis of their C_V (max) or DWTT energies.

In order to determine whether the 2 in. flaw propagates below or above the general yield stress (i.e., whether $\sigma_F \gtreqless \sigma_Y$), and at what plastic-strain level if $\sigma_F > \sigma_Y$, it is necessary to correlate DWTT values with the gross plastic-strain level at which tear fracture takes place. This can be done by the *explosion tear test* (ETT) [33], in which a 2 in. long, 1 in. deep (through thickness) crack is introduced into a 1 in. thick plate, about 2 ft². A mild explosive charge is set off which produces high stresses across the face of the plate. If the tear fracture (or cleavage fracture, in a temperature range where this is a possibility) can propagate below the yield stress (i.e., $\sigma_F < \sigma_Y$), a flat break is obtained; if $\sigma_F > \sigma_Y$, so that gross plastic strain precedes fracture, then the plate will bulge before the crack runs through it (Fig. 3.27).

A grid method can be used to correlate the plastic-strain level with C_V (max) or DWTT energies. For example, Figs. 3.23 and 3.24

indicate that in steel if C_V (max) > 30 ft-lb or if the DWTT energy is greater than 1250 ft-lb, the 2 in. crack cannot propagate at stresses below the yield stress. Consequently $\sigma_F > \sigma_Y$. Similarly, for titanium and aluminum alloys (Figs. 3.25 and 3.26) a DWTT energy that is less than 1600 and 300 ft-lb, respectively, indicates that $\sigma_F < \sigma_Y$ in these materials. This fact is important for design purposes because it indicates that designs based on keeping the working stress below the yield stress will be safe only if the material exhibits a certain value of DWTT energy and if no flaws greater than 2 in. long are present. Similarly, higher values of DWTT energy specify the amount of plastic strain (plastic overload) that can exist before the 2 in. flaw becomes unstable. This is a reasonable estimate for design

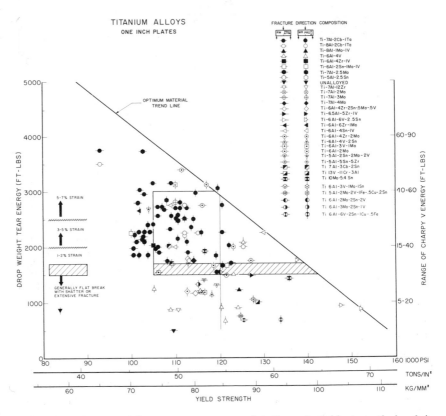

Fig. 3.25. Drop weight tear energy as a function of yield strength level in titanium alloys. *After W. Pellini* et al [21].

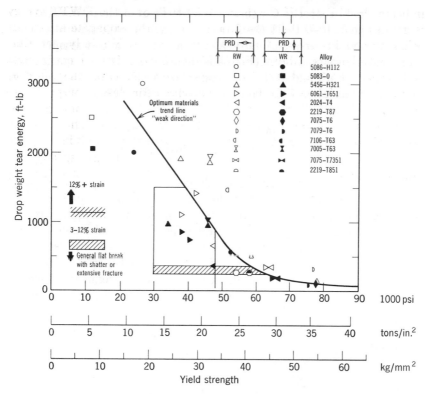

Fig. 3.26. Drop weight tear energy as a function of yield strength level in aluminum alloys. *After W. Pellini* et al [21].

purposes since any flaws greater than even 1 in. long should be detectable by nondestructive testing procedures [63, 64, 65].

The notch tensile test

The notch tensile test is a third type of comparative test that can be used to measure the tendency of a material to fracture in a catastrophic fashion. It is extremely useful for evaluating the toughness of sheet specimens (e.g., of many refractory metals) which are too thin to be tested in a Charpy or drop weight tear test, and useful also for evaluating the fracture resistance of materials that are exposed to specific environments (see Section 3.4). Typically, a 60° notch having a root radius 0.001 in. or less is introduced into a round (circumferential notch) or flat (doubly notched) tensile specimen to a depth such that the notched cross-sectional area is less than or equal to one-half of the load-bearing area in the gage length. Loading is carried out at a standard rate, at a particular test temperature, and the net-section ultimate stress is determined. By comparing this value with the ultimate tensile strength, as measured on an unnotched sample, we can obtain the notch sensitivity ratio (NSR)

$$\text{NSR} = \frac{\sigma_{\text{net}} \text{ (for notched specimen at maximum load)}}{\text{UTS of unnotched specimen}}$$

The NSR is a function of notch geometry as well as of yield strength of the material being evaluated. When notches of low elastic stress concentration factor are present, or in ductile materials, the NSR can actually be greater than 1.0 (Figs. 3.28a and 3.28b). That is, the net-section load-carrying capacity of the notched specimen can actually be greater than that of the unnotched section. The reason for this behavior is that an external notch introduces an *elastic*

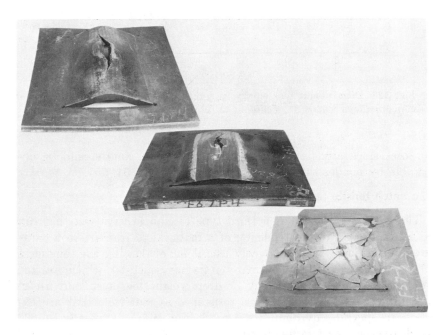

Fig. 3.27. Fractures observed in explosion tear test of 1 in. thick specimens. Top, extreme resistance to propagation; center, fracture requiring 5-7% nominal strain; bottom, shattering at elastic stress levels (flat break). *Courtesy W. Pellini* et al [21].

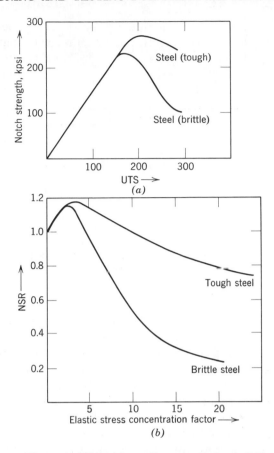

Fig. 3.28. Effect of tensile strength level [35] (a) and elastic stress concentration factor [36] (b) on the notch tensile strength and notch sensitivity ratio for tough and brittle steels.

constraint effect† (Section 1.3, pp. 20–22) which causes the yield and tensile strength of the material to be raised [37, 39] (assuming the material does not fracture below the net-section yield stress).

Suppose, for example, that a very deep and symmetrical external notch exists in a thick plate (Fig. 3.29) of a completely ductile material. Each half piece of a plate (above and below the line AA') can be treated as a flat punch [37] that moves either downward or upward, depending on whether a compressive or tensile stress is applied in the longitudinal direction. The condition for plastic yielding all across

† Also referred to as plastic constraint.

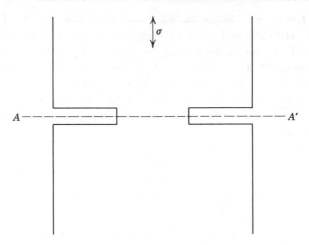

Fig. 3.29. Schematic diagram of cross section of specimen used in notch tensile test.

the notched cross section is then the same as that given by Eq. 1.17 for the indentation of a semi-infinite solid by a punch; the applied tensile yield stress acting across the notched (net) cross section must be (theoretically) 2.57 times the tensile yield stress of an unnotched plate (Tresca criterion), and 2.82 times the tensile yield strength if the von Mises criterion applies. This elastic constraint factor (ECF) is approximately taken to be 3 so that if P/st is the tensile yield load of the notched plate, $3P/st$ is the tensile yield load of the notched plate having load-bearing area st. In a similar fashion the ideal notch causes the UTS of an ideally ductile material to be about three times higher than the value for the unnotched sample (NSR = 3).

In practice this value is rarely achieved for two reasons. First, because most notches are not "ideal" and their constraint factors are less than 3 (e.g., it is 1.5 for the notch used in obtaining Fig. 3.28a). The constraint factor depends primarily on flank angle [39] and root radius, on the degree of plane-strain deformation (i.e., on thickness), and on the condition that the notched area fractures (or necks) before yielding occurs across the unnotched portion of the gage length.

Second, there is the fact that most engineering materials are not completely ductile and fracture occurs if the critical tensile displacement is obtained at the root of the notch before the UTS of the net section is achieved. The decrease in NSR with increasing strength level, and variation of the NSR between two steels of the same strength level, are consistent with the variation of DWTT energy with yield strength and alloy composition shown in Figs. 3.23 to 3.26. Similarly if the stress concentration factor of the notch is increased by decreasing the tip radius ρ, a decrease in NSR is observed.   This occurs because G_{Ic} is lowered (Eq. 2.62) so that fracture occurs at a lower gross stress (Eq. 3.1) and hence at a lower net stress for a notch of given depth.   This effect is shown in Fig. 3.28b.   At the present time it is not possible to use values of the NSR for design purposes because these values have not been correlated with the gross plastic-strain levels required for fracture in various materials.   In general, if the NSR is less than about 0.70 there is a strong probability that an unstable fracture would develop in service [39].

3.4   Testing Procedures Used in Obtaining Specific Information That is Directly Useful in Structural Design

Evaluation of the crack-arrest temperature and the NDT

The crack-arrest temperature is the highest temperature at which unstable cleavage crack propagation can occur under a given level of applied stress.   This temperature can be evaluated directly by means of the Robertson-type test [5].   A uniform stress is applied across a plate of given thickness (0.5–1.0 in.) containing a notch at one end. A temperature gradient is applied across the plate such that the notch resides in the cold end (Fig. 3.30).   This end is impacted by a blow from a bolt gun which causes the formation of a cleavage crack that spreads into the hotter (tougher) end of the plate.   The crack spreads

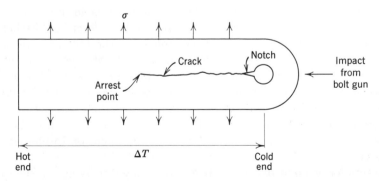

Fig. 3.30. Schematic diagram of specimen used in Robertson crack-arrest test.

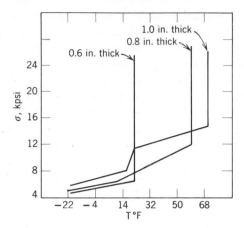

Fig. 3.31. Effect of plate thickness on the Robertson crack-arrest temperature for mild steel. *After T. S. Robertson* [5].

until it becomes arrested at some point (temperature) that varies with applied stress and plate thickness (Fig. 3.31). Propagation at higher temperatures can only occur if the applied stress is raised above the yield stress. The Robertson test (and its subsequent modifications) [40] is the *only* test that *directly* measures the crack-arrest curve. The difficulty with it is that it is complicated and expensive to set up and impractical for use except as a research tool.

Pellini and his coworkers at the United States Naval Research Laboratory have spent the last fifteen years developing simple crack-arrest tests that would correlate well with service failures, give reproducible results, *and* be practical for general use outside the laboratory. They developed the *explosion-bulge test* [33] (described earlier in connection with tear fractures) which measures the temperature range that a starter crack, which initiates at one side of a plate under explosive loading, can propagate across the plate without being stopped. This test, while ideal for measuring the resistance to crack propagation in large structures and for obtaining correlations for various classes of materials, is also complicated and expensive and, unlike the Robertson test, does not easily relate crack propagation to measurable stresses.

A simpler version of this test, which is now receiving considerable attention and use on ferritic steels, is the *Pellini drop weight test* [1, 2, 41]. This test is similar in principle to the explosion-bulge test and DWTT but utilizes a falling weight rather than an explosive charge or pendulum to trigger off the small "starter" crack. Consequently it is considerably simpler to set up and operate than the explosion-bulge test.

Briefly, a bead of "brittle" weld metal is deposited on the surface of a plate (typically $3\frac{1}{2}$ by 14 in. square and 1 in. thick). Smaller plates (5 x 2 x $\frac{5}{8}$ in.) also are used. A small notch ($\frac{1}{4}$ in. long and $\frac{1}{2}$ in. deep) is introduced into the bead and the plate is then immersed in a constant-temperature bath. Once the desired test temperature is reached the plate is removed from the bath, placed in a holder, and impacted with a standard weight (100 or 60 lb) dropped from a height of 6 to 12 ft (Fig. 3.32). A back-up plate is placed underneath the specimen in order to restrict the deflection of the plate after impact. Since the plate is, in effect, a wide "beam" that is loaded in three-point bending, the stress across the tension face of the plate will not exceed the yield stress if the overall deflection of the plate is restricted such that the center of the plate deflects elastically. *Consequently any crack propagation along the tension face will be occurring at the yield stress σ_Y.* Since, by definition, the NDT temperature is the highest temperature at which a small ($\frac{1}{4}$ in. long or less) flaw can initiate fracture at the yield stress σ_Y, it will also be the highest temperature at which the falling weight causes the initial crack to spread to the outer edges of the plate, *along the tension face* where $\sigma = \sigma_Y$. The plates are impacted at various temperatures until a temperature is reached where propagation across the complete tension face no longer occurs; this is the NDT temperature. If the value of the NDT for

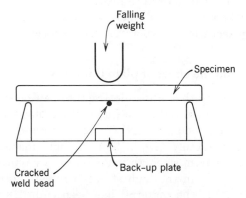

Fig. 3.32. Specimen configuration used in drop weight test.

the particular steel is known, it is possible to use one of the design criteria described on p. 92 in Section 3.1 in order to determine the crack-arrest conditions for given values of σ/σ_Y (i.e., in steels, crack propagation cannot occur at $\sigma \leq 0.5\sigma_Y$ if $T > \mathrm{NDT} + 30°\mathrm{F}$; propagation cannot occur at $\sigma \leq \sigma_Y$ if $T > \mathrm{NDT} + 60°\mathrm{F}$; and propagation cannot occur at $\sigma < \mathrm{UTS}$ at $T > \mathrm{NDT} + 120°\mathrm{F}$. These values were obtained by correlating NDT data with that obtained from crack-arrest tests conducted on the same steels at relative stress levels σ/σ_Y).

Correlations of the measured NDT temperatures for steels have also been made [2] with Charpy V notch curves. In low strength (low Mn to C ratio) mild steels the NDT correlates well with the C_V 8 ft-lb impact temperature. In semikilled steels of higher Mn to C ratio the correlation is with the 20 or even 35 ft-lb impact temperature. In the higher strength quenched and tempered steels (HY-80) the correlation is with the 55 ft-lb temperature. This indicates that if NDT temperatures of other materials are to be established on the basis of Charpy curves, it is first necessary to conduct some drop weight tests to establish the correlation range for the class of materials.

Recent correlations have also been made [21] between DWTT fracture appearance (on $\frac{5}{8}$ in. thick plates) and NDT temperatures. These indicate that cleavage crack arrest can be observed on the fracture surfaces of steel plates broken 40°–60°F above the NDT temperature. The correlation of DWTT energies in the transition temperature range is still in the formative stage, but it appears likely that this test, besides being useful in evaluating the tendency towards low-energy tear fracture, can also be used for obtaining the CAT curve. The DWTT will therefore prove especially beneficial for conducting full thickness tests on thick structures (e.g., line pipe) that have nonhomogeneous microstructures with respect to structure thickness.

G_c and G_{Ic} tests and the effect of plate thickness

It was shown above that unstable fracture can occur in the elastic region ($\sigma_F < \sigma_Y$) in materials containing small flaws if: (1) the operating temperature is below the NDT; (2) the material is susceptible to low-energy tear fracture (i.e., a high strength material at ambient temperature or a medium strength material at low temperature, where its yield strength falls in the high strength range); (3) the structure is extremely thin, (4) the material has been embrittled by the presence of corrosive environments, neutron irradiation, improper heat treatment, and so on.

If a structural material is to be operated under one of these conditions, it is essential that the applied and residual stress be less than σ_F (or $K < K_c$) so that a catastrophic failure does not initiate. σ_F and K_c are given† by Eqs. 3.1 and it is apparent that an accurate determination of G_c is a prerequisite for fracture-safe design.

This determination can best be made by means of a test developed by Irwin and his coworkers [43, 44, 45, 46] at the United States Naval Research Laboratory and the ASTM Committee on High Strength Materials [7, 47]. The details of this test are well documented in the literature [46, 47, 48, 49] and only the more significant aspects will be considered here. Typically a sharp slot is introduced into a tension specimen or a bend specimen whose dimensions are shown in Fig. 3.33. These specimens are then loaded in reverse bending to produce a small amount of fatigue crack growth at both ends of the slot. The radius of curvature of the fatigue crack is quite sharp (less than 0.001 in.) to simulate the sharpness of a crack that might be formed in service. Some brittle materials (e.g., beryllium) are so resistant to fatigue that this approach is not possible. In these cases the sharp crack is introduced by electric discharge machining or by partial cleavage.

The cracked specimen is loaded in a testing machine equipped with a cryostat or heating bath to obtain desired test temperatures other than ambient. The sharply cracked brittle specimens break without necking and the fracture stress σ_F is recorded as the maximum load divided by the gross cross-sectional area Wt. Knowing σ_F and the flaw length at the point of unstable fracture (see below), we can then determine K_c and G_c from Eqs. 3.1, having computed a value for α by the methods of reference [68].

Except in very thick plates or at temperatures much below the NDT, some slow crack growth usually occurs prior to unstable crack propagation. This means that the value of $2c$ used in computing K_c will be greater than the size of the flaw introduced by slotting and fatiguing. The length of the crack during and after slow growth can be obtained by a variety of techniques involving the use of compliance gages [58] or electrical resistance measurements [59]. In the

† At the present time there are no exact relations that can be used to determine σ_F for plastically induced fracture that occurs after gross yielding (i.e., at $\sigma_F > \sigma_Y$) although there are indications that linear elastic fracture mechanics may, under certain conditions, be used to determine an approximate value of σ_F [42]. Also, these relations have been shown not to apply to the case of vessels stressed by internal pressure [63]. For these structures $\sigma_F{}^3 c^2 = \text{const.}$ appears to be in better agreement with experimental data than $\sigma_F{}^2 c = \text{const.}$ This may result from bending stresses produced by bulging of the vessel [51].

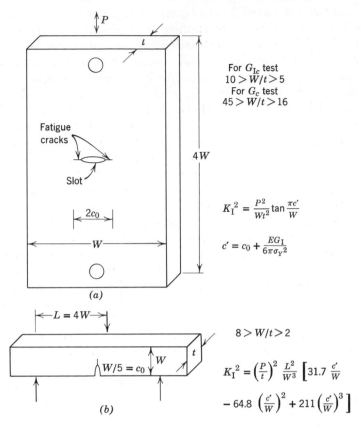

For G_{Ic} test
$$10 > W/t > 5$$
For G_c test
$$45 > W/t > 16$$

$$K_I^2 = \frac{P^2}{Wt^2} \tan \frac{\pi c'}{W}$$

$$c' = c_0 + \frac{EG_I}{6\pi\sigma_Y^2}$$

$$8 > W/t > 2$$

$$K_I^2 = \left(\frac{P}{t}\right)^2 \frac{L^2}{W^3} \left[31.7 \frac{c'}{W} - 64.8 \left(\frac{c'}{W}\right)^2 + 211 \left(\frac{c'}{W}\right)^3 \right]$$

Fig. 3.33. (*a*) Center-notched plate tension specimen used in G_{Ic} test. (*b*) Edge-cracked bend specimen used in G_{Ic} test. *After Brown and Srawley* [49].

latter, for example, the change in electrical resistance of the cracked specimen is measured during loading. By also making a calibration curve of the crack length as a function of electrical resistance, it is possible to relate crack extension directly to applied load and hence to stress intensity. Crack extensions as small as 0.0025 in. can be detected by this technique [49].

Figure 3.34*a* schematically indicates the variation of crack length with applied stress or stress intensity. Two types of slow crack growth occur before fracture. Region 1 is a small region of *microscopic* slow crack growth which precedes plane-strain instability. In this region the stable processes of void coalescence (Fig. 2.14) and crack-tip sharpening take place until the stress builds up to the point where $K = K_{Ic}$ and pop-in occurs. This region of growth prior to pop-in is generally so small that it goes undetected unless extremely sensitive techniques (e.g., acoustic emission [60]) are used to observe it. It is absent in cases where the first microcrack formed near a stopped crack triggers off the instability (e.g., in mild steel at $-321°F$).

In *very thick specimens* the plane-strain instability that occurs when $K = K_{Ic}$ can cause the entire structure to fail so that $K_{Ic} = K_c$ as in Fig. 3.34a. However, in many structures complete failure does

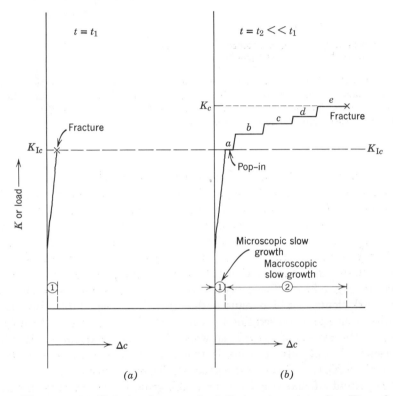

Fig. 3.34. (a) Relation between load P (or stress intensity K) and change in electrical resistance ΔR (or crack length $2c$) when microscopic slow crack growth precedes plane-strain fracture in very thick plate ($t = t_1$). Extent of slow crack growth in this region is extremely small and is usually not detected. (b) Relation between load P (or K) and Δc when microscopic (region 1) and macroscopic (region 2) slow crack growth precede unstable fracture in thinner plate ($t_2 \ll t_1$). The letters a, b, c, etc., correspond to positions of the crack front in Fig. 3.35.

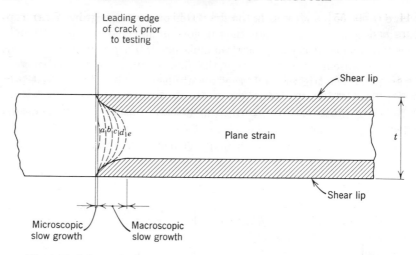

Fig. 3.35. Schematic diagram of combined modes of crack propagation in a plate under mode I tensile loading. $2c_0$ is the initial length of crack; $2c_F$ is the crack length at the onset of unstable fracture. Letters $a, b, c, . .,$ refer to successive positions of crack front.

not occur at $K = K_{Ic}$ and *macroscopic* slow crack growth precedes complete failure [56, 57] (region 2, Fig. 3.34b). In this instance the plane-strain fracture tunnels ahead in the central portion of the structure where the degree of plane-strain loading is greatest (Fig. 3.35). The material on either side of the tunnel is then loaded in plane stress and it eventually fractures by shear rupture. Each of the several "pops" lettered in Fig. 3.34b result from the jumps taken by the crack front in Fig. 3.35a. The increase in fracture resistance with increasing crack length, which causes the growth to be stable, probably results from the increased line tension of the leading edge as the difference in the extent of growth between the center and outer edges increases.† Eventually a point is reached where $\sigma = \sigma_F$, $c = c_F$ (i.e., $K = K_c > K_{Ic}$), and the entire fracture becomes unstable.

The extent of macroscopic slow crack growth, the relative amounts of plane-strain fracture (S) (cleavage or normal rupture), and plane-stress fracture $(1 - S)$ (shear rupture), and the values of K_c and G_c are all strongly dependent on the thickness t as shown in Fig. 3.36

---

† The situation is somewhat analogous to that which occurs when a slingshot is drawn into firing position or a slip dislocation bows between two pinning points.

[44, 47, 50, 55].   At $t < t_0$ the fractures occur entirely by shear rupture and

$$G_c = G_c(45°) \cong \sigma_{YE_f}t \qquad (t \leq t_0) \tag{3.2}$$

as shown in the preceding chapter (Eq. 2.63).   When $t > t_0$ G_c represents an average of the work done in the combined plane-strain propagation and plane-stress propagation (shear lip formation). Thus

$$G_c = SG_{Ic} + (1 - S)G_c(45°) \qquad (t > t_0) \tag{3.3}$$

G_{Ic} will always be less than G_c (45°) since the shear lip size is at least equal† to t_0 and G_c ($t = t_0$) $\gg G_{Ic}$ (Fig. 3.36).   Since S increases with t, G_c *will decrease with increasing thickness*, as seen in Fig. 3.36. Increasing strength level σ_Y or decreasing test temperature causes the shear lip size to decrease [24, 47, 50] (S to increase) and G_{Ic} to decrease, and hence G_c to decrease, in a structure of constant thickness t.

The plastic deformation that initiates unstable fracture can be either plane-strain, plane-stress, or mixtures of the two.   Plane-strain

† There are indications that the shear lip size is equal to t_0, at all $t > t_0$ [55] and other indications [24] that t_0 increases with t.   Further investigation is required to clarify this point.

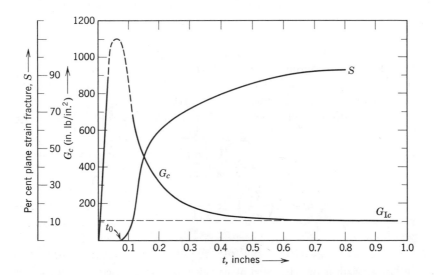

Fig. 3.36. Per cent plane-strain fracture S and G_c as a function of plate thickness.   Schematic diagram based on data obtained on 7075-T6 aluminum alloy.

conditions exist when the root radius ρ and the ratio of plastic zone size to thickness R/t are both small. Plane-stress deformation occurs when $R \geqq 2t$. In high strength alloys plane-strain fracture (normal rupture) occurs under a plane-strain state of deformation and plane-stress fracture (shear rupture) occurs under plane-stress loading†[7, 46]. Thus the ratio of the plastic zone size at fracture R^* to the thickness t is a measure of the state of deformation at the onset of fracture and hence a measure of the fracture toughness, since the latter increases with decreasing t (for $t > t_0$), Fig. 3.36.

A good rule of thumb [7, 46] is that plane-stress fracture initiates under plane-stress deformation when $R^* \geqq 2t$ or when

$$R^* = \frac{K_c^2}{\pi \sigma_Y^2} \geqq 2t$$

or

$$\beta_c \geqq 2\pi \qquad (3.4)$$

where

$$\beta_c = \frac{1}{t}\left(\frac{K_c}{\sigma_Y}\right)^2$$

$$\beta_{Ic} = \frac{1}{t}\left(\frac{K_{Ic}}{\sigma_Y}\right)^2$$

For all practical purposes this is a good criterion for high toughness because sheets that are much thinner than this ($\beta_c \gg 2\pi$) are rarely used in structural design.

Figure 3.36 indicates that maximum toughness is achieved at a thickness t_0 where plane-strain fracture *just begins to develop*. This thickness can be estimated for high strength materials where plane-strain instability occurs under plane-strain loading. Since plane-strain deformation occurs when $R \leqq t/2$ (Chapter 2), $(R^*)_{\text{pop-in}} \leqq t_0/2$. Thus

$$(R^*)_{\text{pop-in}} \cong \frac{1}{2\pi}\left(\frac{K_{Ic}}{\sigma_Y}\right)^2 \leqq \frac{t_0}{2}$$

$$t_0 \geqq \frac{1}{\pi}\left(\frac{K_{Ic}}{\sigma_Y}\right)^2 \geqq \frac{EG_{Ic}}{\pi(1-\nu^2)\sigma_Y^2} \qquad (3.5)$$

or

$$\beta_{Ic} < \pi$$

† In low strength materials tested at or below the NDT, where cleavage is the plane-strain mode of propagation, G_{Ic} is so small relative to G_c (45°), even when plane-stress loading conditions exist, that plane-strain fracture can develop under plane-stress loading [52].

t_0 will vary from one material to another, and for the same material tested at different temperature. For example [50], for X-200 steel, $G_{Ic} = 100$ in. lb per in.^{2}, $\sigma_Y = 240$ ksi, $E = 30 \times 10^6$ psi so $t_0 = 0.02$ in. For 7075-T6 aluminum, $G_{Ic} = 115$ in. lb per in.^{2}, $\sigma_Y = 72$ ksi, $E = 10.5 \times 10^6$ psi so $t_0 \cong 0.09$ in. For 2024-T4 aluminum, $G_{Ic} = 327$ in. lb per in.^{2}, $\sigma_Y = 50$ ksi, so $t_0 \simeq 0.50$ in. *Because G_{Ic} decreases as σ_Y increases, a small increase in σ_Y produces a large decrease in the thickness required to cause plane-strain pop-in.*

Two other points should be noted. First, since G_c and K_c are functions of plate thickness, the plane-strain fracture parameters G_{Ic} and K_{Ic}, which are independent of thickness,† provide lower limiting and hence more reliable measures of a material's toughness than K_c and G_c. The variation of K_{Ic} with yield strength for high strength steels is shown in Fig. 3.37. The decrease in toughness as σ_Y increases is again noted, as in the case of Charpy V and DWTT (Figs. 3.23 and 3.24). It appears, however, that the K_{Ic} test is able to better discriminate between varying degrees of brittleness in the ultra-high strength range than the Charpy V or DWTT and consequently is recommended for comparative testing purposes in this strength range.

Secondly, G_{Ic} (hence G_c) decreases with decreasing temperature [4, 46, 54], especially when fracture occurs by cleavage. (This accounts for the temperature dependence of σ_F that is shown in Fig. 3.1; an explanation of this effect is deferred until Chapter 7.) Thus the measurement of G_c or G_{Ic} that is incorporated in a structural design has to be made at the *minimum temperature at which the material is expected to perform in service.* Similarly, if high loading rates are anticipated, the G_{Ic} value should also be obtained at a high rate of loading.

3.5   Time-Dependent Fracture

In certain instances fracture can occur when materials are subjected to alternating loading (fatigue) or stressed in particular environments for long periods of time. These failures occur because one or both of two effects are taking place: (a) because slow crack growth occurs so that c (and hence K) increases while σ and G_c are maintained constant, or (b) because the value of G_c has been decreased. When

---

† Except in sheets so thin that they would not be used for test pieces anyway, or below the NDT in a low strength material. In general, the test piece thickness should be at least twice t_0 (i.e., $B_{Ic} < \pi/2$) to insure a valid K_{Ic} measurement [56].

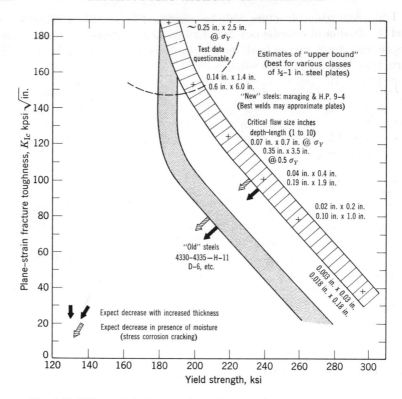

Fig. 3.37. Effect of yield strength level on upper bound value of plane-strain fracture toughness K_{Ic} for various medium and high strength steels. *After W. Pellini* et al [21].

embrittlement results primarily from extensive slow crack growth (a) (e.g., stress corrosion cracking), then *standard rapid tests,* such as a Charpy impact, DWTT, or even a K_c or K_{Ic}, *may fail to detect any susceptibility for time-dependent failure.* Consequently, in these instances a *slow test must be used* to allow the time-dependent embrittling processes sufficient time in which to operate.

Although a standard tensile, notch tensile, or K_c test *may* detect susceptibility, constant stress tests performed for long durations of time (e.g., a stress rupture test) provide more reliable results. In these instances, a smooth, notched, or precracked specimen is loaded in tension or bending or cyclically (fatigue) in various chemical environments, and the time to fail is measured as a function of applied stress.

Time-dependent fracture will be discussed in detail in Chapters 8 and 9. It should only be pointed out here that six basic types of time-dependent fracture, as well as combinations of them, can occur in structural materials. These are:

1. Fatigue. Fatigue occurs in materials under alternating load. G_c appears to be relatively unaffected by the process; the primary effect is slow crack growth which causes c to increase to the point where unstable fracture can occur. Alternating stress tests on smooth or notched specimens provide most reliable information concerning susceptibility as well as some information that can be used in design.

2. Creep rupture. Creep rupture occurs in materials under static load at temperatures greater than 0.5 T_M, where T_M is the absolute melting point. Crack growth takes place as the material creeps under constant stress; in addition, void formation at grain boundaries can lower G_c for subsequent rapid growth. Constant stress test on smooth or notched specimens performed at high temperature provides most reliable information concerning susceptibility and also some information that can be used in design.

3. Stress corrosion cracking. Stress corrosion cracking occurs in materials under static or slowly increasing load, in the presence of corrosive environment. The primary effect is to cause slow crack extension. In some instances G_c may be lowered. Static loading of precracked specimens (see below) gives indication of susceptibility and some information that is useful for design purposes.

4. Liquid metal embrittlement. Liquid metal embrittlement occurs in materials under static or slowly increasing load, in the presence of particular liquids. The primary effect is to lower G_c and to cause some slow crack extension. Slow tensile test on smooth or notched specimens is best for determining susceptibility of given metal in given liquid.

5. Hydrogen embrittlement. Hydrogen embrittlement occurs in certain materials containing hydrogen. The nature of embrittlement varies from one material to another. In steels, for example, embrittlement is favored by static or slowly increasing loads. Primary effect is to cause slow crack extension. In some instances G_c may be lowered. Static loading of notched tensile specimens or slow tensile test on smooth or notched specimens are most useful.

6. Neutron irradiation embrittlement. Neutron irradiation embrittlement occurs in certain materials after exposure to neutron irradiation. The nature of the embrittlement varies from one class of material to another. In steels used in reactor pressure vessels, for example, the primary effect of irradiation is to lower G_{Ic} and hence to increase the NDT. Embrittlement is most pronounced at fast strain rates and C_V, DWTT, or K_c tests are most useful for determining susceptibility or predicting safe operating lifetime.

In cases 1 to 5, where the primary reason for time-dependent fracture is the slow growth of pre-existing flaws, the *rate of slow crack growth* will obviously be an important factor in determining the lifetime of a part. The low-power microscope can be used to measure the crack length in specimens containing through-thickness cracks as a function of time (or number of cycles in fatigue). The slope of the curve of c versus t, dc/dt, can then be plotted as a function of stress intensity. For example, for the stress corrosion of cold-rolled α-brass in ammoniacal copper sulfate [70],

$$\frac{dc}{dt} = A'\sigma^2 c \cong AK^2 \qquad (\sigma < \sigma_Y)$$

where A' and A are materials constants.

Similarly, the rate of fatigue crack growth in sheet specimens of aluminum alloys is given by [71, 72]

$$\frac{dc}{dt} = \beta'\sigma^4 c^2 \cong \beta K^4 \qquad (\sigma < \sigma_Y)$$

where σ is the peak tensile stress applied during a loading cycle and β and β' are materials constants.

If the value of the constants and the growth rate power are known, it is possible to estimate the failure time at a constant value of applied stress† $\sigma < \sigma_Y$. This can be done by integration or through the use of figures similar to Fig. 3.38. For a given σ and G_c the critical crack length at instability c_F can be obtained from Eqs. 3.1. If we know the maximum value of c_0, the size of the starting flaw or the largest flaw that can go undetected, and the growth rate, the time t_F at which $c = c_F$ can be obtained from the plot. If G_c is also affected by these processes, then this must be taken into account in evaluating c_F.

Figure 3.39 shows the failure time of 4340 steel in water as a function of the initial stress intensity factor (i.e., at the time when the stressed specimen was first exposed to the environment). It is interesting to note that even for these sharply cracked specimens there is a lower limit of K_I, labeled K_{ISCC}, below which stress corrosion cracking does not occur [21, 73]. This value is about one-fifth as large as

---

† Some estimates can also be made when $\sigma > \sigma_Y$, as shown in Chapter 8.

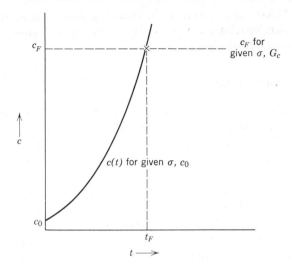

Fig. 3.38. Schematic curve of crack length as a function of time when slow crack growth occurs under constant peak stress σ by fatigue, stress corrosion, etc. Fracture occurs at the time $t = t_F$ at which $c = c_F$, the largest stable crack that can exist for a given σ and G_c.

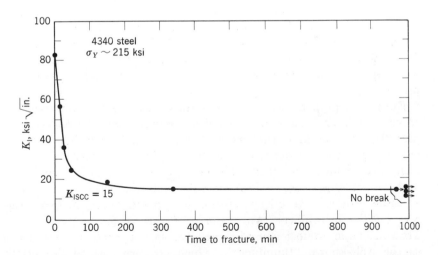

Fig. 3.39. The variation of the time for failure of 4340 steel ($\sigma_Y = 215$ ksi) in water, as a function of the stress intensity K_I of the starting flaw and stress level. Failure does not occur when $K_I < K_{ISCC}$. *After W. Pellini et al [21].*

the value of K_{Ic} for fracture in a dynamic test. It is obvious that stress corrosion cracking can severely limit the reliability of certain materials in certain environments.

3.6 Service Failure Analysis

Analysis of service failures is a procedure which probably depends as much on common sense and experience as on a detailed knowledge of fracture. Basically, three parameters are of interest: (a) the size of the flaw that initiated the failure, (b) the stress acting on the flaw, and (c) the value of G_c or the NDT temperature of the material. Once these parameters have been determined, it is possible to assess the cause of the failure from (1) the fracture analysis diagram (for low strength steels), (2) exact solutions based on Eq. 3.1 (when $\sigma_F < \sigma_Y$), or (3) estimates based on Charpy V or drop weight tear tests. Several excellent reviews [1, 9, 53, 61, 62, 74] of the detailed procedures are now available and only the outlines will be reviewed here.

Basically the idea is to determine whether σ or c were too large or G_c too small for the particular application in which failure occurred. For low strength materials a low G_c appears as a high NDT or Charpy transition temperature. If specifications of the allowable values of σ, c, G_c (NDT) have been made, it is possible to calculate or measure each one of these parameters and determine which ones were out of line. Occasionally failure results from improper design and materials selection; σ is so high and G_c so low (or NDT so high) that the critical flaw size is below the detectable limit. High values of σ can also result from overloading or residual stress (which is partially a design problem).

Even with proper design and materials selection, fracture can still occur if defective material is used in construction. Two cases are possible. In the first, the material (base metal or weld) contains processing defects (such as blow holes, pores, stringers, and large inclusions) which are either long enough to trigger off directly the unstable fracture (e.g., in high strength materials where c_F is small) or, more generally, to initiate slow crack growth which, in turn, triggers off the instability. An example of this is shown in Fig. 3.40 where a casting defect caused large-scale fatigue crack growth (seen as the well-known "thumbnail") which, in turn, caused the brittle fracture of a huge extrusion press. Initiation sites such as these can usually be detected by visual analysis (i.e., by tracing back the radial

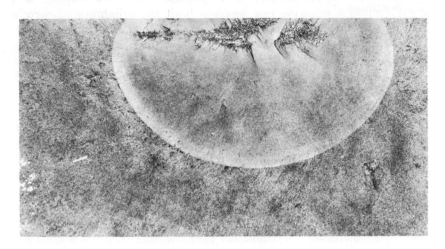

Fig. 3.40. Initiation site of fracture in 80-ton extrusion press cylinder. Casting defect has initiated the fatigue crack (giant thumbnail) which, in turn, has initiated the brittle fracture. *After W. Pellini and P. Puzak* [1].

markings on the fracture surface), although this may be difficult if the structure has shattered into many pieces.

In the second case, the material is defect-free, but improper heat treatment has caused G_c to be much lower than anticipated. Strain-age embrittlement of mild steels, temper embrittlement of high strength steels, and excessive grain coarsening are just a few of the possible ways by which this can occur. The reasons for this behavior are discussed in Part V of the text.

The *mode of failure* is often of interest, particularly in assessing liability. In this case, the characteristic markings discussed in Section **3.2** serve as "finger-prints" which identify the fracture. Visual analyses, by the naked eye, by optical microscope, and, where necessary, by electron fractography, can then be used for identification and classification purposes.

Finally, service failures can result from the environmental effects discussed in the preceding section which tend to increase c, lower G_c, or do both of these things. Thus there are many possible causes of embrittlement that can be wrapped up in permutations of the three parameters σ, c, and G_c and each service failure must be analyzed individually.

3.7 Summary

1. Plane-strain modes of fracture (cleavage or normal rupture) tend to operate in thick structures, at low temperature, and in high strength materials. Plane-stress fracture tends to occur in thin sheets. At any given temperature there is a particular thickness at which the toughness of a given material G_c has a maximum value.

2. In *low strength materials* ($\sigma_Y < E/300$) cleavage becomes progressively less pronounced as the temperature increases (for a structure or test bar of given size); the cleavage to shear transition is a brittle-to-ductile transition. Since FCC metals do not fracture by cleavage, even at cryogenic temperatures, low strength FCC metals are tough at all temperatures. In *medium strength materials* ($E/300 < \sigma_Y < E/150$) the shear fractures are tough (high energy) at ambient temperature but are brittle (low energy) at low temperature. The ductile-brittle transition results primarily from a decrease in tear energy with decreasing temperature. This results from the fact that at low temperature the yield strength is high so that the materials behave as though they were high strength. In *high strength materials* ($\sigma_Y > E/150$) the tear fractures are brittle at all temperatures; a change in fracture mode from cleavage to normal rupture is not a brittle-to-ductile transition.

3. Tough materials absorb large amounts of energy in a Charpy V notch test or DWTT, have high G_{Ic} and K_{Ic} values, and have high notch sensitivity ratios. In brittle materials these parameters are low.

4. In mild steel the NDT and Charpy V notch appearance transition temperatures may be used as the basis for safe design criteria. It will be possible to use NDT and Charpy V notch appearance as design criteria for other materials once correlations with crack-arrest data have been established.

5. The various modes of fracture may be determined by the characteristic markings these produce on the fracture surface. These markings can be detected by optical and replica electron microscopy.

6. Fracture tests are either comparative or specific. Comparative tests are used to determine whether there is a high probability that a material will fail in service, to choose between potential materials for a particular application, and for quality control purposes by the manufacturer. Specific tests are used to determine why a failure has taken place and to obtain data that can be used for design purposes. Recommended comparative and specific tests, for low, medium, and high strength materials, are listed in Table 3.1.

7. Fracture occurs when the stress σ and the flaw depth $2c$ are sufficiently large and the toughness G_c is sufficiently low to satisfy the criteria for unstable crack propagation. Consequently service failures typically result from improper design, overloading, or residual stress (which lead to a high value of σ), fatigue, corrosion, or defective welds (which lead to a high value of c), or improper heat treatment, improper materials selection, and unexpected changes in temperature and loading rate (which lead to a low value of G_c).

References

The asterisk indicates that published work is recommended for extensive or broad treatment.

*[1] W. S. Pellini and P. P. Puzak, *NRL Report 5920* (March 1963).

[2] W. S. Pellini and P. P. Puzak, *NRL Report 6030* (November 1963); also *Trans. ASME (J. Eng. Power)* (1964) p. 429.

[3] G. T. Hahn et al, in *Fracture*, B. L. Averbach et al, eds., M.I.T., Wiley, New York (1959), p. 91.

[4] J. F. Knott and A. H. Cottrell, *J. Iron Steel Inst.*, **201**, 249 (1963).

[5] T. S. Robertson, *J. Iron Steel Inst.*, **175**, 361 (1953).

[6] P. P. Puzak and W. S. Pellini, *Welding J.*, **35**, 275 (1956).

*[7] *Fracture Testing of High Strength Sheet Materials*, ASTM Bull. (January 1960).

*[8] A. Phillips, V. Kerlins and B. V. Whiteson, *Electron Fractography Handbook*, Air Force TR ML TDR-64-416 (January 1965).

[9] F. R. Larson and F. L. Carr, *Metal Progress* (1964), pp. 74, 109.

[10] C. F. Tipper, *J. Iron Steel Inst.*, **185**, 4 (1957).

[11] T. L. Johnston, C. H. Li and R. J. Stokes, *Strengthening Mechanisms in Solids*, ASM, Cleveland (1962), p. 341.

[12] R. Honda, *Int. Conf. on Fracture*, Sendai, Japan (1965).

*[13] J. R. Low, Jr., in *Fracture*, B. L. Averbach et al, eds., M.I.T., Wiley, New York (1959), p. 68.

[14] J. R. Low, Jr., *Relation of Properties to Microstructure*, ASM, Cleveland (1954), p. 163.

[15] C. Crussard et al, *J. Iron Steel Inst.*, **183**, 146 (1956).

*[16] C. D. Beachem and R. M. N. Pelloux, *Fracture Toughness Testing*, ASTM, Philadelphia, STP 381 (1965), p. 210.

[17] C. D. Beachem and D. A. Meyn, *NRL Report 1547* (June 1964).

[18] C. D. Beachem, *Trans. ASM*, **56**, 318 (1963).

[19] C. Crussard et al, in *Fracture*, B. L. Averbach et al, eds., M.I.T., Wiley, New York (1959), p. 524.

[20] A. J. Edwards, *NRL Report* (November 1963).

*[21] W. S. Pellini et al, *NRL Report 6300* (June 1965).

[22] F. R. Larson and F. L. Carr, *Trans. ASM*, **55**, 599 (1962).

[23] F. L. Carr and F. R. Larson, *Proc. ASTM*, **62**, 1210 (1962).

[24] T. S. DeSisto, F. L. Carr and F. R. Larson, *Proc. ASTM*, **63**, 768 (1963).

[25] P. J. E. Forsyth, *Cranfield Symposium on Crack Propagation*, College of Aeronautics, Cranfield, Eng. (1962), p. 76.

[26] W. D. Biggs, *Brittle Fracture of Steel,* MacDonald and Evans, London (1960).

[27] E. R. Parker, *Brittle Behavior of Engineering Structures,* Wiley, New York (1957).

[28] R. Wullaert, Ph.D. thesis, Stanford University (1967).

[29] A. H. Cottrell, *Proc. Roy. Soc.,* **282,** 2 (1964).

[30] J. Hodgson and G. M. Boyd, *Trans. Inst. Nav. Arch., London,* **100,** 141 (1958).

[31] R. J. Eiber and G. M. McClure, *Oil and Gas J.* (September 23, 1963).

[32] E. H. Brubaker and J. D. Dennison, *Seventh Mech. Working Conf.* AIME, Pittsburgh (January 1965).

[33] P. P. Puzak, E. W. Eschbacher and W. S. Pellini, *Weld. J. Res. Supp.,* **31,** 561s (1952).

*[34] J. D. Lubahn, *Fracturing of Metals,* ASM, Cleveland (1948), p. 90.

[35] G. E. Dieter, Jr., *Mechanical Metallurgy,* McGraw-Hill, New York (1961), p. 261.

[36] K. H. Abbott, *Fracture of Structural Metals,* Watertown Arsenal Monograph MS-48 (June 1962).

[37] E. Orowan, *Repts. Prog. Phys.,* **XII,** 185 (1948).

[38] R. Hill, *Mathematical Theory of Plasticity,* Oxford, London (1950).

[39] F. A. McClintock, *Weld. J. Res. Supp.,* **26,** 202s (1961).

[40] F. J. Feeley, Jr., *Weld. J. Res. Supp.,* **34,** 596s (1955).

[41] ASTM *Designation Report E208-63T* (1963).

[42] A. A. Wells, *Brit. Weld. J.* (1963), p. 855.

[43] G. R. Irwin, *NRL Report 5486* (July 1960).

[44] G. R. Irwin, J. A. Kies and H. L. Smith, *Proc. ASTM,* **58,** 640 (1958).

[45] J. M. Krafft, A. M. Sullivan and R. W. Boyle, *Cranfield Symposium,* source cited in [25], p. 8.

[46] G. R. Irwin, *Materials for Missiles and Spacecraft,* McGraw-Hill, New York (1963), p. 204.

[47] "Fracture Testing of High Strength Materials," *Materials Res. and Stds.,* **1,** 389 (1961); **1,** 877 (1961); **2,** 196 (1962); **4,** 107 (1964).

[48] J. E. Campbell, *DMIC Report 207* (August 1964).

*[49] W. F. Brown, Jr., and J. E. Srawley, *Fracture Toughness Testing,* source cited in [16], p. 133.

[50] G. R. Irwin, *J. Basic Eng.,* **82D,** 417 (1960).

[51] W. H. Irvine, A. Quirk and E. Bevitt, *J. Brit. Nuc. Eng. Soc.,* **3,** 31 (1964).

[52] S. Yukawa, *Metallic Materials For Low Temperature Service,* ASTM, Philadelphia, STP No. 302 (1961), p. 193.

[53] J. Sliney, *Fracture of Structural Metals,* Watertown Arsenal Monograph, source cited in [36].

[54] D. H. Winne and B. M. Wundt, *Trans. ASME,* **80,** 1643 (1958).

[55] J. Bluhm, *Proc. ASTM,* **61,** 1324 (1961).

[56] R. W. Boyle, A. M. Sullivan and J. M. Krafft, *Weld. J. Res. Supp.,* **41,** 428s (1962).

[57] A. M. Sullivan, *Mat. and Res. Stds.,* **4,** 20 (1964).

[58] J. E. Srawley, M. H. Jones and B. Gross, *NASA Report* TN D-2396 (August 1964).

[59] E. A. Steigerwald and G. L. Hanna, *Proc. ASTM,* **62,** 885 (1962).

[60] M. H. Jones and W. F. Brown, Jr., *Mat. and Res. Stds.,* **4,** 120 (1964).

*[61] C. F. Tiffany and J. N. Masters, *Fracture Toughness Testing,* source cited in [16], p. 249.

[62] R. W. Nicols et al, *Int. Conf. on Fracture,* Sendai, Japan (September 1965), **D-II,** 19 (1965).

[63] K. M. Entwistle, *Physical Examination of Metals,* Arnold, London (1960), p. 487.

[64] G. Bradfield, *Physical Examination of Metals,* Arnold, London (1960), pp. 559, 605.

[65] J. H. Lamble, *Principles and Practice of Non-Destructive Testing,* Wiley, New York (1963).

[66] M. Hetenyi, *Handbook of Experimental Stress Analysis,* Wiley, New York (1950).

[67] M. M. Frocht, *Photoelasticity,* Wiley, New York (1941).

[68] P. Paris and G. Sih, *Fracture Toughness Testing,* source cited in [16], p. 30.

[69] B. E. Hopkins, *Met. Reviews,* **1,** 117 (1956).

[70] A. J. McEvily and A. P. Bond, *J. Electrochem. Soc.,* **112,** 131 (1956).

[71] H. F. Hardrath and A. J. McEvily, *Cranfield Symposium, Crack Propagation,* source cited [25], p. 231.

[72] P. C. Paris, *Proc. Tenth Sagamore Army Mat. Res. Conf.* (1963), Syracuse Univ. Press.

[73] B. F. Brown and C. D. Beachem, *Corrosion Sci.,* **5,** 745 (1965).

[74] J. E. Srawley and J. B. Esgar, *NASA TM X-1194* (January 1966).

*[75] A. J. Brothers, D. L. Newhouse and B. M. Wundt, *Gen. Electric Report,* GER-2218 (1965).

Problems

1. Show, by means of schematic diagram, the macroscopic (naked eye) appearance of the fracture surfaces of sharply notched, round tensile bars of low strength, medium strength, and high strength steels broken at $-196°$, $-140°$, $-100°$, $-50°$, and $+25°C$.

2. Show, by means of schematic diagram, the effect of temperature on the fracture strength of very thin $(t < t_0)$ and very thick $(t > t_0)$ sheets of (1) low strength steel, (2) high strength steel.

3. Show, by means of schematic diagram, the (naked eye) appearance of the fracture surfaces of round bars, fractured in torsion, of (1) pure silver at room temperature, (2) mild steel at liquid helium temperature, (3) mild steel at room temperature.

4. The fracture strength of tungsten at $-50°F$ is 40 ksi when 2.0 in. long cracks are present. The yield strength at this temperature is 100 ksi. Suppose that G_{Ic} decreases linearly with decreasing temperature, dropping 3 in. lb per in.^{2} per $10°F$. What is the maximum safe working stress at $-100°F$ if the minimum detectable flaw size in a welded structure is 1.0 in. and residual tensile stresses of 10 ksi are known to exist within 0.5 in. from welds? $E = 59 \times 10^6$ psi and $v = 0.3$. You may assume that the residual stresses are parallel to the axis of the applied tensile stress.   (44 ksi)

5. The G_{Ic} value of a low alloy steel plate is 60 in. lb per in.^{2} at −80°F and $\sigma_Y = 110$ ksi. Service failures do not occur unless defects greater than 0.4 in. long are present. What is the maximum strength level this plate can sustain without failing at −50°F, in the presence of a flaw 0.4 in. long, if σ_Y decreases by 1 ksi per 5°F increase in temperature in this temperature range? NDT temperature = −110°F.   (104 ksi)

6. For a quenched and tempered low alloy steel the Charpy energy is 10 ft-lb at −20°F, the highest temperature at which the fracture appearance is 100% crystalline, and 60 ft-lb at +80°F, the lowest temperature at which the fracture appearance is 100% fibrous. Assuming that the energy absorbed in crystalline fracture is temperature-independent and that both the Charpy energy and Charpy appearance vary linearly with temperature in the transition range, determine an approximate value for the Charpy energy at the NDT temperature.   (19 ft-lb)

7. (a) A high strength ($\sigma_Y = 230$ ksi, $\epsilon_f = 0.20$) steel structure, 2 in. thick, fractures without showing any sign of a shear lip in an Irwin-type test and $G_{Ic} = 100$ in. lb per in.^{2}. What is the maximum increase in toughness that can be achieved if this structure is built out of n thin laminated sheets, bolted together, such that the total thickness remains equal to 2 in.? What is the value of n? Assume that there is negligible interaction of the laminates when they fracture.

(b) If the same load-carrying capacity before fracture is to be maintained with the laminate structure as with the homogeneous structure, what is the saving in material that can be realized by using the laminate?   [(a) 820 in. lb per in.^{2}; (b) 200%]

PART TWO

Microscopic Aspects
of Fracture

4

Microscopic Aspects of Plastic Deformation

In Part 1 we have treated the problem of fracture from an engineering point of view. We have seen that fracture initiates at the tip of a flaw when a critical amount of plastic work

$$\gamma_P{}^* = G_c/2 \cong \sigma_Y V^*(c)$$

is done there, work that is necessary to produce a critical displacement $V^*(c)$ at the tip. This critical displacement, in turn, results from the fact that a critical plastic strain $\epsilon_f(c)$ is required to fracture the region adjacent to the tip.

This approach is useful for designing against the initiation and catastrophic propagation of cracks in large structures. It is, however, a phenomenological approach that does not answer any of the fundamental (i.e., smaller scale) questions about fracture, such as: why FCC metals do not fracture by cleavage, why most nonmetallic crystals are very susceptible to cleavage, why γ_P and G_c for cleavage fractures are temperature dependent, why heat treatment has such a large effect on G_c, and so on. In order to answer these questions, and to understand the fracture problem with a view to making better predictions about the behavior of individual materials, it is necessary first to understand (1) how plastic deformation can physically initiate unstable fracture at the tips of flaws, and (2) how it can also initiate unstable fracture when flaws are not even present (i.e., below T_S in Fig. 3.1).

Before we consider this problem (Chapters 5 and 6) it is first necessary to understand the nature of the plastic deformation process from a microscopic point of view. The classical theory of plasticity developed by Hill [1] and others is based on the assumption that a material is a homogeneous body and that the plastic strains (or strain

153

rate) in any volume element of the body are constant under a constant applied shear stress. This is essentially the case when the strains are large (i.e., more than 10%) and when the volume element that is deforming is large (i.e., 1 mm for most polycrystalline materials). Consequently the theory has been successfully used for analysis of problems involving large plastic strains, such as metal-forming operations. However, when the deformations are small and the volume element is small (i.e., one grain in a polycrystalline aggregate), then the plastic strains in the volume element are not homogeneous. In fact, *it is the inhomogeneous distribution of strain on the microscopic scale that leads to the initiation of fracture; if the microscopic strains were completely homogeneous then G_c would be so large that catastrophic fracture would rarely be able to initiate.*

In the following section we describe the nature of this inhomogeneous plastic deformation from a phenomenological point of view and then consider the atomistic aspects of the plastic processes (dislocation theory). It is assumed that the reader has had an introductory course in materials science and that he is familiar with the elementary concepts of crystal structure. Only those elements of dislocation theory that are related to the processes of fracture will be presented here; for readers interested in a more extensive treatment there are the excellent texts by Cottrell [2], Friedel [3], and Weertman and Weertman [4]. In Section 4.3 we apply these concepts in order to understand the microscopic aspects of the flow stress of crystalline solids. We are then in a position to compare the atomistic and engineering approaches to crack propagation; this is the subject of the following chapter.

4.1   The Inhomogeneous Nature of Plastic Deformation

Plastic deformation is inhomogeneous because it occurs by the shearing of whole blocks of crystal† over one another (Fig. 4.1*a*) rather than by continuous and homogeneous deformation where *all* atoms are displaced more or less the same amount from their equilibrium lattice positions as in elastic deformation (Fig. 4.1*b*). The shear displacements occur by either the process of *slip* (glide) or *twinning.* Slip always occurs by the displacement of blocks of crystal (a whole number of lattice spacings) in specific crystallographic directions called *slip directions.* It usually takes place on particular lattice planes called *slip planes* or *glide planes.* In some crystals, particularly those having BCC crystal structure, slip is not always

---

† A crystal is one grain in a polycrystalline aggregate.

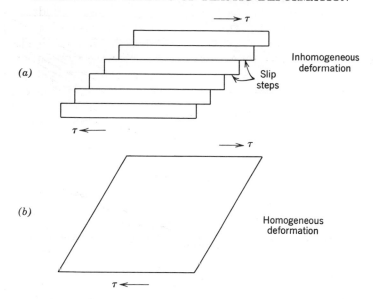

Fig. 4.1. Shear displacements associated with: (a) Inhomogeneous deformation. (b) Homogeneous deformation.

confined to one set of planes. The *slip lines*† (or slip bands) that are observed [because of slip steps emerging from the crystal surface, Fig. 4.2] appear straight when the glide is confined to one plane (Fig. 4.3a) but are corrugated when deformation occurs on many planes. This behavior is termed *wavy* or *pencil glide* (Fig. 4.3b).

Plastic deformation in crystals can also occur by *twinning* (Fig. 4.3c), where layers of atoms slide in such a manner as to bring the deformed part of the crystal into a mirror-image orientation relative to the undeformed part of the crystal. The plane *AB* across which twinning occurs is called the *twinning plane* and twinning, like slip, occurs in a specific direction called the *twinning direction* (Fig. 4.4).

In *single crystals,* macroscopic plastic deformation (yielding) occurs when the applied tensile stress σ, resolved as a shear stress τ on a particular slip plane and in a particular slip direction, is equal to a critical value ($\tau = \tau_Y = k$). The yield criterion is therefore

$$\sigma \cos \chi \cos \lambda = \tau_Y = k$$

where λ is the angle between the tensile axis and the slip (or twin)

---

† These are not the same slip lines used to describe surfaces of maximum shearing stress in plasticity theory, but are actual traces of crystallographic planes on which plastic flow has occurred.

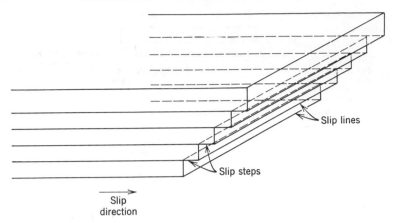

Fig. 4.2. Slip lines and slip steps formed by inhomogeneous plastic deformation.

direction and χ is the angle between the normal of the slip (or twin) plane and the tensile axis. In general, $\cos \chi \cos \lambda \cong 0.5$ so that yielding occurs when $\sigma = \sigma_Y = 2\tau_Y$. This relation also holds for polycrystals and the material is assumed to obey the Tresca criterion (i.e., $k = \tau_Y = \sigma_Y/2 = Y/2$). For simplicity it is assumed in the following discussion that the only stress applied to the solid is uniaxial tension ($\sigma \cong 2\tau$) unless otherwise stated.

The particular mode of deformation which operates will be determined by the criteria that are satisfied at the lowest value of τ_Y. The yield stresses for slip or twinning are not a constant for a given material, but vary with test temperature, strain rate, alloy content, grain size, and other extrinsic and intrinsic variables. In most materials having a BCC structure (iron, refractory metals) the yield stress for slip increases sharply with decreasing temperature, whereas the twinning stress is relatively independent of temperature. At most temperatures τ_Y (slip) $< \tau_Y$ (twin) and slip is the preferred mode of deformation; but at very low temperatures ($T < T_t$, Fig. 4.5a) the stress necessary for twinning is less than that required for slip, and twinning is the preferred mode of deformation. In FCC metals such as copper and aluminum the twinning stress is so much higher than the stress required for slip that twinning is generally not observed. HCP metals such as zinc, magnesium, and beryllium deform quite readily by twinning at most temperatures. Twinning is only observed in a few nonmetallic solids. Table 4.1 lists the various slip and twin systems

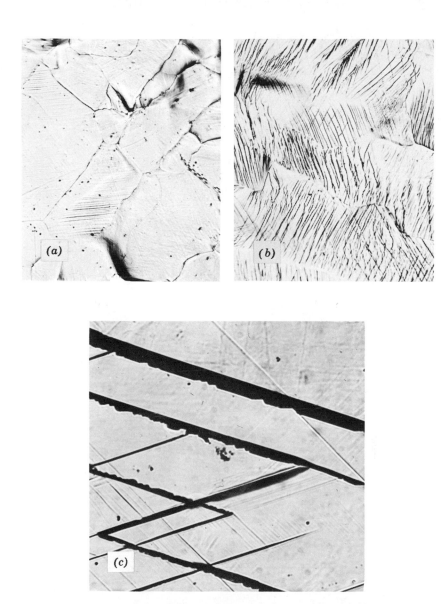

Fig. 4.3. Examples of plastic deformation observed on the polished surface of deformed metals. (a) Straight slip lines and microcracks in ordered FeCo, deformed 4% at 25°C. 70×. (b) Wavy slip in disordered FeCo, deformed 4% at 25°C. 70×. *Courtesy N. Stoloff and R. Davies* [6], *and Acta Met.* (c) Twins in Fe-3% Si deformed to the yield point at low temperature. 350×. *Courtesy D. Hull.*

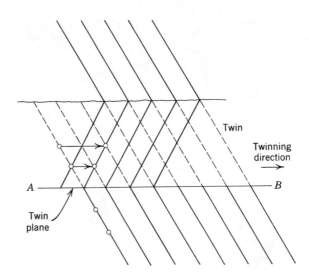

Fig. 4.4. Atomic displacements associated with twinning.

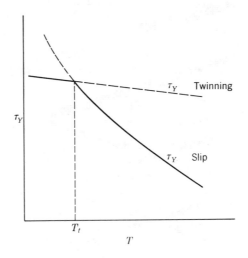

Fig. 4.5. Effect of test temperature on the critical resolved shear stress τ_Y for slip and twinning in a typical BCC metal. Yielding occurs by twinning when $T < T_t$ and by slip at $T > T_t$.

Table 4.1
Crystallographic Modes of Deformation

Crystal Structure	Metal or Base Alloy	Possible Slip Plane	Slip Direction	Twin Plane	Twin Direction
FCC metals	Cu, Ag, Au, Al, Ni, 300 stainless steels, α-brass, bronze	{111}	⟨110⟩		
HCP metals	Be, Mg, Zn, Ti, U, Co, Ti, graphite	{0001} {1010} {1122}	⟨1120⟩ ⟨1120⟩ ⟨1123⟩	{1012}	⟨1011⟩
BCC metals	Li, Na, K, Fe, most steels, V, Cr, Mn, Cb, Mo, W, Ta	{110} {112}	⟨111⟩	{112}	⟨111⟩
Diamond	Diamond, Si, Ge	{111}	⟨110⟩		
NaCl	NaCl, LiF, MgO, AgCl	{110}	⟨110⟩		
Zinc blend	ZnS, BeO	{111}	⟨110⟩		
Fluorite	CaF₂, UO₂, ThO₂	{001} {110}	⟨110⟩		{110}

Note: Some of the materials listed above are allotropic (can have different crystal structures depending on temperature). The most common form in structural use is listed here.

(crystallographic planes and directions) for a variety of materials of engineering and scientific interest [5].

4.2 Elements of Dislocation Theory [2, 3, 4]

Since all atoms on the slip plane are not vibrating in phase with one another, the shearing force that is necessary for atomic displacements will vary from point to point on the plane. Consequently slip cannot occur *simultaneously*, with the entire slip plane $ABCD$ moving simultaneously over $EFGH$ (Fig. 4.6a), but it occurs *consecutively* (Fig. 4.6b), beginning in a minute region on the glide plane and spreading outwards. The boundary between the regions where slip has and has not occurred is called a *dislocation* and is commonly represented as a line in this plane called a *dislocation line*. This line cannot end inside the crystal, so it either forms a closed loop or extends out to a grain boundary or an external surface. A unit of slip has occurred when a dislocation has spread over the entire slip plane (Fig. 4.6c).

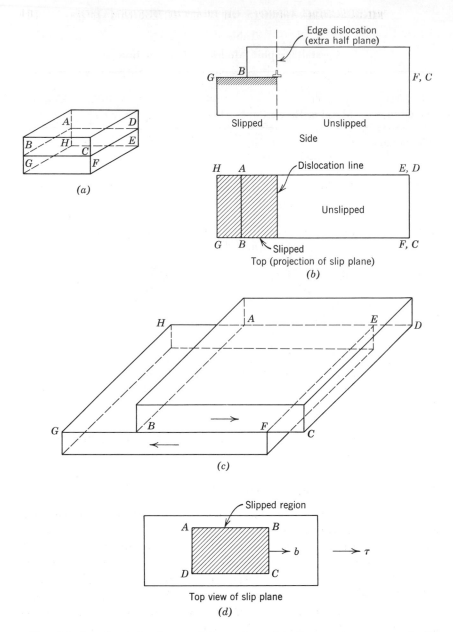

Fig. 4.6. (a) Crystal prior to deformation. (b) Consecutive nature of slip which begins when line AB moves from left to right. Edge-dislocation line AB forms boundary of slipped and unslipped portions of crystal. Side view and top view are shown. (c) Displacement of top and bottom parts of crystal after dislocation has completed motion and formed slip step FC. (d) Slipped region of a crystal bounded by edge-dislocation lines AD and BC and screw dislocation lines AB and CD.

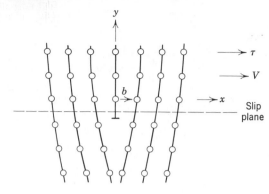

Fig. 4.7. Idealized atomistic picture of a positive edge dislocation. Dislocation moves in direction V that is parallel to displacement vector b. Dislocation line is perpendicular to plane of the paper.

After the dislocation line has moved past a certain point in the crystal, the atoms adjacent to this point are returned to their normal equilibrium positions. Because the atoms in the unslipped portion of the crystal are also at their equilibrium positions, all of the displacement associated with a unit of slip is concentrated around the dislocation line itself. If the displacements from equilibrium occur over one or two lattice spacings, the dislocation is said to be narrow, whereas the dislocation is said to be wide if the displacement extends over several lattice spacings.

A dislocation is commonly defined in terms of two quantities—its line dl and its displacement or *Burgers vector b,* which is constant along the dislocation line. If the Burgers vector is perpendicular to the dislocation line, the dislocation is called an *edge dislocation;* if the Burgers vector is parallel to the dislocation line, the dislocation is of the *screw* type. In practice, most dislocation lines in crystals are neither pure edge nor pure screw but are mixtures of the two. For the displacement shown in Fig. 4.6d the parts AB and CD are screw because they are parallel to the displacement while BC and AD are edge.

The atomic structure of a positive edge dislocation is conveniently represented as an extra half plane of atoms (Fig. 4.7) that lies above the slip plane and glides from left to right, producing the slip displacement shown in Fig. 4.6c. This displacement is the same as that produced by an extra half plane of atoms below the slip plane which moves from right to left. By convention, a negative edge dislocation

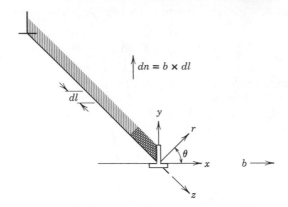

Fig. 4.8. Coordinate system used in defining slip plane of edge dislocation.

would be one whose extra half plane lay above the slip plane and which moved from right to left or whose extra half plane lay below the slip plane and which moved from left to right. The symbol ⊥ or ⊤, symbolizing an extra half plane, is commonly used to represent positive and negative dislocations. Theoretically it is possible to produce an edge dislocation in a crystal by slicing the crystal and inserting an extra half plane of atoms.

The insertion of this extra half plane will cause atoms on either side of the half plane to be strained about their equilibrium positions. The atoms above the slip plane will be put in compression, those below the slip plane will be put in tension. Consequently the edge dislocation will have a stress field around it that is compressive above the slip plane (negative) and tensile below the slip plane (positive). Suppose that the extra half plane lies in the plane $x = 0$, the Burgers vector b lies along the $+x$ axis in the plane $y = 0$, and the dislocation line lies along the z axis (Fig. 4.8). Then the stress components around the dislocation are, in plane strain,

$$\sigma_{xx} = -Dy\,\frac{(3x^2 + y^2)}{(x^2 + y^2)^2} = -\frac{D}{r}\sin\theta(1 + 2\cos^2\theta) \qquad (4.1a)$$

$$\sigma_{yy} = Dy\,\frac{(x^2 - y^2)}{(x^2 + y^2)^2} = \frac{D}{r}\sin\theta(\cos 2\theta) \qquad (4.1b)$$

$$\tau_{xy} = Dx\,\frac{(x^2 - y^2)}{(x^2 + y^2)^2} = \frac{D}{r}\cos\theta(\cos 2\theta) \qquad (4.1c)$$

$$\sigma_{zz} = \nu(\sigma_{xx} + \sigma_{yy}) \qquad (4.1d)$$

where $x = r \cos \theta$, $y = \sin \theta$ in polar coordinates, and

$$D = \frac{Gb}{2\pi(1 - \nu)}$$

G is the shear modulus and ν is Poisson's ratio.

The elastic strain energy U_D of the edge dislocation per unit length of line is obtained by integrating the stress field over the distance r that it extends:

$$U_D = \frac{Gb^2}{4\pi(1 - \nu)} \ln\left(\frac{r_1}{r_0}\right) \qquad (4.2)$$

where r_0 is the core radius of the dislocation (approximately equal to the Burgers vector b) and r_1 is the outer limit of integration, usually taken to be about 1 μ. Taking $b = 2.5 \times 10^{-8}$ cm gives $U_D \cong 7eV$.

The most stable state of any system is that state of lowest free energy. Since entropy effects are usually negligible compared with elastic strain (internal) energies, dislocations will try and arrange themselves in such a way as to minimize the strain energy of the deformed solid. For example, Eq. 4.2 shows that the energy of a dislocation is proportional to the square of its Burgers vector. Since a dislocation of Burgers vector $2b$ has an energy proportional to $4b^2$, while two individual dislocations, each having Burgers vectors b have an energy $b^2 + b^2 = 2b^2$, a dislocation of Burgers vector $2b$ can reduce strain energy by splitting into two individual dislocations. Consequently a *repulsive force* exists along the glide plane between two dislocations of the *same sign*, whereas two dislocations of *opposite sign* are *attracted* toward one another because their stress fields tend to cancel out. Another result of the fact that the energy is proportional to b^2 is that slip usually occurs in closest-packed directions since b is then smallest. If a is the lattice parameter, then $b = a\sqrt{2}/2$ for FCC crystals ($\langle 110 \rangle$ slip) and $b = a\sqrt{3}/2$ ($\langle 111 \rangle$ slip) for BCC crystals. Since $a \approx 3 \times 10^{-8}$ cm $b \cong 2 \times 10^{-8}$ cm for most crystals.

The edge dislocation we have been discussing is confined to glide in a slip plane that contains both its Burgers vector and its line.† The dislocation moves parallel to its Burgers vector under a force F (per unit length) that is equal to

$$F = \tau \cdot b$$

where τ is the applied shear stress. The force required to make a dislocation move from one lattice position to the next depends on the

† The unit vector dn that is perpendicular to the slip plane is $dn = b \times dl$.

inherent resistance of the lattice to the breaking of atomic bonds (called the *Peierls' stress* τ_P) and the internal *stresses* $\tau_D = F_D/b$ that are set up by lattice defects (such as another dislocation of the same sign which repels the motion of the dislocation). The Peierls' stress is least when the slip-plane spacing is greatest since there is then less "friction" involved during the slip process. Thus slip tends to take place on closest-packed (most widely spaced) planes as well as in close-packed directions. The Peierls' stress is also least when the dislocation is wide and the bonds between individual atoms are weak and nondirectional. In inherently soft metals such as aluminum this friction stress may be practically zero, and the high strength of these materials is obtained by increasing the number of defects opposing dislocation motion. In covalently bonded materials, such as diamond, silicon, and germanium, the Peierls' stress is quite large and these materials have a high intrinsic strength, irrespective of the presence of defects.

The only way that an edge dislocation can move out of its glide plane is by *climb*, a process in which the length of the extra half plane is either increased or decreased. For example, suppose that an atom in an interstitial position (Fig. 4.9a) is able to diffuse to a positive edge dislocation and join it. Then, in the plane of the paper P the length of the dislocation has been increased and the glide plane has been changed from AA' to BB'. This is called negative climb and may be easily observed from a side view (Fig. 4.9c). Negative climb can also occur if vacant lattice sites (vacancies) are created at the bottom of the extra half plane and diffuse away from it. Alternatively, the dislocation can climb upwards (positively) by the absorption of vacancies or the emission of interstitial atoms. These processes are not important at low or moderate temperature $T < 0.5T_M$, where vacancy or interstitial diffusion is so slow that the amount of climb is insignificant. However, at higher temperature, sufficient dislocation climb can occur to affect significantly the process of creep and recovery.

The screw dislocation† (AB, CD, Fig. 4.6) does <u>not</u> have an extra half plane associated with it. It is formed by again slicing the crystal but now displacing the two halves of the cut crystal *parallel* to the cut by a distance b at the outer surface (Fig. 4.10a). Slip has then occurred over the shaded area QRS. At the surface the slip displacement QR is equal to the Burgers vector. At S the displacement is

---

† The symbol for the screw dislocation is $\oplus$.

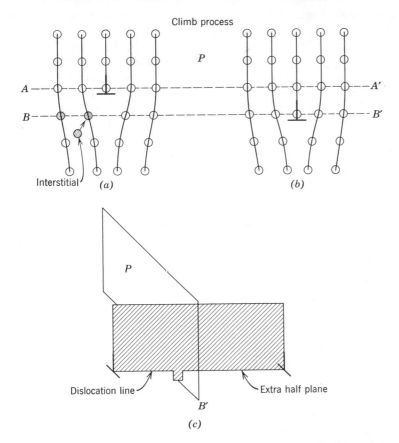

Fig. 4.9. Climb motion of an edge dislocation, perpendicular to its Burgers vector and hence perpendicular to its slip plane.

zero. When the dislocation moves from S to T, additional shear (STU) occurs, and finally a complete unit of slip is achieved (Fig. 4.10b) when the dislocation moves out the back side of the crystal. Thus the screw dislocation line AB is parallel to its Burgers vector and slip occurs when the line moves perpendicular to its Burgers vector. Since the force on a dislocation is $\tau \cdot b$, the dislocation moves perpendicular to the applied force.

The screw dislocation is not confined to glide on a particular plane dn, as is an edge dislocation, because its Burgers vector is parallel to its line (i.e., $dn = b \times dl = 0$). Consequently the screw will follow

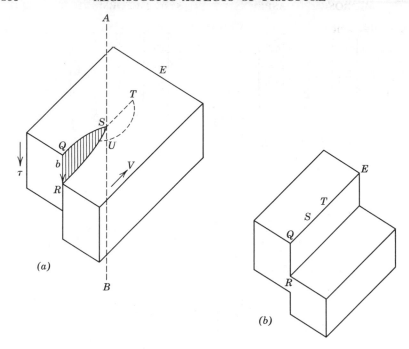

Fig. 4.10. (*a*) Slip step associated with motion of screw dislocation in direction *V*. Slip has occurred over shaded areas *QRS*. (*b*) Unit of slip formed after dislocation has passed through crystal.

a path of least resistance and avoid obstacles that lie in its path by gliding off the plane on which the obstacle lies. The successive motion of a perfect screw dislocation from one plane to the next (position 1 to position 3, Fig. 4.11) is called *cross slipping*.

Because the energy of a dislocation is proportional to b^2, it is possible for some unit dislocations† to decompose into two partial dislocations that have a lower energy than their parent. For example, in the FCC crystal where slip occurs in the $\langle 110 \rangle$ direction on $\{111\}$ planes, the reaction

$$b_1 \rightarrow b_2 + b_3$$

$$\frac{a}{2}[\bar{1}01] \rightarrow \frac{a}{6}[\bar{2}11] + \frac{a}{6}[\bar{1}\bar{1}2]$$

† Those that leave atoms in positions equivalent to the ones that they occupied before the passage of the dislocation. Partial (imperfect) dislocations can cause atoms to lie on different positions after they move through the lattice.

is favorable since

$$b_1{}^2 > b_2{}^2 + b_3{}^2 \qquad \text{(Frank's rule)}$$

$$\frac{a^2}{4}\,[1+1] > \frac{a^2}{36}\,[4+1+1] + \frac{a^2}{36}\,[1+1+4]$$

$$\frac{a^2}{2} > \frac{a^2}{3}$$

The two partial dislocations b_2 and b_3 will try to repel each other, as their Burgers vectors have some parallel components. However, the

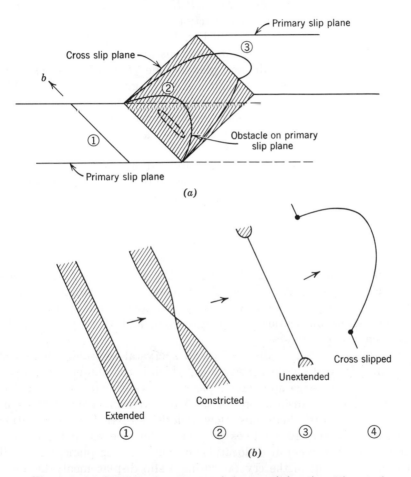

Fig. 4.11. (a) Motion of an unextended screw dislocation (1) out of primary glide plane, onto a cross slip plane (2), and then a return to primary glide plane (3) that is parallel to its original plane. (b) Stages leading to the cross slip of an extended screw dislocation.

atoms between the two partials are not in their normal positions and a *stacking fault* is formed. The separation distance between the two partials is governed by the energy of the fault. When the stacking fault energy (SFE) is low, the partials are widely extended; in materials of high stacking fault energy the partials are close together.

The SFE plays a large role in determining the plastic properties of a crystal. The reason is that while perfect screw dislocations b_1 can cross slip (Fig. 4.11a), as they are not confined to glide on a particular plane, an extended pair of dislocations b_2 and b_3 *is confined to glide on the plane containing the fault that lies between them.* Cross slip can only occur when a constriction is made in the extended dislocation and the process shown in Fig. 4.11b takes place. In crystals having low SFE the extended dislocations are wide, constriction and cross slip are thus difficult, and planar glide (Fig. 4.3a) is observed. High SFE implies easy cross slip and wavy glide (Fig. 4.3b). The ease of cleavage crack and fatigue crack formation is related, in part, to ease of cross slip, and hence the SFE also plays a large role in determining the fracture characteristics of a material at the microscopic level (Chapters 5, 6, and 8).

The screw dislocation has essentially no hydrostatic stress field around it and all strains and stresses are pure shear. The stress field around the screw is symmetrical and is given by

$$\sigma_{\theta Z} = \frac{Gb}{2\pi r} \tag{4.3}$$

in polar coordinates. The energy of a screw dislocation is essentially the same as that of an edge dislocation (Eq. 4.2), except for a numerical factor. Since perfect screw dislocations are not confined to moving on a glide plane, they will always be able to annihilate with a parallel screw dislocation of opposite sign, and thereby reduce the strain energy of a crystal.

The total number of dislocations in a crystal is usually expressed in terms of a dislocation density N_{tot}, which is the number of dislocation lines (both edge and screw) per square centimeter of plane that cuts through the dislocation lines. Most well-annealed crystals of intrinsically "soft" materials (low Peierls' stress) have dislocation densities of about $10^7 \pm 1$ lines per cm². This means that there are about 3×10^3 per cm of dislocations on any one slip plane, and if all of them move out of the crystal during a slip displacement, the average step height (Fig. 4.2) would be $(3 \times 10^3) \times (2 \times 10^{-8}) \cong 6 \times 10^{-5}$ cm. Experiments have shown that the average slip-line step is 100 times larger than this. Furthermore, it is known that crystals

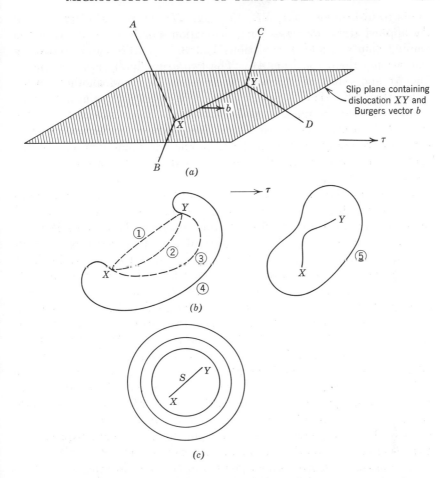

Fig. 4.12. Multiplication of dislocations by Frank-Read source generation.

containing initially very few dislocations are able to be strained plastically and that the dislocation density increases with strain. *Consequently dislocations must be able to multiply when they move.*

Figure 4.12 shows one way in which this occurs. Dislocations in annealed crystals are usually arranged in some sort of a network† and not all of the dislocations lie in planes on which slip can occur easily or have Burgers vectors that are favorably oriented with respect to the applied stress. Suppose that a dislocation line of length $l' = XY$, which can move easily under applied stress, is connected at nodal

† To minimize their strain energy.

points to dislocations XA, XB, YC, and YD, which cannot glide. As the applied stress increases, the dislocation will bow out between the pinning points X and Y (positions 1, 2, 3, Fig. 4.12b) and eventually bow around them (position 4). The two arms finally coalesce (position 5) and pinch off, forming a complete loop and another source dislocation XY. The loop then glides away under stress at a velocity V and the process repeats itself, with $V/\pi l'$ loops forming per second. This process is called *Frank-Read* source multiplication and can produce many dislocation loops in one slip plane if X and Y are coplanar, or loops in parallel slip planes if X and Y are on parallel but separate slip planes.

Another process for dislocation multiplication involves the cross slip of screw dislocations. We have seen earlier that an unextended screw dislocation is not confined to one slip plane because its Burgers vector is parallel to its line. If part of a moving screw dislocation leaves its glide plane (Fig. 4.13), a *jog*† will form (b) and trail behind the moving screw dislocation (c). This is so because the line XY will not be able to glide along with the screw portions AX and YB since its Burgers vector b is perpendicular to the direction of motion of the screw. Consequently a dipole (two dislocations of opposite sign) XC and YD is formed. If the height $r = XY$ is large, the attractive stress between XC and YD is small and these will be able to glide past one another and operate as single-ended dislocation sources (Fig. 4.13d), since the points Y and X are essentially fixed. When the dislocation BC cross slips again, the process of *multiple cross glide* is repeated, and in this manner *slip-line broadening* can occur. Because each moving screw dislocation is, in effect, a single-ended Frank-Read source, the multiplication process can be very extensive. If δ is the number of loops formed per cm of screw dislocation moving at a velocity V, then $dN/dt = 2\delta VN$ and $e^{(\delta 2Vt)}$ loops are formed in a time t from each of the initial sources. Thus, one source on one slip plane can be responsible for a band of active slip planes containing many moving dislocations.

Dislocations can be observed by a variety of techniques. A large portion of recent investigations of crystal plasticity has been concerned with observing the arrangements of dislocations in various crystals after these have been deformed under a variety of conditions [8, 9, 10]. The most direct way to observe dislocation arrangements is by passing a high-energy beam of electrons through a very

---

† A step in the dislocation line that moves part of the dislocation to another slip plane. An example of a one-unit jog (adsorbed atom) is shown in Fig. 4.9c.

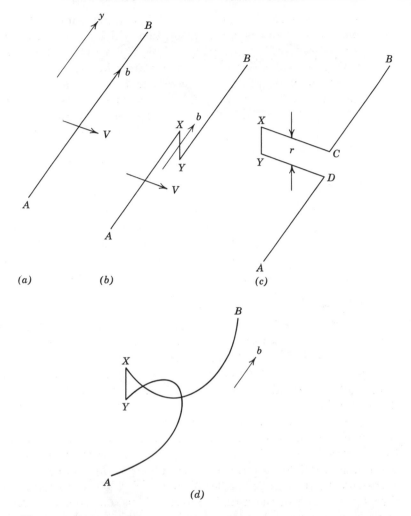

Fig. 4.13. Dislocation multiplication resulting from cross slip of part of moving screw dislocation.

thin (about† 10^3 A) section of a crystalline solid; the strain field around the dislocations diffracts the beam and produces a high-resolution, high-magnification ($\times 20,000$–$100,000$) image of the dislocations.

At very small strains, edge-dislocation dipoles are observed (Fig. 4.14a) to trail behind moving screw dislocations, as schematized in Fig. 4.13c. Dislocations are primarily long and straight (Fig. 4.14b)

† One A or angstrom unit is 10^{-8} cm $\approx 4 \times 10^{-9}$ in.

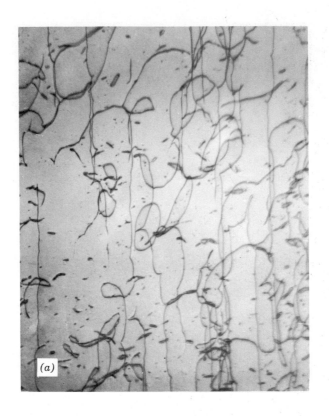

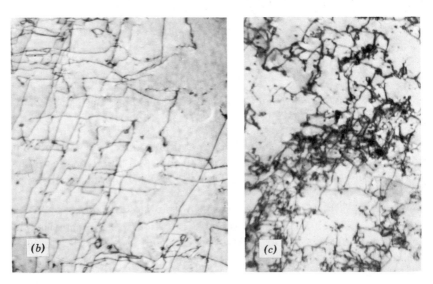

172

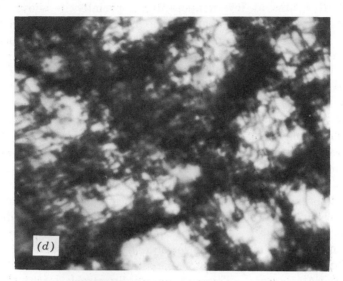

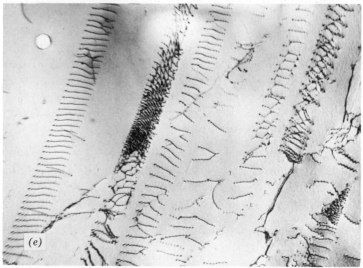

Fig. 4.14. Typical arrangements of dislocation loops observed by transmission electron microscopy. (a) Edge-dislocation dipole trails behind screw dislocations in Fe-3% Si deformed to the yield point at room temperature. 19,000×. *Courtesy A. Turkalo and J. Low* [11]. (b) Long dislocation loops in polycrystalline iron deformed 7% at —78°C. (c) Tangles formed in polycrystalline iron deformed 13% at —78°C. *Courtesy A. Keh* [12]. (d) Cells formed in polycrystalline tantalum deformed 25% at room temperature. 25,000×. *Courtesy T. Barbee*. (e) Planar arrays of dislocation in lightly deformed, nitrogen-bearing austenitic stainless steel. *Courtesy J. Embury and P. Swann* [13].

173

after small strains at low temperature, especially in alloys of low stacking fault energy. The surface slip lines then appear straight or planar. At larger strains or at higher temperatures the dislocations arrange themselves first in tangles (Fig. 4.14c) and then in cells (Fig. 4.14d). Cell formation requires extensive cross slip and hence occurs more easily (lower strain at a given temperature, lower temperature at a given strain) as the SFE is increased. Surface slip lines become more wavy when cell formation occurs in the interior of the crystal. Occasionally planar arrays of dislocations are observed in crystals having extremely low stacking fault energy, such as nitrogen-bearing austenitic stainless steel (Fig. 4.14e).

Thin film microscopy, as this technique is called, is useful for studying the deformation process on the "micron level." Of more immediate interest to the problem of fracture is the *distribution* of dislocations in the vicinity of a micro- or macrocrack or in a slip band.

This can best be observed at lower magnification (and hence in a larger field of view) by etching the crystal (or polycrystal) with certain chemical reagents. These preferentially attack (etch) the sites where dislocation lines emerge on the exposed surface or prepared cross section and the positions of the dislocations are observed by the so-called "etch pits." Figure 4.15 shows the dislocation distribution in slip bands in a single crystal (a) and polycrystal (b) of iron-3% silicon that was lightly deformed at room temperature. The curvy nature of the slip lines near the grain boundaries in (b) results from the complicated set of shears which are set up to maintain continuity at the boundary during deformation.

4.3   The Dislocation Theory of Yielding and Plastic Deformation

Phenomenological aspects of yielding [15, 16, 17]

The total plastic (shear) strain ϵ_p in a crystal whose dislocation density is N_{tot} is given by the relation

$$\epsilon_p = N_{tot}b\bar{x} \qquad (4.4)$$

where b is the Burgers vector and $\bar{x}$ is the average distance that each of the dislocations has moved. Optical and electron microscopic investigations of a variety of deformed crystals have shown that

$$N_{tot} = N_0 + \alpha\epsilon_p{}^\beta \qquad (4.5)$$

The initial or grown-in dislocation density is N_0, about 10^3 per cm^2

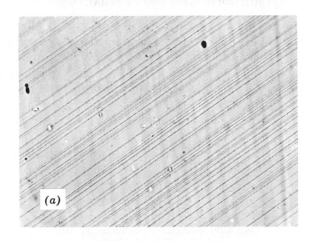

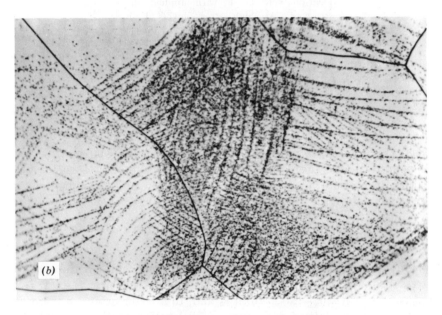

Fig. 4.15. Etch pits delineating dislocation distribution in slip lines in lightly deformed: (a) Single crystals of iron-3% silicon. 100×. (b) Polycrystalline iron-3% silicon. 175×. *Courtesy J. R. Low, Jr., and R. Guard* [14].

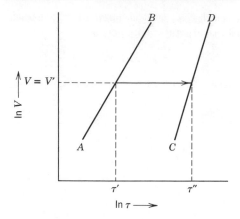

Fig. 4.16. Schematic variation of dislocation velocity V with applied shear stress τ. At high stress levels V approaches the speed of sound V_0.

in ionic and covalent crystals and $10^{7\pm1}$ per cm^2 in most metals; $\alpha \approx 10^{8\pm1}$ per cm^2 and β typically varies from 0.5 to 1.5.

The average plastic strain rate in a crystal $d\epsilon_p/dt = \dot\epsilon_p$ is, from (4.4),

$$\dot\epsilon_p = NbV \qquad (4.6)$$

where N, approximately $0.1N_{tot}$, is the density of dislocations that are *moving* at an average velocity $V = d\bar{x}/dt$.

The dislocation velocity is a strong function of the applied shear stress component τ. Numerous investigations have shown that

$$V = \left(\frac{\tau}{\tau_0}\right)^n = \left(\frac{\sigma}{\sigma_0}\right)^n \qquad (4.7)$$

where n is called the dislocation velocity exponent and τ_0 is the applied stress required to produce a unit dislocation velocity (i.e., $V = 1$ cm per sec, when $\tau = \tau_0$). Both τ_0 and, to a lesser extent, n increase as the inherent lattice resistance to dislocation motion τ_P and the resistance produced by other defects τ_D are increased, for any particular crystal. For example, lowering the temperature of deformation of a BCC metal causes τ_P and hence τ_0 to increase (Fig. 4.16) so that the applied stress must be raised from τ' to τ'' to maintain the same dislocation velocity V'. The maximum value of V, irrespective of τ_0 and n, is about the speed of sound V_0.

Uniaxial tensile tests are performed at a constant applied rate of elongation which leads to a constant applied strain rate $\dot\epsilon_A$ when the elongations are small. $\dot{\epsilon}_A$ is the sum of the elastic strain rate in the crystal $\dot{\epsilon}_e$ and the plastic strain rate $\dot{\epsilon}_p$;

$$\dot{\epsilon}_A = \dot{\epsilon}_e + \dot{\epsilon}_p \tag{4.8}$$

If there is no dislocation motion so that $\dot{\epsilon}_p = 0$, then

$$\dot{\epsilon}_A = \dot{\epsilon}_e = \frac{1}{E}\frac{d\sigma}{dt}$$

and Hooke's law is obeyed (i.e., $\sigma = E\epsilon_e$), curve AB, Fig. 4.17. In an ideal plastic solid which yields at a stress denoted by point C, and which does not strain-harden, plastic strain occurs with no increase in stress (curve CD), and consequently $\dot{\epsilon}_l = 0$. Thus $\dot{\epsilon}_p = \dot{\epsilon}_A$ and the crystal is able to deform at a rate imposed by the tensile machine. Most crystalline solids do strain-harden and their flow curves are given by CE, Fig. 4.17. Since the resistance to dislocation motion σ_0 increases with strain, the applied stress σ must be increased to maintain ϵ_p. Consequently $d\sigma/dt$ and hence $\dot{\epsilon}_e$ do *not* equal zero, and the measured slope of the flow curve falls in between the two extreme cases, $\dot{\epsilon}_p = 0$ or $\dot{\epsilon}_A = 0$.

In Chapter 1 we defined the tensile yield stress as that applied tensile stress at which permanent plastic deformation (i.e., dislocation motion) is produced. For most crystalline solids this definition is an arbitrary one since dislocations can move, and to a certain extent mul-

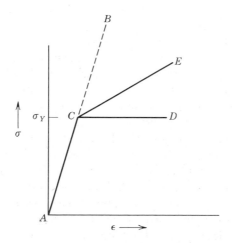

Fig. 4.17. Idealized stress-strain curves for different materials. See text for explanation.

tiply, at stresses far below the stress at which deviations from elastic behavior are observed in a *conventional* tensile test. These strains, of the order of 10^{-6}, are commonly referred to as *microstrains* and can only be detected by special techniques. *The macroscopic flow stress or yield stress* ($\sigma_Y = 2\tau_Y$) is that stress at which *large-scale* dislocation multiplication and extensive slip-band broadening takes place. Then $\dot{\epsilon}_p$ exceeds, becomes equal to, or becomes a large fraction of $\dot{\epsilon}_A$, so that the slope of the stress-strain curve is changed to such a degree that deviations from the initial behavior (essentially all elastic) are easily discernible on the curve obtained from the conventional tensile tests.

In well-annealed, pure FCC metals most of the grown-in dislocations are mobile and τ_0 is small ($\tau_P \approx 0$). Dislocation motion begins at the onset of load application (applied elongation), and the stress-strain curve always deviates from Hooke's law (Fig. 4.18a). Consequently there is no well-defined yield stress (or yield point) and the yield stress is taken to be the stress required to produce 0.2% plastic strain.

In BCC metals and nonmetallic crystals this type of behavior is usually not observed because most of the grown-in dislocations are immobilized (pinned) by impurity atoms (e.g., carbon and nitrogen in iron). These impurities are attracted by the stress fields of the dislocations and segregate around them. In order to move the grown-in dislocations at low temperatures where the impurities cannot diffuse easily, it would be necessary to tear the dislocations away from their impurity atmospheres. The stress required to do this is extremely high and this process rarely occurs. Instead, new (fresh) dislocations are created near stress concentrators, such as inclusions, surface steps, or grain corners, where the *local* stress is raised up to the value of about $G/30$, the stress required to create a dislocation.† Very little dislocation motion and multiplication take place in the microstrain region, so that the stress-strain curve shows only slight deviations from elastic behavior. Eventually a stress σ_{UY} is reached (the upper-yield stress, Fig. 4.18b) where a few dislocations can be created. Once they begin to move and multiply rapidly, N, hence $\dot{\epsilon}_p$, increases sharply. When $\dot{\epsilon}_p$ suddenly becomes greater than $\dot{\epsilon}_A$, stress is relaxed (i.e., $d\sigma/dt$ and hence $\dot{\epsilon}_e$ become negative) and a sharp yield drop is observed. The stress drops to $\sigma = \sigma_Y$ until V, and hence $\dot{\epsilon}_p$, has decreased to the point where $\dot{\epsilon}_p = \dot{\epsilon}_A$. A comparison of measured velocities of dislocations as a function of applied stress

---

† The high strength of whiskers results from the fact that they contain no defects that can trigger off yield at applied stresses much below $G/30$.

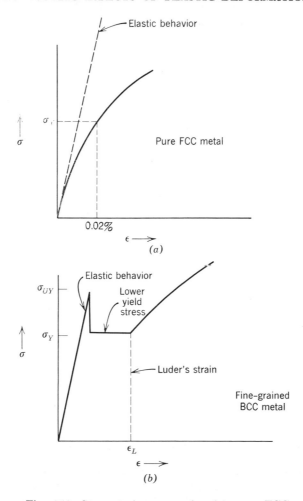

Fig. 4.18. Stress-strain curves for (a) pure FCC metal and (b) fine-grained BCC metal showing sharp yield point and Luder's strain ϵ_L.

with conventional tensile tests performed at rate $\dot{\epsilon}_A \approx 10^{-3}$ per sec in a variety of crystals indicates that $V \cong 10^{-3}$ cm per sec at $\sigma = \sigma_Y$. That is, $N \cong 3 \times 10^7$ per cm² at the yield stress.   Thus we can define *the yield stress as the stress that is required to produce a dislocation velocity of 10^{-3} cm per sec in a given crystal.*† Note that if n is very

† Assuming that the applied strain rate is $\approx 10^{-3}$ per sec as is usually the case. At higher applied strain rates the dislocation velocity at yielding is, of course, higher.

large (e.g., copper, $n > 80$) the stress required to achieve a dislocation velocity of 10^{-2} or 10^{-1} is about the same as that required to
achieve a velocity of 10^{-3} cm per sec. The yield stress of these
crystals is therefore independent of the applied strain rate. However, in silicon or germanium $n \approx 2$, so that a change in applied strain
rate will cause a large change in flow stress. The behavior of BCC
metals and ionic crystals falls in between these two extremes.

Sharp yield points are indicative of inhomogeneous plastic deformation and are observed in crystals when a few dislocations are able
to multiply quickly and form a glide band (or twin). Slip usually
begins at one or two places in the specimen and then spreads across
the entire gage length (as a Luder's band) at a constant stress σ_Y
(i.e., $\dot{\epsilon}_p = \dot{\epsilon}_A$) that is called the lower yield stress; strain hardening in
the deformed region *behind* the Luder's front prevents the specimen
from shearing in two. When the entire gage length has been deformed
($\epsilon = \epsilon_L$), homogeneous strain hardening takes place throughout the
gage length and the flow stress again begins to rise.

If the specimen is unloaded during the lower yield (or Luder's)
elongation $\epsilon < \epsilon_L$ and then *immediately* reloaded, the dislocations at
the Luder's front resume their motion at the same stress level at
which they had previously been moving (Fig. 4.19a) and a yield
point is not observed.† However, if the specimen is *aged* for a short
time at 100°C, or allowed to rest at room temperature before reloading, the impurity atoms will be able to diffuse back to the dislocations
and repin them; consequently a sharp yield point is observed upon
reloading (Fig. 4.19b). This effect is called *strain aging*.

Microscopic parameters that affect the yield stress

Since the yield stress is that stress ($\tau = \tau_Y = \sigma_Y/2$) which must be
applied to bring the dislocation velocity up to a value of about 10^{-3}
cm per sec, $\tau_Y \cong (10^{-3})^{1/n} \tau_0$ from Eq. 4.7. It is convenient, therefore,
to discuss the yield process in terms of the factors that increase the
resistance to dislocation motion τ_0. Thus far we have considered two
of these factors; the Peierls' stress τ_P which increases sharply with
decreasing temperature in all crystals (except FCC and HCP metals)
and the interaction of dislocations with impurity atoms. This process
raises τ_0 even when pinning does not take place. The dislocation
multiplication mechanism also requires that an internal stress be
overcome. In the Frank-Read source process it is necessary that the

---

† Yields points are also not observed if the annealed specimen is accidentally
deformed before the test begins, and large numbers of fresh dislocations are
introduced, or if the tensile machine is "soft."

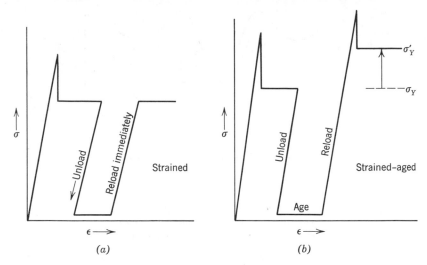

Fig. 4.19. Yield point removal after small plastic strains and its return after low-temperature aging.

dislocation increase its line energy in bowing out into a semicircle, and this requires a stress $\tau = Gb/l'$ where l' is the distance between pinning points. The multiple cross-glide process will only occur when $\tau > Gb/r$, so that the interaction force between segments XC and YD (Fig. 4.13) can be overcome and these dislocations can pass over one another. *In alloys* (crystals composed of two or more elements) two other strengthening mechanisms can contribute to τ_0; solid solution hardening and second-phase or dispersion hardening. If a solute element is dissolved in the lattice, some resistance to dislocation motion arises because of the stress field set up around the solute atom which interacts with the stress field of the moving dislocation. This effect is small if the solute atoms are randomly arranged in the lattice, but can be large if the distribution of solute atoms is not random— for example, impurities segregated around dislocations or arranged in a regular pattern in certain regions of the crystal (short-range order) —because a certain amount of work will have to be done by the applied stress in moving dislocations through the particular pattern and destroying it.

Of most importance from an engineering point of view (high strength steels, aluminum alloys, titanium alloys) is *second-phase* or *dispersion hardening,* where the solute atoms are not dissolved in the matrix material but form second-phase particles (precipitates)

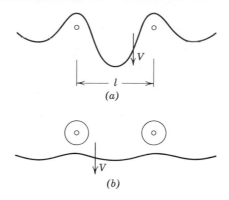

Fig. 4.20. Top view of dislocation lines bowing between hard particles (a) and pinching off on the back side of the particle (b). Dislocation loops are then left around the particles.

dispersed in the matrix. When a moving dislocation encounters one of these precipitates it will not, in general, be able to cut through it because the precipitates are generally stronger than the matrix. Consequently it will have to bow between it (Fig. 4.20a) and around it, leaving a dislocation ring around the particle and a freely expanding loop (Fig. 4.20b). The stress required for this process is essentially

$$\tau = \frac{Gb}{l} \tag{4.9}$$

where l is the distance of closest approach between the particles. In mild steels, carbide and nitride particles precipitate on dislocations during strain aging. In addition to producing the yield-point effect described above, these particles also increase the lattice resistance to dislocation motion so that Luder's extension occurs at a stress $\sigma'_Y > \sigma_Y$ (Fig. 4.19b). When fine dispersions of strong particles ($l \approx 50b$) are present, such as alloy carbides in steel, very large stresses must be applied before dislocation motion can occur and the material has a high yield strength.

All or some of the strengthening mechanisms described here can operate simultaneously in crystals and contribute additively to their yield strength. Table 4.2 summarizes the mechanisms and the manner in which they are achieved for a variety of crystal structures. The potential value or effectiveness of a particular mechanism is also

Table 4.2
Summary of Factors Effecting the Yield Strength of Crystalline Materials

Yield stress $\sigma_Y = \sigma_i + k_y d^{-1/2}$

(flow stress $= \sigma_Y + \Delta\sigma$)

Strengthening Mechanism	Results from	Crystal Structure	Relative Applicability for Achievement of Strength	Temperature Dependence
(1) Peierls' stress (σ_i)	Breaking of atomic bonds	FCC and HCP metals	little	none
		BCC metals	moderate	moderate
		nonmetallics	large	strong
(2) Impurity hardening (σ_i)	Addition of interstitial impurity atoms such as C, N, O which dissolve in between atoms of matrix	FCC metals	none	none
		HCP metals	some	some
		BCC metals	large	strong
		Nonmetallics	none	strong
(3) Solid solution hardening (σ_i)	Addition of substitutional atoms which occupy normal lattice positions	FCC and HCP alloys	large	small
		BCC alloys	moderate	small
		Nonmetallics	large	small
(4) Dispersion hardening (σ_i)	Fine dispersion of 2nd-phase particles in matrix	FCC and HCP alloys	large	none
		BCC alloys	large	none
		Nonmetallics	small	none
(5) Grain boundary hardening ($k_y d^{-1/2}$)	Production of fine grain size	FCC and HCP alloys and metals	small	none
		BCC metals and alloys	large	usually none
		Nonmetallics	Not applicable	
(6) Strain hardening $\Delta\sigma_i \cong (d\sigma/d\epsilon)\epsilon$	Increase in dislocation density	FCC and HCP metals and alloys	some	none
		BCC metals and alloys	some	none
		Nonmetallics	not used	none

listed. Some of the mechanisms (those that depend on long-range stresses) are not affected by deformation temperature. However, when the breaking of short-range bonds is involved in dislocation motion, thermal energy can assist the applied stress and the mechanism will be less effective at elevated temperatures. This fact has also been indicated in Table 4.2. Figure 4.21 shows the (approximate) variation of yield strength with homologous temperature and strain rate for a variety of materials.

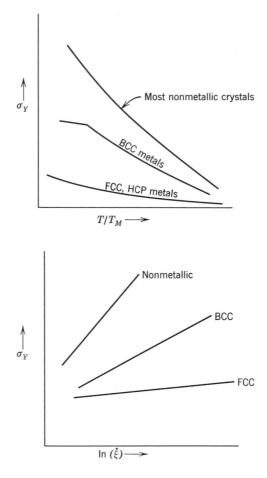

Fig. 4.21. (*a*) Yield strength σ_Y as a function of T/T_M, where T is the absolute value (°K or °R) of the test temperature and T_M is the absolute melting point. (*b*) Variation of σ_Y with strain rate $\dot{\varepsilon}$.

Almost all structural materials are polycrystalline, that is, they are composed of grains (crystals) of varying misorientation with respect to one another. The region between grains, which is a few atom diameters wide and where atoms are not on the lattice of one crystal or another, is called a *grain boundary*. In general, because of the misorientation of the grains, plastic deformation cannot spread easily from one grain to the next. Furthermore, the deformation of any one grain cannot occur independently of its neighbors because these exert a constraint effect on the deformation, in order that continuity be maintained at the grain boundary. Continuity will be achieved if five independent slip systems can operate near the boundary in each of two adjacent grains. Consequently the yield stress of a polycrystalline material will be composed of two parts: the yield stress of a constrained single crystal $\sigma_i = 2\tau_i$, which depends on the strengthing effects (1 to 4) in Table 4.2; and the strengthening effects of grain boundaries.

To estimate the latter, suppose that a dislocation source S in grain A, Fig. 4.22, emits a dislocation, which moves out along its slip plane until it reaches a grain or phase boundary where it becomes blocked by the boundary. The second dislocation emitted by the source will be blocked by the stress field of the obstacle and by the repulsive stress field of the first dislocation. When n dislocations have been emitted by the source and "pile up" against the obstacle, an equilibrium distribution of dislocations will be set up such that the net force acting on any particular dislocation, owing to the applied stress field, the stress field of the obstacle, and the stress field of all the other dislocations in the array, is equal to zero. Theoretically these conditions lead to the following conclusions:

1. The plastic displacement near the tip of a piled-up group of n dislocations of Burgers vector b is nb. The concentrated stress on the leading dislocation τ', owing to the n dislocations piled up behind it, is

$$\tau' = n\tau \qquad (4.10a)$$

2. When a slip band of radius L forms and dislocations pile up against an obstacle, all the applied shear stress τ is relaxed inside the band, except for the friction stress τ_i which opposes dislocation motion. The elastic shear displacement $(\tau - \tau_i)L/G$ that had been homogeneously distributed in the band of length L is replaced by the plastic displacement nb, so that

$$n = \frac{L(\tau - \tau_i)}{Gb} \qquad (4.10b)$$

3. The concentrated *shear stress* τ_L at some point (r, ϕ) from the tip of the pile up is (Fig. 4.22b)

$$\tau_L \cong (\tau - \tau_i) \left(\frac{L}{r}\right)^{\frac{1}{2}} F'(\phi) \qquad (b \ll r \ll L) \qquad (4.11)$$

and the concentrated tensile stress σ_L is

$$\sigma_L \cong (\tau - \tau_i) \left(\frac{L}{r}\right)^{\frac{1}{2}} F(\phi) \qquad (b \ll r \ll L) \qquad (4.12)$$

where $F'(\phi)$ and $F(\phi)$ are two different functions of ϕ. $F'(\phi)$ is a maximum at $\phi = 0°$ and $F(\phi)$ is a maximum at $\phi = 70°$.

The lower yield stress σ_Y is the stress at which slip can propagate across the gage length. Physically this means that the concentrated shear stresses at the tip of a slip band at the Luder's front (grain A, Fig. 4.22a) are able to trigger off a source in grain B or create a new dislocation in the boundary which can move into grain B, depending on which of these processes occurs at a lower applied stress. Suppose that either of these processes will occur at a (different) critical distance r^* from the pile up (Figs. 4.22c and 4.22d) when the concentrated shear stress τ_L reaches some critical value τ^* at $r = r^*$. Then the "local" yield criterion is, from (4.10),

$$(\tau - \tau_i) \left(\frac{L}{r^*}\right)^{\frac{1}{2}} \cong \tau^* \qquad (4.13)$$

and, since $\tau = \tau_Y$ at yielding,

$$\tau_Y = \tau_i + \tau^*(r^*)^{\frac{1}{2}} L^{-\frac{1}{2}}$$

Since L is about one-half the grain size $2d$,

$$\tau_Y = \tau_i + k'_y d^{-\frac{1}{2}} \qquad (4.14a)$$

and, with $\sigma \approx 2\tau$,

$$\sigma_Y = \sigma_i + k_y d^{-\frac{1}{2}} \qquad (4.14b)$$

where $k_y = 2k'_y = 2\tau^*(r^*)^{\frac{1}{2}}$.† Equation 4.14 is known as the *Petch equation* [18], and it applies to a variety of materials (see Fig. 6.2, for example). It indicates that the yield strength (for slip or twinning) increases as the grain size is decreased. This form of strengthening is one of the most important in commercial materials, particularly for steels and refractory metals.

---

† Note that k_y is a microscopic stress intensity factor and has the same dimensions as the macroscopic stress intensity factor K.

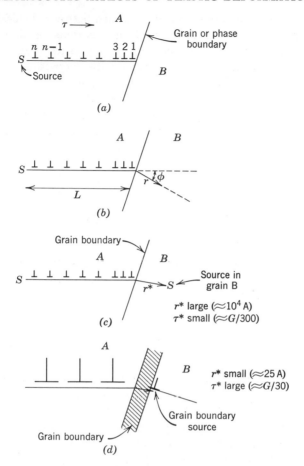

Fig. 4.22. (a) Pile up of n edge dislocations at a grain or phase boundary. (b) Coordinates used in describing stress field of pile up of length L. (c) Propagation of slip across boundary by the operation of a source S in grain B. (d) Propagation of slip across boundary by the creation of a new dislocation at a grain boundary source.

Strain hardening

Once the yield stress has been reached (in either single- or polycrystalline materials) large-scale dislocation motion and multiplication will occur, new slip bands will form and slip bands will broaden. As the amount of plastic strain increases, the dislocation density will increase also and hence the average distance between dislocations will decrease. As pointed out previously, dislocations of like sign will repel one another whereas dislocations of opposite sign will attract one another. This interaction increases with decreasing distance between dislocations. Thus the internal stress field set up by dislocations acting on other dislocations increases with increasing amounts of plastic strain and an increased stress $\Delta\sigma_i \cong 2\Delta\tau_i$ is required for dislocation motion and multiplication.† This phenomena is termed *strain hardening*. At relatively low strains

$$\Delta\tau_i = \Omega Gb \sqrt{N_{tot}} \qquad (\Omega = 0.2 - 0.4) \qquad (4.15)$$

so that

$$\Delta\tau_i = \Omega Gb \sqrt{N_0 + \alpha\epsilon_p{}^\beta} \qquad \text{from (4.5)}$$
$$\cong \Omega Gb\alpha^{1/2}\epsilon_p^{\beta/2} \qquad\qquad (4.16)$$

once yielding has begun. At larger strains k_y is decreased as the deformation is more homogeneous and dynamic recovery sets in. The strain-hardening rate $d\sigma/d\epsilon$ is then reduced.

4.4 Summary

1. Plastic deformation in crystalline solids is an inhomogeneous form of strain which occurs by the processes of slip or twinning. In slip, whole blocks of crystal are sheared over one another. In twinning, the shears are such that the atoms in the sheared regions are in a mirror-image orientation relative to the underforming regions of the crystal. Very high displacements are produced along the planes on which slip or twinning occur.

2. Both of these processes occur by the motion of crystalline defects known as dislocations. Because plastic flow occurs consecutively rather than simultaneously, a dislocation is a boundary between a deformed and undeformed portion of a crystal.

3. Dislocations are described by the orientation of their Burgers vector b and their line l. When b and l are perpendicular, the dislocation is pure edge; when b and l are parallel, the dislocation is pure screw. Most dislocations are mixtures of edge and screw. An edge dislocation is confined to glide on a plane that contains both its

---

† At small strains k_y is not greatly changed by strain hardening so that the relation $\sigma_{flow} = \sigma_i + \Delta\sigma_i + k_y d^{-1/2}$ is applicable. At larger strains k_y is decreased (since the deformation is more homogeneous).

Burgers vector and its line; it moves in the direction of its Burgers vector. A pure screw dislocation always glides perpendicular to its line. When unextended, it is not confined to any particular glide plane.

4. Since large atomic displacements exist in the vicinity of a dislocation line, a dislocation has a long-range strain and stress field associated with it. The stress field of an edge dislocation has both dilatation and shear components; the stress and strain fields of a screw dislocation are (essentially) pure shear.

5. Almost all dislocations in a crystal are thermodynamically unstable defects by virtue of their high strain energy or line tension. Consequently they will move under the action of a force in a manner that will reduce the total energy of the crystal. Dislocations of like sign repel one another, those of opposite sign attract one another.

6. The motion of a dislocation can lead to the creation of new dislocations, a process called dislocation multiplication. The dislocation density increases with increasing plastic strain.

7. The plastic strain rate inside a deforming crystal is proportional to the density of moving dislocations and their velocity. The dislocation velocity increases with increasing applied stress. Yielding in a tensile test occurs when the plastic strain rate is equal to the applied strain rate (nonstrain-hardening material) or is almost equal but less than the applied strain rate (strain-hardening material).

8. The yield or flow strength of a crystal is increased when the mobility of dislocations or the density of mobile dislocations is reduced. This may be accomplished by impeding the motion of dislocations through the use of other crystalline defects. These include impurity atoms (dislocation locking or solid solution hardening), small particles (dispersion hardening) or other dislocations (strain hardening). The intrinsic resistance of the lattice to the shearing of atomic bonds is also important, so that the yield strength is high in covalently bonded crystals and low in pure metals.

9. Thermal energy can assist the motion of a dislocation. Consequently the yield strength decreases as the temperature increases. The temperature and strain-rate dependence of the yield strength is highest in covalently bonded crystals and weakest in FCC metals.

10. Grain boundaries and phase boundaries also provide obstacles to large-scale dislocation motion. Piled-up groups of dislocations can occur at boundaries. The stress concentration factor of a pile-up increases with the square root of the pile-up length. The yield strength varies inversely with the square root of the grain size.

References

The asterisk indicates that published work is recommended for extensive or broad treatment.

[1] R. Hill, *The Mathematical Theory of Plasticity,* Oxford, London (1950).

*[2] A. H. Cottrell, *Dislocations and Plastic Flow in Crystals,* Oxford, London (1956).

*[3] J. Friedel, *Dislocations,* Addison-Wesley, Boston (1964).

*[4] J. Weertman and J. R. Weertman, *Elementary Dislocation Theory,* Macmillan, New York (1964).

[5] C. S. Barrett, *Structure of Metals,* McGraw-Hill, New York (1952).

[6] N. Stoloff and R. G. Davies, *Acta Met.,* **12,** 473 (1964).

[7] D. Hull, *Fracture of Solids,* Wiley, New York (1963).

[8] G. Thomas and J. Washburn, *Electron Microscopy and the Strength of Crystals,* Wiley, New York (1962).

[9] J. Newkirk and J. Wernick, *Direct Observation of Imperfections in Crystals,* Interscience, New York (1962).

[10] G. Thomas, *Transmission Electron Microscopy of Metals,* Wiley, New York (1962).

[11] A. Turkalo and J. R. Low, Jr., *Acta Met.,* **10,** 215 (1962).

[12] A. Keh and S. Weissmann, in G. Thomas and J. Washburn, *Electron Microscopy and the Strength of Crystals,* Wiley, New York (1962), p. 231.

[13] J. Embury and P. Swann, *High Strength Materials,* Wiley, New York (1965).

[14] J. R. Low, Jr., and R. Guard, *Acta Met.,* **7,** 171 (1959).

*[15] J. J. Gilman, "Plasticity," in *Symposium on Naval Structural Mechanics, ONR* (1960).

*[16] A. H. Cottrell, NPL Conf. on *Structure and Mech. Prop. of Metals,* Teddington, Eng. (1963), p. 456.

[17] G. T. Hahn, *Acta Met.,* **10,** 727 (1962).

[18] N. J. Petch, *J. Iron Steel Inst.,* **173,** 25 (1954).

Problems

1. Dislocation A has a Burgers vector $\frac{1}{2} a[110]$ and a line which lies along $[1\bar{1}0]$. Dislocation B has a Burger's vector $a[001]$ and a line which also lies along $[1\bar{1}0]$.

(a) Draw the two dislocations in the standard cube.

(b) What are the glide planes of the two dislocations?

(c) What stress component of dislocation A (σ_{xx} or σ_{xy} or σ_{yy}) tends to make dislocation B glide in its glide plane, in the glide direction?

2. Suppose that dislocation multiplication occurs *only* by multiple cross glide and that dislocations are moving at a constant velocity V. If δ is the multiplication constant (i.e., the number of loops produced

per cm of moving screw dislocation) show that the dislocation density increases linearly with plastic strain and that this increase depends on δ.

3. (a) Determine the average velocity of a dislocation in a single crystal of copper if it is being strained at a rate of 10^{-3} cm per cm sec under steady state creep conditions at a stress of 2,000 psi. $N = 10^8$ lines per cm^2 and remains constant during straining. All slip is on {111} planes in $\langle 110 \rangle$ directions. The lattice parameter of copper is 2.5×10^{-8} cm.   (5.65×10^{-4} cm per sec)

(b) The dislocation velocity exponent $n = 70$ for copper. What would be the new strain rate if the stress were instantaneously raised to 2100 psi?   (3×10^{-2})

4. Assume that strengthening of a low carbon (BCC) steel is *only* achieved by grain size refinement and by fine carbide dispersions. Uniaxial tensile tests show that the tensile yield stress $\sigma_Y = 55,000$ psi when grain size $2d = 0.32$ mm., and that $\sigma_Y = 40,000$ psi when $d = 1$ mm. What is the average distance of closest approach between the particles? The shear modulus $G = 10^{12}$ dynes per cm.^{2}, lattice parameter $a = 2.84$ A$^\circ$.   ($1,250$ A$^\circ$)

5. Suppose that two thermally activated processes of yielding can occur in a BCC metal and that the shear stress τ_Y required to operate each of them decreases linearly with temperature. Torsion measurements of the (shear) yield stress of single crystals indicate the following results:

T (°K)	τ_Y (psi)
100	80,000
200	60,000
500	25,000
600	15,000

What will be the tensile yield stress at 300°K in a polycrystalline specimen of this metal having a grain size $2d = 0.32$ mm if $k_y = 10,000$ psi $\sqrt{\text{mm}}^{1/2}$?   ($115,000$ psi)

6. Suppose that for a certain material having BCC structure the dislocation density varies with strain ϵ as

$$N_{\text{tot}} = 10^{10} \, \epsilon \qquad \text{(loops per cm}^2\text{)}$$

If strain hardening occurs exclusively by the intersection and recombination of $\langle 111 \rangle$ dislocations to form lengths of $\langle 100 \rangle$ so that $\Omega = 0.22$, what is the largest tensile stress the metal can withstand without necking? Assume that the elastic limit is negligible and that all deformation is described by a power law. $G = 10^{12}$ dynes per cm^2, $b = 3 \times 10^{-8}$ cm.   (9.3×10^8 dynes per cm^2)

7. Alloys of metals A and B can be strengthened by precipitation hardening. Metal A melts at 1200°K, metal B at 1400°K. A eutectic is formed at 800°K between liquid of 60% B, an α-phase containing 14% B, and a β-phase containing 84% B. At room temperature (300°K) the maximum solubility of B in A is 4% and A in B is 6%. All liquidus, solidus, and solvus lines are straight. The yield strength of precipitation-hardening alloys is the sum of the stress required to move dislocations through the α-matrix and the stress required to bow the dislocations between the dispersed particles of the β-phase. The distance of closest approach l is given by the approximate relation $l = d/(V_f)^{1/3}$, where d is the particle size and V_f is the volume fraction of dispersed phase. Alloys containing 7 and 10% B were solution-annealed at 700°K, quenched to 300°K, and aged at room temperature for 100 hours to produce optimum strength. The particle size d after this aging treatment was 250 A°. Tensile tests indicated that the yield strength of the 7% alloy was 99,400 psi, and the yield strength of the 10% alloy was 116,000 psi. The shear modulus $G = 10^7$ psi, the Burgers vector $b = 2.5$ A°.

(a) Assuming that the density of the α- and β-phases is the same, determine the strength of the saturated α-phase at 300°K.

(35,000 psi)

(b) Assuming that the yield strength of the α-phase varies with the square of the concentration of dissolved B, determine the strength of pure A. The strength of the 2% alloy is 11,000 psi.   (3,000 psi)

5

Microscopic Aspects
of Crack Propagation

The dislocation theory described in the previous chapter has been very successful in explaining the atomistic aspects of yielding and strain hardening in crystalline solids. It can also be used to explain many of the microscopic aspects of fracture, particularly those aspects related to the *intrinsic* properties of the solid. These include: (1) the question of inherent resistance to crack propagation, (2) the way that *moving* cracks are initiated below the root of a preinduced notch or in flaw-free material, and (3) the reason for the occurrence of a transition temperature region where the cleavage mode of fracture changes over to the tear mode. The latter two questions are dependent, in part, on an understanding of the microscopic aspects of toughness and, consequently, will be deferred until Chapters 6 and 7. In the present chapter we consider physical processes that contribute to a material's resistance to crack propagation. We shall treat a crack as a moving group of dislocations because this analogy is useful in discussing the various processes that blunt the crack tip and hence relax the concentrated stresses and strains there.

In Section 5.1 we consider the cleavage mode of crack propagation in terms of edge-dislocation arrays. In Section 5.2 we discuss the process of normal rupture in terms of these arrays and consider the microscopic aspects of the process of ductile fracture in tensile specimens. In Section 5.3 we consider the problem of antiplane-strain (shear) fractures (mode III deformation) and show how they can be treated as a moving sheet of screw dislocations. We shall show that the distribution of dislocations near the crack tip is the most important factor affecting the resistance to unstable fracture, and that fracture can only take place when the strains are inhomogeneously distributed near the tip of the crack.

5.1 Microscopic Aspects of Cleavage Crack Propagation (Mode I Propagation)

The elastic crack

The elastic cleavage crack described in Chapter 2 is an elliptically shaped discontinuity in a homogeneous elastic solid, where the tensile displacements are greater than the homogeneous displacements produced by the applied tensile stress in the rest of the solid. The crack displacements are approximately equal to zero at the crack tip and vary up to a maximum value $D = 2h$ at the center of the crack; D is the crack height. Because edge dislocations are characterized by tensile displacements (their Burgers vectors), a crack can be formally represented (Fig. 5.1) as a continuous distribution of n edge dislocations [1, 2] of infinitesimal Burgers vector b' such that

$$D = nb' = \int \frac{\partial u}{\partial l} dl \qquad (5.1)$$

The displacement vector of each dislocation is u, and the integration is taken over a path AQA' that starts at the midpoint of one of the crack faces (point A) and finishes at the corresponding midpoint of the opposite face A'. It is important to recognize that the *dislocations do not literally "exist" inside the crack but are employed for representative purposes only.* Friedel [2] has shown that this representation facilitates an understanding of the atomistic and microscopic aspects of fracture toughness which we shall consider shortly.

Figure 5.1b indicates that the crack extends when the dislocation groups move along the positive and negative x axis, perpendicular to their Burgers vector, which lies along the y direction. This type of motion, called *climb*, is shown schematically in Fig. 4.10 for an isolated dislocation. Because the frictional resistance to dislocation motion is zero inside the crack, most of the dislocations will be concentrated (piled up) near the crack tips. At the crack tip the atomic planes Q and Q' have separated to a distance $2a_0$ that is about twice the normal lattice spacing a_0, thereby breaking the atomic bonds. Consequently the leading dislocation d at the crack tip has a Burgers vector b about equal to the lattice spacing. This dislocation (and hence the crack) cannot advance until the force acting on it, from all of the n dislocations piled up behind it, is equal to

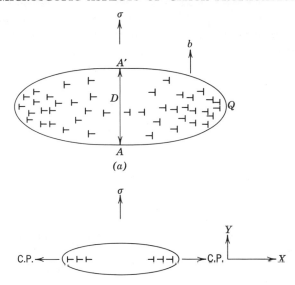

(a)

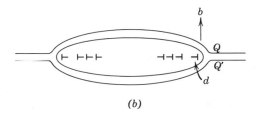

(b)

Fig. 5.1. (a) Representation of an elastic cleavage crack as a climbing group of edge dislocations. (b) Growth of the crack by the climb motion of the edge-dislocation groups. *After J. Friedel* [2].

the resistance $2\gamma_s$ to the breaking of atomic bonds (i.e., the surface energy).

Since the crack length $2c$ is much greater than D, its stress and strain distribution are similar to those around a piled-up group of glide dislocations [2]. By analogy with Eq. 4.10a the force acting on d is equal to $\sigma nb' = \sigma D$, so that the criterion for crack extension is

$$\sigma nb' = \sigma D \geqq 2\gamma_s \tag{5.2}$$

Furthermore, by analogy with Eq. 4.10b the number of dislocations in the pile up is obtained by setting their back stress equal to the applied stress σ.

$$D = nb' = \sigma \frac{(1 - v)(2c)}{G}$$

$$= \sigma \frac{(1 - v^2)(4c)}{E} \tag{5.3}$$

where $E = 2G(1 + v)$. Substituting Eq. 5.3 into 5.2 shows that the crack extends at a stress $\sigma = \sigma_F$ when

$$\sigma_F{}^2 = \frac{E\gamma_s}{2c(1 - v^2)} \tag{5.4}$$

which is the same as the plane-strain Griffith equation $\sigma_F{}^2 = 2E\gamma_s/[\pi c(1 - v^2)]$ except for a factor of $4/\pi$.

Cottrell has defined this type of fracture as *cumulative* [3]; that is, once $\sigma = \sigma_F$ at $c = c_F$, there will be n_F dislocations accumulated (piled up) near the crack tip, where

$$n_F = \frac{\sigma_F(1 - v^2)4c_F}{b'E} \tag{5.5}$$

These dislocations will be able to cause complete fracture of the solid merely by spreading across the fracture plane. No additional ones need be created during crack extension. If σ is maintained constant at σ_F, then additional dislocations will be created during propagation as c becomes greater than c_F in Eq. 5.5. These additional dislocations help push the first group along, but they always remain behind this group and do not contribute to the work required to move d, unless the crack becomes blunted by plastic deformation.

The river patterns observed on cleavage fracture surfaces (Figs. 3.9 to 3.12) result from the propagation of a crack on more than one level. Suppose that the crack tip is moving in the x direction (Fig. 5.2a) and can be represented by a single, climbing-edge dislocation AB which approaches and crosses a screw dislocation CD whose Burgers vector is perpendicular to the cleavage plane. This screw dislocation produces a jog of height b in AB (Fig. 5.2b), which forms a step S on the cleavage plane as the leading edge AB advances [4]. Since additional surface area will be created when the crack advances, the effective surface energy of the crack is increased from γ_s to about $\gamma_s(1 + b/l)$ for steps of height b formed at distances l along the leading edge. These steps tend to be perpendicular to the crack front [2],

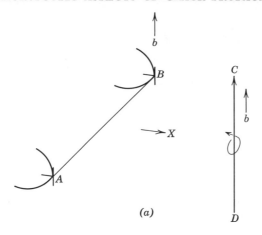

(a)

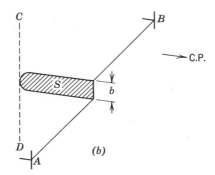

(b)

Fig. 5.2. (a) Cleavage crack approaching screw
dislocation CD that is perpendicular to cleavage
plane. Step S is formed on cleavage plane. (b)
After crack intersects CD. *After J. Friedel* [2].

in order to minimize their energy (area), and hence are roughly parallel to the direction of crack propagation CP.

As the crack advances, it crosses more and more dislocations whose
Burgers vectors have components perpendicular to the cleavage plane,
and additional steps form on the cleavage surface. The steps are
able to coalesce [2] by gliding along the crack front, as shown in Fig.
5.3. Most of the steps annihilate by coalescing with steps of opposite
sign (Fig. 5.3a); a few steps coalesce with others of the same sign
(Fig. 5.3b) and multiple steps of height mb are formed. When their

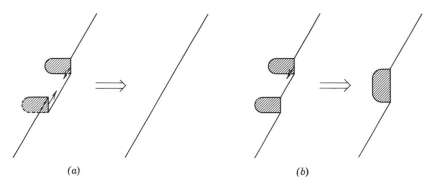

Fig. 5.3. (*a*) Annihilation of cleavage steps of opposite sign. (*b*) Coalescence of steps of the same sign.

height becomes sufficiently large they can be resolved by electron and optical microscopy. These are the river patterns described earlier.

River patterns are formed in numerous ways. (1) After a crack crosses a grain boundary,† a large number of steps of the same sign will be formed on the cleavage surface because the crack has crossed a large number of dislocations of the same sign which lie in the boundary. These steps coalesce into multiple steps as the crack advances, so that the number of steps decreases but the height of each step increases (Fig. 3.10) as the crack moves away from the boundary [5]. (2) When a crack spreads from its nucleation site in the grain in which it was initiated, rivers are observed after the crack has spread some way from the nucleation site and sufficient coalescence of multiple steps has taken place such that the step height is resolvable [2] (Fig. 3.9). (3) When a cleavage crack is stopped and then restarted, large numbers of rivers are formed [5, 6, 7] because of the dislocations produced around the crack during its deceleration and subsequent acceleration. Rivers are also formed when (4) a crack crosses a void (i.e., gas bubble, cracked-particle interface) that lies in its path [8, 9], and (5) when a new crack is nucleated ahead of the main crack (e.g., by stress waves emitted from the main crack) on a different but parallel plane to that on which the main crack is propagating [10]. Steps are produced when the main and new cracks join together, by tearing of material between the cracks [5] or, in very brittle solids, by secondary cleavage [11] on planes inclined to the primary cleavage plane (Fig. 5.4). This is the primary cause of river forma-

---

† Except one that is pure tilt and hence contains no screw dislocations.

tion on the fracture surfaces of noncrystalline solids (e.g., glass) which do not contain screw dislocations [10].

In general, the annihilation of steps is such an effective process for reducing the step density that the steps formed when a crack propagates through a grain are too few in number to affect significantly the work done in propagation. When a crack crosses a grain boundary or dense slip band, however, the step density is sufficiently large that the energy expended in crossing the boundary [2] or band is about equal to $2\gamma_s$. In addition, if the crack crosses from one grain (grain A, Fig. 5.5) to another (grain B), whose cleavage planes are badly misoriented with respect to the first grain, the local stress resolved

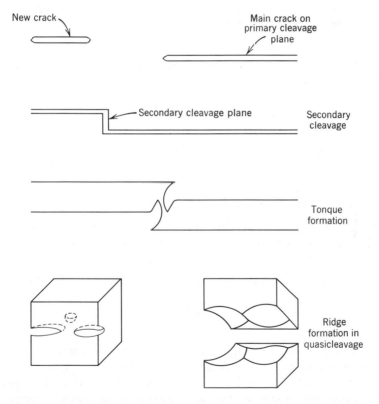

Fig. 5.4. Joining of main crack front with secondary (new) crack on plane parallel to it. Joining can occur by secondary cleavage or tearing (tongue formation). When numerous secondary cracks are present, the high density of ridges formed when they join produces a fracture surface referred to as "quasicleavage."

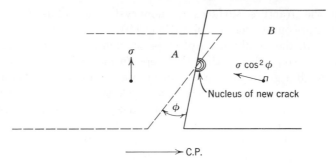

Fig. 5.5. Elastic crack propagation across a grain boundary in polycrystalline material occurs by the nucleation of a new crack at the boundary. The cleavage planes in the two grains are rotated about an axis, parallel to the direction of crack propagation, by an angle ϕ.

normal to the cleavage planes in grain B may be much less than that resolved perpendicular to the cleavage planes in grain A. For a given level of applied stress, this means that energy expended in crossing the grain boundary γ_B will be greater than γ_s. If ϕ is the angle between the cleavage planes in the two grains, then [12] $\gamma_B = \gamma_s/\cos^4 \phi$. In cubic crystals $\phi_{max} = 45°$ so that γ_B can be about $4\gamma_s$. When plastic relaxation occurs during propagation, the work done in advancing the crack increases to $\gamma_m > \gamma_s$ (p. 208) so that $\gamma_B = \gamma_m/\cos^4 \phi$.

Plastic relaxation at the crack-tip

Stopped Cracks. Both the dislocation and elastic stress models predict that the local shear stress a distance r from the tip of a sharp (elastic) crack τ_L is approximately $\sigma \sqrt{c/r}$. Since σ can be $\sqrt{2E\gamma_s/\pi c(1 - v^2)}$ before the crack spreads and $\gamma_s = Eb/20$,

$$\tau_L \cong \frac{G}{3} \sqrt{\frac{b}{r}} \tag{5.6}$$

These stresses are very high and, as we saw in Chapter 2, yielding can extend a distance $r = R \cong (\sigma/\sigma_Y)^2 c$ from the tip of a *stopped* crack in a material that is not *completely* brittle. The plastic deformation can occur by the activation of dislocation sources in a volume element near the crack tip or by the creation of new dislocations at the crack tip [2]. As we shall presently see, the degree of brittleness of a material is intimately connected with the type of relaxation process that occurs at the crack tip.

When a large number of mobile† sources S exist near the tip of a stopped crack (i.e., $\sigma < \sigma_F$), these sources will emit loops, some part C of which, having opposite sign from those of the crack dislocations, are attracted to the crack tip (Fig. 5.6a). The loops C move into the region near the tip until their total Burgers vector just compensates for that of the crack tip D; this neutralizes the large stress field at the tip, both the back stress acting on the center of the crack and the forward stress that acted ahead of the crack. Because the back stresses are relaxed, dislocations can move from the center of the crack to the crack tip, and this blunts the crack (Fig. 5.6b). The crack, therefore, is no longer elliptical but has parallel sides, separated

† Unpinned or lightly pinned.

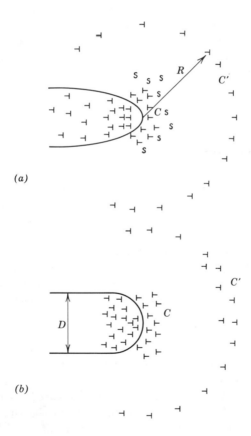

Fig. 5.6. Dislocation representation of crack blunting [2].

by a distance D, and a tip radius ρ that is about equal to $D/2$. Since the stress concentration factor of the crack is reduced from $2(c/b)^{1/2}$ to $2(2c/D)^{1/2}$, the stress required to propagate the crack *directly* as a Griffith crack is raised by a factor of $(D/2b)^{1/2}$, and the toughness of the crystal is increased.

If the stress is removed, the crack will no longer heal (i.e., $D \neq 0$ at $\sigma = 0$) because the stress field of the loops C holds the dislocations D in place. Consequently relaxed cracks can be observed (see Fig. 5.8) in the absence of applied tensile stress. These cracks will partially heal (sinter) if the crystal is annealed at a high enough temperature to allow large-scale diffusion to take place in a measurable period of time [13].

The parts C' of the emitted loops also have a total Burgers vector $C = C' = D$. Each of these no longer feels the elastic stress of the crack tip, which has been relaxed, but only the stresses of the other loops C'. The loops C' are repelled from the crack to a distance R such that their mutual interaction stresses are about equal to σ_Y. In the Friedel model [2] these loops lie along a cylinder (Fig. 5.6a) having radius R

$$R = \frac{GD}{\sigma_Y} \tag{5.7}$$

Physically this means that the plastic relaxation has replaced the large dislocation (displacement) near the tip with a fairly continuous distribution of dislocations (displacements) along a cylinder of radius R that is much greater than D. That is, the relaxation has made the strain distribution much more *homogeneous* by accommodating the displacement† over a distance R that is much greater than D.

If the mobile dislocation density around the stopped crack is very low or zero (heavily pinned dislocations), the complete relaxation described above will not be able to take place. The small amount of relaxation that does occur results from the heterogeneous nucleation of loops C' which are "punched" from the crack tip [2]. This punching can only take place at the tip since it requires a very high stress, of the order of $G/30$. When loops C' are punched from the crack, steps C are created on its surfaces (Fig. 5.7). These do not significantly blunt the tip because their density is low. The reason for this low density is that a step C can only be created when a dislocation C' forms in the slip band that emanates from the tip. If the loops C' are not pure screw dislocations, they will not be able to multiply

---

† Note that Eq. 5.7 is the same as Eq. 2.35, with $D = V(c)$.

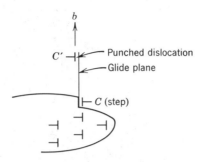

Fig. 5.7. Plastic deformation at crack tip by the punching of a dislocation out of the crack.

by cross gliding and the dislocation density in the band is low. If extensive multiplication takes place, then the band will broaden and C will be greater, but the number of steps will still be small. Extentive blunting by this process could only take place if large numbers of slip bands and large amounts of multiplication take place, but this is rarely observed.

The difference between the two types of relaxation processes is shown by the dislocation distribution near the ends (tips) of stopped cracks in iron-3% silicon and lithium fluoride (Fig. 5.8). Both cracks lie along {100} cleavage planes. The crack in Fe-3% Si was produced during cathodic charging of hydrogen (Chapter 9) and resulted from crack expansion under internal hydrogen pressure P. Each time the crack expands the pressure inside the crack drops (because of the fact that $PV = $ const),† and when P drops down to the Griffith value the crack must stop. It does not restart until additional hydrogen atoms diffuse into it, recombine to form molecular hydrogen, and increase the pressure to the point where fracture can again initiate [14, 15]. Each of the large group of etch pits shown in Fig. 5.8a represents a position at which the crack stopped during its symmetrical, discontinuous growth out from its nucleation site at the crack center. The continuous distribution of pits between the large groups was formed while the crack was moving. The crack shown in Fig. 5.8b was produced [7] by driving a sharp wedge partway through the crystal of LiF. The crack stopped only at its final position. The cracks shown in both crystals were formed at room temperature.

† The volume of the crack V is proportional to Pc^3.

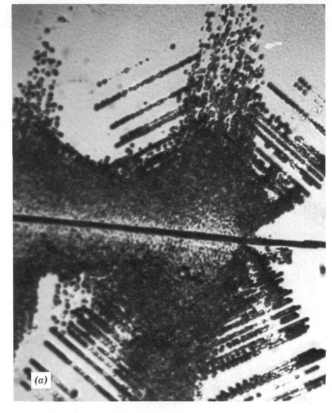

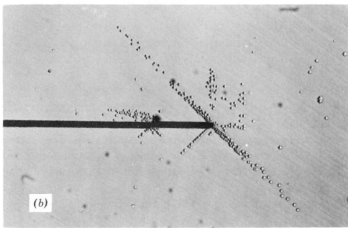

Fig. 5.8. Microcracks observed in single crystal of: (*a*) Iron-3% silicon [14]. 750×. (*b*) Lithium fluoride. 190×. Dislocations revealed by etch pitting technique. *Courtesy J. J. Gilman* [7] *and AIME*.

A comparison of the region near the ends of the two cracks indicates that most of the deformation lies along crystallographic slip planes. These planes are those that have the greatest amount of shear stresses resolved across them, along the respective slip directions [15]. The volume of deformed material is much greater around the crack tip in Fe-3% Si than around the tip of the crack in LiF. In Fe-3%, Si a relatively small amount of deformation has occurred on each of a large number of slip planes, whereas in LiF extensive dislocation multiplication has taken place in a few slip bands. This indicates that a large number of sources have operated in Fe-3% Si, whereas most of the (small amount of) relaxation in LiF has occurred by the motion of dislocations that were punched from the crack tip. At room temperature the tensile yield stress of Fe-3% Si, 25,000 psi, is about an order of magnitude *greater* than the tensile yield stress of LiF, 1000–3000 psi. Consequently the distribution of plastic relaxation is *not related* to a particular numerical value of the yield stress, and in the example cited here more relaxation actually occurs in the material of higher yield stress. The reason for this is that the yield stress is primarily related to the stress required to produce a *particular dislocation velocity* (about 10^{-3} cm per sec) (Chapter 4), whereas the distribution† of the plastic relaxation at the tip of a stopped crack depends on the *density of dislocation sources* available for the relaxation [16]. High densities of mobile sources allow large amounts of deformation, inhomogeneously confined to a few slip planes. In the first case the large amounts of plastic flow reduces the longitudinal stresses at the tip to a value about equal to the tensile yield stress (Fig. 2.9), and the mechanics calculations presented in Chapter 2 are applicable. These cracks cannot reinitiate directly by the process shown in Fig. 2.12a. However, when only a small number of dislocations exist at the tip, as in LiF and other nonmetallic crystals, the amount of blunting is small ($b < \rho \ll D/2$) and the tensile stresses at the tip are higher than σ_Y. Consequently these cracks can begin to propagate *directly* as Griffith cracks, with $\gamma_P \cong \gamma_s(\rho/b)$. We can, therefore, distinguish between "brittle" and "completely brittle" materials at the microscopic level. Brittle materials are those in which unstable fracture can take place, usually at or below the yield stress, when a new, moving crack is *initiated by plastic deformation*. Completely brittle materials are those in which the relaxation at the tip of a stopped crack is so small that unstable fracture can occur by the *direct reinitiation* of a stopped crack, as schematized in Fig.

† Because the extent of plastic deformation around the crack R depends on $\sigma^2 c$, the zone sizes cannot be compared for the two cracks shown in Fig. 5.8.

(2.12a). For example, sharply cracked crystals of iron-3% silicon undergo a transition from completely brittle to brittle as the temperature of testing is raised from $-196°C$ to $18°C$ [20].

Moving Cracks

The *total work* done in propagating a continuous cleavage crack γ_P is greater than γ_s in all but the completely brittle solids. γ_P is composed of two terms: γ_m, which measures the plastic work done around a microcrack or crack which moves through a crystal, and $\gamma_B = \gamma_m/\cos^4 \phi$, which measures the work done when the microcrack or crack crosses a boundary of large misorientation. This is usually a grain boundary, but in some cases twin boundaries [17] and phase boundaries [18] also provide good obstacles to crack propagation.

The intrinsic toughness of various materials is closely related to the value of γ_m. To understand this behavior it is first instructive to compare the plastic relaxation around moving cleavage cracks in Fe-3% Si and LiF. From Fig. 5.8 we observe that essentially no deformation occurred around the crack in LiF until it slowed down and finally stopped, whereas plastic deformation is observed† to occur in Fe-3% Si while the crack was moving. This deformation is much less extensive than that which occurred while the crack was at rest. The reason is that any volume element is under stress for a much shorter period of time when the crack is moving [7, 16, 19], and hence the complete relaxation process described by Fig. 5.6 will not have time to occur.

The distance X that a dislocation near the tip moves in a time t' is

$$X = \int_0^{t' = r'/v_c} V \, dt \tag{5.8}$$

where

$$V = \left(\frac{\tau_L}{\tau_0}\right)^n = \left[\frac{G}{3}\left(\frac{b}{r'}\right)^{\frac{1}{2}}\frac{1}{\tau_0}\right]^n \tag{5.9}$$

τ_L is the concentrated shear stress at the crack tip and is a function of the distance r from the dislocation to the tip, as given‡ by Eq. 5.6.

Consider first the relaxation in Fe-3% Si, which occurs by the operation of dislocation sources near to the moving crack. The time t' that any one dislocation source will be under a particular stress τ

---

† Between the large group of etch pits along the crack profile.

‡ τ_L is also a function of the angle ψ, Fig. 2.3a, and decreases (at $r = $ const) when ψ becomes greater than $\pi/2$. For simplicity, these considerations have been omitted from this discussion.

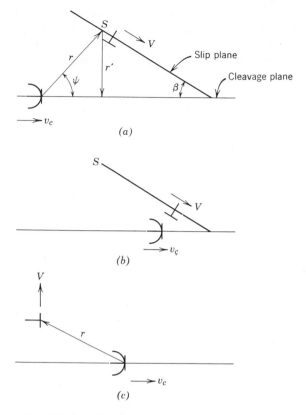

Fig. 5.9. (a) Coordinate system used in calculating interactions between slip dislocation and crack tip. S is a dislocation source. (b) Attraction of slip dislocation to crack tip. (c) Situation that occurs when crack tip sweeps by a dislocation that had been punched from it [16].

is about r'/v_c, where r' is the distance from the source S to the cleavage plane (Fig. 5.9a) and v_c is the crack velocity. In order for a source to produce significant amounts of plastic relaxation, it is necessary that the loops which it emits be able to move to the crack tip before the crack sweeps by the volume element [16]. Since the loops are confined to move on a slip plane that makes an angle β with the cleavage plane, this relaxation can only take place if the loops are moving at a velocity $V = v_c/(\cos \beta)$ at the time that the crack approaches the volume element. As the loops approach the crack tip

(Fig. 5.9b), r decreases, V increases (Eq. 5.9), and the loop will be able to reach the tip. If, on the other hand, $V < v_c/(\cos \beta)$ when the crack approaches, then the distance between dislocation and crack increases as the crack advances, V decreases, and the loop cannot get to the tip. Therefore, τ' and r' are defined by the condition that

$$V = \left[\frac{G}{3} \left(\frac{b}{r'} \right)^{\frac{1}{2}} \frac{1}{\tau_0} \right]^n = \frac{v_c}{\cos \beta}$$

$$t' = \frac{r'}{v_c} = \frac{G^2 b}{9\tau_0^2} (\cos \beta)^{2/n} \left(\frac{1}{v_c} \right)^{2/n+1} \tag{5.10}$$

In all crystals but FCC metals, τ_0 decreases as the temperature increases. *Consequently r' and t' increase with increasing temperature and decreasing crack velocity.*

Similar considerations apply for loops that are punched from the crack tip in "completely brittle" crystals such as LiF. It is obvious that in this case the relaxation will not be extensive because the distance r between the crack tip and the loop C' is continually increasing (Fig. 5.9c). Thus the *loops decelerate from the time that they are created;* if the crack is moving fast, they will only expand to a small final diameter by the time that they no longer feel the stresses of the crack tip. In this case the line tension of the loops will be able to overcome the friction stress of the lattice and the loops will shrink back to the fracture surface and "pop-out" of the crystal. Gilman [7] has shown that stable loops are only produced when the crack velocity is less than $10^{-3}v_0$, where v_0 is the velocity of sound.

Intrinsic variables that affect cleavage fracture resistance

The plastic work done per unit area of *continuous cleavage* crack propagation is a measure of the intrinsic toughness of a solid (i.e., its resistance to the passage of a moving crack). This term, γ_m, is equal to the product of the work done in a unit volume element of material when the crack advances and the distance r' perpendicular to the crack in which the deformation is extensive.

$$\gamma_m = r' \int_0^{t' = r'/v_c} \tau_L \left(\frac{d\epsilon}{dt} \right) dt \tag{5.11}$$

Because $d\epsilon/dt = NVb$, γ_m will depend on the temperature dependence of the dislocation velocity, the initial density of mobile sources N_0, and the multiplication process (Frank-Read source or multiple cross glide) which takes place.

When loops are punched out from the crack tip, γ_m is negligible because V, hence N, is always small. For the case of Fe-3% Si, where relaxation occurs by the operation of dislocation sources and the dislocation velocity varies with applied stress and temperature as $V \cong v_0 \exp (\text{const}/\tau_L{}^2 T)$, it has been shown that [16], by Frank-Read source multiplication (Fig. 4.12),

$$\frac{\gamma_m}{\gamma_s} = \frac{\rho}{b} = \text{const } N_0^{\frac{3}{2}} \left(\frac{v_0}{v_c}\right)^2 T^{\frac{5}{2}} \tag{5.12}$$

and, by multiple cross glide multiplication (Fig. 4.13),

$$\frac{\gamma_m}{\gamma_s} = \frac{\rho}{b} = \text{const } \frac{T^{\frac{1}{2}}}{\delta} N_0 \left[\exp \left\{30 \; \delta T b \frac{v_0}{v_c}\right\} - 1\right] \tag{5.13}$$

δ is the multiplication constant—the number of loops produced per cm of cross-slipping screw dislocation.

These relations indicate that large amounts of plastic blunting (high intrinsic cleavage resistance) will be obtained when the density of mobile dislocation sources N_0 and the temperature T are high and when the crack velocity v_c is low.

Because of the exponential term involved in cross-glide relaxation, greater amounts of relaxation can be achieved, at a given N_0, T, and v_c, by this process than when relaxation occurs by the activation of Frank-Read sources. However, multiple cross glide will only occur if the screw dislocations can cross slip, and this, in turn, requires a high temperature and a high stacking fault energy.† We should thus expect that the variation of γ_m with temperature would appear as shown in Fig. 5.10. Below T' the dislocation multiplication is essentially that of the Frank-Read type and γ_m increases slowly with temperature; this has been experimentally observed in iron silicon [20] and a variety of nonmetallic single crystals [21] containing relatively long and extremely sharp preinduced cracks. Above T', extensive cross slip takes place and γ_m, as given by Eq. 5.13, increases sharply with increasing temperature. Unless the crack is extremely long, continuous cleavage crack propagation is then no longer possible and the crack will stop. Therefore, T' corresponds to the crack-arrest temperature for *single crystals*.

In *polycrystalline materials* the situation is more complicated since, except at very low temperatures, cleavage crack propagation occurs discontinuously [3, 5, 22, 23, 24], by the formation of

---

† So that the screws are not extended into partial dislocations which confine the glide to a particular lattice plane.

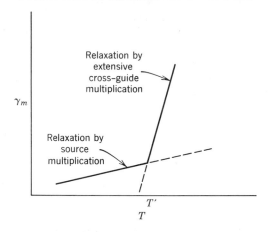

Fig. 5.10. Effect of temperature on the work done in propagation of cleavage crack in a single crystal.

microcleavage cracks one or two grain diameters in length, ahead of the main crack and the subsequent tearing of material between the main crack and the microcracks (Fig. 5.11). Because the formation of microcracks depends, in turn, on the value of γ_m (see Chapters 6 and 7), and some grains ahead of the advancing crack front are oriented more favorably for cleavage than others, the microcrack density and hence the density of cleaved grains decrease with increasing temperature. It appears that unstable propagation, by discontinuous cleavage, is no longer possible when less than 70% of the grains are cleaved [3], and this temperature corresponds to the crack-arrest temperature T_R in polycrystals [25]. The crack-arrest temperature for polycrystalline material is probably much less than that for single crystals T', but this point has never been investigated since 6 ft long, 1 in. thick single crystals are difficult to obtain.

As pointed out in Chapter 2, discontinuous cleavage propagation is actually a form of plastically induced fibrous fracture (hole formation and coalescence) which occurs when a critical plastic zone size is established at the crack tip. Since the plastic zone does not form instantaneously, but requires time for dislocation motion [26, 27], the crack velocity will be limited to some constant value approaching the speed of sound. This, in turn, implies a steady state value of the stress intensity factor $K = \{\sigma^2 \pi c\}^{1/2}$ according to Eq. 2.21, with γ_s replaced by γ_P for plastically induced cleavage. Recent experimental work [23] indicates that K does not increase with c when the length of the moving crack is greater than about 15 in., in mild steel. The reasons for this behavior are unclear. However, it is consistent with the lower limiting stress of 5–8 ksi that is required to cause fracture in mild steel (Fig. 3.1), regardless of the length of the large crack [23].

Equations 5.12 and 5.13 are useful for understanding why certain classes of materials are more susceptible to cleavage than others. In simple *ionic crystals* (LiF, MgO, NaCl) the dislocation mobility is usually quite high, and consequently the yield stress is low, at a particular value of T/T_M, compared with other classes of materials. However, most if not all of the grown-in dislocations in these crystals are pinned by impurity atoms [28] so that N_0 is very small; this is one of the main reasons for the low value of γ_m. In addition, large-scale dislocation cross slip is difficult, and extensive glide band broadness and wavy glide do not occur until $T = 0.6 - 0.7T_M$ (T_M is the absolute melting temperature). This is the temperature range where γ_m rises sharply [29].

Alloying is important in determining the probability of cross slip and hence the fracture behavior as well. Small additions of KBr

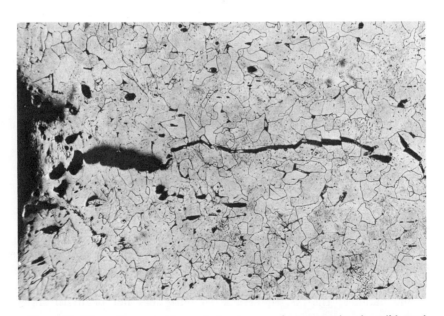

Fig. 5.11. Discontinuous nature of cleavage crack propagation in mild steel at a temperature near the NDT. 60×. *Courtesy J. F. Knott* [56] *and the Iron and Steel Institute.*

promote planar glide in KCl and raise the ductile-brittle transition temperature [30]. Removal of impurities (magnesium ions) lowers the ductile-brittle transition in LiF, and this effect is probably due to changes in slip mode as well as to increases in N_0. Similarly, *covalent crystals* (Ge, Si) also contain low density of mobile dislocations and these also fracture by cleavage at $T < 0.6T_M$. With few exceptions,† little relaxation occurs around stopped cracks in covalent and ionic crystals below $0.6T_M$. Consequently when reloaded at or below this homologous temperature, these specimens fracture by cleavage that initiates *directly* at the tip of stopped cracks, as shown in Fig. 2.12a.

BCC metals such as iron-base alloys and the refractory metals show a cleavage to tear transition at about $0.2T_M$, the temperature at which extensive relaxation by cross slip can take place. N_0 and the probability of cross slip are moderately high. Both can be increased by the removal of impurities. Consequently zone-refined iron is ductile [31] in a tension test at 4°K, whereas cleavage crack propagation can occur in structural steel at 273°K and in iron-3% silicon [32] at 400°K. Except in unusual circumstances the stopped cracks in these materials are sufficiently blunt that fracture cannot reinitiate directly, as in Fig. 2.12a, but does so by the formation of a new, moving cleavage crack ahead of the stopped one, as in Fig. 2.12b.

FCC metals do not fracture by cleavage. The reasons for this are: (1) dislocation pinning is so weak (or nonexistent) that N_0 is always very large and the deformation is homogeneous; (2) unlike the situation in other crystalline solids, the dislocation velocity in FCC metals is independent of temperature and is always high, except in the high strength alloys. Consequently r' will be large. Both effects cause γ_m to be extremely large at all temperatures.

5.2   Microscopic Aspects of Normal Rupture and Ductile Fracture

Stable and unstable crack propagation

At temperatures greater than the crack-arrest temperature, blunting and relaxation of the crack tip are so extensive that unstable cleavage fractures cannot develop. In *low strength materials* (e.g., mild steel and pure copper) fracture in a notched specimen then occurs in a stable manner; that is, the crack (notch) does not begin to spread

---

† AgCl is a notable exception, but the bonding in this compound may be more metallic than ionic.

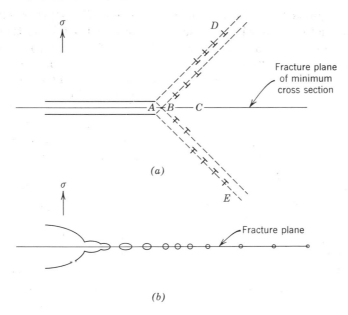

Fig. 5.12. Difference between (a) stable (noncumulative); (b) unstable (cumulative) types of normal rupture. *After A. H. Cottrell* [33].

rapidly when the stress reaches a critical value, as in cumulative fractures, but grows slowly, under an increasing stress that is usually (even in notched specimens) above the yield stress σ_Y.† Cottrell [33] has given a good explanation of this behavior. These ductile (stable) cracks grow by plastic deformation when dislocations are injected into slip lines (the plastic zones) ahead of the tip, as shown in Fig. 5.12a. When the applied stress is less than the general yield stress, the dislocations are localized near the tip and their back stresses opposes further plastic deformation at the tip. Thus the crack cannot grow appreciably until the general yield stress is reached and the plastic zones have spread completely across the specimen. This type of fracture is *noncumulative* because the plastic deformation required for one stage of growth (from A to B) does not help the crack in growing from B to C, and so on. New dislocations must be injected into yield zones at each stage of growth. Consequently the two halves

† This type of microscopic, slow crack growth should not be confused with macroscopic slow crack growth in sheet materials that results from relaxation due to plane stress conditions at the sheet surface (Chapter 3).

of the fracture surface will be heavily deformed. The stress required to cause crack extension is not a simple function of a critical tip displacement and crack length, as in cumulative fractures. In stable fractures, the propagation stress depends on the mechanical history of the specimen as a whole. While the crack is growing at a stress $\sigma \gtrless \sigma_Y$, the material ahead of it is continually strain hardening. Thus it is more difficult to spread the yield zones at later stages of growth than when the crack was initiated and *the crack grows under an increasing applied stress*.† These processes also are importantly involved in the growth of fatigue cracks under cyclic loading, to be discussed in Chapter 8.

On the other hand, low-energy tear fracture can occur in thick notched plates of *ultra-high strength materials* and also in materials of lower strength if a high density of hard particles has segregated at grain boundaries. These fractures are unstable and begin propagating when the applied stress has produced a critical plastic zone size ahead of the notch such that the critical tensile displacement $V^*(c)$ is produced at the notch tip. Physically the critical displacement is reached when a large number of voids have formed at the crack tip and the material between the voids has necked down (Fig. 5.12*b*). Because the deformation is confined to a particular path, such as an embrittled grain boundary, these fractures are *cumulative* and can formally be represented as a climbing group of edge dislocations (Fig. 5.1), as in the case of continuous cleavage crack propagation. The high void density leading to these unstable fractures will exist when a high density of brittle particles or brittle particle-matrix interfaces are present, and in materials of high yield strength where the stress level in the plastic zone will be sufficiently high to cause the particles or their interfaces to crack easily and open up into voids (see Chapters 6 and 7).

Since the fracture strength of the particles or their interfaces is essentially temperature-independent, and since the yield strength of the ultra-high strength materials is not strongly dependent on temperature, low-energy tear fractures occur essentially independent of temperature. Unlike discontinuous cleavage,‡ there is, for all practical purposes, no crack-arrest temperature above which they do not

---

† The applied *load* will probably be decreasing because these fractures usually occur after necking has begun.

‡ Strictly speaking, discontinuous cleavage is also a form of low-energy normal rupture, which propagates by the mechanism shown in Fig. 5.12*b*, but it is convenient to regard it as a form of cleavage fracture because of its strong temperature and strain-rate sensitivity.

occur.† However, in medium strength materials ($E/300 < \sigma_Y < E/150$) the yield strength does reach $E/150$ *at low temperature,* and low-energy tear can then occur. The ductile-brittle transition in these materials results from a transition in the energy required to operate the one mode (rupture), as well as from a change in fracture mode.

Ductility of unnotched tensile specimens

At temperatures where cleavage fracture does not occur, unnotched tensile specimens of commercial materials exhibit the well-known cup-cone appearance at fracture. Except in the ultra-high strength materials, this is a very ductile form of fracture, with final separation occurring after the cross-sectional area of the fracture surface has been reduced by more than 50%. Service failures are rarely a problem in ductile materials, unless the material has been overloaded beyond its ultimate strength and mechanical instability (necking) has taken place. However, the large deformations that occur during mechanical processing (e.g., rolling and deep drawing) can lead to ductile fracture, and for this reason the problem has more than academic interest [35].

Unlike the cleavage fracture which occurs at a fairly well-defined stress level (see Chapter 6), ductile fracture develops slowly, with final separation occurring at a much higher strain and nominal stress than that at which it began. Macroscopically the fracture begins as a normal rupture when a few voids are formed and coalesce in the necked region of a tensile specimen or overloaded member [1, 36–41, 57, 58]. On a microscopic scale voids form at strains much less than the necking strain in OFHC copper [37], and even at the onset of yielding [39] in an aluminum-silicon alloy containing coarse silicon particles, so that the question of when fracture initiates depends on the definition of the phrase "fracture initiation." Voids can form around grain boundary triple points [37], but usually they are nucleated when a nonmetallic inclusion [1, 36] (e.g., an oxide in copper, a sulfide in steel) or its interface is cracked (Fig. 5.13) and the

---

† In high strength materials, the yield stress is determined primarily by the stress required to bow dislocations between finely dispersed particles [34]. Since this process is temperature- (and strain-rate) independent, the yield strength decreases slowly with increasing temperature until high (T $\cong 0.5T_M$) temperatures are reached and particle coarsening occurs. At this point the alloys soften and the tear energy increases sharply. Since this temperature is usually so much greater than normal service temperature (e.g., it is about 500°C in maraging steel) there is, for all practical purposes, no crack-arrest temperature in these materials.

(a)

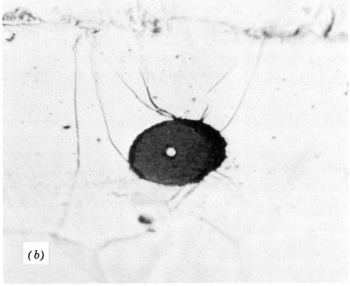

(b)

Fig. 5.13. Void formation associated with particles and inclusions. (a) Cracked silicon particle in Al-13% Si. 400×. (b) Cracked sulfide inclusion interface in Armco iron. 400×. *Courtesy J. Plateau* [39] *and American Society for Metals.*

crack opens up into a void. In two-phase materials (e.g., tempered steels and aluminum-silicon alloys) containing relatively large particles of second phase, the particles themselves or their interfaces can crack and lead to the formation of voids.

The actual process by which the voids coalesce to form the initial crack, while the material as a whole is deforming, is extremely complicated. The coalescence occurs by the elongation of the voids and coincident necking of material between them. It has been shown theoretically [40] that under constant transverse stress ratios $\sigma_{xx}/\bar{\sigma}$, $\sigma_{yy}/\bar{\sigma}$ the ductile fracture strain (in plane strain) is

$$\epsilon'_f = \frac{(1 - n) \ln (l_0/2b_0)}{\sinh \left[(1 - n)(\sigma_{xx} + \sigma_{yy})/2\bar{\sigma}/\sqrt{3}\right]} \tag{5.14}$$

for a material whose stress-strain curve is $\sigma = \sigma_0 \epsilon^n$. The initial radius of the (cylindrical) holes is b_0 and l_0 is the spacing between them. Fracture occurs when $\epsilon = \epsilon'_f$ over a region whose size is of the order of l_0. This relation indicates that the ductility decreases as: (1) the void density increases (i.e., as the particle density and yield strength increase (Chapter 7)), (2) the strain-hardening exponent n decreases, and (3) as the stress state changes from uniaxial to triaxial, as in the neck of a tensile specimen.

Consequently, while isolated voids are formed throughout the deformed specimen, initial crack formation occurs at the center of the specimen, across the plane of minimum cross section, where the tensile stress and hydrostatic components of stress are greatest (Fig. 5.14). At this stage the originally unnotched tensile specimen has become an internally and externally notched bar.

Except in the ultra-high strength materials, this crack extends slowly by the noncumulative process schematically illustrated in Fig. 5.12a. Plastic strain is concentrated in two thin "void sheets" [37] that make an angle of about 50°–60° to the fracture plane [42], and numerous voids form in these highly deformed regions (Fig. 5.15). The crack extends along one of the sheets (e.g., AD, Fig. 5.12a) by the necking of bridges between the cavities.† It cannot extend far, however, because it is moving out of the plane of minimum cross section. Plastic strain is then concentrated again (e.g., along AE, Fig. 5.12a) and the crack advances and moves back towards the plane of minimum cross section. Thus the normal rupture proceeds in a zigzag

---

† This leads to the observation at the high magnification obtained by electron microscopy of dimples on the fracture surfaces (see Figs. 3.17 and 3.18).

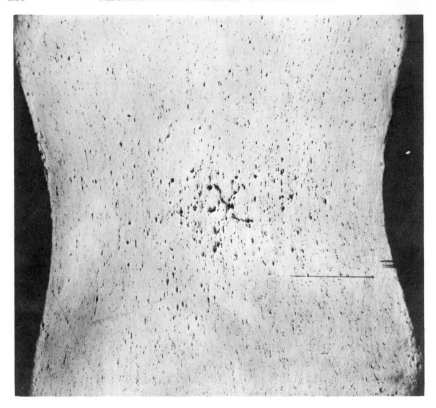

Fig. 5.14. Initial stages of normal rupture in the neck of a tensile specimen of copper. 8½×. *Courtesy J. Bluhm* [41].

fashion, but on the average it lies in the plane of minimum cross section. In relatively pure materials this process continues until the specimen as a whole has fractured with a double cup appearance [37]. Usually final fracture occurs by separation along a 45° plane of maximum shear strain concentration,† and the shear lip so formed gives the cup-cone appearance. The fracture stress is dependent on the strain in the necked region at the onset of final separation and has little relation to the stress level at which voids or fibrous cracks actually form.

It is apparent then that the most effective way to increase resistance to shear fracture (normal or shear rupture) is to decrease the tendency

---

† When the crack approaches the surface, the plane of maximum strain concentration changes from 60° to 45° to the tensile axis.

for void formation and to increase the strain-hardening rate to hinder internal necking between the voids [1]. In fact, if all particles are removed and no external notch is present, then the fibrous crack, which normally initiates by void coalescence, is not able to develop. Large plastic strains are developed and final separation produces a chisel point or slant type of appearance.

In this connection Bridgman has shown [43] that very large increases (more than an order of magnitude) in tensile ductility can be obtained in commercial materials if a large hydrostatic pressure is superimposed on the tensile stress during deformation. For example, a pressure, of the order of 300,000 to 400,000 psi, is able to prevent the formation of voids which otherwise would have formed in a 1045 steel, and change the fracture appearance from cup-cone to chisel point. Similarly, large increases in ductility can be obtained if a specimen is drawn through a die [37] rather than pulled in uniaxial tension, because of the compressive stresses set up by the drawing operation, or formed under high hydrostatic pressure.

In the absence of an external compressive stress the most significant way to increase ductility *is to remove the hard particles* (especially nonmetallic inclusions) that lead to void formation. The increase in ductility that can be obtained by removing second-phase

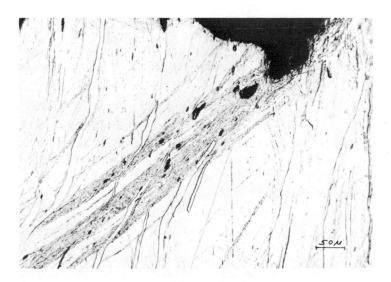

Fig. 5.15. Void sheets associated with stable crack propagation in copper. *Courtesy H. C. Rogers* [37] *and AIME.*

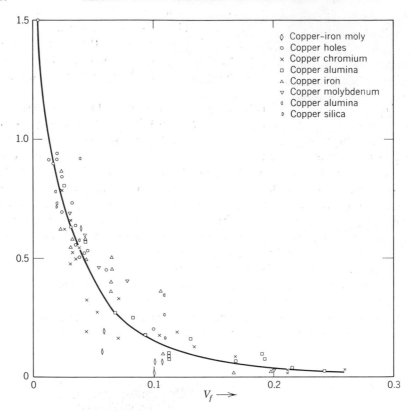

Fig 5.16. Tensile ductility of various low strength two-phase copper-base alloys as a function of volume fraction V_f of hard, second phase. *After B. Edelson and W. Baldwin* [38].

particles from copper-base alloys [38] is shown in Fig. 5.16, and similar results have been observed in other systems as well [39]. When the yield strength is maintained constant, but the particle content V_f is varied, the ductile fracture strain ϵ'_f is given by the empirical relation

$$\epsilon'_f = \frac{\lambda(1 - V_f)}{V_f} \tag{5.15}$$

where λ is a constant for each particular alloy system. Similarly ϵ'_f varies with particle shape and orientation. Lenticular-shaped particles, whose major axis is perpendicular to the stress axis, will open up in the form of sharp cracks which can concentrate strain at their tips more readily than particles whose major axis is parallel to the tensile axis. Consequently the ductility of rolled sheet tends to be *anisotropic* [44], with ϵ'_f least for specimens whose tensile axis is perpendicular to the rolling direction, since rolling tends to align the inclusions parallel to the rolling direction. If second-phase particles are randomly oriented then ϵ'_f is greatest, for the same volume fraction of particles, when the particles are in the shape of spheres rather than plates. Thus quenched and tempered steels are more ductile than ferritic-pearlitic steels at the same volume fraction of carbide [45].

For materials of given composition the tendency for void formation decreases with *decreasing strength level*, as mentioned earlier, so that ductility can be increased by a decrease in strength level. Thus quenched and double-tempered H-11 steel shows [46] 39% RA after tempering at 400°F (UTS = 313,000 psi), but 53% RA after tempering at 1200°F (UTS = 158,000 psi).

In many *high strength aluminum alloys* that are hardened by precipitation processes, fracture tends to be intergranular [47] as a result of the presence of precipitate particles segregated at grain boundaries and the soft, denuded zone adjacent to the boundary [61]

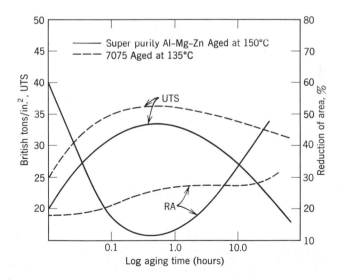

Fig. 5.17. Effect of aging on the UTS and % RA, of a super purity aluminum alloy and the commercial 7075 aluminum alloy [47].

which allows void coalescence to occur easily.   As shown in Fig. 5.17, the ductility of a high purity aluminum-magnesium-zinc alloy decreases as the alloy age-hardens up to its maximum strength but then starts to increase if overaging is carried out.   In the commercial 7075 alloy a fiber-type structure is produced during extrusion which is retained after solution treating.   When specimens are tested in a direction transverse to the extrusion direction, brittle intergranular fracture occurs which is similar to that in the high purity alloy described above.   However, when testing is carried out in the longitudinal direction, the low number of grain boundaries oriented perpendicular to the tensile axis and/or the smaller amount of precipitation in these boundaries allow the fracture to occur transgranularly.   In the solution-treated alloys, discontinuous yielding and a low rate of strain hardening cause the deformation to be concentrated in heavy, localized bands, inclined at 45° to the tensile axis, within which void formation, coalescence, and fracture take place.†   The increase in ductility obtained after aging to maximum strength (Fig. 5.17) results from the presence of a more homogeneous form of deformation (fine conjugate slip) in the hardened condition, which implies a *higher rate of strain hardening,* and makes void coalescence more difficult.   As in the case of the high purity alloys, further increases in ductility and a cup-cone appearance will occur if the alloy is overaged.

The temperature and strain-rate dependence of the yield and tensile strengths of aluminum alloys are relatively low.   Consequently, below about $0.5 \ T_M$ the ductility and toughness are relatively independent of temperature.   Figure 5.18 indicates, however, that the notch toughness of the high strength, 7178-T6 alloy does decrease between ambient and cryogenic temperatures.   From the tensile strength values shown on the figure, it appears that the decrease is associated (for this particular notch geometry) with tensile strengths above about 85 ksi.

5.3   The Dislocation Representation of Elastic-Plastic Antiplane Strain Fractures

In Chapters 2 and 3 it was shown that thin sheets of ductile materials are susceptible to unstable fracture on planes inclined at 45° to the tensile axis when long notches are present.   Similarly 45° fractures can develop in unnotched, high strength aluminum alloys, as shown

---

† Similar types of fracture are observed [59] in precipitation-strengthened single crystals where the fracture stress obeys a critical resolved shear stress law.

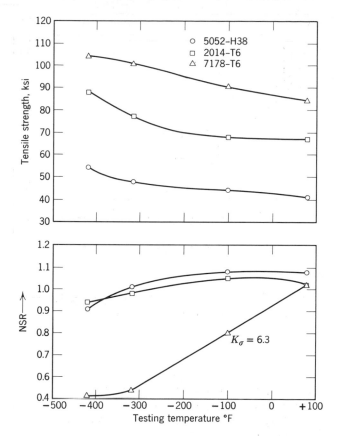

Fig. 5.18. Effect of temperature on the yield strength and toughness of some aluminum alloys [52].

above. To obtain a physical picture of these types of fracture it is convenient to consider first the case of the shear crack loaded in anti-plane strain (mode III deformation). Fracture occurs in the sheet shown in Fig. 5.19 when the two halves of the sheet slide over one another, in the z direction, while the crack is propagating in the x direction. The fracture displacements are similar to those which occur around a pure screw dislocation in a single crystal (Fig. 4.10a), and consequently they can be formally represented by a group of screw dislocations which glide across the fracture plane. Cottrell [48] has pointed out that each dislocation produces an increment of separation on every element of the fracture plane that it traverses. Consequently the initial group of dislocations, which was sufficient to

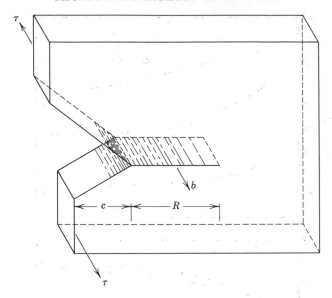

Fig. 5.19. Mode III type of crack propagation represented by a continuous distribution of screw dislocations
[48].

produce complete separation when the fracture initiates, is able to
cause complete separation of the entire sheet by merely gliding forward.
The fracture is *cumulative* and can spread under a decreasing stress
as the crack length increases, once a critical displacement $\phi^*(c)$ is
produced at the crack tip. Like the other cumulative fractures described previously, τ_F, the shear fracture stress, can be less than τ_Y,
the yield stress in pure shear, provided that $\phi^*(c)$ is small and the
initial crack length c is large. Bilby et al [49, 50, 51] and Smith
[52, 53] investigated the relation between $\phi^*(c)$, c, and τ_F, using the
continuum theory of dislocations. Suppose that a uniform shear stress
$\tau_{yz} = \tau$ is applied at infinity to an isotropic elastic solid containing
the distribution of dislocations shown in Fig. 5.20a. The region†
$x < c$ represents the shear crack and the region $a > c$ is the plastic
zone ahead of the crack. $R = a - c$ is again the radius of the plastic
zone. The dislocations do not literally "exist" inside the crack, but
they do exist in the plastic zone $a > x > c$.

† For simplicity we consider the behavior of one-half of the symmetrical crack
of length $2c$.

The resistance to the motion of dislocations is zero inside the crack and equal to the yield stress $\tau_Y > \tau$ outside the crack $(x > c)$. We define a quantity $\Delta\tau = \tau_Y - \tau$ which is the net stress that must be applied to move a dislocation over large distances outside of the crack. Suppose that there are $f(x)\,dx$ screw dislocations of Burgers vector b in any incremental distance dx along the fracture plane $y = 0$. These dislocations, like those in the pile ups described earlier, will assume, at equilibrium, a distribution such that the net stress tending to move any one of them along the fracture plane is equal to zero. The shear stress acting on a dislocation at position x, owing to another dislocation at x', is

$$\frac{Gb}{2\pi}\,\frac{f(x')\,dx}{(x - x')}$$

so that, at equilibrium,

$$\Delta\tau = \frac{Gb}{2\pi} \int_{-a}^{a} \frac{f(x')\,dx'}{(x - x')}$$

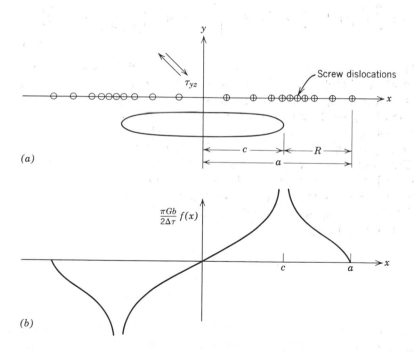

(a)

(b)

Fig. 5.20. Distribution of screw dislocations (a) and displacements (b) in the vicinity of a shear crack (mode III deformation [49].

If we use the boundary condition that $f(x') = 0$ at $x = \pm a$, this equation can be solved to give

$$\frac{c}{a} = \sin\left\{\frac{\pi}{2}\left[\frac{(\tau_Y - \tau)}{\tau_Y}\right]\right\} = \cos\left(\frac{\pi}{2}\frac{\tau}{\tau_Y}\right) \tag{5.16}$$

and for $\tau \ll 0.6\,\tau_Y$,

$$R = a - c = \frac{\pi^2}{8}\left(\frac{\tau}{\tau_Y}\right)^2 c$$

which is similar to relations 2.30 and 2.47 obtained by methods of applied mechanics for the tensile crack in plane stress and the shear crack. For $\tau > 0.6\,\tau_Y$, R can be approximated by the relation

$$R = \frac{\pi^2}{8}\left(\frac{\tau}{\tau_Y}\right)^2 c\left[1 + \frac{\pi^2}{24}\left(\frac{\tau}{\tau_Y}\right)^2\right] \tag{5.17}$$

In the region $-a < x < a$ the distribution function $f(x)$ is

$$f(x) = \frac{2\Delta\tau}{\pi Gb}\left\{\cosh^{-1}\left(\left|\frac{m}{c-x} + n\right|\right) - \cosh^{-1}\left(\left|\frac{m}{c+x} + n\right|\right)\right\} \tag{5.18}$$

where $m = (a^2 - c^2)/a$ and $n = c/a$. Figure 5.20b shows the variation of $(\pi Gbf(x)/2\Delta\tau)$ with x, for $a = 2c$. It is observed that most of the dislocations in the plastic zone are concentrated near the crack tip and that, as the distance from the tip $r = x - c$ increases, the displacements continually decrease. Since each dislocation represents a unit of strain, this behavior is similar to that predicted by Eq. 2.50.

The total number of dislocations in the region $x < a$, $N(a)$, may be obtained by integrating the distribution function over this region. The relative displacement at the crack tip $\phi(c)$ is then equal to $b\{N(a) - N(c)\}$, where $N(c)$ is the total displacement associated with the region inside the crack. This gives

$$\phi(c) = \frac{4c\tau_Y}{\pi G}\ln\left(\frac{a}{c}\right)$$

$$= \frac{4c\tau_Y}{\pi G}\ln\left[\sec\left(\frac{\pi}{2}\frac{\tau}{\tau_Y}\right)\right] \tag{5.19}$$

This relation is similar to those derived by the applied mechanics approach for the tensile crack under plane stress and the shear crack in antiplane strain (Eqs. 2.33 and 2.51).

Fracture occurs at $\tau = \tau_F$ when $R = R^*$ such that $\phi(c)$ attains the critical value $\phi^*(c)$. As shown previously, we can put Eq. 5.16

into 5.19 to obtain

$$\tau_F = \sqrt{\frac{2G}{\pi c}\tau_Y\phi^*(c)} = \sqrt{\frac{2G\gamma_p}{\pi c}} = \sqrt{\frac{GG_c}{\pi c}} \qquad (\tau_F < 0.6\tau_Y) \quad (5.20)$$

where

$$\gamma_p = \frac{G_c}{2} = \tau_Y\phi^*(c)$$

For $\tau > 0.6\tau_Y$ the yield zones spread further from the crack, at a given level of applied stress, and R is given by Eq. 5.17. This again gives a Griffith-type equation for τ_F but with c increased by a factor $[1 + (\pi^2/24)(\tau/\tau_Y)^2]$. This is similar to the Irwin correction for high stresses given in Eq. 2.29 and physically means that the "effective" length of the Griffith crack has been increased.

These results indicate that the dislocation model for the propagation of a shear crack is in excellent agreement with calculations obtained by the methods of applied mechanics. This agreement is also found for plane-strain shear fractures in plates of finite width [50].

At present, none of the models can handle the important problem of blunting of the crack tip. Physically we can picture this blunting to take place when the dislocations have spread on parallel planes ahead of the crack, instead of on one plane (Fig. 5.21). As a result of this

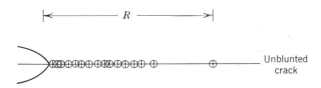

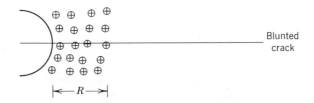

Fig. 5.21. Dislocation distribution associated with sharp and blunted shear cracks (mode III deformation).

lateral deformation the plastic zone radius R is decreased, for a given number of emitted dislocations that must be accommodated ahead of the crack, and fracture does not initiate unless the applied stress is raised considerably. Lateral deformation will be favored when the *strain-hardening rate is high,* since later slip processes will occur parallel to the initial ones in order to avoid the strain-hardened regions where the flow stress is high. A high strain-hardening rate can be achieved at large strains when the stacking fault energy is low,† so that resistance to shear fracture increases with decreases in stacking fault energy and in materials where large numbers of sources are present on parallel planes; these can relax the concentrated shear strains in the distribution, as described previously in connection with the cleavage crack. *Thus high toughness is achieved when plastic deformation is homogeneous on a microscopic as well as macroscopic scale.*

On the other hand, concentrated deformation and unstable shear fracture are favored when deformation is very inhomogeneous. This could be the case, for example, under high loading rates when adiabatic softening of material [54] ahead of the crack takes place during the spreading of the plastic zone, by the heat generated during plastic deformation. This is a particular problem in cold-worked

† So that recovery by cross slip does not occur.

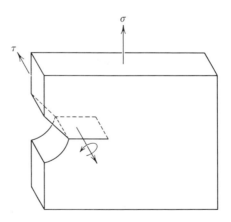

Fig. 5.22. Dislocation representation of crack that has both opening (mode I) and shear (mode III) displacements associated with it. *After A. H. Cottrell* [3].

alloys where small amounts of heating cause recovery and locally decrease τ_Y.

Finally, it should be noted that the plane-stress tensile fractures (45° slant fractures) spread under the combined modes I and III of deformation applied to the crack. The tensile component causes the formation and plastic expansion of the cavities while the shear component causes the fracture surfaces to slide apart. The crack front is then composed of a group of dislocations that are part edge, part screw [3], as shown in Fig. 5.22.

5.4 Summary

1. A plane-strain fracture can be represented as a climbing group of edge dislocations to facilitate an understanding of the microscopic aspects of fracture. Unstable fracture occurs by the motion of these dislocations across the fracture plane, as in cleavage or low-energy tear. Stable cracks (e.g., in low strength, FCC metals) grow by plastic deformation when dislocations are injected into plastic zones ahead of the tip.

2. The intersection of a cleavage crack tip with inhomogenities leads to the formation of cleavage steps which appear as river patterns. These inhomogeneities include grown-in screw dislocations, grain boundaries, screw dislocations that form when a crack decelerates, voids, and cracked particles. Steps are also formed when primary and secondary cracks join together. Step formation increases the work done in crack propagation and hence increases the intrinsic resistance to cleavage.

3. Plastic relaxation at the tip of a cleavage crack occurs by the operation of dislocation sources near the crack tip or by the punching of dislocations out of the crack. The former is much more efficient than the latter in producing relaxation, especially around moving cracks. The relaxation blunts the tip of the crack by neutralizing the crack dislocations at the leading edge and by distributing the displacements of the crack more homogeneously. Large amounts of relaxation produce a high intrinsic cleavage resistance.

4. Extensive plastic relaxation around moving cleavage cracks is facilitated when the crack velocity is low, when the mobile dislocation density is high, when cross-glide multiplication can occur easily, and when the dislocations have a high mobility. This implies that relaxation will be favored in materials of high stacking fault energy, when impurity pinning does not occur and when the test temperature is high. The intrinsically high cleavage resistance of FCC metals and low cleavage resistance of covalently bonded solids and polyvalent oxides can be explained on these grounds, as can the temperature dependence of the toughness of BCC metals.

5. Except at very low temperature, cleavage fracture in polycrystalline solids propagates discontinuously, by the formation of microcleavage cracks ahead of the macroscopic crack and the subsequent tearing of material between the main crack and the microcracks.

6. In the absence of cleavage, normal rupture occurs by the formation of voids near the tip of an advancing crack and by the coalescence of these voids owing to plastic strain concentration. The ductility and resistance to this mode of fracture increase as the void density decreases (i.e., as the distribution of hard particles which cause void formation becomes more homogeneous, as the particle content decreases, as the yield strength is lowered, and by the presence of hydrostatic compressive stresses) and as the strain-hardening rate increases.

7. The ductility of fabricated products tends to be anisotropic since working tends to align the nonmetallic inclusions which play a large role in ductile fracture. Heat treatment also affects ductility by affecting the size and distribution of precipitate particles at grain boundaries and the size of a denuded zone adjacent to the boundaries.

8. Antiplane-strain (mode III) fractures can be represented by a group of screw dislocations which glide across the fracture plane. This approach leads to a fracture criterion that is similar to that derived by the method of fracture mechanics (Chapter 2). The resistance to this type of fracture increases with increasing strain-hardening capacity (i.e., with decreasing stacking fault energy).

References

The asterisk indicates that published work is recommended for extensive or broad treatment.

*[1] A. H. Cottrell, *Fracture,* Wiley, New York (1959), p. 20.
*[2] J. Friedel, in *Fracture,* B. L. Averbach et al. eds., M.I.T., Wiley, New York (1959), p. 498.
*[3] A. H. Cottrell, *Proc. Roy. Soc.,* **A285,** 10 (1965).
 [4] J. J. Gilman, *Trans. AIME,* **203,** 1252 (1955).
*[5] J. R. Low, Jr., in *Fracture,* B. L. Averbach et al. eds., M.I.T., Wiley, New York (1959), p. 68.
 [6] J. J. Gilman, C. Knudsen and W. P. Walsh, *J. Appl. Phys.,* **29,** 601 (1958).
 [7] J. J. Gilman, *Trans. AIME,* **209,** 449 (1957).
 [8] T. L. Johnston, R. J. Stokes and C. H. Li, *Trans. AIME,* **221,** 792 (1961).

[9] C. T. Forwood and A. J. Forty, *Phil. Mag.* **11**, 1067 (1965).

[10] A. H. Cottrell, *Mechanical Properties of Metals,* Wiley, New York (1964), p. 349.

[11] J. M. Berry, *Trans. ASM,* **51**, 556 (1959).

[12] D. Barnett and A. S. Tetelman, to be published.

[13] D. DeSante and A. S. Tetelman, to be published.

[14] A. S. Tetelman and W. D. Robertson, *Trans. AIME,* **224**, 775 (1962).

[15] A. S. Tetelman and W. D. Robertson, *Acta Met.,* **11**, 415 (1963).

*[16] A. S. Tetelman, *Fracture of Solids,* Interscience, New York (1963), p. 461.

[17] C. J. McMahon, Jr., and M. Cohen, *Acta Met.,* **13**, 591 (1965).

[18] M. Baeyertz, W. F. Craig and E. S. Bumps, *Trans. AIME,* **185**, 481 (1949).

[19] J. J. Gilman, *Cranfield Symposium on Crack Propagation,* College of Aeronautics, Cranfield, Eng. (1962), p. 95.

[20] A. S. Tetelman and T. L. Johnston, *Phil. Mag.,* **11**, 389 (1965).

[21] J. J. Gilman, *J. Appl. Phys.,* **31**, 2208 (1960).

[22] C. F. Tipper, *J. Iron Steel Inst.,* **185**, 4 (1957).

[23] P. L. Pratt and T. A. C. Stock, *Proc. Roy. Soc.,* **A285**, 73 (1965).

[24] C. Crussard et al, *J. Iron Steel Inst.,* **183**, 146 (1956).

[25] J. Hodgson and G. M. Boyd, *Trans. Inst. Nav. Arch, London,* **100**, 141 (1958).

[26] A. H. Cottrell, *Trans. AIME,* **212**, 192 (1958).

[27] G. T. Hahn, A. Gilbert and C. N. Reid, *Battelle Inst. Rept. N. G.,* **18**, (1963).

[28] J. J. Gilman, *J. Appl. Phys.,* **30**, 1584 (1959).

[29] T. L. Johnston, C. H. Li and R. J. Stokes, *Strengthening Mechanisms in Solids,* ASM, Cleveland (1962), p. 341.

[30] N. S. Stoloff, D. K. Lezius and T. L. Johnston, *J. Appl. Phys.,* **34**, 3316 (1963).

[31] R. L. Smith and J. L. Rutherford, *Trans. AIME,* **209**, 357 (1957).

[32] M. Gell and W. D. Robertson, to be published.

[33] A. H. Cottrell, *Properties of Reactor Materials,* Butterworth, London (1962), p. 5.

*[34] A. McEvily et al, *Trans. ASM,* **56**, 753 (1963).

*[35] H. C. Rogers, *Fundamentals of Deformation Processing,* Syracuse Univ. Press (1964).

[36] K. E. Puttick, *Phil. Mag.,* **4**, 964 (1959).

[37] H. C. Rogers, *Trans. AIME,* **218**, 498 (1960).

[38] B. Edelson and W. Baldwin, *Trans. ASM,* **55**, 230 (1962).

[39] J. Gurland and J. Plateau, *Trans. ASM,* **56**, 442 (1963).

[40] F. McClintock, to be published.

[41] J. I. Bluhm and R. J. Morrissey, *Int. Conf. on Fracture,* Sendai, Japan **D-II,** 73 (1965).

[42] M. L. Williams, *J. Appl. Mech.,* **24**, 109 (1957).

[43] P. W. Bridgman, *Studies in Large Plastic Flow and Fracture,* McGraw-Hill, New York (1952).

[44] F. DeKazinczy and W. A. Backofen, *Trans. ASM,* **52**, 55 (1960).

[45] E. C. Bain and H. W. Paxton, *Alloying Elements in Steel,* 2nd ed., ASM, Cleveland (1962), p. 38.

[46] J. H. Bucher, G. W. Powell and J. W. Spretnak, *Trans. AIME,* **233**, 884 (1965).

[47] D. A. Ryder and A. C. Smale, *Fracture of Solids*, Interscience, New York (1963), p. 237.

[48] A. H. Cottrell, *Proc. Roy. Soc.*, **A276**, 1 (1963).

*[49] B. A. Bilby, A. H. Cottrell and K. H. Swinden, *Proc. Roy. Soc.*, **A272**, 304 (1963).

[50] B. A. Bilby et al, *Proc. Roy. Soc.*, **A279**, 1 (1965).

[51] B. A. Bilby and K. H. Swinden, *Proc. Roy. Soc.*, **A285**, 22 (1965).

[52] E. Smith, *Int. J. Eng. Sci.*, **2**, 379 (1964).

[53] E. Smith, *Proc. Roy. Soc.*, **A285**, 46 (1965).

[54] C. Zener and J. H. Holloman, *J. Appl. Phys.*, **15**, 22 (1944).

[55] C. D. Beachem and R. M. N. Pelloux, *Fracture Toughness Testing*, ASTM, Philadelphia, STP No. 381 (1965), p. 210.

[56] J. F. Knott and A. H. Cottrell, *J. Iron Steel Inst.*, **201**, 249 (1963).

[57] C. D. Beachem, *Trans. ASM*, **56**, 318 (1963).

[58] C. Crussard et al, in *Fracture*, B. L. Averbach et al. eds., M.I.T., Wiley, New York (1959), p. 524.

[59] C. J. Beevers and R. W. K. Honeycombe, *Fracture*, Wiley, New York (1959), p. 474.

[60] E. Orowan, *J. Appl. Phys.*, **26**, 900 (1955).

[61] G. Thomas, *Electron Microscopy and Strength of Crystals*, Wiley, New York (1963), p. 849.

[62] A. Hurlich and J. F. Watson, *Metal Progress*, **79**, 65 (1961).

Problems

1. By treating a cleavage crack which propagates in a [010] direction on a (001) plane in an ionic crystal as a giant dislocation, and using the Peach-Kohler relations for the forces between dislocations, show that the shear force exerted by the crack on an edge dislocation $b = a$ [0$\bar{1}$1] slipping on a (011) plane is zero directly in front of the crack.

2. Assuming the Friedel model for plastic relaxation along a cylinder of radius R is applicable, determine the length of crack required to produce a plastic zone of radius 0.01 cm when a stress $\sigma = 0.5\sigma_Y$ is applied to a cracked material. $\nu = 0.3$.   (0.0285 cm)

3. Suppose that a cleavage crack of length 0.01 cm propagates at all temperatures below 300°K, at a velocity $v_c = 10^{-2} \, v_0$, when $\sigma_F = 30$ ksi. What is the highest temperature at which a crack this length can propagate at this velocity if $\sigma_F = 20$ ksi. Dislocation multiplication occurs only by Frank Read source operation in this material.   (217°K)

4. The shear yield stress of a particular alloy is $\tau_Y = (G/44)\,T^{-\frac{1}{2}}$, where G is the shear modulus and T is the absolute temperature. Derive an approximate expression for the temperature dependence of the plastic work done per unit area of crack propagation, assuming it

to be equal to the elastic energy stored inside a region (greater than one Burgers vector from the tip), where the local stresses at the crack tip are greater than the yield stress. You may assume that the crack is at rest.

5. Suppose that the distance $a = R + c$ is fixed at 10 in. while the half-length c of a shear crack is varied. What is the value of R at which the maximum plastic displacement is produced at the crack tip and what is the value of this displacement when $\tau_Y = 10^{-3} G$?

$$(6.32 \text{ in.}, 4.7 \times 10^{-3} \text{ in.})$$

6

Cleavage Crack Nucleation and the Ductile-Brittle Transition in Low Strength, Flaw-Free Materials

In previous chapters it was shown that in "completely brittle" solids, where essentially no plastic relaxation occurs around moving or stopped cracks, preinduced flaws can initiate fracture by propagating *directly* as cleavage cracks when the local tensile stress at the flaw tip $2\sigma_F \sqrt{c/\rho}$ reaches the theoretical cohesive stress, about $E/10$. The work done in initiating fracture $\gamma_P{}^*$ is then the same as the work done while the crack is spreading, γ_m, and both of these parameters are about equal to the true surface energy γ_s. This type of behavior is observed when precracked, nonmetallic solids, particularly ionic and covalent crystals, are fractured at very low temperatures.

In metals the plastic relaxation at the tip of a stopped crack or flaw is usually more extensive, and the local stresses at the crack tip are then relaxed to about the value of σ_Y, the tensile yield stress. Because σ_Y is usually much less than $E/10$, *direct* initiation of fracture by the acceleration of the stopped crack is not possible; unstable fracture can only begin when a new dynamic (moving) crack is initiated in the deformed region at the tip of the stopped crack. In these cases the new crack is formed when a critical displacement $2V^*(c)$ is produced at the tip of the stopped crack. Since $V(c)$ is proportional to $\sigma^2 c$ (for $\sigma < 0.6\sigma_Y$), a Griffith-type expression again predicts the relation between fracture stress and crack length. The total work done around the moving crack, γ_P, is usually much less than the work done in fracture initiation, $\gamma_P{}^* = \sigma_Y V^*(c)$, and the crack, once formed, spreads in an extremely brittle manner.

Similarly in *flaw-free* (unnotched) materials tested in tension below

234

or at the brittleness-transition temperature T_D (Fig. 3.1), cleavage fracture occurs upon yielding and the measured fracture strain is essentially zero. The fracture stress of the flaw-free material, subsequently labeled σ_f†, is equal to σ_Y and is, again, much less than the theoretical value $E/10$.

In order to understand (1) why a critical displacement $2V^*(c)$ is required for fracture when flaws are present, (2) why the fracture stress of a material containing flaws σ_F is less than the fracture stress of flaw-free material σ_f, and (3) how cleavage cracks are formed at tensile stress levels σ_f of the order of $\sigma_Y \ll E/10$, it is necessary to focus our attention on the microscopic processes which lead to the nucleation and initial growth of a crack. These processes occur within one or two grains of the polycrystalline aggregate; once a cleavage crack has formed in a grain and spread through the boundaries which surround it, it will be able to continue spreading, at no increase in applied stress, because the stress required to keep a crack in motion σ_M decreases with increasing crack length. Thus in flaw-free material the measured cleavage fracture stress is that stress which is required for the nucleation of a crack and its initial growth through the grain in which it formed and through the surrounding grain boundaries.

Since crack nucleation and initial growth occur at applied stress levels $\sigma_f \ll E/10$, it is apparent that on a microscopic level some mechanism exists which can raise the applied stress to the theoretical strength. In the following section it will be shown that inhomogeneous plastic deformation, in the form of planar arrays of dislocations, provides the stress concentration effect for crack nucleation and initial crack growth. *Therefore, σ_f is the intrinsic fracture stress in a material where some dislocation motion and multiplication (plastic deformation) occurs before fracture.* The various dislocation reactions which produce fracture in specific materials and under particular environmental conditions will be described. The dislocation theory of inhomogeneous deformation will then be used to derive an expression for σ_f that depends on microstructural variables, particularly grain size.

Brittle fracture occurs upon yielding when $\sigma_f \lesssim \sigma_Y$. For if σ_Y were less than σ_f, the material would yield instead of fracture. Since both σ_f and σ_Y are strong functions of temperature, in all but FCC metals, there will be some temperature T_D above which $\sigma_f = \sigma_Y$. This

---

† To distinguish it from the fracture stress when flaws are present, σ_F; for reasons which will be discussed in the following chapter, $\sigma_F \neq \sigma_f$.

is by definition† the brittleness-transition temperature T_D (Fig. 3.1). Thus the relation between σ_f and σ_Y determines T_D and also, as shall be shown, the transition from cleavage to tear fracture (ductile-brittle transition) that occurs above T_D in flaw-free material.

6.1   The Nucleation of Cleavage Cracks by Plastic Deformation

General considerations

It is now established that plastic deformation, specifically, inhomogeneous arrangements of dislocations, is responsible for cleavage crack nucleation in materials that do not contain preinduced "elastic" cracks. The evidence for this statement is both indirect and direct. For example, at $T \leqq T_D$, yielding is a prerequisite for fracture and the fracture stress of flaw-free materials σ_f is never less than σ_Y. This is shown in Fig. 3.1 for steel. It is also observed [1] in NaCl-$CaCl_2$ alloys (Fig. 6.1) where the addition of $CaCl_2$ produces solid solution hardening and raises σ_Y and σ_f by the same amount. The

† Analogous to the definition of the NDT temperature, where $\sigma_F = \sigma_Y$ in a material containing a small (0.25 in.) flaw, and to the definition of the CAT temperature where $\sigma_F = $ UTS in a material containing a very large flaw.

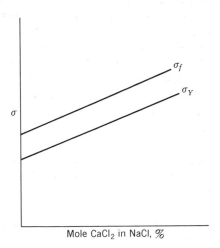

Fig. 6.1. Effect of additions of $CaCl_2$ on the yield strength σ_Y and fracture strength σ_f of NaCl [1] at room temperature.

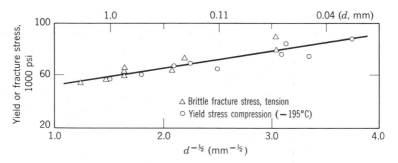

Fig. 6.2. Effect of grain size on yield stress (x) measured in compression and the fracture stress (0) measured in tension, of polycrystalline iron at $-196°C$. *After J. R. Low, Jr.* [2].

equivalence of σ_f and σ_Y has also been shown [2] in iron at 196°C where the yield stress, as measured in compression, varies with grain size in the same manner as the fracture stress measured in tension (Fig. 6.2). In both cases the variation is with $d^{-\frac{1}{2}}$, where $2d$ is the specimen grain size. One other indirect example is worth noting. In HCP single crystals the {0001} basal plane is both the primary slip plane and the cleavage plane. If ϕ is the angle between the pole of this plane and the tensile axis, and λ is the angle between the tensile axis and the slip direction, then the normal stress resolved perpendicular to the cleavage plane is $\sigma_N = \sigma \cos^2 \phi$, while the resolved shear stress for slip is $\tau_S = \sigma \cos \phi \cos \lambda$. Figure 6.3 shows that the fracture stress of zinc single crystals is strongly dependent on ϕ and that σ_f is highest [3] for those crystals which have the greatest amount of normal stress σ_N resolved across the cleavage plane ($\phi = 0°$). A small deviation from this orientation leads to a large decrease in σ_f. The reason for this behavior is that at $\phi = 0°$, λ is equal to $\pi/2$, so that $\tau_S = 0$. Since there is no shear stress component resolved for slip, plastic deformation cannot take place, and hence fracture cannot be nucleated. The high strength of *whiskers* (Table 2.2) results from the absence of dislocation sources, which inhibits plastic deformation (and hence fracture) until the stresses are high enough to cause homogeneous nucleation of dislocations.

Recently a large number of direct observations of crack nucleation by plastic deformation have been made in a variety of materials and we shall discuss some of these shortly. It is instructive to illustrate here one observation made in a magnesium oxide bicrystal. This is

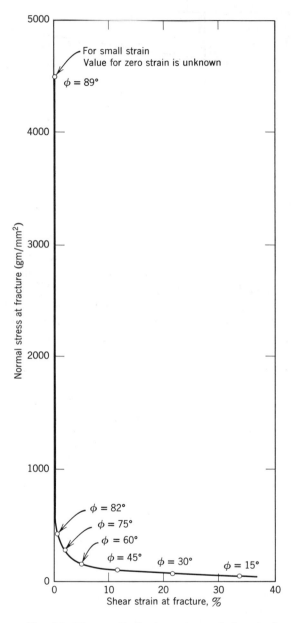

Fig. 6.3. The tensile fracture stress of zinc single crystals at −196°C as a function of the angle ϕ between the tensile axis and the pole of the basal plane. *After J. J. Gilman* [3].

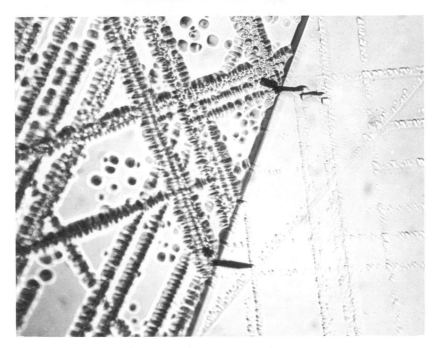

Fig. 6.4. Transgranular microcracks formed at a dislocation pile up at a grain boundary in MgO at room temperature. 1500×. *Courtesy R. Stokes* [74] *and Phil. Mag.*

shown in Fig. 6.4 where a microcrack has been formed at the tip of a blocked slip band, revealed by etching as a piled-up group of dislocations.†

The criterion for crack nucleation

Zener was the first [5] to point out why inhomogeneous plastic deformation could lead to crack formation. He proposed that a glide band containing many moving dislocations, emitted from a source S, Fig. 4.22a, could be treated as a viscous inclusion in an elastic matrix; since a viscous medium cannot support a shear stress, all of the stresses must be concentrated at its tip. As shown in Chapter 4, large tensile stresses as well as shear stresses will be set up at the tip of the glide band, if the band is blocked by a strong obstacle. A grain

† In polycrystalline materials a microcrack is any crack whose length is less than one or two grain diameters. In materials made from single crystals, microcracks are defined as those cracks which are too small to be observed by the naked eye.

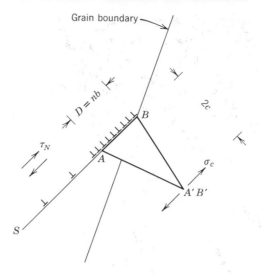

Fig. 6.5. Microcrack formation at the tip of a piled-up group of edge dislocations.

boundary or phase boundary is sufficient for this because, in general, slip planes in adjacent grains (or phases) are not coplanar, so that slip cannot propagate continuously across the boundary (Fig. 4.22).

Because edge dislocations are extra half planes of atoms that have been introduced into a crystal, a group of n adjacent (blocked) dislocations represent a wedge of height $D = nb$ (Fig. 6.5) which has split the neighboring planes AA' and BB' apart and formed a cleavage crack of length $2c$. A pile up of n edge dislocations produces a crack in exactly this manner [6], provided that its concentrated stresses are not relaxed by some of the plastic deformation processes described below. *Crack nucleation* occurs when the concentrated tensile stresses at the tip of blocked edge-dislocation band, approximately $(\tau - \tau_i)$ $(L/r)^{1/2}$ (from Eq. 4.12), are equal to the theoretical cohesive stress $\sigma_c = \sqrt{E\gamma_s/a_0} \approx E/10$ (from Eq. 2.4).† Thus

$$(\tau - \tau_i)\left(\frac{L}{r}\right)^{1/2} \geqq \sqrt{\frac{E\gamma_s}{a_0}} \tag{6.1}$$

† For the "ideal" planar array of edge dislocations shown in Fig. 4.22a. Pure screw dislocations do not have a tensile stress field and hence do not cause cleavage crack formation.

and crack nucleation occurs at a shear stress $\tau \approx \tau_N$

$$\tau_N = \tau_i + \sqrt{\frac{E}{L}\frac{r\gamma_s}{a_0}} \qquad (6.2)$$

where L is again the length of the blocked slip band and r is the distance from the tip of the band to the region where the crack is forming. If r is set equal to the lattice parameter a_0, the shortest unit distance in the crystal, and $E \approx 2G$, the nucleation criterion becomes

$$\tau_N = \tau_i + \sqrt{\frac{2G\gamma_s}{L}} \qquad (6.3)$$

Previously we have shown (Eq. 4.10b) that the length of the band and the "effective" shear stress $(\tau - \tau_i)$ determine the number of dislocations in the band; that is,

$$nb \cong L\frac{(\tau - \tau_i)}{G} \qquad (6.4)$$

Thus Eq. 6.3 may be written

$$(\tau_N - \tau_i)nb \cong 2\gamma_s \qquad (6.5)$$

The physical significance of this equation is that a crack will form when the work done by the applied shear stress in producing a displacement nb, $\tau_N nb$, equals the work done in moving the dislocations against the friction stress $\tau_i nb$, plus the work $2\gamma_s$ in making the new fracture surfaces.

Because these relations for crack formation do not contain the crack length $2c$, they will be satisfied for all lengths if they are satisfied for one length [6]. This means that the crack grows continuously,† *by plastic deformation*, as long as the source S continues to force dislocations into it, provided that $2c \ll L$. When the crack length becomes large, relative to the slip length, Eq. 4.12 is no longer applicable and the stress concentration effect of the slip band decreases. The crack may, however, be long enough at this point to propagate by the release of elastic energy alone (as a regular Griffith crack) under the action of the tensile component of the applied stress. This problem will be discussed separately. It should be emphasized here that the nucleation of a crack by plastic deformation results *only* from the shear stress τ_N which has forced the n dislocations together; *the tensile stresses play no role in the nucleation process.* In fact, cleavage

---

† Recent calculations by Smith [38] do in fact show that the energy of the system can decrease continuously even during sub-critical crack growth.

cracks can be formed in compression [7] (as shown in Fig. 6.6), when Eq. 6.5 is satisfied. These cracks will not be able to propagate unless the compressive stress is removed and a sufficient tensile stress is applied.

The validity of the form of Eq. 6.3 has recently been proved by Ku and Johnston [4]. They heat treated bicrystals of magnesium oxide in such a way as to pin all grown-in dislocation loops with impurities and then introduced fresh (mobile) loops at various distance L from the grain boundary by making indentations on the crystal surface. They found that the stress required for crack nucleation was indeed proportional to $L^{-\frac{1}{2}}$ and that the extrapolated value of τ_i was the same as that stress which is required to produce large-scale dislocation multiplication and glide band broadening (yielding) in single crystals of MgO.

Three points concerning the dislocation model for crack nucleation should be elaborated. First, the model is a real departure from the continuum approach to fracture, but in no way contradicts it. The continuum approach is not concerned with the microscopic *distribution*

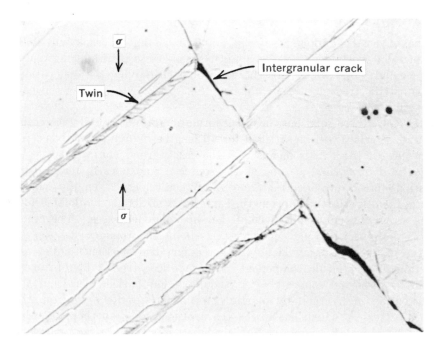

Fig. 6.6. Intergranular microcrack formed by twinning under compressive loading in Mo. 750×. *Courtesy A. Gilbert et al [7] and Acta Met.*

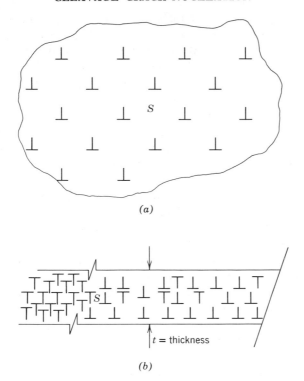

(a)

(b)

Fig. 6.7. (*a*) Homogeneous and (*b*) inhomogeneous distributions of dislocations emitted by a source *S*.

of plastic strains, but only with their *magnitude*. Since it assumes that plastic deformation (in flaw-free material) is completely homogeneous, it does not predict the existence of blocked arrays of dislocations and hence cannot predict that large *internal tensile stresses* can be produced on a *local level*, by an *applied shear stress* τ. The dislocation model (Fig. 6.5) requires that the plastic strain be inhomogeneously distributed and exist in the form of dense glide (or twin) bands. If the n edge dislocations emitted by the source had been homogeneously distributed throughout the crystal (Fig. 6.7a), the local stress would nowhere be sufficient for crack nucleation. The cohesive stress will only be reached† when the dislocations are closely packed in slip planes and the deformation is inhomogeneous.

† Over a distance greater than the lattice parameter (i.e., outside the dislocation core).

Second, the distribution of dislocations in the slip bands will influence the ease (probability) of crack nucleation, since the stress concentration factor at the tip of the band K_σ depends on the shear strain in the band. The isolated slip line used in the calculation presented above gives an oversimplified picture of the crack-nucleation process. In practice, isolated slip lines such as these, which form in the microstrain region, do not cause crack nucleation, as seen from the fact that τ_i is the stress required to produce *glide band broadening* rather than just *slip-line formation*. Thus thick glide bands, containing a *high density of dislocations*, appear to be required for crack nucleation [4].

The stress concentration factor of a band of finite thickness t (Fig. 6.7b) cannot be calculated explicitly because, apart from the difficulty of performing the calculation, the band undoubtedly contains dislocations of positive and negative sign so that all dislocations are not contributing to the concentration of stress. It is more reasonable, therefore, to treat the bands on a more macroscopic scale, as an inclusion or shear crack of length L within which the stresses have been reduced from τ_N to τ_i. Since the stress concentration factor for the relaxed inclusion will be proportional to $L^{-\frac{1}{2}}$, the experiments on MgO [4] provide excellent confirmation for the slip-band model of crack nucleation.

On a microscopic scale, K_σ depends on the shear strain concentrated in the band. For bands of fixed length L, K_σ will increase as the dislocation density in the band increases. In this connection it should be noted that increasing temperature and stacking fault energy facilitate dislocation cross slip and hence cause nb per plane, and therefore K_σ to decrease [8]. This is shown in Fig. 6.8 for LiF deformed at -196 and $25°C$. It is also indirectly observed in metals, in the form of increasing amounts of wavy (compared with planar) glide at high-deformation temperatures and in alloys of high stacking fault energy.

Finally the value of γ involved in forming the crack will only be equal to γ_s if the newly formed crack is completely brittle (i.e., if no relaxation occurs around the blocked glide band and crack nucleus before the crack spreads unstably). In nonmetallic materials it is conceivable that the work done in crack formation would be equal to γ_s, but in metals, where grain boundaries provide sources for dislocations (Fig. 4.22) which can relax the stresses in the pile up, this will not be the case. By analogy with the situation described for crack propagation in the preceding chapter, we should use γ_m rather than γ_s for the work done in crack nucleation. γ_m decreases when few sources

Fig. 6.8. Effect of temperature of deformation on the degree of homogeneity of distribution of slip dislocations in LiF. (a) −196°C. (b) 25°C. 375×. *Courtesy W. G. Johnston and J. J. Gilman* [9] *and J. App. Phys.*

exist around the incipient crack (heavy pinning), when the slip or twin band which concentrates the stress forms *rapidly* [10], so that relaxation does not have time to occur, and when the crack is formed at low temperature. Thus temperature plays a dual role in crack nucleation; it affects the value of the stress concentration in the band (nb per plane), and it affects the amount of relaxation around the band (γ_m). At low temperatures, plastic deformation is sufficiently inhomogeneous so that crack nucleation can take place; at high temperature the homogeneous nature of the deformation prevents crack formation. *So it is not plastic deformation per se, but rather the distribution of the deformation that determines the ease of crack formation; planar glide favors the initiation of cleavage, as compared to wavy glide.*

Crystallographic mechanisms of crack nucleation

Nucleation of Cracks by Slip

1. POLYCRYSTALLINE AND POLYPHASE MATERIALS. In single-phase *polycrystalline* materials, grain boundaries can serve as sufficiently strong obstacles to block the motion of a slip (or twin) band, and this produces the high tensile stress concentration required for crack nucleation. Cracks can then be formed, as shown in Fig. 6.4, if the tensile stresses are not relaxed by plastic deformation before the crack spreads unstably [6]. When hard nonmetallic phases are dispersed in grain boundaries of the matrix material, such as carbides in steels of low manganese content [11], the process of crack nucleation at the boundary takes place more easily. The reason is that the second-phase particles, being brittle,† do not permit much relaxation at the tip of the blocked slip band so that the work done in crack nucleation is small. If the cleavage planes in the matrix grain are well oriented with respect to those in the particle, there will be little or no misorientation energy expended when the crack crosses the particle-matrix boundary AA' (Fig. 6.9). When the cleavage planes between particle and matrix are badly misoriented, the misorientation energy expended in crossing the boundary makes crack nucleation in the matrix more difficult. Consequently not all second-phase particles are equally effective in causing crack nucleation, and the probability of nucleating a crack at a given stress level increases with increasing numbers of grain boundary particles.

The value of γ_m for any particular particle-matrix orientation depends on the size and shape of the particle. When the particle length

---

† Probably because they contain a low density of mobile sources and high friction stress for motion of glide dislocations.

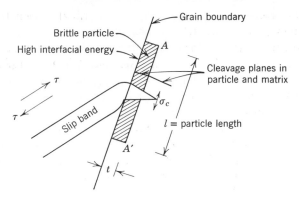

Fig. 6.9. Transgranular microcrack initiation associated with brittle second-phase particle.

l is short, dislocations in the pile-up may be able to activate sources above and below it and γ_m is not decreased. The particle will be able to prevent this relaxation and keep γ_m small only when it is longer than the thickness of the blocked slip band. A second relevant factor is the thickness of the particle t. When t is large (Fig. 6.9), a substantial part of the critical-sized crack† nucleates in the brittle second phase, whereas when t is small, most of the nucleation takes place in the softer matrix. Consequently γ_m increases as particle thickness decreases.

These considerations show why the *distribution of second phase* has such a large effect on cleavage nucleation. A given amount of phase, homogeneously distributed throughout the matrix in the form of fine particles, has no effect on γ_m because the particles are too short to prevent much relaxation around the blocked slip band. Furthermore, the slip distance L between particles will be so decreased that the stress-concentrating effect of the pile up is small. The particles raise τ_i but have no direct effect on γ_m. On the other hand, when the same amount of second phase is segregated at grain boundaries, then crack nucleation is facilitated, as described above. The amount of second phase required for crack nucleation by this process varies from one material to another, but need not be large. In ferrite [11] 0.1 volume per cent Fe_3C has a large effect; 0.01 wt per cent oxygen, segregated at grain boundaries in iron on a scale too fine to be observed by electron microscopy [12], embrittles the boundaries to such a degree that

---

† That can propagate by release of elastic energy alone, as a regular Griffith crack.

intergranular fracture is produced under conditions where crack formation cannot take place in the matrix (ferrite) grains.

This indicates that a *homogeneous microstructure*, as well as a homogeneous form of deformation, increases the resistance to crack nucleation. We shall see shortly that the presence of inhomogeneous distributions of grain boundary particles can have a large effect on the tensile and impact properties of various materials. We now define *dirty materials* as those which contain inhomogeneous distributions of particles† and impurity atoms (usually at grain boundaries), and *clean materials* as those in which the material is single phase only or, if it does contain a second phase, those in which the second phase is *homogeneously distributed*.

When larger volume fractions (2 to 10%) of second-phase particles are homogeneously distributed in the matrix, cleavage crack nucleation will be more difficult (than in the matrix material alone) if the particles are spherical and if the particle-matrix bond strength is high. The reason is that dislocation pile ups will crack the particle-matrix interfaces [13, 14] (Fig. 6.10a) and cavity formation rather than cleavage crack nucleation will take place. Fracture then occurs

† Nonmetallic inclusions are of course included in this definition. Note that even high purity materials may be regarded as *dirty* in this sense.

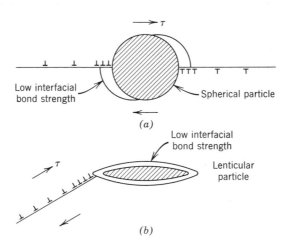

Fig. 6.10. (a) Cavity formation when slip band impinges on spherical particle, weakly bound to the matrix [13]. (b) Microcrack formation when slip band impinges on lenticular particle that is weakly bound to the matrix [14].

by the higher-energy tear mode and, as will be discussed in the following chapter, the impact-transition temperature can be lowered. This beneficial effect will only be observed if the particles are spherical [14]; for if the particles are lenticular, they will open up in the form of cleavage cracks rather than voids (Fig. 6.10b), and the ductility transition temperature T_D will be unaffected or increased. When large (greater than about 15%) amounts of hard phase are dispersed in the matrix, the particles will, in general, be so large that critical-sized Griffith cracks can be formed in them alone, or unstable fibrous cracks will be nucleated. These cracks can propagate through the material at temperatures both below and above the crack-arrest temperature of the matrix and the toughness and ductility of the material is lowered. In this connection it must again be emphasized that in *all* materials containing hard particles dispersed in a softer matrix (including FCC metals which do not fracture by cleavage), the tendency for void formation and low energy fibrous fracture increases with increasing particle content. This will be so, regardless of particle distribution, although the propagation of the fibrous crack requires less energy as the inhomogeneity of the distribution increases. Consequently the tensile and impact properties of polyphase materials are a function of the volume fraction of hard phase. As shown previously, the tensile (ductile) fracture strain is given by the relation

$$\epsilon'_f = \frac{\lambda(1 - V_f)}{V_f} \tag{6.6}$$

where λ is a constant and V_f is the volume fraction of hard phase.

2. SINGLE CRYSTALS. In single crystals, strong intrinsic barriers such as grain boundaries and phase boundaries do not exist† and hence crack nucleation cannot occur by the processes depicted in Fig. 6.5. The fracture strength of single crystals is, however, much less than the theoretical strength (except in whiskers). In fact, Fig. 6.2 shows that σ_f is lower in single crystals than in polycrystalline materials.‡ Consequently inhomogeneous deformation does occur and high tensile stresses are set up when the *deformation processes themselves* are able to provide strong barriers which can block further dislocation motion, on a local level. The specific process for doing this varies from one crystal structure to another.

† In iron single crystals, a few small included grains are often present. Cleavage is often nucleated at the boundary surrounding one of these grains [73].

‡ In some refractory metals this is not always the case, for reasons which will be discussed in Chapter 11.

In *hexagonal crystals* such as zinc and graphite the basal plane is both the primary slip plane and the cleavage plane. Crack nucleation occurs if a subboundary† wall terminates inside the crystal [13], because the stresses at the bottom of a long wall can be greater than the theoretical cohesive stress. The criterion for crack nucleation is [15]

$$\theta \ln \left[\left(\frac{H}{h} \right) \theta \right] > 1.2 \qquad (6.7)$$

where $\theta = b/h$ = angle of misorientation of the wall of length H, containing n dislocations separated by a height h. For $H \cong 1$ mm, $\theta > 5°$ is the required misorientation. Crack nucleation by this process has been observed [16] in a zinc bicrystal where $\theta = 8°$.

Such a wall can be formed [17] if part (BC, Fig. 6.11a) of a moving "infinite" wall of dislocations is held up by an obstacle, such as a symmetric, low-angle grain boundary. This obstacle need not be strong because a large number of dislocations are being blocked *simultaneously* along its length, so that there is no concentrated shear stress at any point which could break the obstacle and relax local tensile stresses. The critical step in the process of crack nucleation is the separation of the wall. Since the attractive force between AB' and BC will try to reunite the wall, a certain shear stress τ_N (greater than the friction stress τ_i) will be required. It has been shown [15, 17] that

$$\tau_N = \frac{Gb}{H} \exp \frac{1}{\theta} \qquad (6.8)$$

which is of the order of $10^{-3}G$ for typical values of H and θ. Since this stress is easily attained, the separation of the wall and the resulting nucleation of a crack should not be difficult processes.

Although this model can account for some, but not all [36], of the fractures which occur in crystals where the slip plane and cleavage plane are coincident, it is not applicable to the case of fractures in *BCC crystals* where slip occurs on {110} and {112} planes and fracture occurs on {100} planes. Suppose, following Cottrell [18], that edge dislocations having Burgers vectors $a/2$ [$\bar{1}11$] and $a/2$ [111] are gliding on orthogonal (101) and (101) planes and that they intersect

---

† A vertical array of parallel edge dislocations having the same sign. The wall has no long-range stress field, except at its ends, because the stress fields of dislocations separated a distance h in the wall cancel out at distances perpendicular to the wall that are greater than h.

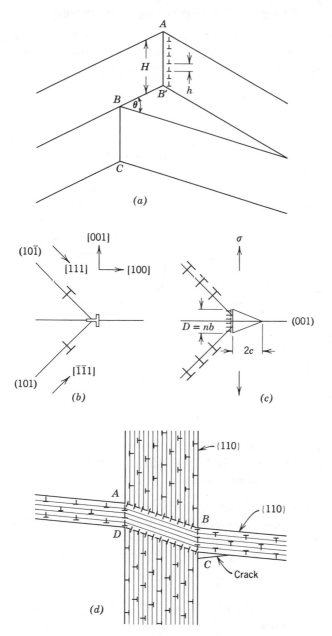

Fig. 6.11. (a) Dislocation mechanism responsible for microcrack nucleation in HCP single crystals [16, 17]. (b), (c) Dislocation mechanism of microcrack nucleation in BCC single crystals [18]. (d) Dislocation mechanism of microcrack nucleation in ionic single crystals [20].

251

along the direction of their lines [010] (Fig. 6.11b). These dislocations can lower the energy of the system (according to Frank's rule) by coalescing to form an $a[001]$ dislocation

$$\frac{a}{2}[\overline{1}\overline{1}1] + \frac{a}{2}[111] \rightarrow a[001] \tag{6.9}$$

The $a[001]$ dislocation lies in the (001) cleavage plane and therefore acts as a cleavage knife. A crack of height $D = nb$ is formed after n dislocations of each Burgers vector have reacted according to Eq. 6.9, Fig. 6.11c. Evidence for this mode of crack initiation was obtained [19] in iron silicon single crystals strained along a [110] direction at $-321°F$, where cracks formed on the (001) plane, which is parallel to direction of the applied stress for this orientation. Since there is no tensile component of the applied stress acting normal to this cleavage plane, the cracks must have been formed by shear stresses only, and subsequent analysis of the slip traces on the specimen surface showed that the operative slip planes were consistent with the Cottrell model.

In *ionic crystals,* crack nucleation occurs when two {110}, ⟨110⟩ glide systems intersect one another. Crack nucleation does not occur when the glide bands pile up against one another, as in the Cottrell model, but only when they *interpenetrate* [20]. The cracks (or slits) form on {110} planes, parallel to one set of glide planes before transferring over to the {100} planes on which large-scale propagation occurs. The strains set up when the glide bands interpenetrate are equivalent to four dislocation walls, one of which is able to nucleate a crack as described above [21]. Figure 6.11d is a schematic drawing of the nucleation process.

The Nucleation of Cleavage Cracks by Twinning

Under high loading rates or at low temperature, twinning is the preferred mode of deformation in BCC metals. The yield stress for twinning obeys a Petch relation of the form given by Eq. 4.14,

$$\sigma_Y = \sigma_i + 2k_y d^{-\frac{1}{2}} \tag{6.10}$$

where σ_i is the friction stress that opposes the growth of a twin in a crystal, k_y is a measure of the stress concentration factor at the tip of the twin,† and $2d$ is the grain diameter. As shown in Fig. 4.5, $d\sigma_Y/dT = d\sigma_i/dT$ is smaller in magnitude for twinning than for slip,

---

† Or the stress at which local yielding can occur at the tip of the blocked twin band.

so that σ_i (twinning) $<\sigma_i$ (slip) at low temperature. k_y (twinning) $> k_y$ (slip), and both of these are essentially independent of temperature at low $(T < 0.1T_M)$ temperature. Consequently σ_Y (twinning) $< \sigma_Y$ (slip) in coarse-grained materials (Fig. 6.12) and twinning is the preferred mode of yielding. Slip is favored in materials of finer grain size. Lowering the temperature from T_1 to T_2 increases the range of grain sizes in which yielding can occur by twinning.

Twinning usually takes place in materials containing few mobile dislocation sources. Consequently it is rarely observed after strains greater than a few per cent and slip is the primary deformation mode in deformed materials. Similarly small amounts of prestrain at higher temperatures prevent twinning from occurring at temperatures where it normally occurs in unstrained materials.

There is a large body of evidence [7, 18, 22–26], both direct and indirect, that twinning is an extremely effective deformation mode for nucleating cleavage cracks. In BCC metals twinning occurs on {112}

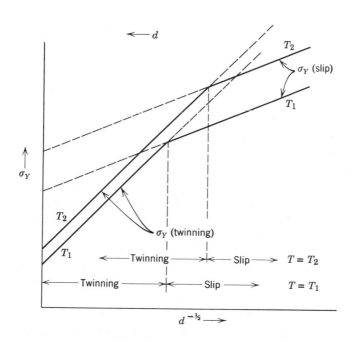

Fig. 6.12. Effect of grain size on the yield stress for slip and for twinning. Decreasing test temperature from T_1 to T_2 increases the range of grain size within which yielding occurs by twinning.

planes and in $\langle 111 \rangle$ directions [27]. The twinning shear is $1/\sqrt{2}$ which is the same as a displacement of $a/6 \langle 111 \rangle$ on every successive $\{112\}$ plane in the twin of thickness t'. Thus a twin can be represented as a moving sheet of dislocations having $nb = t'(1/\sqrt{2})$ as the shear displacement at its tip [25]. For a typical value of $t' = 10^{-4}$ cm, nb is of the order of 10^4 A, which is much greater than the average displacement at the tip of a slip band. Because twins spread more rapidly than slip bands, small amounts of relaxation should occur (at low temperatures) when a twin is blocked by a hard obstacle, such as a grain boundary or hard second-phase particle, and cracks will be nucleated (Fig. 6.6). These cracks can be either intergranular or transgranular, depending on whether the grain boundary is clean or has been embrittled by impurity atoms or a film of second phase. When thick particles lie in the grain boundaries, cleavage cracks are formed in the particle and these spread into the matrix grains, in the same manner as for slip-nucleated fractures. When the boundaries are clean, the cracks are formed directly in the matrix grains by the stresses set up at blocked twin bands.

Twinning is especially important for crack nucleation in single crystals of BCC metals. The orientation-dependence of the cleavage fracture stress of iron alloy single crystals is the same [25] as that required to obtain twinning in alloys tested above T_D where fracture does not occur upon yielding. Both the fracture and yield stress increase with increasing angle ϕ between the tensile axis and the pole of the $\{001\}$ cleavage plane. Decarburization or prestrain at ambient temperature in iron single crystals prevents twinning [22], and when these crystals are subsequently strained to fracture at low temperature, the fracture mode changes from cleavage to shear (chisel point).

There are direct observations that crack nucleation can occur at twin intersections [19, 23] (Fig. 6.13a) and along a single twin interface [28]. When sheet specimens of Fe-3% Si single crystals were strained at $-196°$C with the tensile axis chosen perpendicular to the (010) cleavage plane [23], the crystal orientation was such that only four of the twelve possible $\langle 111 \rangle \{112\}$ twin systems were operative; the other systems did not have sufficient stress resolved on them to satisfy a resolved shear stress criterion. These four twinning systems produced six possible types of twin intersections; two of the six possible intersections occurred along a [101] line which lay in the (010) cleavage plane. Cracks were nucleated only when these types of intersections occurred (Fig. 6.13b). The twin intersections able to nucleate cracks in these experiments are those which produce [29] a large concentration of normal strain across the favored (010) cleavage

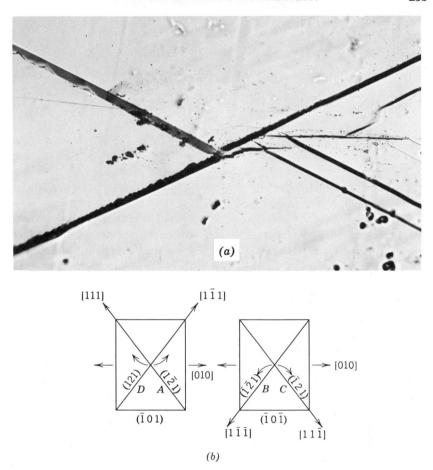

Fig. 6.13. (a) Transgranular cleavage crack formed by the intersection of twins in Fe-3% Si. 250×. *Courtesy D. Hull* [23] *and Acta Met.* (b) Shears associated with twin intersections.

plane (Fig. 6.13b). Another possible [19] reason why crack nuclea-tion is favored by the $(1\bar{2}1)$ [111], (121) [$1\bar{1}1$], and $(\bar{1}\bar{2}1)$ [$1\bar{1}\bar{1}$], $(\bar{1}21)$, [$11\bar{1}$] type intersections is that, of the six possible intersections, these are the only two in which the intersecting twins cannot penetrate one another because of the unfavorable reorientation of the twinning planes in the twinned material. These twins can thus act as effective barriers for one another and permit the formation of pile ups which lead to crack formation.

Plastically Induced Growth of Subcritical Size Cleavage Cracks

The Griffith criterion discussed in Chapter 2 shows that the smallest length of crack $2c$ that can propagate under a stress σ is given by the relation

$$c_{min} = \frac{2E\gamma_s}{\pi\sigma^2(1 - \nu^2)} \qquad (6.11)$$

where γ_m is the work done in propagation. The lowest value of γ_m occurs in completely elastic solids where $\gamma_m = \gamma_s$, so that

$$c_{min} = \frac{2E\gamma_s}{\pi\sigma^2(1 - \nu^2)} \qquad (6.12)$$

Ionic crystals are extremely weak when small microcracks are present on their surfaces. Fracture initiates at the tip of these preinduced cracks in NaCl [30] and MgO [31], when $\sigma = \sigma_Y$. This is not surprising, except for the fact that the initial crack lengths are observed to be an order of magnitude *smaller* than the value of c_{min} given by Eq. 6.12. The subcritical-sized cracks (with respect to the Griffith length) grow slowly, *by plastic deformation*, until their length is equal to c_{min}, at which point they can spread unstably *by the release of elastic energy* [32]. Figure 6.14 shows one way by which this might occur. Dislocations are created at the crack tip and spread into the body of the crystal. When double cross slip and dislocation multiplication occur (at D), large numbers of dislocations are formed which are attracted back toward the tip of the crack, and a large, inhomogeneous strain is produced there. The Griffith criterion is satisfied *locally*, within the highly strained region [33], and the crack jumps forward from A to B but it cannot advance further. Slip is again nucleated at the crack tip and the process repeats itself; the crack grows discontinuously in this manner until its length is equal to c_{min}, at which point it can advance by unstable elastic propagation.

Similar processes have been observed in single crystals [34] and polycrystals [35] of iron-3% silicon. When microcracks, one grain diameter in length, are introduced into polycrystalline iron-3% silicon by the cathodic charging of hydrogen† described in the preceding chapter, the cleavage fracture stress at $-196°C$, σ_f $(-196°C)$, is decreased.

It can be decreased still further if the cracked specimens are strained small amounts at ambient temperatures before they are fractured at $-196°C$ (Fig. 6.15). The microcracks behave as short notches and

† Followed by low temperature degassing of the specimens.

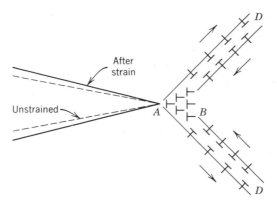

Fig. 6.14. Dislocation mechanism of plastically induced, slow growth of a cleavage crack.

the prestraining process concentrates plastic strain at their tips, which raises the stress level there. This allows the fracture criterion to be satisfied at lower nominal stresses and strains (see Section 6.3, p. 275) when the specimens are subsequently restrained at low temperatures.

When cracked specimens are prestrained at higher temperatures T_{ps} or greater amounts at ambient temperature, $\sigma_f(-196°C)$ increases (Fig. 6.15). The reason is that at higher temperature or after larger strains the plastic deformation at the crack tip is much more *homogeneous* so that the amount of crack-tip blunting increases.

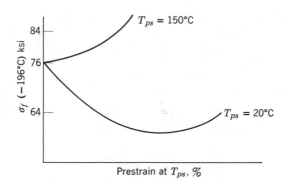

Fig. 6.15. Effect of prestrain at 20°C and 150°C on the cleavage fracture stress of premicro-cracked, polycrystalline iron-3% silicon broken at −196°C [35].

This blunting decreases the stress- and strain-concentrating effect of the crack and increases the low temperature strength and ductility. The example emphasizes the *fact that plastic deformation can assist or inhibit the cleavage fracture process, depending on the homogeneity of the plastic deformation; this choice in turn is strongly influenced by the temperature at which the deformation is performed.*

6.2   The Initial Growth and Propagation of Microcleavage Cracks

The critical step in fracture

The three important stages associated with the slip or twin nucleated cleavage fracture of a polycrystalline aggregate are illustrated in Fig. 6.16a.   They are:

1. The *nucleation* of a crack, by inhomogeneous plastic deformation, which occurs under the action of *shear stresses* that cause the dislocations to coalesce, forming a crack nucleus,

2. The *initial growth* of the cleavage crack, under *tensile stresses,* through the grain in which it was nucleated,

3. The traversing of the first strong barrier (e.g., a grain boundary) that the crack encounters.   This step also occurs under *tensile stress.*

Since these stages occur in series, *the measured fracture stress σ_f will be that nominal stress at which the most difficult of the three steps is overcome.*   If the yield stress σ_Y is less than the fracture stress, the material will be partially or completely ductile; if the yield stress is greater than the fracture stress, then the material will fracture rather than deform to any significant degree (i.e., $\epsilon_f \cong 0$).

In *clean materials,* stages 1 and 2 are more difficult than 3, for reasons which will be discussed shortly.   The critical step in fracture will be either nucleation or initial crack growth, and it is difficult to decide between 1 and 2 on purely theoretical grounds [6, 18, 37, 38].

Consider a polycrystalline specimen strained in uniaxial tension such that $\sigma \cong 2\tau$.   Suppose that the length L of the slip band which nucleates the crack is one-half the grain size $2d$.   According to Eq. 6.3, crack nucleation occurs at $\tau = \tau_N$ when

$$\tau_N = \tau_i + \sqrt{\frac{2G\gamma_m}{d}} \qquad (6.13)$$

where γ_m is written in place of γ_s to account for any relaxation which accompanies fracture.   Suppose that the nucleation step is the more difficult and that the fracture stress is therefore determined by the

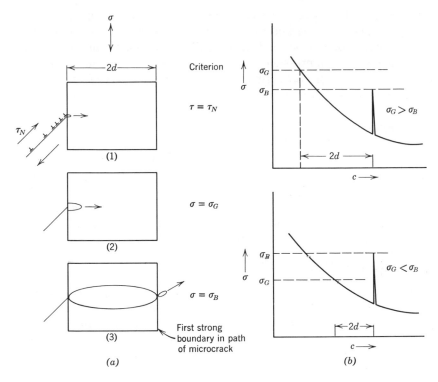

Fig. 6.16. (a) Three stages associated with the development of an unstable cleavage fracture in a polycrystalline metal. In single crystals only stages (1) and (2) are present. (b) Conditions under which nonpropagating microcracks are not formed (1) and are formed (2) in a polycrystalline metal.

nucleation stress. Then $\tau_N = \tau_f$. Fracture will only occur (instead of yielding) if τ_N is less than or equal to τ_Y, that is, if

$$\tau_N = \tau_i + \sqrt{\frac{2G\gamma_m}{d}} \leqq \tau_i + k'_y d^{-\frac{1}{2}} = \tau_Y$$

Thus the fracture criterion would be that

$$k'_y > \sqrt{2G\gamma_m} \tag{6.14}$$

Because the ratio $k'_y/\gamma_m^{\frac{1}{2}}$ increases as the temperature decreases, for reasons discussed previously, this relation might account for the fact that a brittleness-transition temperature T_D exists, below which cleavage is observed and above which yielding is observed. However, it cannot account for the facts that (1) grain-size refinement increases ductility [39, 47] (Fig. 6.19) since d does not appear in Eq. 6.14; (2) at $T > T_D$, cleavage can occur after strain hardening (τ_i is increased), since τ_i does not appear in Eq. 6.14; (3) nonpropagating cracks can form in simple compression [7] (Fig. 6.6) or parallel to the tensile axis in specially oriented single crystals [19]. If crack nucleation alone determines the fracture criterion, the first crack that forms should produce cleavage fracture, regardless of the sign or magnitude of the hydrostatic stress system.

These facts indicate that tensile stresses are required for cleavage fracture [18, 47], in order that the incipient crack nucleus be able to spread elastically as an unstable Griffith crack. Thus it seems more realistic to base a fracture criterion on the stresses required for initial crack growth rather than for crack nucleation. *If the yield stress is less than the growth stress, the material yields; if the yield stress is greater than the growth stress, the material fractures.*

The criterion for initial crack growth

Suppose, following Cottrell [18], that two slip bands of length L, oriented symmetrically at 45° to the applied tensile stress axis, intersect to produce a crack by the nucleation model† described by Fig. 6.11c. This crack can be treated as a giant dislocation of Burgers vector (crack height) nb, and length $2c$. Its energy W in the two-dimensional (plane-strain) case is

$$W = \frac{Gn^2b^2}{4(1 - \nu)} \ln\left(\frac{2R}{c}\right) + 4\gamma_m c - \frac{\pi(1 - \nu)\sigma^2c^2}{2G} - \sigma nbc \quad (6.15)$$

The first term is the self-energy of the giant dislocation, which extends over an effective radius R, the second is the effective surface energy of fracture $\gamma_m = \gamma_s\rho/a_0$, the third is the elastic energy of the crack in the applied stress field, and the last is the work done by the applied stress field owing to the increase in volume when the crack opens. By defining

$$c_1 = \frac{Gn^2b^2}{8\pi(1 - \nu)\gamma_m} \qquad c_2 = \frac{8G\gamma_m}{\pi(1 - \nu)\sigma^2} \qquad \left(\frac{c_1}{c_2}\right)^{\frac{1}{2}} = \frac{\sigma nb}{8G} \quad (6.16)$$

W can be written

$$W = 2\gamma_m \left[c_1 \ln\left(\frac{2R}{c}\right) + 2c - \frac{2c^2}{c_2} - 4\left(\frac{c_1}{c_2}\right)^{\frac{1}{2}} c\right] \quad (6.17)$$

† A similar calculation can be carried out for each of the various nucleation processes and the result (Eq. 6.19) is always about the same. This one is the easiest to visualize and will be used as an example here.

The equilibrium length of the crack, which is the largest crack length that can exist without spreading unstably, is defined by the condition $\partial W/\partial c = 0$ or by the roots of the equation

$$4c^2 - 2\left[1 - 2\left(\frac{c_1}{c_2}\right)^{\frac{1}{2}}\right]cc_2 + c_1c_2 = 0 \qquad (6.18)$$

When both dislocations and applied stress are present, this equation has either two positive roots, in which case the smaller gives the stable crack length, or no real roots, in which case the crack grows in an unstable manner. The transition between the two cases occurs when $(c_1/c_2)^{\frac{1}{2}} = 0.25$, or when

$$\sigma nb \geqq 2\gamma_m \qquad (6.19)$$

which is similar to Eq. 6.5 but with all of the applied stress $\sigma = 2\tau$ tending to spread the crack, not just an effective shear stress $(\tau - \tau_i)$. For L equal to one-half the grain size $2d$,

$$nb \cong (\tau - \tau_i)\frac{d}{G} \qquad (6.20)$$

so that initial crack growth occurs when

$$\sigma(\tau - \tau_i)d = 2G\gamma_m$$

Once yielding begins in one or two grains, $(\tau \cong \tau_Y)$. Thus $(\tau - \tau_i) \cong (\tau_Y - \tau_i) = k'_y d^{-\frac{1}{2}}$ and initial crack growth will occur when

$$\sigma = \sigma_G = \frac{2G\gamma_m}{k'_y}d^{-\frac{1}{2}} = \frac{4G\gamma_m}{k_y}d^{-\frac{1}{2}} \qquad (6.21)$$

using the value of k_y, as normally measured in a tension test, rather than k'_y (see Eqs. 4.14a and 4.14 b). Initial crack growth will therefore occur upon yielding when

$$\sigma_Y \geqq \sigma_G$$

or

$$\sigma_Y \geqq \sigma_G = \frac{4G\gamma_m}{k_y}d^{-\frac{1}{2}} \qquad (6.22)$$

or when

$$\tau_Y = \frac{\sigma_Y}{2} \geqq \frac{\sigma_G}{2} = \frac{2G\gamma_m}{k_y}d^{-\frac{1}{2}} \qquad (6.23)$$

$$(\tau_i + k_y d^{-\frac{1}{2}})k_y \geqq 4G\gamma_m$$

Since yielding is a prerequisite for crack nucleation, these relations *only apply when $\sigma_G \geqq \sigma_Y$. If $\sigma_G < \sigma_Y$ then, as initial crack growth cannot occur until yielding in one or two grains has taken place, the observed $\sigma_G = \sigma_Y$.*

The length of the critical crack which spreads unstably according to Eq. 6.22 is much less than the grain diameter $2d$. This is so because the tensile stresses at the tip of the piled-up group of dislocations are only of the order of the theoretical stress for distances $r = 2c \ll L = d$ [6]. Once the crack starts to grow as a Griffith crack, it will continue spreading through the grain in which it was nucleated until it arrives at a boundary of large misorientation (step 3, Fig. 6.16a), at which point the energy expended in crack propagation rises from γ_m to γ_B. The crack will not necessarily be blocked by the boundary, if it is long, because the stress γ_B required to keep the crack moving through the boundary,

$$\sigma_B = \sqrt{\frac{2E\gamma_B}{\pi c(1 - \nu^2)}} \tag{6.24}$$

can be less than the stress σ_G at which it began to grow initially (Fig. 6.16b). In general, the first grain boundary that the crack encounters will be the critical one (Fig. 6.17a) so that $2c = 2d$, the grain size. However, when twins are present in the grain in which the crack was formed, their boundaries may be strong enough to stop the crack [11] (Fig. 6.17b). Then $2c$ will be equal to $\alpha 2d$, where $\alpha \leqq 1$ is a parameter that measures the mean free path between boundaries that can block the crack. Thus, the "microscopic" Griffith relation for a *moving microcrack* becomes

$$\sigma_B = \sqrt{\frac{2E\gamma_B}{\pi \alpha d(1 - \nu^2)}} \tag{6.25}$$

When $\sigma_G > \sigma_B$ the first crack which forms is able to propagate until the specimen has fractured, so that $\sigma_f = \sigma_G$ and nonpropagating microcracks are not observed. When $\sigma_G < \sigma_B$ microcracks will form at $\sigma = \sigma_G$ but they will be stopped by the boundaries. The fracture stress σ_f is then the stress required to initiate fracture at the tip of a *stopped microcrack* and will be different from the value of σ_B given by Eq. 6.25, as explained in Section 6.3, p. 275. Since numerous microcracks can form before fracture is triggered from the tip of one or two of them (see below), nonpropagating microcracks are observed [2, 7, 11, 63] in fractured specimens. A comparison of Eqs. 6.21 and 6.25

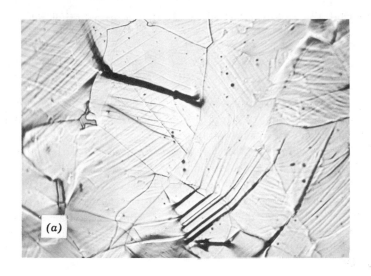

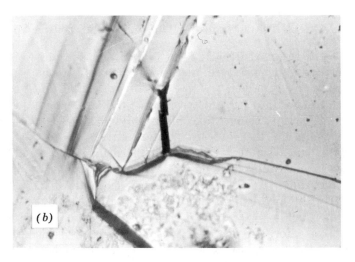

Fig. 6.17. (a) Microcracks stopped by grain boundaries in ferrite. 125×. *Courtesy G. Hahn et al* [63]. (b) Microcracks nucleated at carbide particle in ferrite, stopped by twin boundary in ferrite. 500×. *Courtesy C. McMahon and M. Cohen* [11] *and Acta Met.*

shows that the condition which leads to nonpropagating microcracks (i.e., $\sigma_B > \sigma_G$) can be written

$$\gamma_B > (\gamma_m{}^2\alpha) \left\{ \frac{2G\pi(1 - \nu)}{k_y{}^2} \right\} \tag{6.26}$$

When γ_m is small (dirty material), and if there is no preferred orientation (so that γ_B can be large), then Eq. 6.26 can be satisfied and nonpropagating cracks can be observed [2, 7, 11, 63]. Large amounts of twinning inside the grains reduce α and increase the microcrack density by providing good barriers to propagation [11], as shown in Fig. 6.17. Thus in a *dirty* material (Fe containing grain boundary carbides), where crack nucleation and initial crack growth are easy, twinning can improve ductility [42], in contrast to the situation in *clean* materials where crack nucleation is difficult and twinning, as the most effective deformation mode for forming cracks, reduces ductility.

6.3   The Ductile-Brittle Transition in Unnotched Tensile Specimens

Since σ_Y increases while γ_m/k_y and γ_B decrease (or remain constant) with decreasing temperature, there will be a brittleness-transition temperature T_D at which $\sigma_Y = \sigma_f$ and the material fractures upon yielding. Below T_D, σ_Y is greater than σ_f, but since crack nucleation, and hence fracture, cannot take place until yielding has occurred in at least one grain, the *observed* fracture stress $\sigma_f = \sigma_Y$. Above T_D, σ_Y is less than σ_f so that some strain hardening $\Delta\sigma_f$ is required to bring the stress level from σ_Y up to σ_f; the material deforms plastically before fracturing. This behavior is shown schematically in Fig. 6.18.

Similarly, at any temperature T there will be a transition grain size d^* at which $\sigma_Y = \sigma_f$. For $d > d^*$, fracture occurs upon yielding and $\sigma_f = \sigma_Y$. For $d < d^*$, $\sigma_Y < \sigma_f$ so that strain hardening $\Delta\sigma_f$ is required to bring the stress level from σ_Y up to σ_f before fracture. This behavior is shown schematically in Fig. 6.19.

In actual practice the experimental studies on the effect of temperature and grain size on fracture behavior is consistent [2, 11, 35, 47, 63] with the behavior described by Figs. 6.18 and 6.19. There are, however, certain aspects of the fracture behavior above the brittleness-transition temperature T_D that are significantly different for materials of varying microstructure. These aspects have a significant influence on macroscopic parameters such as G_{Ic} and the shape of the Charpy V notch curve, as will be shown in the following chapter, and it is

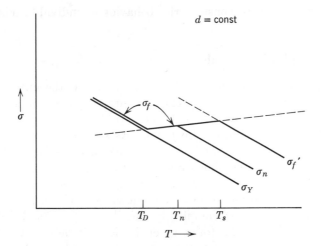

Fig. 6.18. Effect of temperature on the yield stress σ_Y, cleavage fracture stress σ_f, and necking (true stress at ultimate point) stress σ_n for a clean polycrystalline material, or a dirty one in which cleavage nucleates intergranularly. $\Delta\sigma_f$ is the amount of strain hardening that precedes cleavage, σ'_f is the true stress at which ductile fracture occurs and ϵ_f and ϵ'_f are, respectively, fracture strains when fracture is initiated by cleavage and by shear.

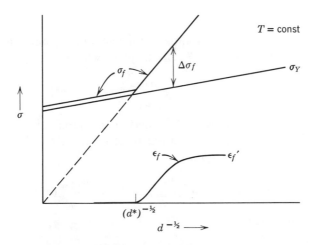

Fig. 6.19. The effect of grain-size variations at constant temperature on σ_Y, σ_f, $\Delta\sigma_f$, ϵ_f, and ϵ'_f for a clean polycrystalline material or a dirty one in which cleavage nucleates intergranularly.

necessary, therefore, to consider the behavior of individual classes of materials.

Clean, single-phase materials

In these materials the critical step for fracture is the initial growth of the crack so that $\sigma_f = \sigma_G$. The fracture stress of *single crystals* is given, with $\sigma = \sigma_f$, by Eq. 6.19 where (nb) is the displacement at the tip of a blocked twin or slip band. As shown above, twinning is an extremely effective means of producing critical-sized cracks in BCC metals and the twinning stress and fracture stress are coincident [25]. If the twinning is prevented (e.g., by prestrain) [22], the crystals of a given orientation ϕ neck down to a chisel point (100% RA) and the fracture is completely ductile [65]. The brittleness-transition temperature T_D for a given angle ϕ is determined by the highest temperature at which twinning can take place. When fracture is slip nucleated (e.g., in ionic single crystals), the brittleness transition is controlled by the distribution of dislocations in slip bands, the intersection of slip bands, and the condition of the specimen surface [43, 44]. These features will be considered in detail in Chapter 11.

In *polycrystalline materials*, nb can be directly related to k_y and grain size $2d$ so that the fracture stress is a function of grain size. Since $\sigma_f = \sigma_G$, the fracture stress is obtained from Eq. 6.21

$$\sigma_f = \frac{4G\gamma_m}{k_y} d^{-\frac{1}{2}} \tag{6.27}$$

At $T \leqq T_D$ yielding is a prerequisite for fracture so that the *observed* $\sigma_f = \sigma_Y$. By definition, $T = T_D$ at the point where

$$\sigma_Y = \sigma_f = \frac{4G\gamma_m}{k_y} (d^{-\frac{1}{2}}) \tag{6.28}$$

When $T > T_D$, σ_Y has decreased, σ_f may have increased (Fig. 6.18) so that strain hardening $\Delta\sigma_f$ is required to bring the stress level from σ_Y up to σ_f. Thus, when $T > T_D$,

$$(\sigma_Y + \Delta\sigma_f) = 2(\tau_y + \Delta\tau_f) = \sigma_f = \frac{4G\gamma_m}{k_y} (d^{-\frac{1}{2}}) \tag{6.29}$$

The fracture strain

$$\epsilon_f \cong \frac{\Delta\sigma_f}{d\sigma/d\epsilon} = \frac{\sigma_f - \sigma_Y}{d\sigma/d\epsilon} \tag{6.30}$$

assuming linear strain-hardening behavior for simplicity.

For $T_D < T < T_n$, ϵ_f is less than the strain ϵ_n at which necking begins, so that only uniform strains are achieved before fracture. For $T_n < T < T_S$, $\epsilon_f > \epsilon_n$ and necking precedes fracture, at the stress σ_n indicated in Fig. 6.18. The highest temperature at which cleavage cracks can be nucleated before fracture begins is T_S. Above T_S the ductile fracture *strain* ϵ'_f is reached before the cleavage fracture *stress* σ_f and the single-phase materials draw down to a chisel point. Since strain-hardening rate is essentially independent of temperature, ϵ'_f is also independent of temperature [18] and the ductile fracture stress σ'_f decreases with increasing $T > T_S$ in the same manner as σ_Y decreases.

All crystals *except* FCC metals show a ductile-brittle transition temperature region. In FCC metals k_y is always small because the deformation is homogeneous and γ_m is large for reasons discussed in Chapter 5. Since σ_Y is independent of temperature, $\sigma_Y \ll \sigma_f$ even at $0°K$. These metals always fracture by tear, except when embrittling chemical environments are present (Chapter 9).

The diagram shown in Fig. 6.18 is of great assistance for understanding why changes in certain variables affect T_D. Grain-size refinement raises σ_Y, but it raises σ_f even more† and the net effect (Fig. 6.20a) is to shift T_D to lower temperatures. On the other hand, if σ_Y is raised by increasing σ_i (by increasing the strain rate [45], by cold-working [46], by alloying to produce solid solution or precipitation hardening [47], or by neutron irradiation [48] in BCC metals) then, since σ_f is not directly affected,‡ T_D will be increased, as shown in Fig. 6.20b. *Thus grain-size refinement is the only means by which both the yield stress and the fracture stress can be increased, and the ductility transition can be lowered.*

The variation with temperature of the fracture appearance and fracture strain ϵ_f [plotted as the % RA = 1-exp $(-\epsilon_f)$] are shown in Fig. 6.21. The fracture strain of the *clean* single-phase material, as given by Eq. 6.30, rises sharply above T_n and behaves differently from that of other classes of materials, for reasons which are discussed below.

A common interpretation of Eq. 6.27 has been that the ductile-brittle (i.e., cleavage—tear) transition arises *only* because of the temperature dependence of σ_i and hence σ_Y. It now appears that this interpretation is too restrictive and that the effect of temperature on the mode of plastic deformation, and hence on (γ_m/k_y), must also be considered [8, 40, 49]. For example, iron and some BCC refractory

† As seen from the slopes of σ_f and σ_Y in Fig. 6.19.
‡ If these changes lower (γ_m/k_y), T_D can be increased still further.

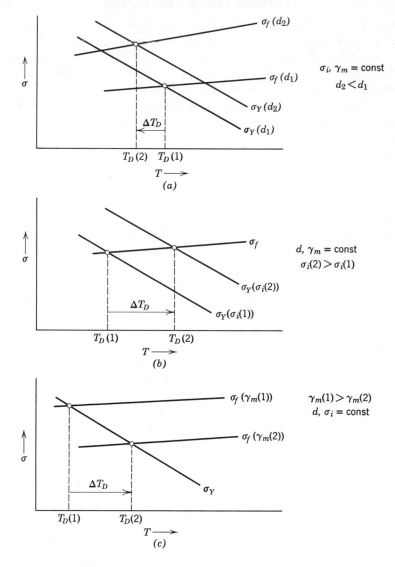

Fig. 6.20. Effect of changes in d, σ_i, and γ_m on σ_Y, σ_f, and the brittleness transition temperature T_D. (a) Grain-size refinement. (b) Increase in σ_i. (c) Lowering of γ_m.

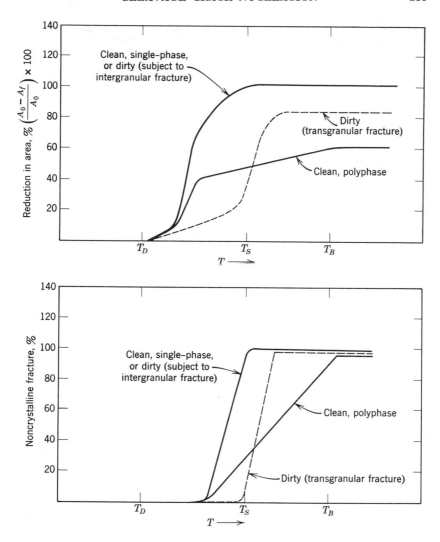

Fig. 6.21. Effect of test temperature on the tensile ductility and tensile fracture appearance of various classes of materials (d = const).

metals deform by twinning at low temperatures and small strains, but by slip at higher temperatures and at higher strains. Since k_y for twinning is greater than k_y for slip [50], the fracture stress increases, according to Eq. 6.27, when the temperature $T_f(t)$ and/or strain is too high for twinning to occur (Fig. 6.22) before the cleavage fracture

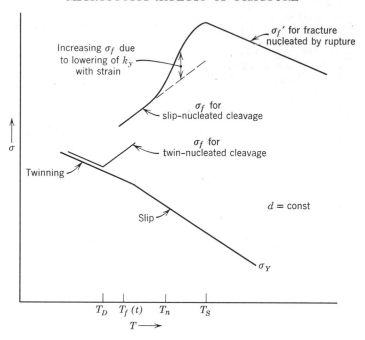

Fig. 6.22. Effect of a change in the deformation mode and degree of homogeneity of slip, caused by a change in temperature or plastic strain, on the cleavage fracture stress of clean materials, or dirty ones in which fracture initiates intergranularly.

stress is reached [51]. Since grain-size refinement also reduces the tendency for twinning (Fig. 6.12), further decreases in T_D, in addition to the effect described previously, can be realized if slip replaces twinning as the deformation process responsible for crack initiation.

Even when deformation occurs exclusively by slip, k_y will be a function of intrinsic variables such as stacking fault energy [8, 40] and, in solid solutions, the tendency for long- and short-range ordering of atoms. In Mg_3Cd, ordering changes the slip mode from planar to wavy and causes the ductile-brittle transition temperature to decrease [62]. However, in Fe-Co-V, ordering changes the slip mode from wavy to planar and k_y is doubled [40]. This raises T_D by 550°C, to the temperature where σ_Y (ordered) is about one-half the value of σ_Y for the disordered material at its T_D temperature, so that σ_Y still is equal to σ_f at T_D. k_y is also a function of temperature since it is proportional to nb per slip plane and nb decreases (Fig. 6.8) at temperatures where extensive cross slip can take place. Finally, k_y

is a function of plastic strain and decreases in metals of high stacking fault energy after initial yielding (Luder's extension) has been completed [52, 53, 72]. In some materials k_y decreases further with increasing strain since deformation is becoming increasingly homogeneous. This means that above T_D, where some plastic strain ϵ_f is required to bring σ_Y up to σ_f, σ_f *will itself be increasing with strain.*

In addition to producing a decrease in σ_Y and k_y, increasing temperature also increases $\gamma_m = \gamma_s(\rho/a_0)$, according to Eqs. 5.12 and 5.13. This effect will be especially strong if (1) the temperature is high enough to allow extensive dislocation cross slip, because the relaxation will then be more extensive (Fig. 5.10), and (2) after large strains since the density of mobile sources N_0 which cause relaxation is then increased. The increase in fracture stress with increasing temperature above T_D (Fig. 6.22) is partially the result of an increasing γ_m. Because it is difficult to separate the k_y and γ_m effects, it is most reasonable to attribute the total increase in σ_f to one of an increasing (γ_m/k_y) ratio (i.e., an increasing degree of homogeneity of deformation).

In addition to the *microscopic* requirement for ductile behavior that the right-hand sides of Eq. 6.27 be greater than the left-hand sides, there is also the *macroscopic* (von Mises) requirement that five independent slip systems be operative near a grain boundary to allow homogeneous deformation in polycrystals [54]. This requirement is always satisfied in FCC and BCC metals. In HCP metals (zinc, titanium, beryllium, magnesium) slip occurs only on the basal plane at low temperatures so that only two independent slip systems are available per grain.† These metals are therefore brittle until the temperature is raised sufficiently to allow the prismatic and pyramidal glide systems to be operative also [6]. Similarly ionic bicrystals and polycrystals are brittle at temperatures where glide is planar (only two independent slip systems are available), but are ductile at higher temperatures where cross slip allows the von Mises requirement to be satisfied [56, 57].

Clean, polyphase materials

At temperatures below T_S the behavior of clean, polyphase materials is qualitatively the same as that for clean single-phase materials described above. Quantitatively there will be differences because the second phase (1) causes dispersion hardening and an increased strain-hardening rate [58], (2) relaxes stress concentration by the formation

† As shown by Armstrong et al [55] this amounts to saying that k_y is large and hence σ_f is small.

of internal voids (Fig. 6.10a), or (3) refines the grain size by hindering grain growth during heat treatment [59]. These effects, once determined, can be explained in terms of the diagrams shown in Figs. 6.20a and 6.20b. Above T_S, single-phase materials draw down to a chisel point and the reduction in area at fracture approaches 100%. When hard second-phase particles are present, such as Fe_3C in iron, this phase, or the particle-matrix interface, may break and cause the formation of voids, which coalesce to form a fibrous crack. This crack propagates relatively slowly, but since the strain rate at its tip, and hence σ_i, have been raised, cleavage cracks can be formed ahead of it (i.e., Eq. 6.29 can eventually be satisfied). Because T_S is much below the crack-arrest temperature (Fig. 3.1), these cleavage cracks can propagate to cause a mixed fracture and the fracture strain is lower than in the case of the single-phase material where fibrous cracks are not formed. Since fracture begins as a fibrous crack, the nominal fracture stress is actually a ductile fracture stress σ'_f and it decreases with increasing temperature above T_S. Similarly the fracture appears as a mixture of shear and cleavage (Fig. 6.21) at all temperatures below $T < T_B$. At $T \geqq T_B$ cleavage cracks can no longer be formed ahead of the fibrous crack and the fracture appears to be 100% fibrous.† The reduction in area at fracture, however, is not equal to 100% because of the formation and coalescence of voids. If the volume fraction of hard-phase V_f is very large (Eq. 6.6), or if the hard phase is inhomogeneously distributed (e.g., along prior austenite grain boundaries in a high strength steel [60, 61], then ϵ'_f can be very small (less than 0.30) and the fibrous crack can propagate unstably to produce a low-energy tear fracture.

Intergranular fracture initiation in dirty materials

The segregation of interstitial solutes at grain boundaries can cause intergranular fracture, and transgranular fracture initiated by an intergranular crack, in BCC metals and polycrystalline ionic solids. In these cases the relationship between fracture stress and grain size (Eq. 6.27) does not always hold true because γ_m can itself be a function of grain size. Single crystals can be more ductile than polycrystals for the simple reason that they do not contain grain boundaries which can be embrittled. Similarly, if only a small amount of impurity is present, in the form of isolated inclusions which trigger off cleavage fracture‡ in the form of an intergranular crack, the

---

† T_B is the propagation-transition temperature for flaw-free materials.

‡ As in refractory metals, discussed in Chapter 11.

probability of having an embrittling inclusion in a grain boundary that is suitably oriented (i.e., perpendicular to the tensile stress axis) decreases as the grain size increases.† Consequently large-grained materials can sometimes be more ductile than fine-grained ones [41]. At temperatures above T_D some strain hardening would be required to bring the stress level from σ_Y up to σ_f. It has been shown [41] that small plastic prestrains at $T > T_D$ prevent intergranular fracture, probably because they relax stress and strain concentrations around the inclusions and introduce fresh dislocation loops. The fracture initiation mode then changes from intergranular to transgranular, σ_f increases, and the material behaves as though it were clean; the fracture stress, fracture strain, and fracture appearance vary with temperature in the same manner as for a clean, single-phase material, but with T_D and hence T_S increased, because γ_m and hence σ_f have been decreased (Fig. 6.20c).

If the grain boundaries are completely embrittled, such that the fracture path is intergranular, then probability considerations do not come into play. $\gamma_m = \gamma_{GB}(\rho/a_0)$ is lowered because the grain boundary energy γ_{GB} is lowered, owing to the presence of the impurity (e.g., oxygen in decarburized iron), and γ_m is independent of d. Refinement of grain size increases the fracture stress [2], and hence the ductility, by reducing the stress concentration set up [7], by plastic flow, at grain boundaries. The fracture behavior is again described by Eq. 6.27 and by Figs. 6.18 and 6.19, but with a low value of γ_m. Consequently T_S and T_D can be quite high.

Transgranular fracture in dirty material

Cleavage crack initiation occurs transgranularly when second-phase particles are strongly bound to the grain boundary. Since γ_m is decreased, compared to a clean material whose composition and microstructure is otherwise the same, σ_G will be decreased. Consequently initiation and initial growth of microcracks can occur upon yielding ($\sigma_Y \geqq \sigma_G$) at much higher temperatures than for the clean material (Fig. 6.23a). Since $\sigma_G < \sigma_B$ for dirty materials, the brittleness transition temperature is defined by the condition that

$$\sigma_f = \sigma_Y = \sigma_B$$

$$\sigma_f = \sigma_Y = \sqrt{\frac{2E\gamma_B}{\pi\alpha(1-\nu^2)d}} \qquad \text{at } T = T_D \qquad (6.31)$$

---

† This is the same as saying that γ_m increases as d^2, since the number of grains and hence the number of suitably oriented boundaries is proportional to $1/d^2$.

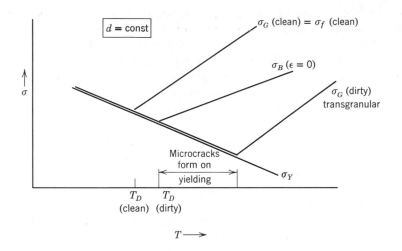

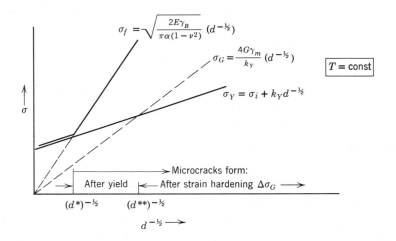

Fig. 6.23. Effect of a change in temperature (a) or change in grain size (b) on the initial growth stress σ_G and microcrack propagation stress σ_B (see text) for a dirty material in which cleavage initiates transgranularly. The temperature and grain-size ranges in which stopped microcracks form upon yielding are indicated.

As shown on the figure, T_D is increased, compared to the clean material, because $\sigma_B < \sigma_G$ (clean).

Similarly at any temperature T there exists a brittleness-transition grain size d^*. For $d > d^*$, $\sigma_Y > \sigma_B = \sigma_f$ and fracture occurs upon yielding (Fig. 6.23b). For $d^* > d > d^{**}$, $\sigma_B > \sigma_Y > \sigma_G$ so that nonpropagating microcracks form upon yielding. For $d < d^{**}$, $\sigma_Y < \sigma_G$ so that cracks are not formed upon yielding.

Since the value of γ_m, and hence that of σ_G, depends on particle size, shape, and orientation, there will be a spectrum of σ_G values at which initial growth of microcracks takes place. Consequently the number of microcracks increases with increasing plastic strain [11, 63] at $T > T_D$.

As shown in Fig. 6.24, the tensile behavior of dirty materials is quite different from that of the clean materials at $T > T_D$. The most important differences are (1) the large increase in the number of nonpropagating microcracks, (2) the *decrease* in fracture stress with increasing temperature between T_D and T^*, and (3) the low value of the fracture strain ϵ_f (Fig. 6.21) compared to the clean material [11, 63].

The decrease in fracture stress with increasing temperature probably results [66] from the fact that nonpropagating cracks, which form at yielding or after some strain hardening, behave as internal notches which concentrate strain at their tips, similar to the hydrogen-charging cracks described previously. This strain concentration and plastic stress concentration (see Chapter 7) cause sufficient local strain hardening, so that the local stress level is raised to the fracture stress of the matrix (i.e., that of the clean material). The nominal strain ϵ_f at which fracture occurs at $T > T_D$ is, approximately, that strain at which the true fracture strain of the matrix (i.e., for the clean material, at the temperature T) is reached over a critical distance r^* in the plastic zone ahead of the crack, (Fig. 6.25). r^* is determined by the condition that the Griffith criterion is satisfied throughout the stress gradient ahead of the stopped microcrack.

The ratio ϵ_f (clean)/ϵ_f (dirty), obtained from curves similar to those shown in Fig. 6.21, is then a measure of the strain concentration factor K_ε, evaluated at $r = r^*$. Consequently the measured fracture stress σ_f is the stress required to produce the nominal fracture strain ϵ_f at $r = r^*$. Since, to a first approximation, $\sigma_f \cong \sigma_Y + (d\sigma/d\epsilon)\ \epsilon_f$, the decrease in fracture stress with increasing temperature simply reflects the fact that σ_Y is decreasing faster with temperature than $(d\sigma/d\epsilon)\ \epsilon_f$ is increasing.

Because nominal plastic strain causes blunting of the microcrack tips as well as strain concentration, K_ε decreases at large nominal

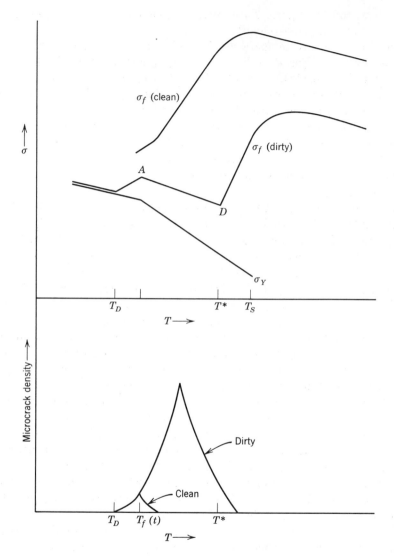

Fig. 6.24. The effect of temperature on the cleavage fracture stress σ_f for clean materials and dirty ones in which fracture initiates transgranularly. A comparison of microcrack densities observed on fractured specimens is shown on the lower part of the figure.

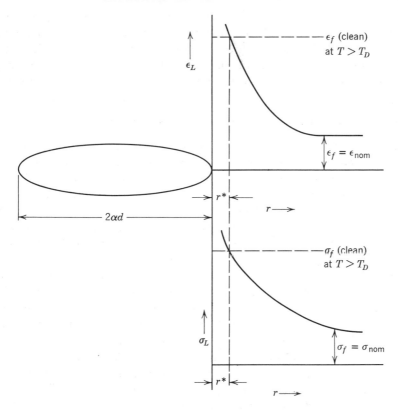

Fig. 6.25. Strain and strain-hardening concentrations in the vicinity of a stopped microcrack. Cleavage initiates on the opposite side of the boundary when the local strains ϵ_L, and hence the local stresses σ_L, reach the values of ϵ_f (clean) and σ_f (clean) over a critical distance r^* such that the new microcrack is sufficiently long to grow unstably.

strains and the nominal fracture strain ϵ_f begins increasing more rapidly as the temperature increases. Above T^*, microcrack formation at particles will itself become more difficult, as σ_G begins to increase sharply above σ_Y. Finally, at $T = T_S$, microcracks are no longer formed and the ductility of the dirty material approaches that of the clean material.

However, the second-phase particles themselves are still able to crack at temperatures much above T_S and these open up to form voids [11]. Even when the fracture surface appears completely noncrystalline, the reduction-in-area at fracture is only about 80% because of this void formation. If the amount of second phase at the boundaries is small, then the partially fibrous crack will not be able to transform over to a cleavage crack and there will be no propagation transition T_B. This temperature will only exist when sufficiently large amounts (greater than about 10%) of hard phase are present, regardless of distribution, so that a predominantly fibrous fracture can be produced (i.e., % RA. = 60% in the fully ductile condition).

The effect of alloying additions and stress state on ductility

It is apparent from the discussion presented above that the fracture stress depends on many variables, both intrinsic with respect to microstructure and extrinsic with respect to operating conditions. The effect of these variables on the parameters γ_m, k_y, γ_B, σ_i, and d must always be considered in order to predict the effect on fracture behavior of changes in temperature and chemical composition of a structure. Since the brittleness transition temperature is determined by the point at which the fracture stress *and* yield stress are coincident, it is also necessary to determine the effect of these variables on σ_Y as well as σ_f. Similarly, to understand how the other transition temperatures T^*, T_S, and T_B are affected by changes in certain variables, it is necessary to know how these variables influence the strain-hardening rate, since $\epsilon_f = (\sigma_f - \sigma_Y)/(d\sigma/d\epsilon)$. For example, manganese additions in iron lower T_D, and, more important, raise the fracture strain and fracture stress above $T_f(t)$ because they prevent segregation of carbides to ferrite grain boundaries [67, 68]. Thus, Mn makes a "clean" material out of ferrite that otherwise would be "dirty." The fracture stress and fracture strain are then shifted, as shown by Figs. 6.24 and 6.21, as γ_m and γ_B are now raised. In addition, manganese raises σ_i slightly [69], lowers d [70], probably by lowering the austenite-ferrite transformation temperature, and also lowers k_y to a certain degree [70]. The *net* effect of all these changes is to lower T_D and T_S as well as to raise σ_f and ϵ_f in the region $T_D - T_S$.

In Chapters 10, 11, and 12 we shall consider the behavior of important engineering materials on the bases of the variables that affect σ_f and σ_Y. It is important to realize that a change in one variable, such as temperature or chemical composition, can affect *more than one parameter* in the fracture and yield equations and consequently can produce large changes in the tensile ductility.

It must also be appreciated that the fracture criteria discussed in this chapter apply only to the case of *uniaxial tensile loading*. It was assumed here that the incipient crack lay perpendicular to the tensile axis, so that all of the applied stress $\sigma_{max} = 2\tau_{max}$ would be available for crack growth. Since plastic deformation depends only on shear stresses, whereas cleavage fracture depends on both shear and tensile stresses, it is also necessary to consider the *stress state* that is applied to a structure to determine whether flow or fracture will occur at a given nominal stress level. This can be done by defining a parameter $1/\beta$ as the ratio of the maximum tensile stress that exists when general yield† occurs under a particular type of loading as compared with the case of uniaxial tension [18]. Fracture will occur instead of plastic deformation when

$$\tau_f = \beta \frac{\sigma_f}{2} \leqq \tau_Y \qquad (6.32)$$

where τ_Y is the yield stress or flow stress if some strain hardening has occurred. For simple tension $1/\beta = 1$. For pure torsion $\sigma_{max} = \tau_{max}$ so $1/\beta = \frac{1}{2}$. Thus the yield stress τ_Y at which fracture occurs upon yielding can be doubled, as compared to uniaxial tension. Consequently the brittleness-transition temperature can be lowered, and the fracture strain can be increased [71], as compared to loading in uniaxial tension.

Alternatively, $1/\beta \approx 3$ near the root of a notch in a thick plate, since plastic constraint raises, by a factor of about 3, the tensile stress level that exists at general yield, as compared to an unnotched tensile specimen. This increases the tendency for cleavage fracture and causes T_D to rise. This effect will be discussed in more detail in the following chapter.

6.4 Summary

1. Inhomogeneous plastic deformation (piled-up groups of dislocations) is able to cause microcrack nucleation in crystalline solids. Since dislocations move under the action of shear stresses, the nucleation process results only from the presence of a shear stress. Tensile stresses play no role in the nucleation process. Crack nucleation cannot occur until the shear stresses are at least equal to τ_i, and usually equal to τ_Y, so that large-scale dislocation motion can occur.

2. The degree of inhomogeneity of the plastic deformation plays a significant role in determining whether cleavage crack initiation can occur. Low stacking fault energy, low test temperature, and a low density of mobile dislocations favor planar glide and the nucleation and initial growth of microcracks.

---

† Yielding all across the load-bearing area.

3. In polycrystalline solids, grain boundaries, hard particles and inclusions are able to act as barriers to slip or twin dislocations and allow crack initiation to occur. The mode of crack initiation and the plastic work done in the process depend on the distribution of impurity particles and the degree of wetting between particles and the matrix or grain boundary. Crack initiation is favored in dirty materials (where the second phases are inhomogeneously distributed) compared with clean materials in which second phases, if present, are homogeneously distributed.

4. In single crystals the strong barriers required for dislocation pile-up formation and crack initiation are not initially present. Instead they are formed by plastic deformation. Numerous dislocation models for crack initiation have been devised and all of them are probably responsible for fracture in one material or another. The general requirements for crack initiation are that an inhomogeneous displacement nb be produced normal to a cleavage plane (or on a plane that shares a common axis with a cleavage plane), and that this displacement be not relaxed by plastic deformation before unstable growth of the microcrack occurs. Twinning is a form of plastic deformation which can very easily lead to crack initiation. In ionic crystals, fracture usually begins at the tip of a surface flaw.

5. There are three important stages associated with the cleavage fracture of a polycrystalline aggregate: (1) the nucleation of a crack by shear stresses τ_N, (2) the initial growth of the crack under tensile stresses σ_G through the grain in which it was nucleated, and (3) the traversing of the first strong barrier that the crack encounters σ_B. The measured cleavage fracture stress σ_f is that stress at which the most difficult of these steps is overcome. Yielding will occur instead of fracture when $\sigma_Y < \sigma_f$. Typically σ_f is the initial growth stress (i.e., $\sigma_f = \sigma_G$), but in some dirty materials (e.g., mild steel) σ_f is the stress at which a new, unstable microcrack is initiated in the plastic zone near the tip of a stopped one.

6. The initial growth stress $\sigma_G = (4G\gamma_m/k_y) \; d^{-\frac{1}{2}}$; the tensile yield stress $\sigma_Y = \sigma_i + k_y d^{-\frac{1}{2}}$. The brittleness-transition temperature T_D is the highest temperature at which $\sigma_Y = \sigma_f = \sigma_G$ (for clean materials). Above T_D, $\sigma_Y < \sigma_f$ and some strain hardening $\Delta\sigma$ is required to raise the tensile stress level from σ_Y up to σ_f. For linear strain-hardening behavior, $\Delta\sigma = (d\sigma/d\epsilon) \; \epsilon_f$ so that $\epsilon_f = (\sigma_f - \sigma_Y)/(d\sigma/d\epsilon)$.

7. T_D is decreased and ϵ_f is increased as d, k_y, σ_i and the ratio of hydrostatic tensile to shear stress are decreased and as γ_m and γ_B are increased. These parameters are, in turn, affected by processing history, stacking fault energy, particle content and distribution, loading rate, and stress state. Brittle fracture is also favored when five independent slip systems are not operative, as in some HCP metals and ionic crystals at low temperatures.

8. In single-phase materials noncleavage fracture occurs by necking down to a chisel point. In two-phase materials, fracture initiates by slow fibrous tearing above a temperature T_s; this mode of fracture can, in turn, develop into cleavage below T_B. The tensile ductility decreases with increasing amounts of second-phase particles or inclusions.

References

The asterisk indicates that published work is recommended for extensive or broad treatment.

[1] A. Edner, Z. für Physik, 22, 286 (1924).
[2] J. R. Low, Jr., Relation of Properties to Microstructure, ASM, Cleveland (1953), p. 163.
[3] J. J. Gilman, Trans. AIME, 212, 783 (1958).
*[4] R. Ku and T. L. Johnston, Phil. Mag., 9, 231 (1964).
[5] C. Zener, Trans. ASM, A40, 3 (1948).
*[6] A. N. Stroh, Advances in Phys., 6, 418 (1957).
[7] A. Gilbert et al., Acta Met., 12, 754 (1964).
[8] A. J. McEvily, Jr., and T. L. Johnston, Int. Conf. on Fracture, Sendai, Japan, B-I, 23 (1965).
[9] W. G. Johnston and J. J. Gilman, J. Appl. Phys., 30, 129 (1962).
[10] C. J. McMahon, Jr., and R. Honda, private communication.
*[11] C. J. McMahon, Jr., and M. Cohen, Acta Met., 13, 591 (1965).
[12] J. R. Low, Jr., Fracture of Solids, Interscience, New York (1963), p. 197.
[13] E. Orowan, Dislocations in Metals, AIME (1954), p. 190.
[14] T. L. Johnston, R. J. Stokes and C. H. Li, Trans. AIME, 221, 792 (1961).
[15] J. Friedel, Les Dislocations, Gautier-Villars, Paris (1956).
[16] J. J. Gilman, Trans. AIME, 200, 621 (1954).
[17] A. N. Stroh, Phil. Mag., 3, 597 (1958).
*[18] A. H. Cottrell, Trans. AIME, 212, 192 (1958).
[19] R. Honda, J. Phys. Soc. Japan, 16, 1309 (1961).
[20] A. S. Argon and E. Orowan, Nature, 192, 447 (1961).
[21] A. S. Argon, Int. Conf. on Fracture, Sendai, Japan, DII, 195 (1965).
[22] W. D. Biggs and P. L. Pratt, Acta Met., 6, 694 (1958).
[23] D. Hull, Acta Met., 8, 11 (1960).
[24] A. W. Sleeswizk, Acta Met., 10, 803 (1962).
*[25] D. Hull, Fracture of Solids, Wiley, New York (1963), p. 417.
[26] A. S. Tetelman and T. L. Johnston, AIME, Conf. on Twinning, Orlando, (1962).
[27] C. S. Barrett, Structure of Metals, 2nd ed., McGraw-Hill, New York (1952).
[28] J. R. Low, Jr., in Fracture, B. L. Averbach et al. eds., M.I.T., Wiley, New York (1959), p. 68.
[29] J. R. Low, Jr., "Fracture of Metals," Prog. in Mat. Sci., 12, (1963).
[30] R. J. Stokes and C. H. Li, Fracture of Solids, Wiley, New York (1963), p. 289.

[31] H. G. Tattersall and F. J. P. Clarke, *Phil. Mag.,* **7**, 1977 (1962).

[32] E. Orowan, in *Fracture,* B. L. Averbach et al. eds., M.I.T.,¯Wiley, New York (1959), p. 147.

[33] E. E. Smith, to be published in *Acta Met.* (1966).

[34] A. S. Tetelman and T. L. Johnston, *Phil. Mag.,* **11**, 389 (1965).

[35] A. S. Tetelman, *Acta Met.,* **12**, 993 (1964).

[36] E. J. Stofel and D. S. Wood, *Fracture of Solids,* Interscience, New York (1963), p. 521.

*[37] A. H. Cottrell, in *Fracture,* B. L. Averbach et al. eds., M.I.T., Wiley, New York (1959), p. 20.

[38] E. E. Smith, to be published in *Acta Met.,* 1966.

[39] L. L. Seigle and C. D. Dickinson, *Ref. Met. and Alloys,* AIME (1962), p. 65.

*[40] T. L. Johnston, R. G. Davies and N. S. Stoloff, *Phil. Mag.,* **12**, 305 (1965).

[41] A. Gilbert, C. N. Reid and G. T. Hahn, *J. Inst. Met.,* **92**, 351 (1964).

[42] B. Rosof, M.S. thesis, M.I.T. Cambridge, Mass. (1964).

[43] T. L. Johnston, C. H. Li and R. J. Stokes, *Strengthening Mechanisms in Solids,* ASM, Cleveland (1962), p. 341.

[44] T. L. Johnston and E. R. Parker, *Fracture of Solids,* Interscience, New York (1963), p. 267.

[45] T. R. Wilshaw and P. L. Pratt, *Int. Conf. on Fracture,* Sendai, Japan, **B-III**, 3 (1965).

[46] W. S. Owen, in *Fracture,* B. L. Averbach et al. eds., M.I.T., Wiley, New York (1959), p. 141.

*[47] N. J. Petch, in *Fracture,* B. L. Averbach et al. eds., M.I.T., Wiley, New York (1959), p. 54.

[48] I. Mogford and D. Hull, *J. Iron Steel Inst.,* **201**, 55 (1963).

[49] E. E. Smith and P. J. Worthington, *Int. Conf. on Fracture,* Sendai, Japan, **A**, 163 (1965).

[50] D. Hull, *Acta Met.,* **9**, 191 (1961).

[51] A. S. Tetelman, *Acta Met.,* **12**, 324 (1964).

[52] J. D. Meakin and N. J. Petch, *Symposium on Substructure,* ASD-TDR-63-324 (1963), p. 243.

[53] B. Lement, to be published.

[54] R. von Mises, *Z. Angew. Math. Meth.,* **8**, 161 (1928).

[55] R. Armstrong et al, *Phil. Mag.,* **7**, 45 (1962).

[56] G. Groves and A. Kelly, *Phil. Mag.,* **8**, 877 (1963).

[57] N. S. Stoloff, D. K. Lezius and T. L. Johnston, *J. Appl. Phys.,* **34**, 3316 (1963).

[58] A. Kelly, *Proc. Roy. Soc.,* **A282**, 63 (1964).

[59] J. L. Ratliff et al, *Trans. AIME,* **230**, 490 (1964).

[60] S. Floreen and G. R. Speich, *Trans. Quart. ASM,* **57**, 714 (1964).

[61] N. P. Allen, C. C. Earley and J. H. Rendall, *Proc. Roy. Soc.,* **285A**, 120 (1965).

[62] R. G. Davies and N. S. Stoloff, *Trans. AIME,* **230**, 390 (1964).

*[63] G. T. Hahn et al, in *Fracture,* B. L. Averbach et al. eds., M.I.T., Wiley, New York (1959), p. 91.

[64] M. Baeyertz, W. F. Craig and E. S. Bumps, *Trans. AIME,* **185**, 481 (1949).

[65] N. P. Allen, B. E. Hopkins and J. E. MacLennon, *Proc. Roy. Soc.,* **A234**, 221 (1956).

[66] A. S. Tetelman, to be published.

[67] N. P. Allen et al, *J. Iron Steel Inst.*, **174**, 108 (1953).

[68] G. T. Hahn, M. Cohen and B. L. Averbach, *J. Iron Steel Inst.*, **200**, 634 (1962).

[69] F. B. Pickering and T. Gladman, *Special Report 81*, Iron and Steel Inst. (1963), p. 9.

[70] J. Heslop and N. J. Petch, *Phil. Mag.*, **2**, 649 (1957).

[71] T. Yokobori, A. Otsuka and T. Takahashi, *Fracture of Solids*, Interscience, New York (1963), p. 261.

[72] D. V. Wilson and D. Russel, *Acta Met.*, **8**, 36 (1960).

[73] R. Honda, *Int. Conf. on Fracture*, Sendai, Japan **B-I**, 109 (1965).

[74] T. L. Johnston, R. S. Stokes and C. H. Li, *Phil. Mag.*, **7**, 23 (1962).

Problems

1. Suppose that the average slip distance L in a polycrystal is equal to one-half the grain size $2d$. What shear stress is required to nucleate a crack in a molybdenum alloy whose grain size is 0.02 mm if 55 ksi are required when $2d = 0.32$ mm and 40 ksi are required when $2d = 2$ mm? The transition grain size is 0.002 mm. (130 ksi)

2. Consider a tungsten single crystal strained along a [110] direction.

(a) Work out the six possible lines of intersection for any two of the operative {110}, ⟨111⟩ slip systems that cross one another.

(b) Show these intersections in a cube, using separate diagrams for each intersecting pair and determine the line of intersection.

(c) Which two of these intersections could conceivably lead to microcrack formation by an intersection mechanism?

3. Derive an expression for the longest stable microcrack that can exist according to the Cottrell-Petch theory.

4. Stroh has shown that a crack of length

$$2c = \frac{\pi \tau^2 L^2}{2G\gamma_m}$$

forms in HCP metals under the action of a shear stress τ, where L is the length of the dislocation walls in Fig. 6.11a.

(a) Derive an expression for the spreading of the crack, in terms of the normal and shear stress acting on the metal and the length L of the dislocation wall. Assume that the friction stress τ_i is negligible.

(b) For a Mg sheet specimen having preferred orientation such that the basal plane is perpendicular to the sheet surface, and the basal slip vector b lies in the sheet surface, calculate the angle between the tensile axis and b which causes the sheet to fracture at the lowest applied stress.   Assume that both slip and cleavage occur only on the basal plane.   (30°)

5. The yield strength of polycrystalline iron varies with absolute temperature T (°K) as $\sigma_Y = (200 - .5T)$ ksi.   The cleavage strength σ_f increases linearly with temperature and is equal to 170 ksi at 200°K.   The ductility-transition temperature $T_D = 100°$K.   What is the new ductility-transition temperature if the alloy is strengthened (by dispersion hardening) by 21 ksi if σ_f and the temperature dependence of σ_Y are unaffected by the presence of precipitates?

(140°K)

6. Consider a clean refractory metal whose tensile cleavage strength $\sigma_f = Ad^{-\frac{1}{2}}$.   If $\sigma_i = B - cT$, derive an expression for the variation of the tensile (unnotched) transition temperature with grain size, in terms of C, A, and k_y.   You may assume that A is temperature-independent.

The Relation Between the Microscopic and Macroscopic Aspects of Plane-Strain Fracture

7

The Physical Meaning
of Fracture Toughness

In Chapter 2 the *macroscopic aspects* of fracture in large structures containing sharp, deep notches or cracks were considered. It was shown that plastically induced fracture begins at the notch root (i.e., crack tip) when a critical plastic strain is produced over a critical distance ahead of the tip. At low nominal stress levels ($\sigma \ll \sigma_Y$) the displacement of the crack faces near the tip $2V(c)$ is proportional to the strain in the volume element directly ahead of the tip so that fracture initiates when $V(c)$ reaches a critical value $V^*(c)$. The work done in initiating a plane-strain fracture is

$$2\gamma_P{}^* = G_{\mathrm{I}c} \cong 2\sigma_Y V^*(c) \tag{7.1}$$

where σ_Y is (approximately) the uniaxial tensile yield stress. The nominal fracture stress σ_F is given by the relation

$$\sigma_F = \sqrt{\frac{EG_{\mathrm{I}c}}{\pi c(1 - \nu^2)}} \qquad (\sigma_F \ll \sigma_Y) \tag{7.2}$$

At higher nominal stress levels ($\sigma_F \rightarrow \sigma_Y$), $V^*(c)$ may itself be a function of geometry [1] and $G_{\mathrm{I}c}$ is no longer a simple function of $\sigma^2 c$, so that the mechanics of the fracture process become considerably more complicated.

The *microscopic aspects* of fracture are primarily concerned with events in one or two grains of a polycrystal where microcracks or voids are formed by inhomogeneous plastic deformation. Cleavage fracture of an unnotched tensile specimen occurs when the microcracks can spread unstably, at a tensile stress σ_f that is influenced by intrinsic variables such as microstructure and extrinsic variables such as temperature and starin rate. Tear fracture occurs when the voids coalesce by localized plastic deformation, and this process is also influenced by intrinsic and extrinsic variables. As shown in the preceding chapter, the choice between these two modes of fracture depends on many factors and is very involved.

Consequently we should not expect that the relation between the microscopic and macroscopic aspects of the problem could be explained, except superficially, in a simple fashion. At this writing only a few bridges have been built between the two approaches to the fracture problem and much work remains to be done. In the following sections we shall describe the principles that have been established and show how, in specific instances, it is possible to determine the physical significance of G_{Ic}. We shall then outline how certain changes in relevant macroscopic and microscopic parameters (e.g., in root radius ρ and grain size $2d$) might be expected to influence G_{Ic} and *why* they would be expected to do so.

The microscopic processes of fracture that take place in an unnotched material also take place below (i.e., ahead of) the notch root in a notched structure. For those cases where the root radius ρ is greater than the smallest structural unit of the fracture process (e.g., grain size in a cleavage fracture) we can, to a first approximation, analyze the problem of notch brittleness by treating the volume of material ahead of the notch as a miniature tensile specimen [2, 3] and determining the effect of the adjacent notch on the mechanical behavior of this specimen.

For example, under certain conditions, microcrack formation in one grain ahead of the notch (Fig. 7.1a) leads to cleavage fracture of the "specimen" (Fig. 7.1b), which leads to unstable fracture of the entire plate (Fig. 7.1c). Fracture of the "specimen," and hence fracture of the entire notched structure, occurs when the tensile stress level in the plastic zone ahead of the notch is raised to σ_f. Physically the notch is able to produce this elevation in tensile stress level by one or more of the following processes: (1) by raising the rate at which the specimen is loaded, thereby raising the shear yield stress $\tau_Y = k$ from the value measured, at the same temperature, in a standard, unnotched tensile specimen loaded at the same crosshead rate [4, 5], (2) by concentrating plastic strain in the "specimen," as the plastic zones spread across the structure, thereby raising τ_Y by strain hardening [5, 6], and (3) by introducing a state of triaxial tension in the specimen [1–7], thereby increasing the tensile stress level (σ_{yy}) required to produce a given amount of shear stress† ($\sigma_{yy} - \sigma_{xx}) = 2\tau_Y$. The tensile stress level ahead of the notch is therefore raised from $\sigma_Y = 2\tau_Y$ to $K_{\sigma(p)}\sigma_Y$, where $K_{\sigma(p)}$ is called the *plastic stress concentration factor*.

To a first approximation, $K_{\sigma(p)}$ can reach a value of about 3 for a deep, sharp, external notch in a thick plate [7, 8]. In the preceding

---

† For simplicity, the Tresca criterion will be used in this chapter.

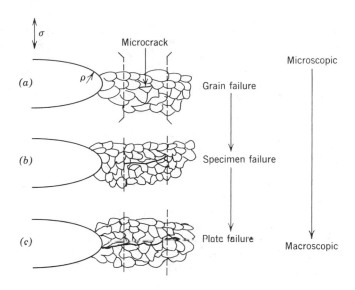

Fig. 7.1. Three stages leading to the development of an unstable cleavage fracture in a notched structure. Failure of a grain (a) leads to failure of a collection of grains (b) (a hypothetical miniature "tensile specimen") which leads to failure of the entire structure (c).

chapter the brittleness-transition temperature T_D was defined as the highest temperature at which fracture would occur in smooth specimens in the absence of macroscopic plastic deformation (i.e., when $\sigma_Y = \sigma_f$). Since σ_f is not strongly affected by temperature, an increase in the tensile stress level in the plastic zone ahead of the notch, from σ_Y to $3\sigma_Y$, significantly increases the temperature (Fig. 7.2) at which cleavage occurs in a large structure before or at general yield ($\sigma_F \leqq \sigma_Y$), from T_D to $T_{D(N)}$.

This is the classic explanation for notch brittleness of mild steel as originally proposed by Orowan [7]. Qualitatively it can account for an increase in ductile-brittle transition temperature in notched as compared with unnotched structures. Other factors, such as the nominal stress level at which fracture occurs, the influence of notch and specimen geometry, and the occurrence of low stress, plane-strain fracture in high strength materials require an understanding of the factors that influence $K_{\sigma(p)}$ and the relation of $K_{\sigma(p)}$ to G_{Ic}.

In Section 7.1 we present a discussion of the plasticity aspects of plane-strain yielding of notched structures. Most of the discussion

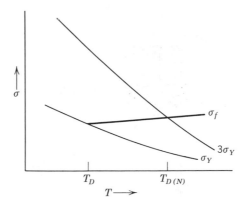

Fig. 7.2. The increase in nil-ductility temperature, from T_D to $T_{D(N)}$, owing to the presence of a sharp notch. The notch raises the tensile stress level from σ_Y to about $3\sigma_Y$ at general yield. *After E. Orowan* [7].

will be concerned with bending, since solutions of this problem are available for various notch and specimen geometries. Although the plane-strain tension problem has not yet been completely worked out, some aspects of the problem have been treated and these will also be presented. We then consider (Section 7.2) the problem of cleavage fracture in notched specimens and illustrate how the shape of the Charpy curve is dependent on microstructural parameters. We also consider the effect of notch and structural geometry on the value of G_{Ic}. In Section 7.3 we consider the physical significance of G_{Ic} for low-energy tear fractures which occur by void formation and coalescence. Finally, in Section 7.4 we discuss some of the means of reducing the danger of notch brittleness in various materials.

7.1 The Distribution of Stress and Strain in the Vicinity of a Notch in a Thick Structure (plane strain)

Constraint in a notched structure

Consider an externally notched plate loaded in tension to a low nominal stress level such that there is no local yielding at the notch root and the entire plate is elastic. The high longitudinal stresses $\sigma_{yy} \cong 2\sigma\sqrt{c/r}$ set up at a distance r ahead of the root (for $r > \rho$) cause the material there to extend elastically and consequently to contract because of the Poisson effect.   This contraction is greatest near the notch root (Fig. 7.3) where the longitudinal stresses are highest. The area A that has been cut by the notch does not want to contract because there are no longitudinal stresses acting across it; all of these are concentrated ahead of the root.   Since the unstressed area A tries to maintain its original dimensions while the material beneath the root is contracting, transverse tensile stresses σ_{zz} are set up in the contracting material [9].   The stress σ_{zz} is a maximum at the center of the plate ($z = 0$).   Because the faces (xy planes) of the plate are not loaded externally, σ_{zz} drops to zero at $z = \pm t/2$, for any value of r (Fig. 7.4a).   Similarly, for any value of z, σ_{zz} decreases with increasing distance r ahead of the root.   In addition, transverse tensile stresses σ_{xx} also are set up ahead of the notch, by the constraint which prevents contraction in the width (x) direction and by the cantilever-type deflection induced by the presence of the notch [9].   The

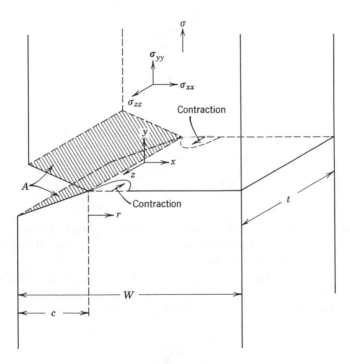

Fig. 7.3. Transverse contractions that occur near the tip of a notch in a thick plate. These contractions are opposed by the unyielding faces A of the notch; consequently transverse tensile stresses σ_{zz} and σ_{xx} are set up ahead of the notch tip.

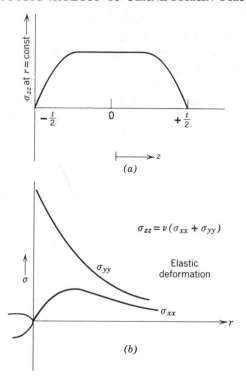

Fig. 7.4. (a) Variation of σ_{zz} across the thickness (z) direction at r = const. (b) Variation of σ_{yy} and σ_{xx} with r at mid thickness (z = 0) of a thick, notched plate. *After E. Parker* [9].

variation of the longitudinal and transverse (elastic) stresses σ_{yy} and σ_{xx}, with distance r in the center of the plate (z = 0), are shown in Fig. 7.4b. At any point $\sigma_{zz} = \nu\,(\sigma_{xx} + \sigma_{yy})$.

Since $\sigma_{xx} = 0$ at the root surface (r = 0), local yielding will occur when the longitudinal stresses σ_{yy} are equal to the uniaxial tensile yield stress. The nominal stress† at which this occurs is then

$$\sigma = \frac{\sigma_Y}{K_\sigma} \tag{7.3}$$

† For pure (four-point) and three-point bending the nominal stress at the notch tip is given by the relation $\sigma = 6M/ta^2$ where M is the applied bending moment and $a = W - c$ is the depth of the notched cross section.

where K_σ is the elastic stress concentration factor of the notch and σ_Y is the uniaxial tensile yield stress, at the same test temperature, and evaluated at the same applied strain rate as that which exists across the material at the root of the notch. Since the effective gage length of the "specimen" is of the order of the root radius (0.010–0.001 in.), the strain rate can be two or three orders of magnitude higher than that which is measured in a conventional test specimen of 1 in. gage length, for the same rate of cross head elongation [4, 5]. In strain-rate sensitive materials, such as mild steel, this may cause σ_Y to be about 25% greater than the conventionally obtained value.

The stress distribution that exists during local yielding in a nonstrain-hardening solid

As the nominal stress increases above the value given by Eq. 7.3, plastic zones spread to a distance R that increases with increasing applied stress and hence with increasing applied load P. Theoretical slip-line field solutions for the plane-strain general yielding of a rigid plastic, nonstrain-hardening material, loaded in pure bending, have been obtained by numerous investigators [10, 11, 12, 13] following the methods of Hill [8]. Basically these solutions predict that under plane-strain conditions the lines of maximum shear stress or "slip lines" assume the form of logarithmic spirals (Fig. 7.5a) which increase in depth R (e.g., from position 1 to position 3, Fig. 7.5c) as the applied load increases. These logarithmic spirals are only observed at the center of the plate where the highest degree of constraint is set up [14]. At the plate surface the plastic zones appear as plastic hinges (Fig. 7.5b), similar to those described in Chapter 2 (Fig. 2.8). The plastic zone size measured on the plate surface (Fig. 2.8) is about twice that measured in the center of the plate, in the plane of the notch.

Since the transverse stresses σ_{xx} and σ_{zz} are both greater than zero ahead of the root, and since the maximum shear stress criterion requires that $\sigma_{yy} - \sigma_{xx} = 2k = \sigma_Y$ inside the plastic zone,

$$\sigma_{yy} = \sigma_Y + \sigma_{xx}$$

Both σ_{xx} and hence σ_{yy} increase with increasing distance r ahead of the root. *Inside the plastic zone* $(r \leq R)$, the longitudinal stresses are given by [8]

$$\sigma_{yy} = 2k \left[1 + \ln \left(1 + \frac{r}{\rho} \right) \right] \qquad (r \leq R_\beta) \qquad (7.4)$$

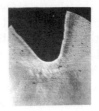

$$\frac{\sigma}{\sigma_{GY}} = 0.4$$

 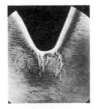

$$\frac{\sigma}{\sigma_{GY}} = 0.7$$

$$\frac{\sigma}{\sigma_{GY}} = 0.8$$

(a) Center   *(b)* Surface

Fig. 7.5. (a) Local plastic deformation observed (in mild steel) in the form of logarithmic spirals at mid thickness $(z = 0)$ of a standard Charpy V notch specimen, under three-point bending, at $\sigma/\sigma_{GY} = 0.4, 0.7, 0.8$. 20×. *Courtesy T. R. Wilshaw and P. Pratt* [14] *and J. Mech. Phys. Solids.* (b) Local plastic deformation observed (in mild steel) as plastic hinges at the surface $(z = t/2)$ of a standard Charpy V notch specimen, under three-point bending, at $\sigma/\sigma_{GY} = 0.4, 0.7, 0.8$. 20×. *Courtesy T. R. Wilshaw and P. Pratt* [14] *and J. Mech. Phys. Solids.*

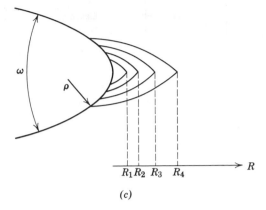

(c)

Fig. 7.5. (*c*) Successive positions of the elastic-plastic interface at mid thickness as the nominal stress σ is increased across a plate containing a notch of given depth.

Ahead of the plastic zone $(\sigma_{yy} - \sigma_{xx}) < 2k$ and the material remains elastic. Both σ_{yy} and σ_{xx} decrease with increasing values of r for $r > R$ (Fig. 7.6).

The plastic stress concentration factor $K_{\sigma(p)}$ is *defined* as the ratio of the maximum tensile stress which exists below the root to the tensile yield stress of an unnotched specimen tested at the same strain rate. At low stress levels $K_{\sigma(p)}$ is a maximum at the elastic-plastic interface $(r = R)$, as seen in Fig. 7.6a. Using a Tresca criterion, we have

$$K_{\sigma(p)} = \frac{\sigma_{yy}^{\max}}{\sigma_Y} = \left[1 + \ln\left(1 + \frac{R}{\rho}\right)\right] \qquad (R \leq R_\beta) \qquad (7.5)$$

As shown in Chapter 2, R increases with increasing ratios of nominal stress σ or stress intensity factor $K = \sigma\sqrt{\alpha\pi c}$ to general yield stress† σ_{GY}. Since $K_{\sigma(p)}$ increases with R, $K_{\sigma(p)}$ also increases with K/σ_{yy} or σ/σ_{GY} (Fig. 7.7) for constant values of ρ.

The maximum possible value of $K_{\sigma(p)}$ is independent of root radius and depends only on flank angle ω for the notch, according to the relation [8]

$$\frac{1}{\beta} = \frac{\sigma_{yy}^{\max}}{2k} = \left[1 + \frac{\pi}{2} - \frac{\omega}{2}\right] = K_{\sigma(p)}^{\max} = \left(\frac{\sigma_{yy}^{\max}}{\sigma_Y}\right)^{\max} \qquad (7.6)$$

† σ_{GY} and P_{GY} are, respectively, the nominal stress and load at which yielding spreads completely across the notched cross section. $\sigma/\sigma_{GY} \cong P/P_{GY}$.

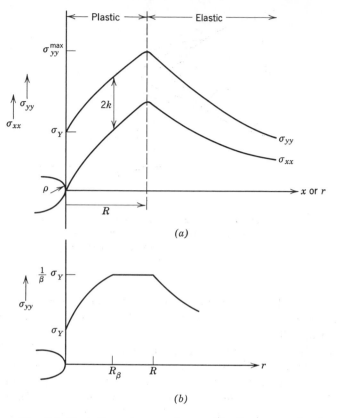

Fig. 7.6. Variation of σ_{yy} and σ_{xx} with distance r ahead
of a notch at a given value of plastic zone size R. (a)
$R < R_\beta$. (b) $R > R_\beta$. R_β is the value of R at which σ_{yy}^{max}
is first raised to its maximum value, $(1/\beta)\sigma_Y$. (a) and
(b) are not on the same scale.

For a parallel-sided notch or crack, $\omega = 0$ and $1/\beta = 2.57$ according to
the Tresca criterion and 2.82 according to the von Mises yield criterion.
For a 45° notch (Charpy specimen) $1/\beta = 2.18$. These relations
also apply for three-point bending [14] and for externally notched
tensile specimens [8].

A comparison of Eqs. 7.5 and 7.6 indicates that for a given flank
angle ω, $K_{\sigma(p)}$ reaches its maximum value of $1/\beta$ when

$$R = R_\beta = \rho[e^{1/\beta-1} - 1] \tag{7.7}$$

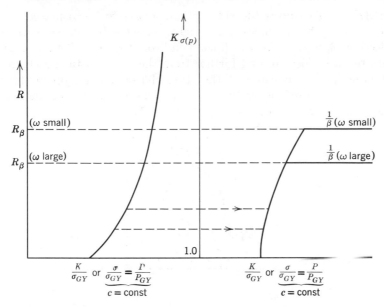

Fig. 7.7. Schematic diagram of the relation between plastic zone size R, K/σ_{GY}, and plastic stress concentration factor $K_{\sigma(p)}$. An increase in K/σ_{GY} causes R to increase, which, in turn, causes $K_{\sigma(p)}$ to increase up to a maximum value $1/\beta$ that depends only on flank angle ω.

Once $R = R_\beta$, further increases in load (i.e., in $K^2/\sigma_{GY}{}^2$) cause R to increase but produce no further increases in $K_{\sigma(p)}$, as shown on Fig. 7.7. For a given value of $K^2/\sigma_{GY}{}^2$, $K_{\sigma(p)}$ remains equal to $1/\beta$ in the region $R_\beta < r < R$ ahead of the notch (Fig. 7.6b) and decreases to unity for $R < R_\beta$, according to Eq. 7.4. Similarly $K_{\sigma(p)}$ decreases to unity in the elastic region $r > R$.

When the applied stress σ is sufficient to cause general yielding (i.e., $\sigma = \sigma_{GY}$ or $P = P_{GY}$), plastic hinges form at the root at midthickness and propagate rapidly across the specimen [5, 10–14], which is then fully plastic. The theoretical slip-line field for the case of three-point bending [11] is shown in Fig. 7.8a and is in good agreement with the experimental observations shown in Fig. 7.8b.† At general yield there is a central pivot $ABCD$ which remains rigid, around which the rigid parts of the bar rotate by shearing over the circular areas AB

---

† For the case of pure bending the solutions are slightly different above general yield.

and AD. This causes the sides of the notch to be displaced by an amount (i.e., bend angle) that increases with increasing crosshead deflections $\delta > \delta_Y$. The slip lines then broaden, spreading into regions that are both closer to and further from the plane of the notch [5, 14, 15] (Fig. 7.8c). Because of the strain hardening set up in the slip lines, the load (and hence the nominal stress) required for increasing deflection increases with δ for $\delta > \delta_Y$ (Fig. 7.8).

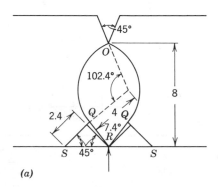

(a)

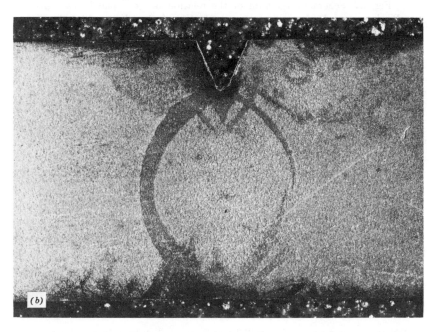

(b)

Fig. 7.8. (a) Calculated [11] and (b) observed [25] slip-line fields at mid-section of standard Charpy specimens of mild steel, loaded to general yield in three-point bending.

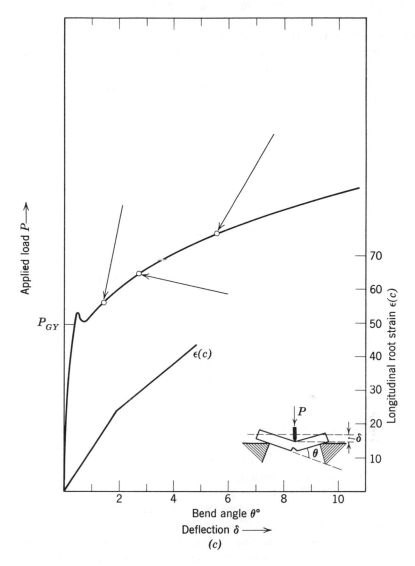

Fig. 7.8. (c) Load (P) deflection (δ) curve for Charpy specimens of mild steel in three-point bending. The variation of root strain ε(c) and bend angle θ are also shown [64], as is the relation between P, δ, and θ.

299

For an *unnotched* beam loaded in *pure bending*, the general yield load P_{GY} is that load required to produce a fully plastic bending moment $M = (\sigma_Y/4)\, a^2 t$, where a is the depth of the beam, corresponding to the depth of the notched cross section in the notched specimen.   In a notched beam the constraint described previously increases the applied loads required for general yielding, and hence the fully plastic bending moment, by a factor L, referred to as the *constraint factor*. L is a function of flank angle ω and root radius, as shown in Table 7.1, for some important notch and specimen geometries and loading conditions.   These factors only apply to the case of extremely thick specimens where the deformation is entirely plane strain and are limited to specimens containing sufficiently deep notches such that there is no yielding in regions removed from the vicinity of the notch [8, 16] (i.e., the stresses acting across the un-notched sections must be below the yield stress).

For the case of *tensile loading*, L is defined as the ratio of the general yield load for a notched bar to the general yield load of an unnotched bar of the same cross-sectional area, tested at the strain rate which exists below the notch.   Thus

$$L = \frac{\sigma_{GY}}{\sigma_Y} = \frac{P_{GY}}{P_Y} \tag{7.8}$$

It is important to emphasize the distinction between L and $K_{\sigma(p)}$; L refers to applied loads and hence to applied (nominal) stresses required for yielding all across the notched cross section.   In tensile loading, L is a measure of the *mean stress* across the bar at general yield. $K_{\sigma(p)}$ refers to the *maximum* longitudinal stresses that exist in the plane of the notched cross section.   Decreasing root radius decreases R_β and hence increases L (Fig. 7.10b).   For example, for a hyperbolic notch in a cylindrical specimen ($\omega = 29°$, $\rho = 0.01$ in.), $L = 1.35$ and $K_{\sigma(p)}^{\max} = 1/\beta = 2.35$.   For a specimen containing two parallel-sided, external notches of infinitely sharp radius ($\omega = 0$, $\rho \to 0$), $L = 2.57 = 1/\beta$.   The slip-line field for this case and the large internal notch are shown in Fig. 7.9.   For pure shear (mode II and mode III loading) there are no constraints set up and both $K_{\sigma(p)}$ and L are one-half.

At present there is no theoretical treatment, and only few experimental observations, of the variation of R and root displacement $2V(c)$ with nominal stress and notch depth for two important cases:

1. Plane-strain deformation, $\rho \to 0$ (e.g., sharp crack).
2. Plane-strain deformation, $\rho =$ finite (e.g., Charpy V notch).

Table 7.1

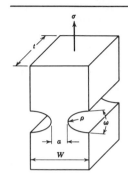

$$L = \left[\left(1 + \left(\frac{\pi - \omega}{2} \right) \right) \right.$$

$$\left. - \frac{\rho}{a} \left(c^{(\pi-\omega)/2} - 1 - \left(\frac{\pi - \omega}{2} \right) \right) \right]$$

$$\frac{1}{\beta} = \left[1 + \frac{\pi}{2} - \frac{\omega}{2} \right]$$

For $\dfrac{t}{a} > 1$ $\dfrac{W}{a} > L$

$$L = 1$$

For $\dfrac{c}{W}$ large

$$\frac{1}{\beta} = 1$$

$$L = 2.9$$

For $\rho = 0$, $\omega = 0$

$$\frac{1}{\beta} = 2.9$$

$$L = 1.26$$

For $\omega < 60°$

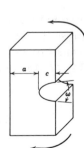

$$\frac{1}{\beta} = 1 + \frac{\pi}{2} - \frac{\omega}{2}$$

Three- or four-point (pure) bending

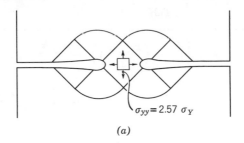

$$\sigma_{yy} = 2.57 \; \sigma_Y$$

(a)

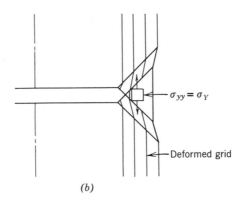

$\sigma_{yy} = \sigma_Y$

Deformed grid

(b)

Fig. 7.9. Slip-line field at general yield under plane-strain tensile loading. (a) Double, external sharp notch (c/W small). (b) Internal sharp notch (c/W large). *After F. McClintock and G. Irwin* [1].

Concerning Case 1, we suppose that to a first approximation R and $V(c)$ are similar functions of σ and c as in plane-stress deformation (Eqs. 2.31 and 2.33) but that σ_Y is replaced by its constrained value σ_{GY}. There is sufficient experimental evidence [5, 14, 17, 18, 19] to indicate that this is a good approximation. Thus

$$V(c) \cong \Omega \frac{4\sigma_{GY}}{\pi E} c \ln \left(\sec \frac{\pi}{2} \frac{\sigma}{\sigma_{GY}} \right) \qquad (7.9a)$$

or

$$V(c) \cong \Omega \frac{4\sigma_{GY}}{\pi E} R \qquad (7.9b)$$

where

$$R = \frac{1}{\psi} \left\{ \left(\sec \frac{\pi}{2} \frac{\sigma}{\sigma_{GY}} \right) - 1 \right\} c \qquad (7.10a)$$

or

$$R = \frac{\pi^2}{8\psi} \left(\frac{\sigma}{\sigma_{GY}} \right)^2 c = \frac{\pi}{8\alpha\psi} \frac{K^2}{\sigma_{GY}^2} \qquad (7.10b)$$

where

$$K = \sigma \sqrt{\alpha\pi c}, \qquad \alpha = \frac{\Omega}{\psi}$$

Ω, ψ, and $\alpha = \Omega/\psi$ are parameters that account for finite specimen and notch geometry. Figure 7.10a shows the reduction in R, at a given K, that occurs in plane-strain deformation as compared with plane stress.

Case 2 is more complicated. When ρ is finite, local yielding cannot occur until Eq. 7.3 is satisfied and the local stresses at the notch tip reach the yield stress (point A, Fig. 7.10). Furthermore, the elastic stress field a distance r from the tip no longer varies [62] as $r^{-\frac{1}{2}}$, but as $(\rho + r)^{-\frac{1}{2}}$. Consequently the plastic zone size will be smaller, at a given stress, in the case of a blunt notch than in the case of a sharp one [63]. When r is large compared with ρ, the effect of variations in ρ become negligible and R is the same as for the infinitely sharp crack (point B, Fig. 7.10a). The curve AB thus describes R as a function of K/σ_{GY} when R is of the order of ρ. No functional form of this relation has yet been established.

Equation 7.5 indicates that $K_{\sigma(p)}$ decreases, at a given value of R, as ρ increases. Figure 7.10a indicates that R decreases with increasing ρ, for a given value of K/σ_{GY}. Consequently $K_{\sigma(p)}$ *strongly decreases with increasing* ρ, as shown in Fig. 7.10b, for constant values of K/σ_{GY} and flank angle ω. Furthermore, since R_β increases with increasing ρ (Eq. 7.7), the stress level K/σ_{GY} required to achieve the maximum degree of stress intensification $[(1/\beta)\sigma_Y]$ also increases with increasing ρ.

Strains at and near the notch root

The root displacement $2V(c)$ can be obtained from measurements of the bend angle $\theta°$ in a bend test or from the change in flank angle of the notch in a bend or tensile test (Fig. 7.8c). It has been shown that [21]

$$V(c) = A\theta° \qquad (7.11)$$

where A is a parameter that depends primarily on notch and specimen geometry (e.g., $A = 1.2 \times 10^{-3}$ in. per degree for Charpy V specimens).

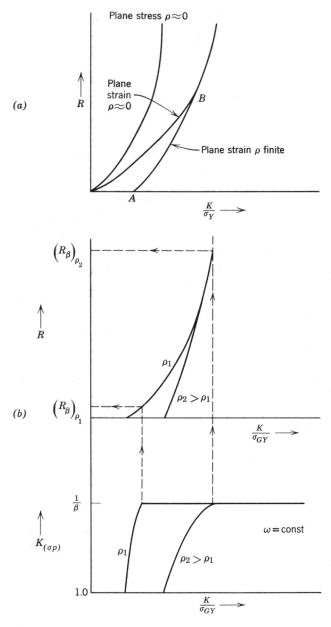

Fig. 7.10. Variation of R and $K_{\sigma(p)}$ with K/σ_{GY} for notches of varying root radii under plane-strain loading. The flank angle ω, and hence $K_{\sigma(p)}^{\max} = 1/\beta$, are constant. R_β is the value of R at which $K_{\sigma(p)}$ first reaches $1/\beta$.

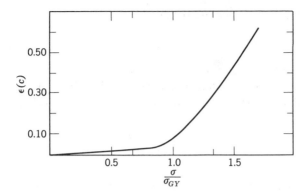

Fig. 7.11. Relation between root strain $\epsilon(c)$ and σ/σ_{GY} for Charpy specimens of mild steel in three-point bending. *After T. Wilshaw* [17].

The root strains $\epsilon(c)$ are also related to θ [22, 64];

$$\epsilon(c) \cong B\theta° \qquad (7.12)$$

where B is a parameter that depends on specimen geometry and the magnitude of the strains. Figure 7.8c indicates that B decreases by a factor of 2 (from 0.12 to 0.06) when $\epsilon(c) > 0.25$, for Charpy V specimens of mild steel. θ is proportional to deflection δ and hence depends on σ/σ_{GY}. Thus $\epsilon(c)$ is a function of σ/σ_{GY}, as shown in Fig. 7.11 [17].

When the plastic zone size is small ($\sigma < \sigma_{GY}$), essentially all of the strains are localized in a "miniature tensile specimen," placed adjacent to the root surface, whose gage length (i.e., the thickness of the logarithmic spirals) is of the order of 2ρ. As shown in Eq. 2.38,

$$V(c) \cong \rho\epsilon(c) \qquad (\sigma < \sigma_{GY}) \qquad (7.13a)$$

At higher values of σ/σ_{GY} the thickness of the logarithmic spirals increases and a large fraction of the displacement tending to open up the notch is accommodated along the plastic hinges, in addition to that concentrated at the root. Thus

$$V(c) \cong 2\rho\epsilon(c) \qquad (\sigma > \sigma_{GY}) \qquad (7.13b)$$

is probably a better estimate of the relation between tip displacement and root strain under these circumstances.

Combining Eqs. 7.12 and 7.13 gives

$$V(c) \cong D\rho\theta \tag{7.14}$$

where $D \approx 0.12$ for Charpy V goemetry.

In Chapter 2 it was shown that the strain distribution ahead of an infinitely sharp crack ($\rho, \omega \to 0$) subjected to mode III deformation is given by [20]

$$\epsilon_{yz} = \frac{k}{G} \frac{R}{r} \tag{2.53}$$

Recent measurements [22] indicate that the strains directly ahead of the notch† in a Charpy V specimen under three-point bending also vary inversely with distance r from the tip

$$\epsilon(r) = \frac{F\theta}{r} \tag{7.15}$$

when the specimens are bent through an angle θ. $F = 0.02$ for Charpy V geometry [22].

By combining Eqs. 7.14 and 7.15 we can determine the root displacement required to achieve a strain $\epsilon(r)$ at any point r ahead of the root

$$V(c) = J\rho\epsilon(r)r \tag{7.16}$$

($J = D/F \approx 6$ for Charpy V geometry) and also the relation between local strain and root strain

$$\epsilon(r) = \frac{\epsilon(c)}{Jr} \tag{7.17}$$

Strain hardening near the notch root

In a strain-hardening material the strain in the vicinity of the root produces strain hardening. To a first approximation we can assume linear strain hardening and that at any point r inside the plastic zone, in the plane of the notch, the yield stress σ_Y is raised by an amount

$$\Delta\sigma(r) = \frac{d\sigma}{d\epsilon} \epsilon(r)$$

$$= \frac{d\sigma}{d\epsilon} \frac{\epsilon(c)}{Jr} \tag{7.18}$$

† In the plane, $y = 0$.

so that

$$\sigma_{yy}(r) = K_{\sigma(p)}[\sigma_Y + \Delta\sigma(r)] \tag{7.19}$$

$$\sigma_{yy}(r) = \left[1 + \ln\left(1 + \frac{r}{\rho}\right)\right]\left[\sigma_Y + \frac{d\sigma}{d\epsilon}\frac{\epsilon(c)}{Jr}\right] \qquad (r < R_\beta) \quad (7.20a)$$

$$\sigma_{yy}(r) = \frac{1}{\beta}\left[\sigma_Y + \frac{d\sigma}{d\epsilon}\frac{\epsilon(c)}{Jr}\right] \qquad (R_\beta < r < R) \tag{7.20b}$$

Although $\Delta\sigma(r)$ decreases with increasing values of r (Fig. 7.12a), $K_{\sigma(p)}$ increases with r so that $K_{\sigma(p)}\Delta\sigma(r)$ reaches a maximum somewhere below the root, rather than at the root surface. It turns out [22] that Eq. 7.20 almost always has a maximum at $r = R_\beta$, as seen on the figure.† At this point $K_{\sigma(p)} = 1/\beta$, so that

$$\sigma_{yy}^{max} = \frac{1}{\beta}\left[\sigma_Y + \frac{d\sigma}{d\epsilon}\frac{\epsilon(c)}{JR_\beta}\right] \qquad (r = R_\beta) \tag{7.21}$$

$$\sigma_{yy}^{max} = \frac{1}{\beta}\left[\sigma_Y + \frac{d\sigma}{d\epsilon}\frac{\epsilon(c)}{J\rho(e^{1/\beta-1} - 1)}\right] \tag{7.22}$$

according to Eq. 7.7. Consequently a decrease in ρ produces a large increase in σ_{yy}^{max} (because R_β is reduced), and hence a greater amount of strain hardening $\Delta\sigma(r)$ is superimposed on σ_Y for a given root strain $\epsilon(c)$ (Fig. 7.12b). If σ_{yy}^{max} is evaluated at constant root displacement $2V(c)$, then the root radius ρ becomes even more important:

$$\sigma_{yy}^{max} = \frac{1}{\beta}\left[\sigma_Y + \frac{d\sigma}{d\epsilon}\frac{V(c)}{\rho^2}H\right] \tag{7.23}$$

where H is a constant that depends primarily on flank angle, specimen geometry, and strain level (i.e., according to whether Eq. 7.13a or 7.13b applies).

Summarizing relations between the maximum tensile stress produced ahead of a notch σ_{yy}^{max} and notch-tip displacement $2V(c)$

At low nominal stress levels (K^2/σ_{GY}^2), $R < R_\beta$, $K_{\sigma(p)} < 1/\beta$, and σ_{yy}^{max} is given by Eq. 7.5. Combining this with 7.9b, we have

$$\sigma_{yy}^{max} = \left[1 + \ln\left(1 + \frac{V(c)\pi E}{4\Omega\sigma_{GY}\rho}\right)\right]\sigma_Y \qquad \left(\sigma_{yy}^{max} < \frac{1}{\beta}\sigma_Y\right), \quad V(c) < [V(c)]_\beta$$

$$\tag{7.24}$$

† The amount of strain hardening that occurs when the plastic zone size $R < R_\beta$ is negligible and can be neglected.

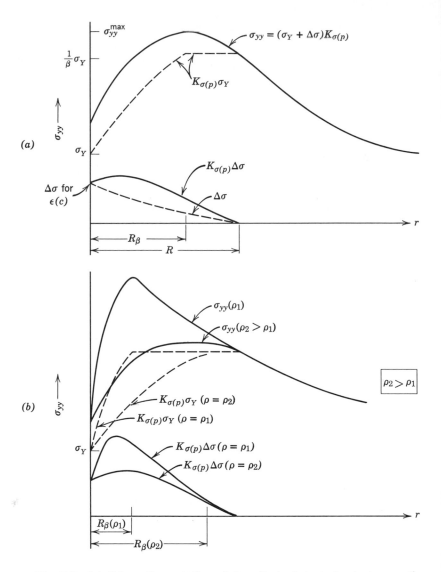

Fig. 7.12. (a) Schematic variation of the effect of strain hardening on the stress distribution ahead of a notch in a thick plate. Linear strain hardening ($\Delta\sigma = \epsilon(d\sigma/d\epsilon)$) and superposition of stresses are assumed for simplicity. For any given value of root strain $\epsilon(c)$, and hence of $\Delta\sigma(r)$, the maximum value of σ_{yy} is still reached at R_β. As shown in (b), a decrease in root radius causes $\sigma_{yy}^{(\text{max})}$ to be increased for a given value of $\epsilon(c)$.

where $[V(c)]_\beta$ is the value of $V(c)$ when $R = R_\beta$ and r is a constant. Once $R = R_\beta$, plastic constraint alone has achieved a maximum degree of hydrostatic stress intensification and

$$\sigma_{yy}^{max} = \left(\frac{1}{\beta}\right)\sigma_Y \qquad V(c) = [V(c)]_\beta \tag{7.25}$$

In a nonstrain hardening material this is the maximum value of σ_{yy}^{max}. When strain hardening occurs $(R > R_\beta, V(c) > [V(c)]_\beta, d\sigma/d\epsilon > 0)$

$$\sigma_{yy}^{max} \cong \frac{1}{\beta}\left[\sigma_Y + \frac{d\sigma}{d\epsilon}\frac{V(c)}{\rho^2}H\right] \qquad V(c) > [V(c)]_\beta \tag{7.23}$$

These relations will be used extensively in the following sections where the relation between microscopic and macroscopic aspects of brittle fracture is discussed.

7.2   The Physical Significance of G_{Ic} for Cleavage Fracture in Low Strength Materials ($\sigma_Y < E/300$)

The temperature dependence of G_{Ic} in clean, single-phase materials

G_{Ic} is a complicated function of many variables such as the inherent properties of the material, the strain rate, and the radius of the notch. It is convenient to start our discussion with the simplest and best understood case and then vary the relevant parameters. Accordingly, we begin by considering the effect of temperature on G_{Ic} for (1) clean material (e.g., pure Fe and W), (2) Charpy V geometry, (3) Tested at moderate loading rate in three-point bending.

In the preceding chapter it was shown that cleavage microcracks are initiated and begin to spread in an unnotched tensile specimen when the applied tensile stress σ reaches a critical value $\sigma_G = (4G\gamma_m/k_Y)\ d^{-\frac{1}{2}}$. For those cases (e.g., *clean materials*) where the first microcrack that forms is able to spread through the surrounding grain boundaries at the same stress level, the fracture stress $\sigma_f = \sigma_G$.

Similarly, at any temperature T there will be some tensile stress level σ_G at which a microcrack will form in a grain *ahead of the root* (Fig. 7.1a) where the tensile stress level σ_{yy} is a maximum [23]. In a clean material this will immediately cause failure of the miniature tensile specimen (Fig. 7.1b). Consequently

$$\sigma_G = \sigma_f{}^*$$

where $\sigma_f{}^*$ *is the fracture stress of the "tensile specimen" imbedded ahead of the notch root.* $\sigma_f{}^*$ can differ somewhat from the fracture

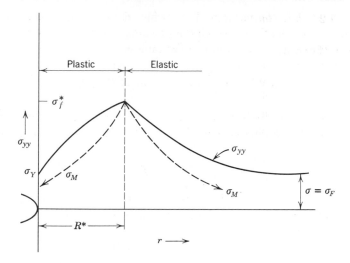

Fig. 7.13. Requirements for cleavage fracture before general yield of a notched, thick plate, in the absence of any stable crack growth. When $\sigma = \sigma_F$, the local stresses ahead of the root $\sigma_{yy} = \sigma_f{}^*$ and a cleavage microcrack initiates. The microcrack can propagate unstably as long as the local stresses $\sigma_{yy} > \sigma_M$. σ_M is the stress required for unstable microcrack propagation and decreases as the microcrack lengthens (spreads away from its initiation site).

stress of a smooth tensile specimen σ_f for reasons which are discussed below.

Once the miniature tensile specimen has fractured, the notched structure will contain an unstable crack a few grain diameters in length. This crack will propagate both backwards, toward the notch, and forwards, into the elastically stressed portion of the structure (Fig. 7.1c). The stress required to keep the crack in motion is σ_M. If the tensile stress level σ_{yy} in the plastic and elastic zones is greater than σ_M (Fig. 7.13), the crack will be able to propagate until the entire structure has fractured. Consequently the microscopic fracture criterion [3, 4, 5]

$$\sigma_{yy}^{\max} = \sigma_f{}^* \tag{7.26}$$

is also the criterion for macroscopic cleavage fracture in completely clean materials where there is no slow crack growth prior to fracture. *The nominal fracture stress σ_F is that stress which is required to sat-*

isfy Eq. 7.26, by plastic constraint and, where necessary, by strain hardening.

Figure 7.14 shows the typical variation of the fracture load and general yield load (measured with the notch in compression) with temperature, for V notch specimens loaded in tension as well as in bending [5, 26, 27, 28]. Also included on the diagram are (1) a plot of the root displacements at fracture $2V^*(c)$, (2) the ratio of the fracture load to general yield load $(P_F/P_{GY} \cong \sigma_F/\sigma_{GY})$, and (3) the load at which slip or twinning occurs in the first grain ahead of the notch. Three regions are of interest for clean materials.

Region I.   Fracture Due to Initial Yielding

At temperatures below the brittleness transition T_D of smooth tensile specimens, the yield stress σ_Y is greater than the extrapolated σ_f and slip or twinning in one or two grains can initiate cleavage. As shown in Fig. 6.18, the observed fracture stress is equal to σ_Y and decreases with increasing temperature for $T \leqq T'_D$. Similar behavior takes place in notched specimens at very low temperatures [5, 29], and fracture occurs upon local yielding when slip bands or twins form in the *first grain* adjacent to the root. Since the deformation must extend over at least one grain (i.e., $R^* \approx 2d$) to produce a pile up which can nucleate a microcrack, the nominal fracture stress

$$\sigma_F = \frac{\sigma_Y}{K_\sigma} \qquad (\sigma_Y \geqq \sigma_f{}^*) \qquad (7.27)$$

where K_σ is the elastic stress concentration factor *at the first grain boundary*, a distance $r = 2d$ ahead of the notch [5]. As in the case of unnotched specimens this relation only applies in the temperature range AB where $\sigma_Y \geqq \sigma_f{}^*$.

Region II.   Fracture Due to Plastic Constraint

The yield stress σ_Y decreases with increasing temperature, whereas $\sigma_f{}^*$ is independent of or increases slightly with increasing temperature. At temperatures above that corresponding to point B, $\sigma_Y < \sigma_f{}^*$ and consequently slip or twinning in one grain ahead of the notch will not, of its own accord, initiate cleavage. Thus Eq. 7.27 no longer is applicable. Instead, some plastic stress intensification is required to raise the tensile stress level in the plastic zone from σ_Y to $\sigma_f{}^*$. This is provided by the triaxial constraint discussed earlier; cleavage fracture initiates at the elastic-plastic interface, where σ_{yy} has its maximum value, when

$$\sigma_{yy}^{max} = K_{\sigma(p)}\sigma_Y = \sigma_f{}^* \qquad (7.28)$$

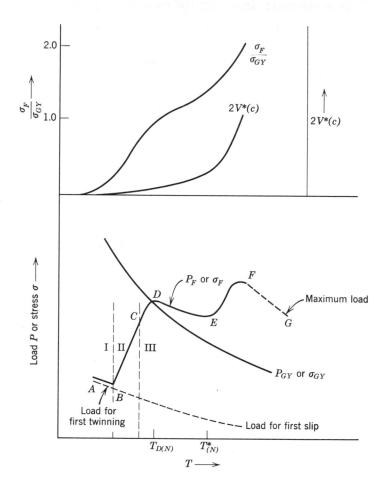

Fig. 7.14. Schematic variation of the effect of temperature on the general yield load P_{GY} and fracture load P_F of notched-bend or tensile specimens of a low strength (non-FCC) material. The variation of the ratio of these parameters $P_F/P_{GY} \simeq \sigma_F/\sigma_{GY}$, the critical root displacement for unstable fracture, $2V^*(c)$, and the loads at which slip or twinning first occur at the notch root are also shown. $T_{D(N)}$ is the temperature at which $\sigma_F = \sigma_{GY}$, and $T^*_{(N)}$ is the temperature at which the ductility and toughness increase rapidly. I, II, and III refer to regions of different fracture criteria.

Consequently, from Eq. 7.24,

$$V^*(c) = \frac{4\Omega\sigma_{GY}\rho}{\pi E}\left[\exp\left(\frac{\sigma_f^*}{\sigma_Y} - 1\right) - 1\right]$$

$$= \frac{4\Omega\sigma_{GY}\rho}{\pi E}R^* \qquad \left(\frac{\sigma_f^*}{\sigma_Y} < \frac{1}{\beta}\right) \qquad (7.29)$$

and

$$G_{Ic} \cong 2\sigma_{GY}V^*(c) = \frac{8\Omega\sigma_{GY}^2\rho}{\pi E}\left[\exp\left(\frac{\sigma_f^*}{\sigma_Y} - 1\right) - 1\right]$$

$$G_{Ic} = \frac{8\Omega L^2\rho}{E}(\sigma_i + k_y d^{-1/2})^2\left[\exp\left(\frac{4G\gamma_m d^{-1/2}}{k_y(\sigma_i + k_y d^{-1/2})} - 1\right) - 1\right]$$

$$\left(\frac{\sigma_f^*}{\sigma_Y} < \frac{1}{\beta}\right) \qquad (7.30)$$

so that the toughness is a sensitive function of the *ratio* σ_f^*/σ_Y.

Similarly $\sigma = \sigma_F$ at $R = R^*$ so that, from Eq. 7.10a,

$$R^* = \frac{1}{\psi}\left\{\left(\sec\frac{\pi}{2}\frac{\sigma_F}{\sigma_{GY}}\right) - 1\right\}c = \rho\left[\exp\left(\frac{\sigma_f^*}{\sigma_Y} - 1\right) - 1\right]$$

$$\left(\frac{\sigma_f^*}{\sigma_Y} < \frac{1}{\beta}\right) \qquad (7.31)$$

Consequently, for a given notch and specimen geometry (ρ, c), the ratio of nominal fracture stress to general yield stress (σ_F/σ_{GY}) is directly related to (σ_f^*/σ_Y). *In fact, for a given notch and specimen geometry, equivalent values of (σ_f^*/σ_Y) lead to equivalent values of (σ_F/σ_{GY}).* This fact is extremely useful in alloy development studies, as will be shown shortly.

Finally it should be noted that if the value of the shape factor α were known, it would be possible to determine directly the nominal fracture strength σ_F in terms of macroscopic (α, c, ψ, ρ) and microscopic (σ_f^*, σ_Y) parameters that are measurable. The relationship

$$\frac{K_{Ic}^2}{\sigma_{GY}^2} = \frac{\sigma_F^2}{\sigma_{GY}^2}\pi^2\alpha c = \frac{8\rho}{\psi(1 - \nu^2)}\left[\exp\left(\frac{\sigma_f^*}{\sigma_Y} - 1\right) - 1\right] \qquad (\sigma \ll \sigma_{GY})$$

$$(7.32)$$

is convenient to use at low stress levels.

Between points B and C (Fig. 7.14) σ_F rises sharply. At point B fracture is usually initiated by twinning (in BCC metals). At C both twin-nucleated and slip-nucleated cleavage has been reported [5, 25, 30]. The sharp rise in σ_F simply results from the fact that (σ_f^*/σ_Y)

increases rapidly with increasing temperature and, although σ_{GY} is decreasing, the exponential term dominates.

A comparison [4, 5, 25, 26, 27, 31] of calculated values of σ_f^* with measured values of the cleavage strength of smooth specimens σ_f indicates that σ_f^* is about 30–50% greater than σ_f. The reasons for this behavior are not well established at present but are probably related to two phenomena. First, calculations of σ_f^* using Eq. 7.31 and measured σ_F/σ_{GY} are sensitive to the value taken for the yield strength in the plastic zone σ_Y. Often the tensile yield strength at the particular temperature is used, but since the strain rate beneath the notch is not known, σ_Y can only be estimated. Furthermore, since the plastic strains at the elastic-plastic interface are less than the Luder's strain [22], and since the tensile yield strength is actually the flow stress at the Luder's strain, the value of σ_Y used in calculating σ_f^*, and hence the value of σ_f^*, are too high [22]. Second, there are probability considerations. In Chapter 6 the value for σ_f was predicted on the basis that fracture would initiate in the most suitably oriented grain (i.e., in the one whose cleavage plane is perpendicular to the tensile axis in a clean material or the one containing the most favorable grain boundary particle–matrix cleavage plane orientation in a dirty material). In a conventional, smooth tensile specimen, containing a large distribution of grains and particles, this assumption is reasonable and in agreement with experiments [32]. In the notched structure, however, the size of the "specimen" will be small and only a few grains will be subjected to high plastic stress intensification. If these grains are not favorably oriented for cleavage, the measured value of σ_f^* will be equal to $m\sigma_f$, where $m > 1$ is a misorientation factor.

Region III.   Fracture Due to Plastic Constraint and Strain Hardening

As the temperature is raised, a point will eventually be reached where $\sigma_f^*/\sigma_Y = 1/\beta$ and plastic stress concentration alone will be insufficient to cause brittle fracture. At this temperature (point C, Fig. 7.14) $V^*(c) = [V^*(c)]_\beta$ and $R^* = R_\beta$, as given by Eq. 7.7. For Charpy V specimens this occurs when $\sigma_F/\sigma_{GY} \cong 0.8$.

Above this temperature (region III, Fig. 7.14) cleavage cracks will only form and spread when a certain amount of strain hardening [5] $\Delta\sigma$ has occurred, so that the tensile stress level ahead of the notch is raised to the point where

$$\sigma_{yy}^{\max} = (\sigma_Y + \Delta\sigma)\frac{1}{\beta} = \sigma_f^* \qquad (7.33)$$

As discussed previously (Fig. 7.12), σ_{yy} reaches its maximum value at R_β. The amount of strain required to initiate fracture there, $\epsilon_{f(R_{\beta'})}$, is

$$\epsilon_{f(R_\beta)} = \frac{\beta\sigma_Y}{d\sigma/d\epsilon}\left(\frac{\sigma_f{}^*}{\sigma_Y} - \frac{1}{\beta}\right) \qquad (7.34)$$

This, in turn, requires a critical root displacement that can be obtained from Eq. 7.23

$$V(c) = V^*(c) = \frac{\rho^2}{H}\frac{\beta\sigma_Y}{d\sigma/d\epsilon}\left(\frac{\sigma_f{}^*}{\sigma_Y} - \frac{1}{\beta}\right) \qquad V^*(c) > [V^*(c)]_\beta \quad (7.35)$$

and consequently

$$G_{Ic} \cong 2\sigma_{GY}V^*(c) = \frac{2\rho^2}{H}\frac{L\beta\sigma_Y{}^2}{d\sigma/d\epsilon}\left(\frac{\sigma_f{}^*}{\sigma_Y} - \frac{1}{\beta}\right) \qquad V^*(c) > [V^*(c)]_\beta \quad (7.36)$$

In relatively narrow specimens containing relatively blunt notches (e.g., the Charpy V) the amount of plastic strain and hence the amount of strain hardening that occurs at R_β *before general yield is reached* is extremely small. Thus the fracture load rises sharply from point C to point D. The temperature corresponding to point D, where $\sigma_F = \sigma_{GY}$, is the macroscopic, brittleness-transition temperature of the notched specimen, $T_{D(N)}$.

At D, plastic hinges (Fig. 7.8) form at the notch root and spread rapidly across the specimen. The fracture load then drops or remains constant until point E is reached, whereupon it again increases sharply. This effect can be explained as follows. As T increases, $\sigma_f{}^*/\sigma_Y$ increases so $V^*(c)$ and G_{Ic} increase. This, in turn, requires larger bend angles at fracture θ_F and hence larger nominal deflections δ_F. These, in turn, require increased values of $(\sigma_F - \sigma_{GY})$ because of general strain hardening. Thus, in region III, σ_F is the stress required to produce a deflection such that Eq. 7.35 can be satisfied. The total effect of a temperature change may be summarized as follows:

$$T\uparrow \rightarrow \frac{\sigma_f{}^*}{\sigma_Y}\uparrow \rightarrow \epsilon_{f(R_\beta)}\uparrow \rightarrow \epsilon_f(c)\uparrow \rightarrow V^*(c)\uparrow \rightarrow \theta_F\uparrow \rightarrow \delta_F\uparrow \rightarrow \epsilon_F \rightarrow \frac{\sigma_F}{\sigma_{GY}}\uparrow \rightarrow (\sigma_F - \sigma_{GY})\uparrow$$

| Metallurgy | Plasticity theory | Fracture mechanics |

where $\uparrow$ indicates an increase in the particular parameter and $\rightarrow$ stands for the word "produces" or the phrase "required for fracture."

Equation 7.37 indicates that three disciplines are involved in the relation between microscopic and macroscopic aspects of fracture:

1. *Metallurgy* explains (Chapter 6) the temperature dependence of $(\sigma_f{}^*/\sigma_Y)$ and the effect of microstructural changes (e.g., grain-size refinement) on this parameter.

2. *Plasticity theory* can be used to relate $V^*(c)$ to $\sigma_f{}^*/\sigma_Y$, as in Eqs. 7.29 and 7.35, and to show that G_{Ic} *increases with* $\sigma_f{}^*/\sigma_Y$ (Eqs. 7.30 and 7.36).

3. *Fracture mechanics* can be used to relate nominal fracture stress σ_F to $V^*(c)$ in specimens of a given geometry. When $\sigma_F < \sigma_{GY}$ (ρ small, c large), the principles of linear elastic fracture mechanics (Chapter 2) can be used to obtain closed form solutions in terms of α, c, and ρ. However, when $\sigma_F > \sigma_{GY}$ this is not possible, at least at the present time.

Between points D and E, $V^*(c)$ and hence $(\sigma_F - \sigma_{GY}) \cong (d\sigma/d\epsilon)\epsilon_F$ increase more slowly than σ_{GY} decreases (for BCC metals, Charpy geometry), and consequently σ_F decreases with increasing temperature.† At a certain temperature $T^*_{(N)}$, known as the *ductility transition*, σ_F increases very rapidly with temperature because $V^*(c)$ increases rapidly with temperature. This behavior, which is extremely sensitive to specimen thickness [42, 43, 65], results from a change in stress state at R_β due to an increasing tendency for plane-stress deformation [22, 35]. In effect, the large strains reduce constraint and cause $1/\beta$ to decrease, which causes $V^*(c)$ to rise rapidly. Eventually a point F is reached where the fracture load is greater than the ultimate load. Since clean, single-phase materials cannot fracture by ductile rupture, they continue to deform by bending and simply wrap themselves into a V shape. In tension, localized necking occurs and produces a chisel-point failure.

The effect of nonhomogeneous and homogeneous distribution of dispersed phases

In the previous chapter it was shown that in dirty materials initial growth of microcracks occurs at a lower stress σ_G than the stress σ_B required to spread these cracks through grain boundaries and cause unstable fracture. In notched specimens the formation of nonpropagating microcracks ahead of the notch at a local stress level

$$\sigma_{yy} \approx \sigma_G < \sigma_f{}^*$$

† On a macroscale, the behavior is similar to that observed on a microscale in unnotched dirty materials (see Fig. 6.24) where nonpropagating microcracks behave like small notches.

considerably complicates the mechanics of the fracture problem since the formation of the stable microcrack will alter the constraint and the local strain distribution ahead of the notch. At low temperatures very few stable cracks form before fracture. To a first approximation, we can assume that the alterations in $K_{\sigma(p)}$ and $\epsilon(r)$ are negligible, so that cleavage will occur when the strains in the cracked miniature tensile specimen ahead of the root are equal to $\epsilon_{f(R\beta)}$. Physically cleavage occurs when a new, unstable microcrack is formed at the tip of one of the stopped ones (or near a few of these linked together by tearing), in a manner similar to that described in Chapter 6. Consequently there are two "notch effects" superimposed on one another—one due to the macroscopic notch and one due to the microcracks (microscopic notches) which operate in the stress and strain field of the former. Under these conditions the temperature dependence of $V^*(c)$ will be similar to the temperature dependence of $\epsilon_{f(R\beta)}$ and hence to the temperature dependence of the tensile ductility ϵ_f for a given notch and specimen geometry. In accordance with Fig. 6.21, $V^*(c)$ and hence $(\sigma_F - \sigma_{GY})$ increases more slowly for dirty materials than for clean ones (Fig. 7.15) in regions I, II, and III.

Region IV. Fracture Initiated by Fibrous Tearing

Fibrous fracture (i.e., rupture) occurs in dirty materials, or in clean ones containing a homogeneous distribution of hard, second-phase particles, when the strains build up to ϵ'_f, where $\epsilon'_f \approx 0.50$–0.80 in low strength materials. Above the temperature $T_{S(N)}$ the root strain $\epsilon_f(c)$ that would be required to initiate cleavage beneath the root is greater than the strain $\epsilon'_f(c)$ required to initiate rupture

$$\epsilon_f(c) = \frac{\rho}{H}\frac{\beta\sigma_Y}{d\sigma/d\epsilon}\left(\frac{\sigma_f^*}{\sigma_Y} - \frac{1}{\beta}\right) > \epsilon'_f(c) \qquad (T > T_{S(N)}) \qquad (7.38)$$

Fracture is therefore initiated by rupture at or close to the root [5, 23], unless ρ is large.† Since the mode of unstable fracture initiation changes from cleavage below to rupture at the root, $T_{S(N)}$ is known as the *initiation-transition temperature*. Typically, about 80–85% of the fracture surface of broken Charpy impact specimens is crystalline at $T_{S(N)}$, which marks the lower end of region IV.

† If ρ is large, the stress level near the root will be small (Eq. 7.4) and there will be an insufficient number of voids (cracked inclusions or particles or their interfaces) to initiate fibrous rupture at the notch. This is the case in unnotched tensile specimens which fracture after necking. The fracture usually initiates at the center of the neck, along the center line of the specimen (see Fig. 5.4).

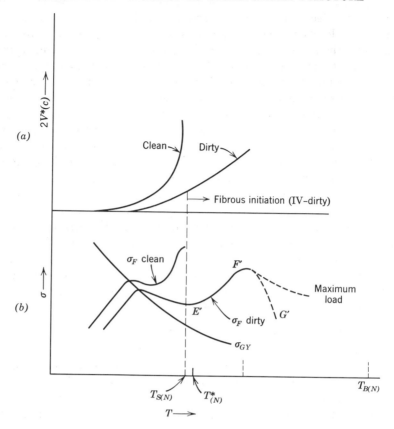

Fig. 7.15. The change in critical notch tip displacement (a), and nominal stress σ_F (b) at which unstable fracture occurs, when a clean material is made dirty and all other variables are held constant.

Once fibrous cracking has occurred, the notch will lengthen and perhaps sharpen, especially when it links up with any nonpropagating microcracks that previously may have formed beneath the root [17]. Initially these processes are noncumulative (stable) and occur slowly, under an increasing load, as additional notch-tip displacements are required. However, as the notch lengthens and sharpens, the strain concentrated near its tip increases and the point of maximum stress intensification moves back toward the tip (Fig. 7.12b). Consequently the stress level in front of the crack σ_{yy} increases with increasing amounts of slow growth and the crack begins to accelerate. This acceleration, in turn, raises the local strain rate and hence σ_Y, causing

(σ^*_f/σ_Y) to decrease. This effect, coupled with sharpening, causes further acceleration; eventually microcrack propagation becomes unstable and the specimen fractures. The value of $2V^*(c)$, as measured at the *tip of the original notch*, is thus the sum of the displacements required to initiate the fibrous tear and that required to cause stable, microscopic slow crack growth and sharpening before unstable fracture.

As in the case of clean materials, there is a ductility transition $T^*_{(N)}$ at which constraint is relaxed and $V^*(c)$ and hence G_{Ic} increase sharply. The increase in G_{Ic} and the occurrence of stable, plane-strain fracture allow shear lip formation to precede instability. Consequently *macroscopic slow crack growth* will precede unstable fracture (Chapter 3) and the fracture resistance will become $G_c > G_{Ic}$, as shown in that chapter.

Above a temperature corresponding to point F' the ultimate load is reached before the fracture load. Whereas clean, single-phase materials do not fracture once the ultimate load is reached (in bending), dirty materials, or clean ones containing hard phases, can still fail by rupture (and some radial cleavage) after the ultimate load is reached. The fracture load then decreases with increasing temperature. Since this load has little physical significance, instability already having been reached at the ultimate load, the curve $F'G'$ is commonly drawn with a dashed line.

At temperatures in the vicinity of $T_{S(N)}$, the load drops immediately to zero once unstable fracture has initiated. At higher temperatures shear lip formation occurs at the sides and back of the specimen *after* the unstable fracture has initiated; this produces the picture-frame effect discussed in Chapter 3. In these instances the load drops sharply from the fracture load to some lower value, but then tapers off slowly [30, 34] since additional nominal deflections are required to cause the final separation along the shear lips. The work done in this process is known as *postbrittle fracture energy*, in contrast to the *prebrittle fracture energy* which was expended in reaching the fracture load. As shown in Chapter 3, the size of the radial zone of fast cleavage decreases and the shear lip size and postbrittle fracture energy increases, until $T = T_{B(N)}$. At this point the fracture is 100% fibrous.

Figure 7.16 illustrates the Charpy V curves for two steels; one is a dirty steel containing a high density of grain boundary carbides because of its low manganese/carbon ratio. The other is a clean steel having a high Mn/C ratio. The shape of the Charpy curve at low temperatures closely parallels the value of $V^*(c)$ shown in Fig. 7.15. It is also noteworthy that the sharp rise in $V^*(c)$ is accompanied by a change

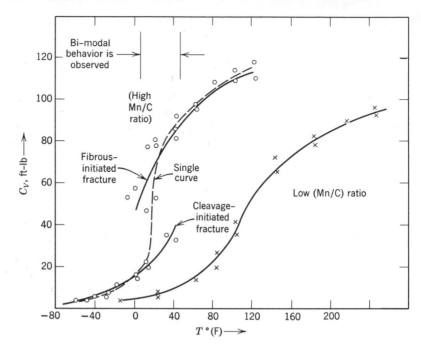

Fig. 7.16. The difference in shape of Charpy V notch impact curves for a clean (high Mn/C ratio) and dirty (low Mn/C ratio) mild steel. *After Pellini and Puzak* [69].

in the mode of fracture initiation. This results from the fact that $T_{S(N)}$ is equal to or just above $T^*_{(N)}$ for this particular steel. In some cases $T^*_{(N)}$ is not well defined because of the probability considerations (e.g., orientation) that determine σ_f^*. The change in mode of fracture initiation may then occur over a range of temperatures and the behavior is termed *bimodal*.

The effect of strain rate

In BCC metals and ionic crystals, increases in applied strain rate $\dot{\epsilon}$ (e.g., from slow bend conditions to Charpy impact conditions) produce three effects which cause the position of the σ_F and σ_{GY} curves to be shifted [25]: (1) σ_Y and hence σ_{GY} are increased; (2) the temperature dependence of σ_Y and σ_{GY} is reduced; (3) the rate of strain hardening $d\sigma/d\epsilon$ is decreased.

These effects, in turn, cause the σ_F and σ_{GY} curves to shift in the manner shown in Fig. 7.17. The shift may be explained as follows.

Since $\sigma_f{}^*$ is independent of strain rate, the increase in σ_Y decreases the value of $\sigma^*{}_f/\sigma_Y$ at temperature T. Consequently, in the low stress range $(\sigma_F < \sigma_{GY})$, R^* and hence σ_F/σ_{GY} will be reduced. Since σ_F/σ_{GY} is decreased more than σ_{GY} is increased, σ_F itself will be lower. A decrease in $d/dT^-(\sigma_F)$ results from a lowering of $d/dT^-(\sigma_f{}^*/\sigma_Y)$, owing to effect 2 listed above. The net effect of lowering σ_F and $d\sigma_F/dT$ and raising σ_{GY} is to increase markedly the value of $T_{D(N)}$, the brittleness-transition temperature where $\sigma_F = \sigma_{GY}$.

Similarly, the initiation-transition temperature $T_{S(N)}$ and the ductility-transition temperature $T^*_{(N)}$ will be raised. Recent experiments have noted [25] that a six-order of magnitude change in the applied strain rate (10^{-3} per min up to 10^3 per min) can produce a 90°C increase in $T^*_{(N)}$ for mild steel. Figure 7.17 indicates that the temperature range $T_{D(N)}$ to $T^*_{(N)}$ is smaller at higher strain rates than at low strain rates. This effect appears to be related to a decrease in strain-hardening rate which occurs at high strain rates [25], at least in mild steel at moderate temperatures. According to Eq. 7.35, a low value of $d\sigma/d\epsilon$ implies a large value of $V^*(c)$, all other factors being held constant. Furthermore, when $d\sigma/d\epsilon$ is small, $V^*(c)$ increases more sharply with

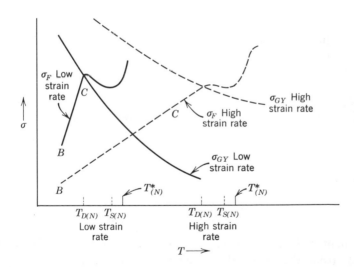

Fig. 7.17. The effect of strain rate on σ_F and σ_{GY} for low strength, strain-rate sensitive (non-FCC) material. Increasing strain rate increases $T_{D(N)}$ and $T_{(SN)}$ and T_N^* but decreases the difference between $T_{D(N)}$ and $T^*_{(N)}$. After T. Wilshaw and P. Pratt [25].

increasing temperature (i.e., increasing $\sigma_f{}^*/\sigma_Y$) than when $d\sigma/d\epsilon$ is large.

The use of Charpy-type specimens for alloy development purposes

In addition to providing information that is useful for design purposes (see Chapter 3), the Charpy impact test is extremely useful in alloy development studies when the effect of alloying additions or processing changes on toughness are to be determined. Usually the change in a particular transition temperature (e.g., 50% fibrous) is taken as a measure of the beneficial (or detrimental) effects of the variation in alloy composition or processing conditions. This is satisfactory for obtaining a first approximation of the magnitude of these effects. A great deal more information can be obtained, at a relatively low cost (less than $2,500 for capital equipment), if the Charpy test is *instrumented* to record the dynamic load-time curve (Fig. 7.18a). Recent investigations have shown [30, 35, 36, 61] that the impact energy can be correlated directly with the area under this curve. Figure 7.18b indicates the effect of temperature on these various parameters for a mild steel.

At temperatures somewhat below $T_{D(N)}$ (e.g., when $\sigma_F < 0.8\sigma_{GY}$), fracture results only from plastic constraint (in Charpy V specimens), and Eq. 7.31 describes the relation between σ_F/σ_{GY} and $\sigma_f{}^*/\sigma_Y$. As stated earlier, equivalent values of $\sigma_f{}^*/\sigma_Y$ lead to equivalent values of σ_F/σ_{GY}. Thus, if a particular heat treatment produces a decrease in $\sigma_f{}^*$ from $\sigma_{f(1)}^*$ to $\sigma_{f(2)}^*$ without producing any other changes in σ_Y, σ_F will be reduced from $\sigma_{F(1)}$ to $\sigma_{F(2)}$ while σ_{GY} stays the same (Fig. 7.19). Suppose that at $T = T_1$, $\sigma_{F(1)}/\sigma_{GY} = 0.6$, and that at $T = T_2$, $\sigma_{F(2)}/\sigma_{GY}$ also equals 0.6. Then, if $\sigma_f{}^*$ is independent of temperature,

$$\frac{\sigma_{f(1)}^*}{\sigma_{Y(T_1)}} = \frac{\sigma_{f(2)}^*}{\sigma_{Y(T_2)}}$$

and hence

$$\frac{\sigma_{f(1)}^*}{\sigma_{f(2)}^*} = \frac{\sigma_{Y(T_1)}}{\sigma_{Y(T_2)}} = \frac{\sigma_{GY(T_1)}}{\sigma_{GY(T_2)}} = \frac{P_{GY(T_1)}}{P_{GY(T_2)}} \tag{7.39}$$

Consequently the relative change in cleavage strength can be obtained directly from the ratio of the general yield loads at temperatures where the σ_F/σ_{GY} *ratios* are equivalent.

The advantage of using Charpy V specimens for this determination is that quite often the alloys are so tough that σ_f cannot be measured directly and easily in a smooth tensile test. That is, even at $-196°C$, the lowest test temperature that can be obtained easily and

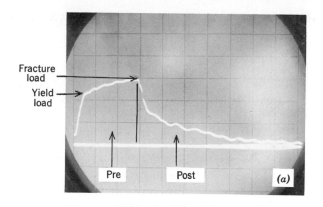

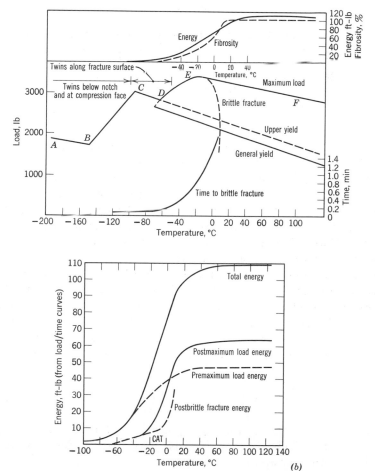

Fig. 7.18. (a) Typical trace from an instrumented Charpy test on mild steel, showing prebrittle and post-brittle fracture energies. *Courtesy R. Wullaert.* (b) Effect of temperature on various fracture parameters, as determined by instrumented Charpy tests on mild steel. The time to brittle fracture is directly proportional to $V^*(c)$ [30].

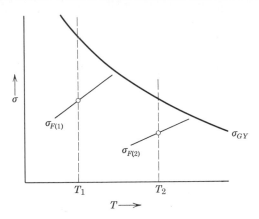

Fig. 7.19. Effect of a decrease in the microscopic fracture stress $\sigma_f{}^*$, from $\sigma_{f(1)}^*$ to $\sigma_{f(2)}^*$ on the temperature dependence of the nominal fracture strength σ_F when σ_Y remains constant:

$$\left(\frac{\sigma_{F(1)}}{\sigma_{GY}}\right)_{T_1} = \left(\frac{\sigma_{F(2)}}{\sigma_{GY}}\right)_{T_2}.$$

inexpensively, smooth specimens are so ductile that fracture initiates by fibrous tearing (i.e., $T_s < -196°C$) and hence variations in the cleavage strength cannot be observed. The method described above may then be easily used for evaluating the effect of microstructure changes on $\sigma_f{}^*$. For example, in medium strength steels, temper embrittlement produces a decrease in γ_m and hence a decrease in $\sigma_f{}^*$ (Chapter 10). This change may be evaluated from the change in σ_F after heat treatment, as shown above.

It should be pointed out that this approach is *not restricted* to the instrumented Charpy test, but may be applied even more easily to the three-point bend test. (The instrumented Charpy simply provides a means of studying the microscopic effects at high strain rates.) For the Charpy V specimen $\sigma_{yy}^{max} = (1/\beta)\sigma_Y = 2.18\sigma_Y$ at $\sigma/\sigma_{GY} = 0.8$, so that $\sigma_f{}^* = 2.18\sigma_Y$ at $\sigma_F/\sigma_{GY} = 0.8$. Since $\sigma_Y = 33.3P_{GY}$ when σ_Y is in psi and P_{GY} is in pounds,

$$\sigma_f{}^* = 72.5P_{GY(0.8)} \qquad \left(\frac{\sigma_F}{\sigma_{GY}} = 0.8\right) \qquad (7.40)$$

where $P_{GY(0.8)}$ is the (estimated) general yield load when $\sigma_F/\sigma_{GY} = P_F/P_{GY} = 0.8$. This relationship is very useful for determining quickly a value for $\sigma_f{}^*$.

At temperatures below the isothermal crack-arrest temperature

(CAT) essentially all of the energy is absorbed in causing the initiation and unstable propagation of microcracks, so that the temperature dependence of the Charpy curve will be similar to the temperature dependence of $V^*(c)$ and hence similar to the temperature dependence of the tensile ductility ϵ_f. Above the CAT a significant amount of postbrittle fracture energy will be absorbed by shear lip formation at the sides and back of the specimen.

At $T = T_{B(N)}$ the fracture is 100% fibrous. This point defines the onset of the Charpy shelf, where the energy $C_V(\max)$ becomes independent of the temperature. $C_V(\max)$ decreases with ϵ'_f, the tensile ductility of unnotched specimens in a temperature range where the fracture is 100% fibrous or shear [37], and hence with: (1) decreasing cleanliness or increasing volume fraction of second phase [38], (2) increasing strength level, and (3) decreasing strain-hardening rate (see Section 5.2, p. 217).

The slope of the Charpy curve in the transition region is related to the values of $T^*_{(N)}$, the energy E^* absorbed in fracture initiation at $T^*_{(N)}$, and the values of $T_{B(N)}$ and $C_V(\max)$. This slope is difficult to define, particularly when bimodal behavior (Fig. 7.16) occurs. To a first approximation

$$\text{Charpy slope} \cong \frac{C_V(\max) - E^*}{T_{B(N)} - T^*_{(N)}} \qquad (T > T^*_{(N)}) \qquad (7.41)$$

The shape and position of the Charpy curve is strongly dependent on microstructural variables. Those factors that decrease $(\sigma_f{}^*/\sigma_Y)$, such as large grain size, an increase in σ_i (owing to cold work, neutron irradiation [40], precipitation hardening [41], etc.), a decrease in γ_m (owing to impurity segregation), and an increase in k_y (owing to a more planar slip mode), all tend to raise $T_{D(N)}$, $T^*_{(N)}$ and $T_{B(N)}$. Since the NDT (as measured in a drop weight test) is about equal to $T^*_{(N)}$, it, too, is raised by these metallurgical processes. Those factors such as low-strain hardening rate [30] that lower $C_V(\max)$ tend to broaden out the transition range and lower the Charpy slope. As shown in Chapters 10 to 13, the effect of alloy content and processing conditions on toughness must be thought of as: (1) What is the effect of processing on microstructure? (2) What is the effect of microstructure on toughness? The latter effects are summarized in Table 10.1.

Effect of variations in notch and specimen geometry

Root Radius ρ. Equations 7.30 and 7.36 indicate that in regions II and III respectively the notch toughness G_{Ic} decreases with decreasing root radius ρ at a given temperature [2, 24, 33] and that the temperature (i.e., $\sigma_f{}^*/\sigma_Y$) dependence of G_{Ic} or σ_F decreases with decreasing ρ (Fig. 7.20). This has been observed for $\rho > \rho_{eff}$, where ρ_{eff} is called the effective root radius [2, 5, 24]. For $\rho < \rho_{eff}$, G_{Ic} is independent of changes in ρ and the structure behaves as though the radius of the notch were equal to ρ_{eff}. The reason for this effect is that the plastic deformation processes (slip or twinning) which initiate cleavage require a minimum volume, of the order of the grain size or twin band spacing, in which to operate. Thus, irrespective of the value of ρ, R^* must be at least equal to the grain size $2d$. Consequently the minimum value of ρ_{eff} is about $d/2$ for a parallel-sided notch and about d for a Charpy notch, according to Eq. 7.7.

When fracture is initiated by fibrous tearing at the root, $V^*(c) \cong 2\rho\epsilon'_f(c)$ will also decrease with decreasing ρ, for ρ greater than ρ'_{eff}. Now ρ'_{eff} is determined by the size of the localized regions of tearing between voids and stable microcracks. For example, if equal elongation of the voids and contraction of material between them occurs during tearing, then the displacement $2V^*(c)$ accompanying rupture is of the order of the distance of the void spacing l. Consequently, when $\rho < l$, the tip will assume a radius of curvature approximately equal to l after tearing has begun. Thus ρ'_{eff} is of the order of l. This value of ρ'_{eff} may decrease during propagation if instability occurs and increasing amounts of cleavage (i.e., decreased tearing) result from the acceleration of the crack.

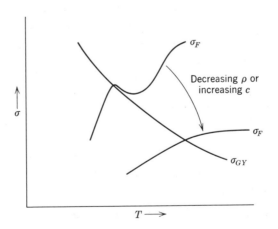

Fig. 7.20. Effect of a decrease in ρ or an increase in flaw depth $2c$ on the nominal fracture strength σ_F.

The decrease in σ_F with decreasing ρ (Fig. 7.20) at low stresses results from two effects, (1) the decrease in G_{Ic} with decreasing ρ, and (2) the increase in α with decreasing ρ which causes σ_F to be lowered (in region II) for a given G_{Ic} (Eq. 7.31).

Thickness t

Figure 7.21 shows the effect of thickness on the fracture load, and hence on $V^*(c)$ and G_{Ic}, for square, V notch specimens loaded in slow bending [42]. Thickness has no significant effect on σ_{GY}, nor on σ_F below general yield, but it has a significant effect on σ_F above general yield. These effects can be explained as follows. Plane-strain conditions will exist in the center of the plate when the plastic zone size is about $t/4$, provided that ρ is small. Since the plastic zone size before general yield is about 1 mm [14], essentially† all of these specimens are loaded in plane strain prior to general yield; consequently σ_F and σ_{GY} will be independent of thickness.

Once general yield occurs, constraints are relaxed in the thickness direction, by an amount that decreases with increasing distance inward from the plate surface. From Fig. 7.21 it appears that this relaxation is so complete in the 0.1 in. thick specimen that $V^*(c)$ (and hence σ_F) must increase sharply at $\sigma = \sigma_{GY}$, to provide the additional strain hardening before fracture which can compensate for the decrease in $1/\beta$. In the 0.2 in. thick specimen, $K_{\sigma(p)}$ retains its maximum value $1/\beta$ until a higher temperature is reached (i.e., until after additional deformation has occurred). σ_F and hence $V^*(c)$ are independent of thickness for $t > 0.3$ in. Similar results have been noted in impact tests [43] on Charpy specimens. In addition to affecting the value of $V^*(c)$ and hence G_{Ic}, a change in thickness will also affect the fracture stress when shear lip formation occurs *prior* to instability. This was discussed in some detail in Chapter 3 and need not be repeated here.

Finally, thickness will have an indirect effect on toughness since gradients in microstructure occur in thick structures after materials processing [44]. Because the center of a plate, for example, cools more slowly than the outer surface, the grain size at the center will be larger than at the outer surface. Consequently there is a gradient in toughness across the plate thickness that increases with increasing thickness.

Plate Width W and Notch Depth 2c

Relatively little is known about the effect of changes in plate width in specimens containing a crack of fixed length, for in most examina-

† The thinnest specimen will be in plane strain until $\sigma/\sigma_{GY} \cong 0.8$.

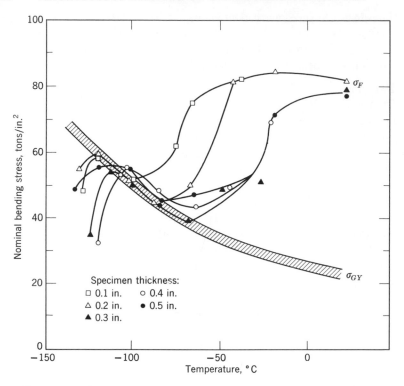

Fig. 7.21. Effect of thickness and temperature on the nominal fracture strength of notched specimens of 0.2% C steel. *After J. Knott* [42].

tions of size effect the ratio c/W is kept constant while W is changed. When the ratio of crack depth to plate width c/W is large, G_{Ic} will be a function of W [1]. The reason is that in this situation the slip-line field ahead of a notch loaded in tension will appear as straight lines inclined at 45° to the tensile axis rather than as logarithmic spirals (Fig. 7.9b). Since there is no constraint associated with this form of deformation, and since the strains directly ahead of the notch are small, larger values of crack-tip displacement are required before unstable cleavage fracture can develop ahead of the notch. This is the reason that shear lips tend to form along the sides and back surfaces of a specimen, except at very low temperatures. When c/W is small, the slip-line field ahead of the tip assumes, at least initially, the spiral shape associated with constrained deformation, and stress intensification and strain hardening increase ahead of the crack. Consequently we should expect that $V^*(c)$ and hence G_{Ic} will decrease as c/W decreases until a point is reached where they become independent of c/W.

When c/W and G_{Ic} are small, $\sigma_F < \sigma_{GY}$, and the variation of σ_F with c is given by $\sigma_F = K_{Ic}/\sqrt{\pi \alpha c}$, where α is a function of c at high stress levels. When $\sigma_F > \sigma_{GY}$, the fracture stress will be related to the compliance of the structure and hence to the ratio of c/W. Increasing values of c in a structure of fixed width cause $(\sigma_F - \sigma_{GY})$ and hence σ_F to decrease as shown in Fig. 7.20.

7.3   The Physical Significance of G_{Ic} for Low-Energy Tear Fracture in High Strength ($\sigma_Y > E/150$) and Medium Strength ($E/300 < \sigma_Y < E/150$) Materials

Unstable fracture in high strength materials occurs by the formation and coalescence of voids at the tip of an advancing crack or notch. This process is similar to that which takes place above the cleavage propagation transition $T_{B(N)}$ in low strength materials, except for the fact that much smaller crack opening displacements are required. Consequently G_{Ic} and C_V (max) are lower for the ultrahigh strength materials, and unstable fracture can develop at much lower nominal stress levels for a given notch and specimen geometry. At the present time only a simplified description of this complicated process can be presented.

For a notch whose radius ρ is small, but greater than ρ'_{eff}, $V^*(c)$ will be the sum of the displacement associated with the formation of a void at the notch tip (Fig. 7.22a) and the displacement required to join the void and the tip together by plastic deformation (i.e., tearing), Fig. 7.22b. Void formation will occur when inhomogeneous plastic deformation is able to crack the interfaces between dispersed particles (or inclusions) and the matrix or the particles or inclusions themselves. This will require a tensile stress of the order of

$$\sigma_p \cong \frac{M\gamma}{nb}$$

where M is a numerical constant, n is the number of dislocations piled up behind the particle, and γ is the work done in cracking the particle or its interface. The value of σ_p will depend on particle size and will be larger for small particles than for big ones. When the particle is small, dislocations will be able to cross slip out of the pile up so that n is reduced and σ_p is increased. γ will be higher when the particle is strongly bound to the matrix, so that σ_p will be lower for inclusions

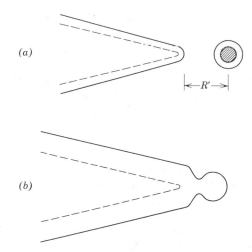

Fig. 7.22. Processes leading to void formation (a) and coalescence of the void and crack tip (b) during normal rupture. $V^*(c)$ decreases with decreasing R'.

than for precipitate particles of the same shape. Consequently σ_p will vary with particle size p in a manner shown schematically in Fig. 7.23a.

All materials, especially those produced commercially, contain a distribution of particle sizes (Fig. 7.23b), where N is the number of particles of a particular size p. In ultra-high strength materials, most of the strengthening is due to very small precipitate particles [45] and $\sigma_Y \approx Gb/\lambda^*$. Consequently, for a given volume fraction of precipitate and a given strength level σ_Y, most of the particles will have a size p^* such that the average distance between them λ is equal to λ^*. Since λ, defined as the average distance between two particles of the same size, is proportional to $1/N$, the average distance between the particles increases with particle size (Fig. 7.23c) for $p > p^*$. Consequently σ_p varies with particle spacing λ as shown in Fig. 7.23d.

Since the average distance from the notch tip to a particle is of the order of the spacing of particles of that particular size, λ also represents the distance ahead of the notch at which the longitudinal stress σ_{yy} must be equal to σ_p before a particle will crack and a void can be formed.

For a given notch geometry and yield strength σ_Y, σ_{yy} increases with increasing plastic zone size until $R = R_\beta$; for $R > R_\beta$, $\sigma_{yy} = (1/\beta)\sigma_Y$

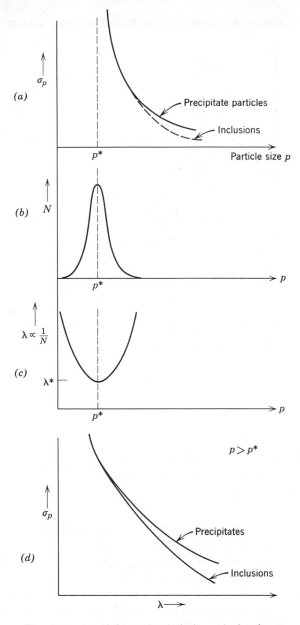

Fig. 7.23. (*a*) Schematic variation of the fracture strength of a hard particle or inclusion σ_p with particle size p. (*b*) Typical distribution of particle sizes about an average value of particle size p^*. (*c*) The variation of λ, the mean distance between two particles of the same size, with particle size p. (*d*) The variation of σ_p with λ for $p > p^*$.

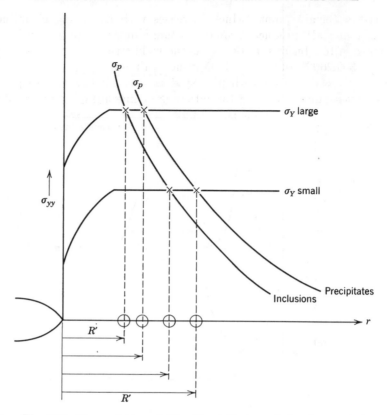

Fig. 7.24. The variation of R' with yield strength level σ_Y and the presence or absence of inclusions, in high strength materials. R' is the distance from the notch tip at which $\sigma_{yy} = \sigma_p$ and a void can be formed.

between R_β and the elastic-plastic interface (Fig. 7.6). As shown in Fig. 7.24a, the plastic zones will have to spread to a distance R' before a particle is encountered which can fracture at the stress level $\sigma_{yy} = (1/\beta)\sigma_Y$ and a void can be formed.† Decreasing particle strength or increasing σ_Y decreases the plastic zone size required for void formation, since a particle that will fracture at the stress level in the plastic zone will exist at a smaller distance from the notch. However, because σ_p increases sharply as λ decreases,‡ increases in

† $d\sigma/d\epsilon$ is very small in these materials and so strain hardening does not raise $\sigma_{yy}^{\max}$ appreciably over $(1/\beta)\sigma_Y$.

‡ Consequently the very fine particles which contribute to the yield strengthening and whose spacing $\lambda = \lambda^*$ are normally not the ones that crack.

σ_Y past a certain point (which increases with increasing cleanliness of the material) produce a smaller change in R'. Consequently the decrease in toughness that occurs as the yield strength increases (Figs. 3.37, 3.23–3.26) begins to level off at high values of σ_Y.

Since $V(c)$ decreases with plastic zone size, smaller root displacements are required for void formation as σ_Y increases (up to an effective value) and as σ_p decreases (higher inclusion content).

The tip displacement required to join the void and the notch together by shearing will be of the order of $\epsilon_f'(c)\rho$, where $\epsilon_f'(c)$ is (approximately) the ductile fracture strain. As shown in Eq. 5.14, ϵ_f' decreases as the spacing between the crack tip and the void $l_0 \approx R'$ decreases. Consequently the same factors which favor small tip displacements during void formation (σ_Y large, σ_p small) also favor small displacements during void coalescence. In addition, the coalescence will be easier in materials having a low rate of strain hardening [66, 67, 68] (Section 5.2, p. 217) if all other variables are held constant.

Thus $2V^*(c)$, the total displacement required for fracture, hence $G_{Ic} \cong 2\sigma_{GY}V^*(c)$ and $K_{Ic} = \{EG_{Ic}\}^{\frac{1}{2}}$, all decrease with increasing yield strength (Fig. 3.37) and decreasing root radius (above a limiting value ρ_{eff}' that decreases with σ_Y). The large difference in toughness between air-melted and vacuum-melted steels, when both are evaluated at the same yield strength level, results from a decrease in the inclusion content (associated with high sulfur and phosphorus content) after vacuum melting [46].

As pointed out in Chapter 3, the ductile-brittle transition in *medium strength materials* results from the fact that σ_Y increases as the temperature decreases. At low temperatures the yield strengths are greater than $E/150$ and the alloys are susceptible to low-energy tear fracture, as well as to cleverage [70].

7.4 Micromechanical Techniques for Improvement of Notch Toughness in Brittle Materials

In the preceding sections it was shown that increases in σ_Y produce a decrease in notch toughness associated with both cleavage fracture[†] and low-energy tear fracture because the high tensile stress level in the plastic zone ahead of the notch is then able to crack grains and particles at lower tip displacements. This means that a high degree of notch toughness and a high load-carrying capacity in the absence of notches (i.e., a high σ_Y) tend to be incompatible. The only way

---

† Except when σ_Y is increased by grain-size refinement.

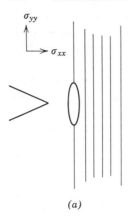

(a)

Fig. 7.25. Splitting produced along a weak interface perpendicular to an advancing crack or notch by transverse stresses σ_{xx}.

that they can be made compatible is when some means exist for reducing plastic stress concentration (i.e., constraint) before the structure fails unstably. This is possible in *anisotropic materials* [47] such as cold-rolled plate or extruded bars, where most of the grain boundaries or phase boundaries are lined up parallel to the working direction and are themselves embrittled (e.g., by precipitation) to a certain degree. If the boundaries are relatively weak, the transverse stresses σ_{xx} set up ahead of a notch lying perpendicular to them may be able to cause them to split open *before* an unstable tear or cleavage fracture can develop in the plane of the notch [48, 49]. When splitting occurs (Fig. 7.25), the crack tip becomes blunted and constraint is relaxed. Then $1/\beta$ drops sharply and larger root displacements are required for microcrack initiation or void formation in the plane of the notch.†

This splitting effect can produce pronounced improvements in materials that normally are brittle. Ausformed steel (Chapter 10) can stop a 240 ft-lb impact hammer, even when $\sigma_Y = 200,000$ psi, by splitting repeatedly (Fig. 7.26) [49]. Similarly, in refractory metals where cleavage is a potential problem even above ambient temperature, the initiation-transition temperature $T_{s(N)}$ can be lowered by large amounts (e.g., 250°C for tungsten) [50] after cold rolling has produced the elongated grain structure.

In order for these processes to be effective, two requirements must be satisfied. First, the distance between the splitting interfaces must be small so that the process can occur repeatedly and prevent an unstable fracture from developing. Although in some materials (e.g., wrought iron) the presence of elongated inclusions may be sufficient, in general, their density is too low. A fine elongated grain structure, embrittled to a certain degree by precipitation, is inherently more desirable. Secondly, although the interfaces must be relatively weak,

† This process is only effective when the nominal stress is parallel to the fiber direction. When it is perpendicular to it, low-energy tear fracture develops along the boundary and the material is extremely weak.

they cannot be too weak in the absolute sense or the structure will simply slide apart under load. Consequently the trick here is to develop materials that have the right balance between high strength and adequate toughness for particular application. In addition to the examples cited above, *composite materials* such as bamboo and fiberglass also fall in this category. These will be discussed in some detail in Chapters 12 and 13.

Another means of increasing resistance to cleavage fracture in certain materials (e.g., AgCl [51], W [52], zinc [53]) is by the addition of small amounts (2–5%) of fine (1–5 μ diameter) hard, spherical particles. This is usually done by conventional powder metallurgy techniques. When these particles are loosely bound to the matrix, they will be easily separated from it during plastic deformation and cavity formation will occur prior to fracture [54]. These cavities will interfere with an advancing microcrack by causing it to decelerate [55] (as it approaches the cavity surface which cannot support a tensile stress) and to become blunt after it runs into the cavity. Both of these processes increase the work done in crack propagation, which then occurs by fibrous tearing rather than rapid cleavage [51]. This increases the notch toughness, as shown in Fig. 7.27 for the case where 2.5 volume per cent spherical particles of Al_2O_3 are dispersed in AgCl [51]. This process is only effective in increasing cleavage resistance of *low strength materials;* the addition of hard particles to

Fig. 7.26. Splitting of an ausformed steel after impact. *After A. McEvily and R. Bush* [49].

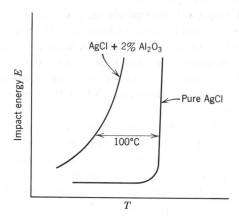

Fig. 7.27. Subsize Charpy curves for pure
AgCl and AgCl containing 2% of spherical
alumina particles. *After T. Johnston et al*
[51].

high strength materials would only serve to decrease the work done in
the already brittle tear fracture and reduce the toughness.

A similar method for increasing the toughness of a particular material containing a given notch and specimen geometry is by the addition of macroscopic cavities (i.e., drilled holes) near the root of the
notch [56]. These holes need not be large and are able to produce
significant improvements in properties when their size is of the order
of root radius. When the holes are placed in the trace of the plastic
hinges that emanate from the notch tip at general yield, $K_{\sigma(p)}$ is
reduced by relaxation of constraint and the local strains are redis-
tributed (Fig. 7.28a). Consequently $T_{D(N)}$ can be lowered, by about
20°C in a bend test, and the material is considerably stronger in a
temperature range where both drilled and undrilled specimens fracture by cleavage. There is a higher temperature range in which
drilled specimens are able to stop completely the 240 ft-lb Charpy impact hammer while standard samples fracture completely by cleavage.
The mechanism by which this occurs is shown in Fig. 7.28b. When
ductile tearing between the notch and the holes occurs *before* cleavage
can initiate, the notch tip becomes extremely blunt, constraint is reduced, and cleavage is then completely suppressed. Consequently,
the holes, like the fibers described above, provide a fail-safe mechanism by causing fracture, when it does occur, to do so by the ductile
tear mode.

The amount of reduction depends on a combination of macroscopic and microscopic parameters. For a given alloy composition the holes cause a maximum reduction in 50% transition temperature when they lie in the plastic hinges that emanate from the tip, since this favors the tearing process shown in Fig. 7.28b, and when they are close to the notch. Holes placed at or in front of the tip have no effect, unless their radius is much greater than that of the notch, as they do not produce large-scale tip blunting. Increasing pearlite content in low carbon steel increases the tendency for fibrous cracking, so that the mechanism shown in Fig. 7.28b can operate more easily. Consequently the 50% transition temperature can be lowered by 55°C in a 0.1% carbon steel as compared with 33°C in a 0.02% carbon steel.

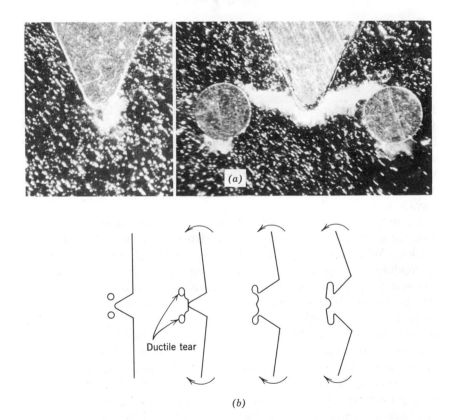

Fig. 7.28. (a) Effect of two drilled holes on the strain (etch pit) distribution at midthickness of a notched specimen of Fe-Si, $\sigma/\sigma_{GY} = 0.50$. 18×. (b) Mechanism by which two holes cause large-scale notch blunting and improve toughness. *After A. Tetelman and C. Rau* [56].

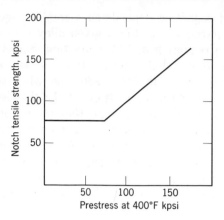

Fig. 7.29. Effect of prestressing at 400°F on the notch tensile strength at room temperature of high strength steel. *After E. Steigerwald* [58].

In 250 grade maraging steel, which fractures by the low-energy tear fracture described above, drilled holes are not able to stop the impact hammer but they are able to double $C_V(\max)$, from 18–37 ft-lb, at ambient temperature. Consequently the beneficial effects of this technique are general and may be applicable to a wide variety of alloys containing notches fixed by design requirements (e.g., a keyway).

Warm prestressing [57, 58, 59] is another technique that can be used for improvement of toughness in structures containing notches or cracks. When ultra-high strength materials are prestressed in tension at moderate temperatures† before being tested at ambient or subambient temperature, an increase in toughness is observed (Fig. 7.29) that increases with increasing amount of prestress and decreasing prestress temperature, within a certain temperature range. This effect is believed to result from the introduction of residual compressive stresses in the plastic zone ahead of the crack. These would, of course, lower the value of σ_{yy} and improve toughness for reasons discussed earlier. Blunting of the crack tip during prestressing may also contribute to the increased toughness. It must be borne in mind that this technique will not be effective if the preloading is carried out at relatively low temperatures since substantial amounts of slow crack growth and crack sharpening may then occur. That prestrain at am-

---

† Recrystallization or softening of the material need not occur.

bient temperatures can reduce the resistance to low temperature cleavage of precracked iron silicon has already been demonstrated [60] (see Fig. 6.15).

7.5 Summary

1. A notch in a thick plate, loaded in bending or tension, introduces a state of triaxial tension (plastic constraint) ahead of it. This raises the tensile stresses in the plastic zone, and hence raises the temperature at which cleavage fracture occurs at or before general yield, and lowers the nominal ductility.

2. During local yielding ($\sigma < \sigma_{GY}$), $\sigma_{yy}^{(max)}$ builds up from a value of σ_Y at the notch root to $K_{\sigma(p)}\sigma_Y$ ahead of the root, where $K_{\sigma(p)}$ is known as the plastic stress concentration factor. $K_{\sigma(p)}$ increases with decreasing root radius ρ and increasing plastic zone size R (and hence with increasing nominal stress σ and increasing flaw size $2c$). The maximum potential value of $K_{\sigma(p)}$ (if fracture does not intervene) is $K_{\sigma(p)}^{max} = 1/\beta$; it depends only on the flank angle ω of the notch. If we assume that yielding is controlled by a maximum shear stress criterion, $1/\beta = [1 + (\pi/2) - (\omega/2)]$.

3. R_β is the value of R at which $K_{\sigma(p)} = 1/\beta$. It is given by the relation $R_\beta = \rho[\exp(1/\beta - 1) - 1]$. Further increases in R produce no further increase in $K_{\sigma(p)}$. Consequently, before general yield, and in the absence of strain hardening, σ_{yy} is given by the relations

$$\sigma_{yy} = \sigma_Y \left[1 + \ln\left(1 + \frac{r}{\rho}\right)\right] \qquad r < R_\beta$$

$$\sigma_{yy} = \sigma_Y \frac{1}{\beta} \qquad R_\beta < r < R$$

σ_{yy} decreases with r when $r > R$ (elastic region).

4. The plastic-constraint factor L is a measure of the increase in applied bending moment or tensile load required to produce general yielding of a notched specimen compared with an unnotched specimen having the same load-bearing area and tested under the same conditions. Physically it is a measure of the mean hydrostatic stress that exists across the notched structure at general yield. As $\rho \to 0$, $R_\beta \to 0$ and the mean value of $\sigma_{yy} \to \sigma_Y(1/\beta)$. Thus $L \to 1/\beta$ as $\rho \to 0$.

5. In plane-strain deformation the plastic zone size R and crack-tip displacement $2V(c)$ increase with nominal stress and crack length, but at a slower rate than in plane-stress deformation.

6. For a notch of given radius ρ, the crack-opening displacement $2V(c)$ increases with increasing bend angle θ, as do the root strains $\epsilon(c)$. The strains at a point r ahead of the root $\epsilon(r)$, and hence the amount of strain hardening there $\Delta\sigma(r)$, increase with increasing root strains (i.e., with increasing C.O.D. and with decreasing values of r).

$$\epsilon(r) = \frac{V(c)}{J\rho r} = \frac{\epsilon(c)}{Jr}$$

7. When strain hardening occurs, the tensile stresses are a maximum at $r = R_\beta$ and can be approximated as

$$\sigma_{yy} = \frac{1}{\beta}(\sigma_Y + \Delta\sigma) = \frac{1}{\beta}\left[\sigma_Y + \frac{d\sigma}{d\epsilon}\epsilon(R_\beta)\right]$$

$$\sigma_{yy} = \frac{1}{\beta}\left[\sigma_Y + \frac{d\sigma}{d\epsilon}\frac{V(c)H}{\rho^2}\right]$$

for linear strain-hardening behavior. A decrease in root radius produces a large decrease in R_β, and hence a large increase in $\epsilon(R_\beta)$ and σ_{yy}.

8. Cleavage fracture initiates (ahead of the notch) when σ_{yy} builds up to a value of $\sigma_f{}^*$, where $\sigma_f{}^*$ is of the order of the cleavage strength of an unnotched tensile specimen. In a deeply notched specimen, under conditions where $\sigma_f{}^*/\sigma_Y < 1/\beta$, the nominal fracture stress σ_F is that stress which is required to spread the plastic zones out to a distance R^* such that $K_{\sigma(p)}$ build up to a value of $\sigma_{yy}/\sigma_Y = \sigma_f{}^*/\sigma_Y$. When $\sigma_f{}^*/\sigma_Y > 1/\beta$, σ_F is the nominal stress at which a critical displacement $V^*(c)$ is produced such that a critical amount of strain hardening occurs at R_β to initiate cleavage there. This requires that root displacement reach a critical value

$$V(c) = V^*(c) = \frac{\rho^2\beta\sigma_Y}{H\,d\sigma/d\epsilon}\left(\frac{\sigma_f{}^*}{\sigma_Y} - \frac{1}{\beta}\right)$$

9. The plain-strain cleavage fracture resistance G_{Ic} increases with increasing root radius ρ and increasing values of $\sigma_f{}^*/\sigma_Y$ (i.e., with increasing temperature T, and the work done in microcrack propagation γ_m and γ_B, and with decreasing strain rate $\dot\epsilon$, grain size $2d$, friction stress σ_i, and yield parameter k_y).

10. In most polycrystalline solids a temperature exists at which the root strains $\epsilon(c)$ build up to the ductile fracture strain $\epsilon'_f(c)$ before they reach the strain required to initiate cleavage $\epsilon_f(c)$. Fracture then initiates by fibrous tearing at the root rather than by cleavage ahead of the root. This fibrous crack gets sharper as it advances, and below a temperature $T_{B(N)}$ it can transform over to an unstable, discontinuous cleavage crack. The amount of crystalline fracture drops to zero at $T_{B(N)}$.

11. The displacements associated with unstable cleavage fracture above general yield (and hence the nominal fracture strength) increase with decreasing plate thickness because a decrease in thickness reduces the plastic stress concentration factor $1/\beta$.

12. The shape and position of the Charpy V notch impact curve can be correlated with the critical tip displacement $V^*(c)$ required to initiate unstable fracture (prebrittle fracture energy) and the shear lip size (postbrittle fracture energy). The instrumented Charpy test is one of the most useful tests for alloy development purposes.

13. In high strength materials and in medium strength materials at low temperature (high σ_Y) unstable fracture occurs by the formation of voids and the subsequent coalescence of these voids with the crack tip. The voids form around cracked particles or inclusions. An increasing yield strength level implies a higher level of tensile stress in the plastic zone ahead of a crack and hence a higher density of broken particles (voids). This, in turn, implies that a smaller displacement occurs during coalescence. Consequently the toughness of these materials increases with decreasing yield strength, decreasing particle or inclusion content and size, and increasing strain-hardening rate.

14. High toughness is only compatible with high yield strength if some means exist for reducing constraint ahead of a crack. This is possible in anisotropic material such as wrought products or in fiber composites where planes of weakness lie perpendicular to the direction of crack propagation. These planes can be opened by the transverse tensile stresses that exist at the crack tip and the crack will then be blunted. Appropriately drilled holes and warm prestressing are also able to increase the toughness of notched structures of brittle materials.

References

The asterisk indicates that published work is recommended for extensive or broad treatment.

[1] F. A. McClintock and G. R. Irwin, *Fracture Toughness Testing*, ASTM, Philadelphia, STP No. 381 (1965), p. 84.
*[2] A. H. Cottrell, *Proc. Roy. Soc.*, **A285**, 10 (1965).
[3] A. A. Wells, *Proc. Roy. Soc.*, **A285**, 34 (1965).
[4] J. R. Hendrickson, D. S. Wood and D. S. Clark, *Trans. ASM*, **50**, 656 (1958).

*[5] J. F. Knott and A. H. Cottrell, *J. Iron Steel Inst.*, **201**, 249 (1963).
 [6] D. C. Drucker, *Fracture of Solids*, Interscience, New York (1963), p. 3.
 [7] E. Orowan, *Repts. Prog. Phys.*, **12**, 185 (1948).
 [8] R. Hill, *Mathematical Theory of Plasticity*, Oxford, London (1950).
 [9] E. Parker, *Brittle Behavior of Engineering Structures*, Wiley, New York (1957).
[10] A. P. Green, *Quart. J. Mech. Appl. Math.*, **6**, 223 (1953).
*[11] A. P. Green and B. B. Hundy, *J. Mech. Phys. Solids*, **4**, 128 (1956).
[12] G. Lianis and H. Ford, *J. Mech. Phys. Solids*, **7**, 1 (1958).
[13] J. M. Alexander and T. J. Komoly, *J. Mech. Phys. Solids*, **10**, 265 (1962).
*[14] T. R. Wilshaw and P. L. Pratt, *J. Mech. Phys. Solids*, **14**, 7 (1966).
[15] R. F. Koshelov and G. V. Ushik, *Izv. Akad. Nauk. SSSR Otn. Mekhanika i Mashinostroenie*, **1**, 111 (1959).
[16] F. A. McClintock, *Welding J. Res. Suppl.*, **26**, 202s (1961).
[17] T. R. Wilshaw, Ph.D. thesis, London University (1965).
[18] A. A. Wells, *Brit. Weld. J.*, (1963), p. 855.
[19] G. T. Hahn and A. R. Rosenfield, *Acta Met.*, **13**, 293 (1965).
[20] J. A. H. Hult and F. A. McClintock, *Ninth Int. Congr. Appl. Mech.*, **8**, 51 (1957).
[21] A. A. Wells, *Cranfield Symposium on Crack Propagation*, College of Aeronautics, Cranfield, Eng. (1962), p. 210.
[22] T. R. Wilshaw, *J. Iron Steel Inst.*, **204**, 936 (1966).
[23] C. Crussard et al, *J. Iron Steel Inst.*, **183**, 146 (1956).
[24] G. Oates, *Proc. Roy. Soc.*, **A285**, 166 (1965).
[25] T. R. Wilshaw and P. L. Pratt, *Int. Conf. on Fracture*, Sendai, Japan, **B-III**, 3 (1965).
[26] R. T. Ault, *Fracture of Molybdenum*, WAFB, TDR-63-4088 (1963).
[27] N. P. Allen, C. C. Earley and J. H. Rendal, *Proc. Roy. Soc.*, **A285**, 120 (1965).
[28] D. H. Winne and B. M. Wundt, *Trans. ASME*, **80**, 1643 (1958).
[29] J. R. Griffiths and A. H. Cottrell, *J. Mech. Phys. Solids*, **13**, 135 (1965).
*[30] G. D. Fearnehough and C. J. Hoy, *J. Iron Steel Inst.*, **202**, 912 (1964).
[31] F. W. Barton and W. J. Hall, *Bur. Ships, U.S. Navy Rept. SSC-147* (1964).
[32] G. T. Hahn et al, in *Fracture*, B. L. Averbach et al. eds., M.I.T., Wiley, New York (1959), p. 91.
[33] T. R. Wilshaw and P. L. Pratt, to be published.
[34] W. D. Biggs, *Brittle Fracture of Steel*, MacDonald and Evans, London (1960).
[35] C. A. Rau, Jr., and A. S. Tetelman, to be published.
[36] J. D. Lubahn, *Weld. J.*, **34**, 518S (1955).
[37] D. L. Newhouse, *Proc. ASTM*, **62**, 1192 (1962); *Welding J. Res. Supp.* (March 1963).
[38] J. A. Rinebolt and W. J. Harris, Jr., *Trans. ASM*, **44**, 225 (1952).
[39] N. J. Petch, *Prog. Met. Phys.*, **5**, 1 (1954).
[40] I. Mogford and D. Hull, *J. Iron Steel Inst.*, **201**, 55 (1963).
[41] N. J. Petch, in *Fracture*, B. L. Averbach et al. eds., M.I.T., Wiley, New York (1959), p. 54.
[42] J. F. Knott, *Proc. Roy. Soc.*, **A285**, 150 (1965).
[43] R. Castro and A. Gueussier, *Rev. Metal.*, **46**, 517 (1949).
[44] W. S. Pellini et al, *NRL Report 6300* (June 1965).

[45] A. J. McEvily et al, *Trans. ASM,* **56,** 753 (1963).

[46] G. E. Gazza and F. R. Larson, *Trans. ASM,* **58,** 183 (1965).

*[47] A. H. Cottrell, *Proc. Roy. Soc.,* **A282,** 2 (1964).

[48] J. Cook and J. E. Gordon, *Proc. Roy. Soc.,* **A282,** 508 (1964).

*[49] A. J. McEvily and R. H. Bush, *Trans. ASM,* **55,** 654 (1962).

[50] R. J. Stokes and C. H. Li, *Trans. AIME,* **230,** 1104 (1964).

*[51] T. L. Johnston, R. J. Stokes and C. H. Li, *Trans. AIME,* **221,** 792 (1961).

[52] A. Gilbert, J. L. Ratliff and W. R. Warke, *Trans. ASM,* **58,** 142 (1965).

[53] W. McCarthy, O. Sherby and J. Shyne, to be published.

[54] E. Orowan, *Dislocations in Metals,* AIME (1954), p. 190.

[55] C. T. Forwood and A. J. Forty, *Phil. Mag.,* **11,** 1067 (1965).

[56] A. S. Tetelman and C. A. Rau, Jr., *Int. Conf. on Fracture,* Sendai, Japan, **B-I,** 161 (1965).

[57] J. E. Srawley and C. D. Beachem, *NRL Report 5460* (April 1960).

[58] E. A. Steigerwald, *Trans. ASM,* **54,** 445 (1961).

[59] A. J. Brothers and S. Yukawa, *Trans. ASME,* **85D,** 97 (1963).

[60] A. S. Tetelman, *Acta Met.,* **12,** 993 (1964).

[61] M. Tanaka and S. Umekawa, *Proc. 1st Japan Congr. Test. Mat.,* 95 (1958).

[62] V. Weiss, *ASME preprint 62 WA 270* (1962).

[63] T. R. Wilshaw, C. A. Rau, Jr., and A. S. Tetelman, to be published.

[64] H. A. Lequear and J. O. Lubahn, *Weld. J.* (1954), **33,** p. 585s.

[65] D. E. W. Stone and C. E. Turner, *Proc. Roy. Soc.,* **A285,** 83 (1965).

[66] F. McClintock, to be published.

[67] J. M. Krafft, *Appl. Mat. Res.,* **3,** 88 (1964).

[68] J. Gurland and J. Plateau, *Trans. ASM,* **56,** 442 (1963).

[69] W. Pellini and P. P. Puzak, *NRL Report 6030* (November 1963).

[70] J. Knott, *J. Iron Steel Inst.,* **204,** 1014 (1966).

Problems

(Assume that $\sigma_f = \sigma_f^*$ in all problems.)

1. Suppose that the uniaxial tensile yield stress σ_Y varies with absolute temperature T as $\sigma_Y = A - BT$ and the uniaxial tensile cleavage fracture stress $\sigma_f = C + DT$. To a first approximation, what is the increase in nil ductility temperature that occurs when a deep, sharp external notch is introduced into a thick plate of this material? If $C = 3 \times 10^5$ psi, $B = 400$ psi per °K and $D = 0$, what is the magnitude of this effect? (500°K)

2. At 300°K the yield stress of a pure BCC metal is $\sigma_Y = 16 + d^{-\frac{1}{2}}$ in units of kpsi and the tensile transition grain size $2d^*$ is 0.0315 in. Suppose that a thick plate of this material contains a deep notch whose root radius $\rho = 0.03$ in. What is the plastic zone size at which cleavage fracture can initiate ahead of the notch if the grain size $2d$ is 0.005 in.? (0.0273 in.)

3. If the yield stress of this material decreases by 0.3 ksi per °K and the cleavage fracture strength is temperature-independent, what

plastic zone size can initiate cleavage at 320°K when the grain size is maintained at 0.005 in. and $\rho = 0.03$ in.? What is G_{Ic} and what is the nominal fracture strength at 320°K if the notch depth is 2.0 in. and the plastic constraint factor is 1.35? $E = 42 \times 10^6$ psi.

(0.051 in., 10 in, lb per in.^{2})

4. Suppose that cleavage fracture initiates by plastic stress concentration alone (i.e., strain hardening is not required) when the grain size $2d$ of polycrystalline iron is less than 0.0089 in. and a notch having a 60° included angle is present. If the tensile yield stress is given by $\sigma_Y = 25 + 2d^{-\frac{1}{2}}$, what is the maximum flank angle that a notch can have and still initiate cleavage without strain hardening when $2d = 0.005$ in.?   (28°)

5. A thick plate of mild steel contains a notch whose dimensions are ($\rho = 0.01$ in., $c = 1.0$ in., $\omega = 65.4°$).

$$\sigma_f{}^* = 100 \text{ ksi} \quad \text{and} \quad \sigma_Y = (100\text{--}0.3T)\text{ksi} \quad L = 1.35$$

(a) What is the highest temperature at which cleavage can be initiated without strain hardening?

(b) What is the nominal fracture strength of the cracked plate at this temperature?

6. Suppose that the flow curve of a material is given by the relation $\sigma = \sigma_0 (B + \epsilon)^n$. Derive a simple relation, analogous to Eq. 7.35, for $V^*(c)$ in terms of σ_0, B, n, $\sigma_f{}^*$, ρ, B, and H.

7. Assuming that $\sigma_f{}^*$ is independent of temperature, estimate its value for the steel shown in Fig. 7.18b.   (220 ksi)

PART FOUR

Time-Dependent Fracture

8

Fracture Under Cyclic Loading (Fatigue)

Fracture, because of repeated rather than unidirectional or static loading, is known as fatigue and is the commonest cause of service failures. Such fractures can occur at stress levels well below the tensile strength, even in normally ductile low strength materials. Since it was first recognized more than a century ago, fatigue has been intensively investigated in order to understand the phenomenon and guard against its occurrence. These studies indicate that it is convenient to subdivide the over-all fatigue process into three main stages; crack initiation, including a consideration of the effects of cyclic loading on bulk properties, crack propagation to critical size, and unstable rupture of the remaining section. An example of a fatigue fracture surface clearly showing the second and third of these stages is given in Fig. 3.40. In this instance a crack initiated at a flaw in a casting and propagated under repeated loading to create a characteristic thumbnail region, the boundary of which marks the onset of the final rupture stage. The appearance of a transition region is shown in greater detail in Fig. 3.18.

It is now recognized that fatigue occurs as the result of plastic deformation, both in the initiation and propagation of cracks. Up to the terminal fracture fatigue is a form of ductile (stable) rupture, although often of an extremely localized nature. Because fatigue failure involves the cumulative effect of small-scale events taking place over perhaps millions of cycles, it is extremely difficult to make an a priori prediction of the fatigue lifetime. However, certain aspects of fatigue such as low cycle fatigue (failure in less than 10^3 cycles) and fatigue crack propagation can be treated quantitatively on a semiempirical basis. Therefore, although our understanding of this complex subject is far from complete, the present state of knowledge is nonetheless useful in dealing in an enlightened manner with a number of problem areas:

1. The planning of fatigue testing and evaluation programs and the statistical analysis of fatigue data.

2. The design of cyclically loaded components involving the prediction of service lifetime under complex loading and environmental conditions.

3. The scheduling of inspections for the purpose of detecting fatigue cracks before they grow to critical length.

4. Failure analysis.

5. The improvement of fatigue resistance by control of surface condition, residual stresses, and metallurgical structure.

6. Provision of a basis for further research, for there are questions still to be answered about fatigue which arise because of the complexity of the process and the large number of parameters that can influence fatigue behavior. These parameters include the nature of the load spectrum, cyclic speed, geometrical effects, size effects, notch sensitivity, surface condition, residual stresses, metallurgical structure, temperature, environment, and the effect of history.

In recent years the subject of fatigue has been extensively treated in review papers, texts, and conferences, and in this chapter only a brief review of the subject including an indication of the effect of the above parameters is presented. The interested reader is referred to the bibliography and references for more comprehensive treatments.

8.1 Effects of Cyclic Loading

The most important factor in determining the fatigue lifetime is the magnitude of the external stress or strain amplitude. Figure 8.1 indicates the usual shapes of fatigue curves as a function of these parameters. These curves differ in appearance in the low cycle region corresponding to amplitudes well into the plastic range in which a small increment in stress leads to a large increment in strain. If a positive mean stress is present during cycling, the allowable amplitude for a given lifetime will be reduced, particularly at long lives. However, as a simplification, we deal chiefly with cyclic loading in the absence of a mean stress. The equation of the steeply sloping part of the curve of Fig. 1a can be expressed as [1]

$$\sigma^a N = \text{constant} \tag{8.1}$$

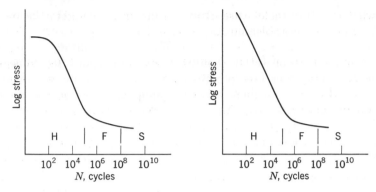

Fig. 8.1. Stress-cycle and strain-cycle fatigue curves, indicating H, F, and S ranges.

where a varies from 8 to 15. Thus a 2% reduction in stress can lead to a 30% increase in life [2], underlining the importance of the stress amplitude in fatigue.

Beyond 10^6 cycles the curves in Fig. 8.1 may be more horizontal than described for such materials as iron which contains carbon as an interstitial element, in which case the material is said to have a *fatigue limit*. This limit is thought to be the result of strain aging during cyclic loading [26, 27]. Fatigue results can also be influenced by the cyclic frequency of testing; for example, the fatigue limit of carbon steels has been found to increase monotonically by a factor of two over the frequency range from 40 to 100,000 cps [3]. However, in the usual range of testing frequencies, that is, below 200 cps, the fatigue limit is not greatly affected. We can expect that for stresses well into the plastic range the effect of cyclic frequency on fatigue lifetime will be more pronounced.

The σ-N and ϵ-N curves of Fig. 8.1 can be subdivided into three regions [4], H, F, and S, as indicated. In a high stacking fault energy material such as copper, tested within the S region, only fine slip, distributed uniformly over the surface, is observed and fatigue failures do not occur. In the F region plastic deformation becomes concentrated within a relatively few slip bands, and this concentration of deformation leads to the formation of cracks within such bands. In the H region large-scale plastic deformation results in surface folding and cracking, and the formation of substructures in materials of high stacking fault energy.

The amplitude of plastic deformation is an important factor in determining the fatigue lifetime. If the stress required to reverse a given plastic strain is plotted against successive reversals, the reversals being added unidirectionally, independent of sign, a curve such as that shown in Fig. 8.2a is obtained for each amplitude of plastic strain [5]. It is observed that at large amplitudes for annealed copper hardening increases progressively with increasing strain until fracture occurs, though more slowly than in unidirectional straining. At smaller amplitudes, hardening is less and tends to a limiting value before fracture. In this "cyclic state" a stable hysteresis loop is developed within a few per cent of the total lifetime at low amplitudes, and Fig. 8.2b shows how the loop shape varies during the first loading cycles before attaining a saturation level of hardening. As indicated in Fig. 8.2c, the curve drawn through the saturation stress for each plastic strain amplitude describes a cyclic stress-plastic strain curve. If the stress amplitude rather than the strain is held fixed, a characteristic value of strain will be achieved, and again a cyclic stress-strain curve can be established [6, 7, 8]. An equivalent type of "cyclic-state" curve established in bending is shown in Fig. 8.3 [9]. Note that in copper the onset of large-scale plastic deformation occurs at about the same load level for both cyclic as well as unidirectional loading. It is observed that for copper the stress corresponding to a fatigue lifetime of 10^7 cycles is about equal to its yield strength. For

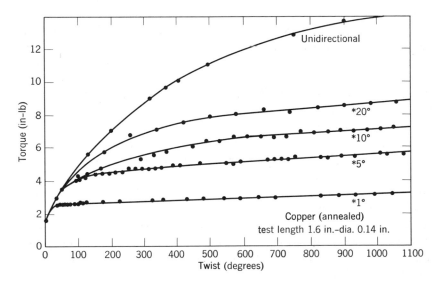

Fig. 8.2. Cyclic stress-strain behavior. (a) Torque versus twist curves for different amplitudes of alternating torsion [129].

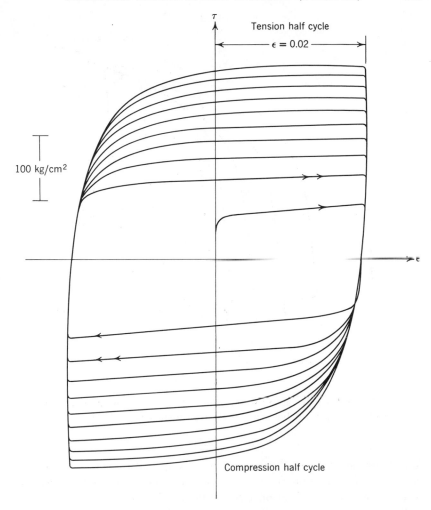

Fig. 8.2. Cyclic stress-strain behavior. (b) The first 10 cycles of a copper crystal subjected to a reversed plastic strain amplitude of $\epsilon = +0.02$ at 200°K [13].

iron, on the other hand, the load for the onset of large-scale plastic deformation is increased considerably by cyclic loading, and in this case the fatigue limit is some 50% higher than the yield strength in accordance with the cyclic hardening behavior. The difference between copper and iron is thought to be due to the aforementioned process of strain aging in iron, which inhibits plastic deformation at

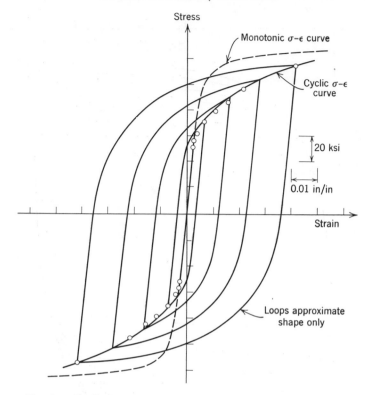

Stress

Monotonic σ–ε curve

Cyclic σ–ε curve

20 ksi

0.01 in/in

Strain

Loops approximate shape only

Fig. 8.2. Cyclic stress-strain behavior. (c) The monotonic and cyclic stress-strain behavior of SAE 4340 steel [130] [126].

low stress levels. The existence of a well-defined knee in the σ-N curve of iron may be related to the relatively sharp break observed in the cyclic load-deflection curve [43, 44, 45, 121].

Transmission electron microscopy studies have been useful in determining the nature and extent of the dislocation arrays introduced by cyclic loading. In the H range it has been observed that cell structures develop in materials of high stacking fault energy such as copper [10], aluminum [40, 41], and iron [42], in which cross slip and climb can occur readily, and that the size of the cells after saturation is directly related to strain amplitude [10]. (The appearance of these cells is similar to those shown in Fig. 4.14d.) As the strain amplitude increases, the cell size (of the order of 1μ) decreases. Furthermore, if the amplitude of straining is altered, a cell size characteristic of the new amplitude is rapidly established [11]. The cell

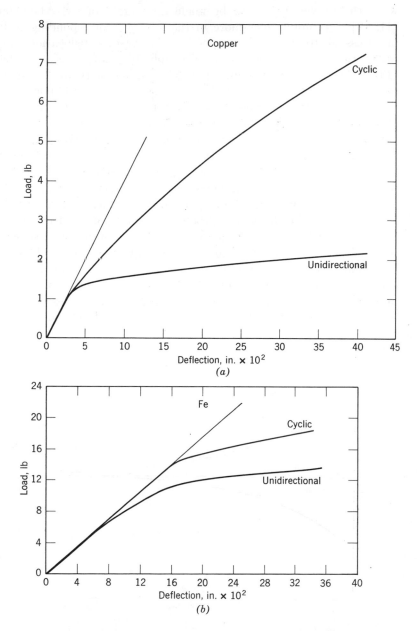

Fig. 8.3. (a) Unidirectional and cyclic load-deflection curves for copper [9]. (b) Unidirectional and cyclic load-deflection curves for iron [9].

walls themselves appear to be made up largely of twist boundaries [11, 12]. At amplitudes more in the F range corresponding to lives in excess of 10^5 cycles for copper single crystals dislocation dipoles rather than cells appear as the principle form of dislocation debris [13]. Vacancies created during cyclic loading may also contribute to cyclic hardening [14]. Current investigations are aimed at determining how these various types of lattice defects are involved in the attainment of the saturation stress during cyclic loading.

In contrast to the behavior of pure metals, engineering alloys often do not exhibit a direct correlation between their fatigue strengths at high cycles and their yield strengths. The reason for this lies in the fact that in such alloys inhomogeneities in metallurgical structure exist, particularly along grain boundaries, which permit localized plastic deformation to occur, although the material is stressed in the nominally elastic range. In addition, certain metallurgical structures such as coherent (Guinier-Preston) zones are not stable when localized, plastic deformation takes place under cyclic loading. In such alloys dislocations can cut through and cause dissolution of the structure [15, 16]. Aluminum alloys are strengthened by these zones, and it is thought that the low resistance of the alloys to cyclic loading, in

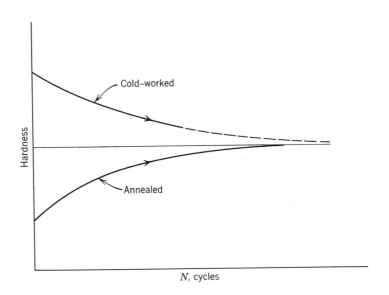

Fig. 8.4. Influence of cycling on the hardness of annealed and cold-worked copper specimens [17].

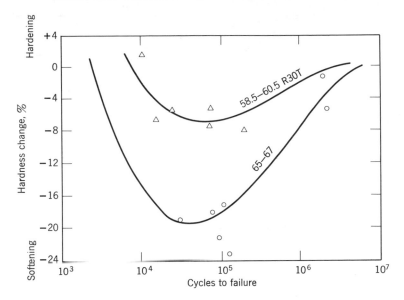

Fig. 8.5. Variation in hardness of cold-worked, low carbon steel as a function of initial hardness and cyclic life [18].

contrast to their high tensile strengths (a ratio of approximately 0.3), is due to the instability of the zones during cyclic loading. In contrast, precipitation-hardened iron-base alloys have a much higher fatigue ratio (0.5), a reflection of the greater stability of their microstructure during cyclic loading.

Cold-worked structures can also be unstable under cyclic load; the result is a cyclic softening process rather than a hardening process. Figure 8.4 indicates the change in hardness of copper during cyclic loading as a function of initial condition [17]. Such a softening process may have serious practical implications. For example, a beneficial residual stress system in a cold-worked surface may be rendered ineffective by cyclic loading into the plastic range. The amount of softening depends, as indicated in Fig. 8.5 [18] for a low carbon steel, on the initial degree of cold work and the amplitude of cyclic straining. Recent electron transmission studies indicate that for materials of high stacking fault energy the cell structure developed in the 10^4 life range is largely a function of the cyclic conditions and relatively independent of the initial condition of the material [11].

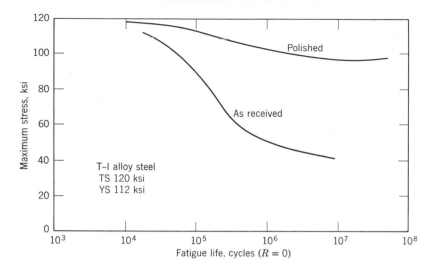

Fig. 8.6. *S-N* curves for T-1 steel in polished [22] and as-received [21] conditions.

8.2   Microscopic Aspects of Fatigue Crack Nucleation and Growth

Nucleation

Fatigue cracks can be initiated in a number of ways, but the important fact is that *they are usually nucleated at a free surface.* Exceptions exist, as in rolling contact fatigue, in which shear stresses are greatest beneath the surface [19], but even so cracks are probably nucleated at an interface between a particle and the matrix [20]. The importance of surface condition in fatigue is indicated in Fig. 8.6 [21, 22]. In this example a high strength structural steel alloy in the polished condition tested in pulsating tension has a fatigue strength at high cyclic life twice that of the as-received material. Because the sensitivity to surface flaws and notches generally increases with tensile strength (Chapter 3), the stronger the material the greater the influence of surface condition. The low fatigue strength of the as-received alloy is due to the presence of millscale which is easily cracked, thereby creating stress concentrators. Decarburization, especially of forgings, is another common cause of the reduction in fatigue strength. Although some of this loss can be avoided by machining off the decarburized layer, even the method of machining is an important factor [31]. Recarburizing, surface rolling, and shot peening can also improve fatigue resistance, and it is significant that the benefits developed in the shot peening of notched specimens are greater than in unnotched specimens [23].

In addition to cracks initiated at stress-raisers in millscale coatings, cracks can appear at a variety of sites, as indicated in Fig. 8.7; (a) and (b) represent crack initiation at particles, either surface or sub-surface. In (c) the soft zone existing along a grain boundary can lead to crack initiation at a triple point. The other cases are fundamentally more interesting, for they represent examples of crack initiation at slip bands created during cyclic loading. In (d) an extrusion [24, 25], or ribbonlike bit of metal, is observed to emanate from the slip bands. In (d), (e) and (f) various other types of slip-band topography are shown. Each of these effects can lead to the localization of plastic strain by the creation of discontinuities on a previously featureless surface. It has been observed that regions beneath these fatigue slip bands are softer than the adjacent matrix [28, 29, 30]. It appears that this recovery facilitates further dislocation motion within the regions. Recent studies of the subsurface dislocation arrays have indicated a correlation with the surface disturbance [32, 12]. The soft zones have been found to penetrate some distance beneath the surface in axially loaded copper single crystals [30]. Substructures have been observed within such persistent slip bands, but their dimensions are an order of magnitude larger than those observed in the H region. The presence of the substructure suggests that cross slip and climb are involved in the recovery process [30].

A factor influencing the nature of the surface topography developed is the degree of planarity of glide. Note in Fig. 8.7 the difference in the appearance of slip bands in copper and in Fe-3 Si. Copper exhibits wavy glide, Fe-3 Si exhibits more planar glide. Alloys of low stacking fault energy, such as Cu-7 Al, also develop slip bands that resemble those in Fe-3 Si rather than those in copper [33, 34]. It has also been observed that the slip vector must lie at a finite angle to the surface in order to develop this topography [35]. The fact that the appearance of slip bands in copper is much the same at 4.2°K as at ambient temperatures is an indication that fatigue results from the motion of dislocations rather than the accumulation of point defects [36].

In materials whose slip systems are such that cross slip does not occur easily, these surface effects are not developed. Ionic crystals (e.g., LiF), which do not contain two glide planes with a common slip direction, are of this type [37]. Another example is zinc [38] tested at low temperatures wherein slip occurs primarily on the basal plane. The resistance to cyclic stressing of materials that cannot form surface stress raisers is quite high, and, in general, fatigue resistance increases with decreasing stacking fault energy.

The geometry developed if screw dislocations cross slip to cause nonreversing plastic deformation, based on the Mott [39] mechanism, is indicated in the simplified model of Fig. 8.8. The merging of

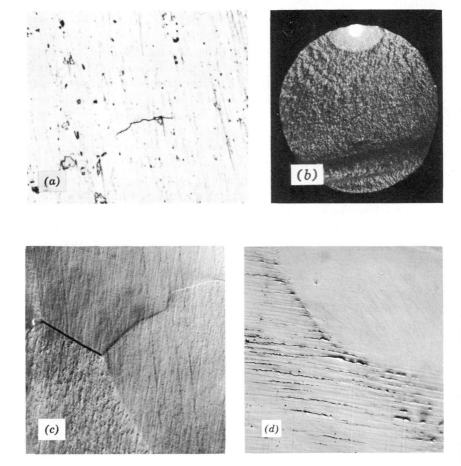

Fig. 8.7. Initiating sites for fatigue cracks. (*a*) Cracked particle [113]. 300×. (*b*) Crack initiated at subsurface particle in ausformed steel [114]. (*c*) Crack in grain boundary, Al-10Mg [115]. 100×. (*d*) Extrusions in Al-10Mg [115]. 640×.

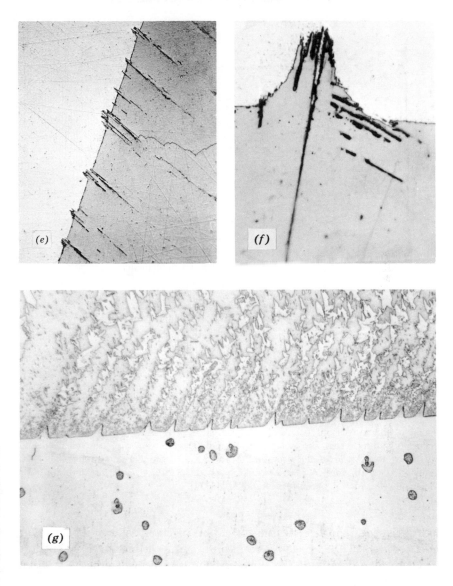

Fig. 8.7. (e) Extrusions in copper [116]. Taper section. (f) Twin boundary extrusion in copper [116]. Taper section. (g) Extrusions in Fe-3 Si [117]. Taper section.

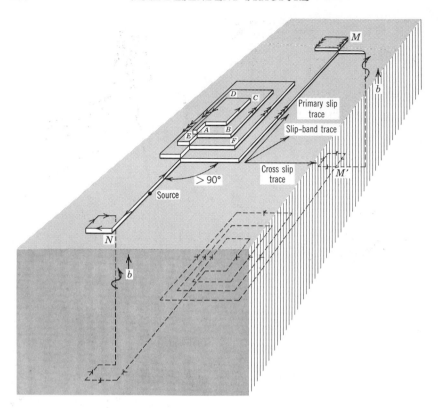

Fig. 8.8. Schematic model for extrusion formation. During loading, screw dislocation moves out from source, cross slips at A and B and moves past C. On unloading, cross slip occurs at C and D and dislocation moves past E. In next loading, cross slip occurs at E and F, etc. Smaller subunits are shown forming at M and N. Subsurface edge dislocations are indicated, and a vacancy-type loop is seen forming at M' [117].

parallel peaks can lead to the formation of cracks shown in the taper section in Fig. 8.9. Cracks are also observed to initiate at re-entrant angles. The details of the dislocation mechanism of crack initiation are not yet clear.

Crack growth

Stage I Growth. Once a crack is initiated at a surface slip band in a single crystal it will continue to advance into the material along the primary slip planes involved in the creation of the slip band before veering onto a plane macroscopically at right angles to the principal tensile stress. Crack growth before the transition is referred to as Stage I growth, that after the transition as Stage II growth [122]. These stages are indicated in Fig. 8.10. The transition is governed by the magnitude of the tensile stress, and the lower the magnitude of this stress, the larger the extent of the first stage of growth. For this reason Stage I growth is favored in torsion testing, for the tensile component at right angles to the Stage I crack is low. If the tensile stresses are high enough, Stage I may not be observed at all, as in sharply notched specimens, and growth occurs entirely in the second mode.

In polycrystalline metals Stage I growth usually terminates when the slip band crack encounters a grain boundary. In polycrystalline brass, for example, if the stress amplitude is high enough to nucleate a Stage I crack in a large-grained specimen, that stress would also be sufficient to cause the crack to propagate through the adjacent grains. On the other hand, in fine-grained specimens cracks may be initiated at a stress that is insufficient for propagation into the adjacent grains.

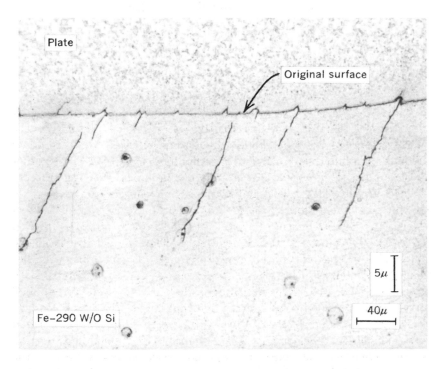

Fig. 8.9. Taper section of a slip bands shown in Fig. 8.7g. Crack formation associated with merging of ridges and with reentrant angles of ridges [117].

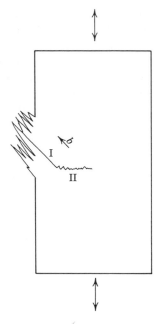

Fig. 8.10. Schematic diagram of stage I and stage II fatigue crack growth.

In one case the event resulting in final failure is nucleation of a crack, whereas in the other propagation is the important factor [123].

Relatively little is known about the mechanism of Stage I growth largely because of the fine scale of the processes involved. It has been proposed that intrusions, a type of reverse extrusion, may develop and facilitate this mode of growth. If so, there would be no clear-cut distinction between initiation and propagation [37]. It has also been proposed that a component of tensile stress may assist in the growth of Stage I cracks [124], in which case the mechanism would be similar to that of Stage II growth to be discussed.

Stage II Growth

Stage II growth can be investigated under conditions of high strain amplitude. As a consequence, the plastic deformation taking place at the tip of a crack can be directly observed [46], a circumstance that contributes greatly to our understanding of the process. One of the most important characteristics is that the crack advances a finite increment in each loading cycle [118]. Another is that a mark, referred to as a striation, is created on the fracture surface in each load cycle and provides a record of the passage of the fatigue crack front. These two aspects are not unrelated, as shown in the schematic diagram of the growth process in Fig. 8.11 [124]. At the start of a loading cycle the crack tip is sharp, but during extension, as the crack advances, it simultaneously becomes much blunter, and the plastic zones at the tip expand. Both effects are involved in establishing a balance between the applied stress and the amount of plastic deformation at the crack tip. It is during the loading stage that new fracture surface is created by this plastic shearing process. During the unloading portion of the cycle as the sharp tip of the crack is re-established the extended material at the tip is heavily compressed and exerts a back stress which causes the deformation marking or striation on the fracture surface as the crack closes.   There appears to be no correlation between the subsurface dislocation arrangements and these striations [47].   The resharpened crack is then ready to advance and be blunted in the next cycle.   The repetition of this blunting and resharpening process is the basic aspect of Stage II growth.   The mechanism can be modified but not altered in principal if the crack advances along only one of the two shear zones at the crack tip, as is sometimes observed at the surface (e.g., in Fig. 8.12 [9]).   Examples of striations formed at high as well as low strain amplitudes are shown in Fig. 8.13 together with the macroscopic appearance of fatigue fracture surfaces.   Although in one case the striations are visible to the naked eye, and in the other they can be observed only by means of electron microscopy, nonetheless they are created basically by the same mechanism.   This process is not limited to crystalline solids; it is also observed as well in noncrystalline polymeric materials (Chapter 12).   Section 8.3 treats some of the quantitative aspects of Stage II growth.

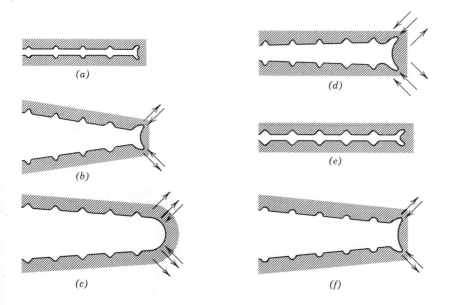

Fig. 8.11. The plastic relaxation process of fatigue crack propagation in the stage II mode. (a) Zero load. (b) Small tensile load. (c) Maximum tensile load. (d) Small compressive load. (e) Maximum compressive load. (f) Small tensile load. The double arrowheads in (c) and (d) signify the greater width of slip bands at the crack in these stages of the process. The stress axis is vertical [124].

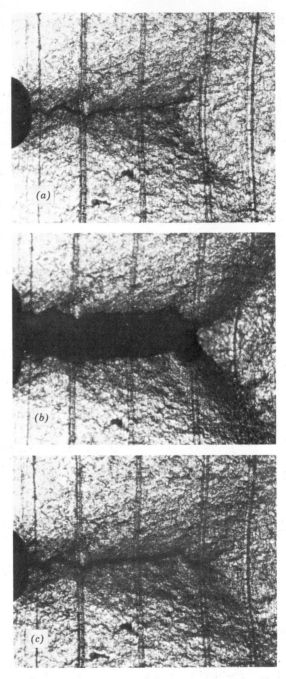

Fig. 8.12. Sequential crack-tip deformation during loading cycle [9]. (a) After compression. (b) After extension. (c) After compression.

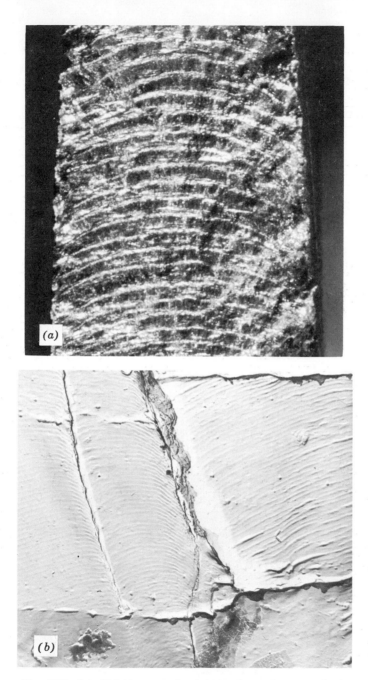

Fig. 8.13. (a) Striations on fracture surface of copper in low cycle fatigue. 20×. (b) Striations on fracture surface of aluminum alloy in high cycle fatigue. 9000×.

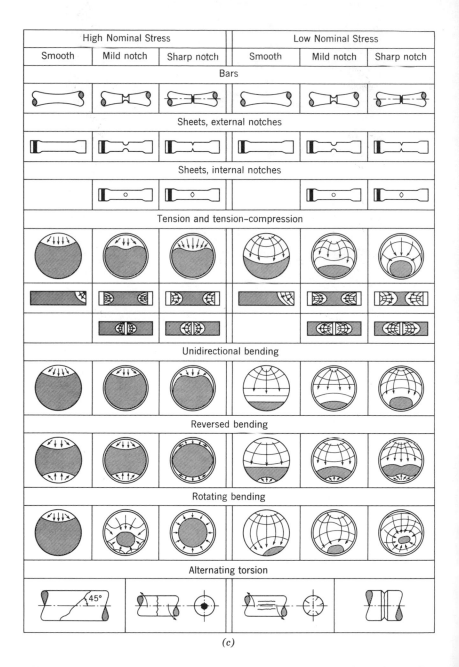

High Nominal Stress			Low Nominal Stress		
Smooth	Mild notch	Sharp notch	Smooth	Mild notch	Sharp notch
Bars					
Sheets, external notches					
Sheets, internal notches					
Tension and tension–compression					
Unidirectional bending					
Reversed bending					
Rotating bending					
Alternating torsion					

(c)

Fig. 8.13. (c) Appearance of fatigue failures on macroscopic scale [119].

366

Stage II growth continues until the crack becomes long enough to trigger off final instability. As discussed in Chapters 2 and 3, *in brittle materials* the instability criterion is simply that a critical displacement $V^*(c)$ be achieved at the crack tip, at which point the crack runs unstably. This implies that the crack length reaches a critical value $2c_F$, for nominal stresses in the elastic range. In *ductile materials* Stage II continues until the remaining cross-sectional area can no longer support the applied load. Fracture usually occurs by shear rupture, on shear planes inclined at 45° to the tensile axis. Thus, the extent of Stage II growth is also governed by the material's toughness G_c, for this determines the critical-sized crack that can exist before causing final instability at a given peak stress σ.

8.3   Macroscopic Aspects of Fatigue Crack Growth

Net section stresses in the elastic range

Two factors that are important in determining the rate of crack growth are the applied stress or strain amplitude and the length of the crack itself, for they determine the stress intensity factor K. To measure the rate of growth as a function of these parameters, we can measure the spacing of adjacent striations on the fracture surface and obtain directly the rate of growth per cycle [48]. Another method is to test sheet specimens and observe the length of the crack as a function of the number of cycles applied and then determine the rate of growth by graphical analysis of a plot of crack length as a function of the number of load cycles applied. These two techniques are equivalent, but it is usually simpler to use the second. In this method cracks are started at a stress raiser in the central portion of the sheet specimen, and their growth along the surface under pulsating tensile loading is followed with a low power microscope. Usually, the crack front beneath the surface is in advance of that at the surface, but the difference in actual length is small in sheet specimens. In notched sheet specimens crack growth occurs initially in the Stage II mode, but after some distance a gradual shift occurs to a plane containing the width direction but inclined 45° to the sheet thickness. This transition region occurs when the radius of the plastic zones at the tip of the crack equals one-half the sheet thickness [49], [51]. In accordance with the principles discussed in Chapter 3, the initial growth before this transition is called "plane strain growth" or "tensile modes" as well as "Stage II," whereas the growth after the transition is referred to as "plane stress growth" or "shear modes." It has been noted that for the same stress and crack length growth in plane stress is slower than in plane strain [9]. This can be attributed to (a) the ability of the larger plastic zones in plane stress to absorb some of the applied strain, whereas in plane strain the deformation is more concentrated at the tip, and (b) the higher tensile component that is set up at the crack tip under plane strain conditions (Chapter 7). Analysis of crack propagation data indicates that the influence of plane stress conditions is felt as soon as a shear lip starts to develop at the surface [50], [131].

A large number of tests with sheet specimens indicate that the results can be correlated by means of the parameter $\sigma \sqrt{c}$, where σ is the peak gross stress and c is one-half the tip-to-tip length of a crack started at a stress raiser in the center of the sheet. The parameter $\sigma \sqrt{c}$ is related to K, the stress intensity factor [52] by

$$\frac{K}{\sqrt{\pi}} = \sigma \sqrt{c}$$

for a small crack in a wide sheet. Within experimental scatter the rate of crack growth is a single-valued function of this parameter for net section stresses in the elastic range. The trend of fatigue crack propagation rates obtained for an aluminum alloy as a function of the parameter $\sigma \sqrt{c}$ is shown in Fig. 8.14 [9]. For rates in excess of $10^{-5} - 10^{-4}$ in./cycle the plane stress mode of propagation is dominant in sheet specimens. For this alloy the rate of crack growth under fully reversed loading does not differ greatly compared with just pulsating tensile loading [53], indicating that compression stresses do little to advance the crack in an alloy of low strain hardening capability.

There is no simple relationship between the rate of crack growth and the stress intensity factor that holds precisely over the *entire range*, but as indicated in the figure a straight-line approximation can be made. The slope of this line is such that the rate of propagation is proportioned to $\sigma \sqrt{c}$ raised to the fourth power [54]. At low values of $\sigma \sqrt{c}$ the rate is much less than given by this relationship, and in fact at very low stresses crack growth may not be detected over a period of observation of more than 10^8 cycles. At high values of the net section stress, approaching or exceeding the yield strength, crack growth occurs at a higher rate than predicted by the fourth-power approximation. This approximation holds for other alloys as well as aluminum as indicated

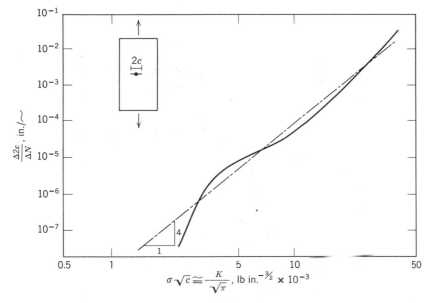

Fig. 8.14. Rate of fatigue crack growth as a function of the stress intensity parameter $\sigma \sqrt{c}$ in a 12-inch-wide sheet specimen of the aluminum alloy 7075-T6 [9].

in Fig. 8.15 [9], but there is a wide variation in the resistance to fatigue crack growth from alloy to alloy. It is interesting to consider the factors that are responsible for this difference.

In contrast to the fatigue strength at 10^7 cycles for a material such as copper (which is related simply to the yield strength), crack propagation, involving as it does large strains at the crack tip in each cycle, depends more on the total plastic response of the material. The result of an analysis [55] of crack growth in an elastic-plastic solid based on the continuum dislocation model of plastic yielding at a notch (Section 5.4) is useful in determining the relative importance of the parameters affecting resistance to crack propagation (for other analyses see [56], [57, 58]). The rate of growth given by this analysis [55] is

$$\frac{dc}{dN} \, \alpha \, \frac{(\sigma \sqrt{c})^4}{\gamma G \sigma_Y^2} \qquad (8.2)$$

where γ is a surface work term, G is the shear modulus and σ_Y is the yield stress. Relation 8.2 has been empirically modified [9] to

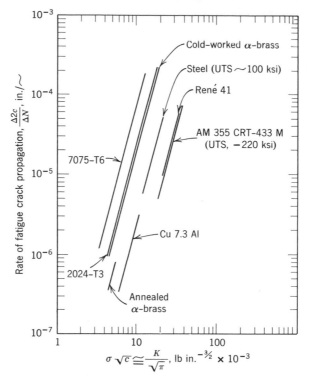

Fig. 8.15. Rate of fatigue crack growth for a number of alloys [9].

incorporate the effects of strain-hardening by replacing γ with the area under the stress-strain curve up to the point of necking instability. This area is given approximately by $(\sigma_Y + \sigma_u)/2\ \epsilon_u$, where ϵ_u is the strain at necking. Young's modulus is used in place of G, and σ_u^2 is used instead of σ_Y^2 to include more strongly the effects of strain hardening. The result is

$$\frac{dc}{dN} \ \alpha \ \frac{(\sigma\sqrt{c})^4}{[(\sigma_Y + \sigma_u)/2]E\sigma_u^2\epsilon_u} \tag{8.3}$$

The data of Fig. 8.15 are plotted in Fig. 8.16 in terms of this modification. It is seen that except for the aluminum alloy 7075-T6, which may be metallurgically unstable under the test conditions, the results fall in a fairly narrow band. From (8.3) it appears that significant improvements in resistance to crack growth by alteration of mechanical properties are not likely.

For cracks initiated at sharp notches Stage II growth occupies most of the fatigue lifetime. Certain predictions about the dependence of the lifetime can therefore be made in terms of the fourth-power relationship. For example, if

$$\frac{dc}{dN} \propto \sigma^4 c^2 \tag{8.4}$$

integration between the limits of c_0, the initial crack length, and c_F, the critical crack length, leads to

$$N_F \propto \frac{1}{\sigma^4}\left(\frac{1}{c_0} - \frac{1}{c_F}\right) \tag{8.5}$$

If c_F is always much larger than c_0, the slope of a log σ-log N curve should be equal to $-\frac{1}{4}$ in the range in which the fourth-power

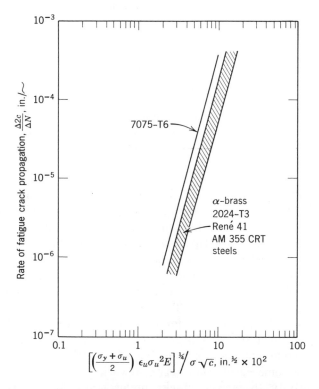

Fig. 8.16. Normalized fatigue crack growth rates for alloys shown in Fig. 8.15 [9].

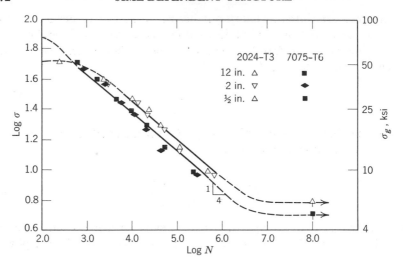

Fig. 8.17. *S-N* curves for notched specimens of the aluminum alloys 7075-T6 [9].

approximation is applicable.   As we can see in Fig. 8-17, the experimental results [59] for sharply notched aluminum-alloy sheet specimens fall along a line of expected slope.   The total lifetime in these tests is independent of specimen width because most of the lifetime is spent while the crack is quite small and insensitive to the width effects. Width effects can be quite important, however, if we wish to know the critical crack length at failure or the rate of growth as a function of crack length.   Note that the fatigue lifetime is greater for 2024-T3 than 7075-T6, a reflection of the higher work-hardening capacity measured by the area under the stress-strain curve of this alloy.   As shown in Chapter 5, work hardening leads to the lateral spread of plastic deformation at the tip of a fatigue crack.   The more rapidly the crack is blunted as the cyclic stress increases, the less it will advance per cycle.   Further, on unloading, the crack may not close so completely [53], with the result that it is less of a geometrical stress raiser in the following cycle.   The lack of work-hardening capacity also appears to be responsible for the fact that a cold-worked brass of 90,000 psi yield strength is less resistant to crack growth in the elastic range than annealed brass of but 15,000 psi yield strength [60]. There is also some correlation between work-hardening capacity and notch sensitivity to fatigue cracks under static loading conditions. For example, in Fig. 8.18 [61] the 2024 alloy is less notch sensitive

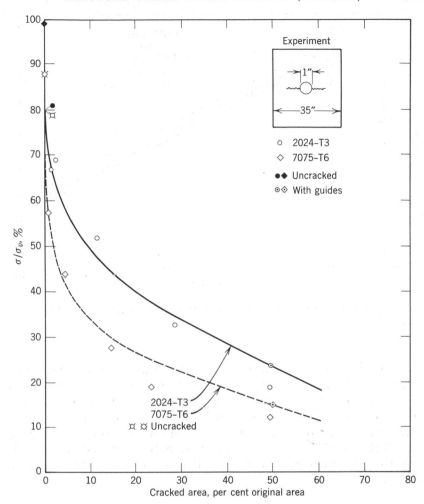

Fig. 8.18. Effect of fatigue cracks on strength of 2024-T3 and 7075-T6 aluminum-alloy-sheet specimens 35 in. wide. σ is the maximum gross stress, σ_u is the tensile strength [61].

than the 7075 alloy of lower work-hardening capacity, but it has not yet been established whether a general relationship between resistance to fatigue crack propagation and fracture toughness exists.

Net Section Stresses in the Plastic Range

In the low cycle range it has been shown that fatigue crack propagation in the Stage II mode occupies most of the lifetime [62]. As

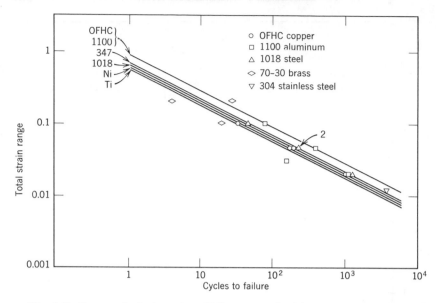

Fig. 8.19. Low cycle fatigue data [65] compared with general trend for such data [120].

shown in Fig. 8.19, the fatigue lifetime of a large number of unnotched ductile metals and alloys are remarkably similar when tested under fully reversed strain range ϵ_r. This correlation serves to emphasize the importance of plastic strain in the fatigue process, for if stresses rather than strains had been used this correlation would be lacking. In addition to the similarity in lifetime, it is noted that the slope is such that the following relationship due to Manson [63] and Coffin [54] holds:

$$N^n \epsilon_r = C \qquad (8.6)$$

where n often has a value of $\frac{1}{2}$.

It is thought that the similarity in lifetime of materials of such dissimilar mechanical properties results from the similarity in the deformation processes at the tip of the growing crack [65]. By counting the striations on the fracture surface it is possible to account for at least 75% of the lifetime in this range. In addition, by measurement of the spacing of the striations it is possible to determine the rate of advance of the crack. Such results are plotted in Fig. 8.20 in terms of a *strain intensity factor* $\epsilon_r \sqrt{c}$, where ϵ_r is the total strain range.   This factor is similar in concept to the $\sigma \sqrt{c}$ parameter described earlier.   The slope of the line through the data points of Fig. 8.20 is close to **2**, and the rate of crack growth can be expressed empirically as

$$\frac{dc}{dN} \, \alpha \, (\epsilon_r \sqrt{c})^2 = \epsilon_r{}^2 c \qquad (8.7)$$

Assuming that Stage II growth starts when the strain intensity factor $\epsilon_r \sqrt{c}$ at the base of a slip-created notch reaches a critical value $(\epsilon_r \sqrt{c})_1$ and final fracture occurs when a second critical value $(\epsilon_r \sqrt{c})_2$ is reached, we can find the total number of cycles spent in Stage II by integrating (8.7) between appropriate values of c for a constant value of ϵ_r.

$$\epsilon^2 N_{\text{II}} = \kappa \ln \frac{c_2}{c_1} = C' \qquad (8.8)$$

where N_{II} denotes the number of cycles spent in Stage II growth and κ and c' are (approximately) constant.   It is known from experimental determinations that for lives less than 10^3 cycles almost the entire

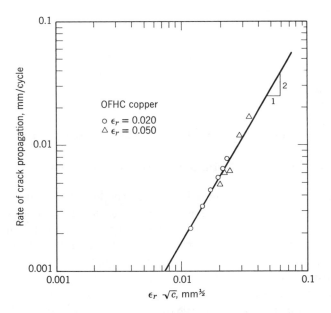

Fig. 8.20. Rate of crack growth in low cycle range as a function of the strain intensity factor $\epsilon_r \sqrt{c}$ [65].

lifetime N is spent in Stage II growth; therefore N can be substituted for N_{11} to obtain the result that

$$N^{\frac{1}{2}}\epsilon_r = C \tag{8.9}$$

which is the expression experimentally found to describe the behavior of ductile materials in the low cycle range.

At lower strain amplitudes, where the net section stresses remain in the elastic range, the characteristics of the individual materials become more important in crack propagation. As already discussed, crack growth in this range can be given in terms of $(\epsilon_r \sqrt{c})^4$ or its equivalent $(\sigma \sqrt{c})^4$. If we compute as above the portion of total lifetime spent in Stage II growth for an unnotched specimen, we find that

$$N_{II} \, \alpha \, \frac{1}{\sigma^2} \quad \text{or} \quad \frac{1}{\epsilon_r^2} \tag{8.10}$$

This result indicates that the number of cycles spent in Stage II crack propagation in an unnotched specimen depends inversely only on the second power of the stress-strain amplitude. However, the total life may vary by as much as the tenth power on the stress (or strain) level. *Hence the life spent in crack growth (Stage II) at low stresses is but a small portion of the total lifetime.* Most of the lifetime of unnotched specimens stressed at a level corresponding to a fatigue lifetime of specimens stressed 10^5 cycles or more is spent in slip-band formation and Stage I crack growth [62].

8.4   Prediction of Fatigue Life in Engineering Structures

Predictions based on plastic and elastic strain amplitudes

The finding that the fatigue lifetime in the low cycle range is related to the magnitude of the plastic strain range has given rise to a number of related proposals for predicting fatigue behavior at longer lifetimes [66, 67, 68, 69]. However, to determine the plastic strain amplitude a cyclic stress-strain curve is required. This stipulation is not so important when the plastic strain range is large with respect to the elastic strain range, and a constant value of the plastic strain amplitude can be assumed [68]. (At longer lifetimes the elastic strain becomes a much larger portion of the total strain and knowledge of the extent of cyclic hardening or softening in affecting the degree of plastic strain per cycle is needed. A rule of thumb is that $\pm 1\%$ strain causes failure in 1000 cycles [70]. It has also been observed that at about 1000 cycles the cyclic plastic strain range equals the elastic strain range [66].) In addition, for materials that exhibit a fatigue limit this fact must also be included in the expression for fatigue lifetime.

Despite the difficulties involved, there is considerable interest in these prediction methods, especially for preliminary design estimates. An example of one approach due to Manson [66] is given. The cyclic plastic strain range $\Delta \epsilon_{\text{pl}}$ is plotted in Fig. 8.21(a) for 29 materials, and the range is found to be related to the number of cycles to failure by the Manson-Coffin law

$$\Delta \epsilon_{\text{pl}} = \left(\frac{N}{D}\right)^{-0.6} \tag{8.11}$$

where D is the ductility defined as

$$D = \ln \frac{1}{1 - RA}$$

RA is the fractional reduction in area in a tensile test. Similarly, as shown in Fig. 21(b), the elastic strain range $\Delta \epsilon_{\text{el}}$ is related to the number of cycles to failure by

$$\Delta \epsilon_{\text{el}} = 3.5 \frac{\sigma_u}{E} N^{-0.12} \tag{8.12}$$

which is a form of Basquin's law (see Eq. 8.1). Here σ_u is the ultimate tensile strength. These expressions can be combined to give the total strain range as

$$\Delta \epsilon = 3.5 \frac{\sigma_u}{E} N_f^{-0.12} + D^{0.6} N_f^{-0.6} \tag{8.13}$$

A comparison of predicted and experimental axial fatigue life for low-alloy and high strength steels is given in Fig. 8.22. Equations 8.11 and 8.12 can be combined [68] to give the cyclic stress-plastic strain relation which is dependent on only four experimental constants:

$$\sigma = \frac{1.75 \sigma_u}{D^{0.12}} \Delta \epsilon_{\text{pl}}^{0.2} \tag{8.14}$$

This is a remarkable and surprising result, for it indicates that the only distinguishing characteristics of a material that determine the

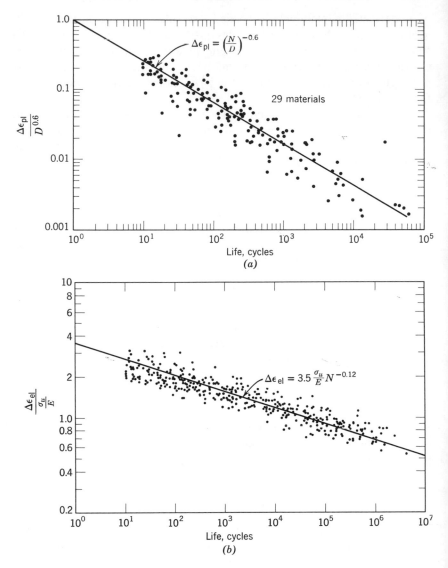

Fig. 8.21. Dependence of fatigue life on strain amplitude [66]. (*a*) Relation between plastic-strain ductility and cycles to failure. (*b*) Ratio of elastic strain range to σ_u/E against cycles to failure (29 materials).

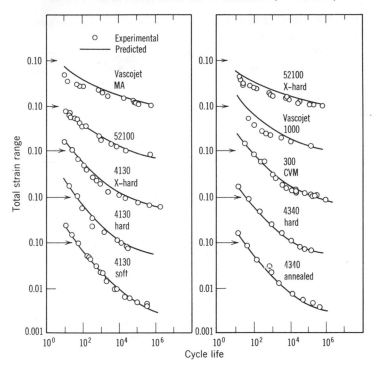

Fig. 8.22. Comparison of predicted and experimental axial fatigue life for low alloy and high strength steels [66].

cyclic stress-strain characteristics, and hence the finite fatigue life, are the tensile strength and the reduction in area. It is implicit therefore that the degree of cyclic hardening or softening must somehow be reflected in these two constants. Where greater accuracy is desired the cyclic stress-strain curve should be established [68].

Other considerations

In addition to use in predicting the fatigue life as a function of strain range, empirical methods have also been proposed for the prediction of the endurance limit of ferrous alloys. The best known is for steels of moderate strength and states that the endurance limit is one-half the tensile strength. A more exact expression is given by [71]

$$\sigma_e = 1.0139^{1/n} K^{0.911} \tag{8.15}$$

where n is the slope of the true stress—true strain curve, and K is the stress at unit plastic strain.

For heat-treated aluminum alloys a similar type of relationship has been proposed to predict the stress corresponding to a life of 10^8 cycles, namely,

$$\sigma_{(10^8)} = 1.0075^{1/n} K^{0.843} \qquad (8.16)$$

An important consideration in any prediction of fatigue life is the amount of scatter of data of nominally identical specimens. An example of the degree of scatter encountered in fatigue testing is given in Fig. 8.23 [72]. This figure is a logarithmic normal probability diagram that shows individual fatigue lifetimes obtained at different stresses for the aluminum alloy 7075-T6. A straight line on this plot corresponds to a log-normal distribution of data, and the steeper the slope of the lives the smaller the standard deviation. At the lowest stress the scatter of data covers more than an order of magnitude, whereas at the highest stress, which corresponds to the yield strength of the alloy, the degree of scatter is much less extensive. One of the factors contributing to the decrease in scatter with increasing stress amplitude is that the crack propagation stage, which constitutes a greater fraction of the total lifetime as the stress level is increased, is affected more by the average properties of the material than by heterogeneities. In fact, tests involving sharply notched specimens in which cracks are quickly initiated usually show very little scatter. In many instances, when the lifetime at a given stress level corresponding to 1 or 10% probability of failure of the group, rather than the 50% level, is of interest, statistical analysis with extreme value distributions is employed [73, 74]. Because of the statistical nature of fatigue, careful planning of test programs is an important consideration [75].

The accurate prediction of the service lifetime of a complex structure from simple tests is not yet a reality, and for this reason components are often tested under simulated service conditions. One reason why predictions are difficult is that reliable analytical methods have not yet been developed to include the effects of variable amplitude loading on fatigue life [77, 78]. Such a task is truly a formidable one, for even the sequence of loading events can influence the total lifetime indicated in Fig. 8.24 [76]. One method used as a guide to the effects of variable amplitude loading is based on Miner's rule, which states that failure will occur when the sum of the ratio of the number of cycles at the ith stress level, n_i, to the constant amplitude

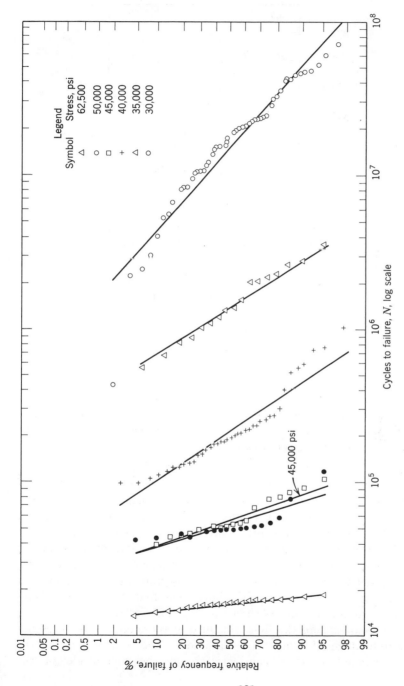

Fig. 8.23. Logarithmic normal probability diagram showing individual fatigue lifetimes obtained at different stresses, 7075-T6 aluminum alloy [72].

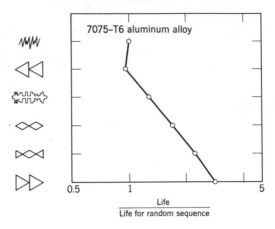

Fig. 8.24. Sequence effect. Type of loading spectrum indicated along ordinate scale [76].

lifetime at that level, N_i, summed over all i stress levels is equal to unity or

$$\sum_{i=i}^{i=i} \frac{n_i}{N_i} = 1 \qquad (8.17)$$

As the results shown in Fig. 8.24 indicate, this method is not always accurate. Therefore more refined, hence more complex, methods are being developed [79]. Meanwhile, periodic inspection provides some insurance against unexpected failure.

Size effects

Three types of size effect influence fatigue behavior. These are (a) a statistical size effect related to the probability of finding a critical flaw within the most highly stressed region, (b) a metallurgical size effect that is associated with a change in metallurgical structure and properties as a function of absolute size, and (c) a notch size effect related to the steepness of the stress-gradient at the root of a notch as a function of the notch radius. The first two of these effects could account for the results shown in Fig. 8.25 [80].

As a consequence of the first and third of these effects, the presence of a notch of small radius does not reduce the fatigue limit to the extent expected from a consideration of the magnitude of the

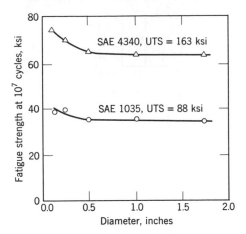

Fig. 8.25. Size effect in unnotched steel specimens in rotating bending [80].

theoretical stress concentration factor K_σ. The experimental fatigue notch sensitivity q can be expressed as

$$q = \frac{K_F - 1}{K_\sigma - 1} \tag{8.18}$$

where K_F, the effective notch stress concentration factor, is the ratio of fatigue strengths determined for unnotched and notched fatigue specimens. The value of q is a function both of the material tested and the radius of the notch ρ, and ranges in value from $q = 0$ for a notch insensitive material to $q = 1$ for a notch sensitive material containing a blunt notch. The effect of notch size on K_F is shown in Fig. 8.26 [81]. To account for this size effect, Neuber [82] proposed that the actual notch factor K_N be given in terms of the theoretical factor $K\sigma$, modified in the following manner:

$$K_N = 1 + \frac{K_\sigma - 1}{1 + \sqrt{\rho'/\rho}} \tag{8.19}$$

where the material constant ρ' is a distance across which a stress gradient cannot exist. As ρ approaches zero, K_N approaches unity. In the analysis of notch fatigue results it has been assumed that K_σ can be similarly modified to give

$$K_F = 1 + \frac{K_\sigma - 1}{1 + \sqrt{\rho'/\rho}} \tag{8.20}$$

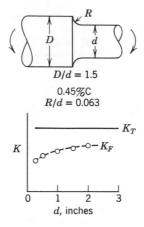

Fig. 8.26. Experimental and predicted notch-fatigue factors for low alloy steel rotating beams with shoulders [81]. K_T is the theoretical stress concentration factor, K_F is the ratio of the endurance limit unnotched to the endurance limit for notched specimens.

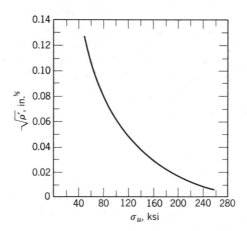

Fig. 8.27. The Neuber constant as a function of tensile strength for low alloy steels [81].

Values of ρ' as a function of tensile strength for steels are given in Fig. 8.27 [81]. As the strength increases, the value of ρ' decreases, and thus there may be no particular advantage in using a high strength steel where notch fatigue properties are a consideration. A comprehensive review of notch effects in fatigue has been given by Peterson [70].

8.5   Effects of Environment and Temperature

The environment

The fatigue properties of a metal can be greatly influenced by the nature of the environment, especially in long time tests at low stresses and at low frequencies. Even normal air has a deleterious effect on fatigue properties, and recent reviews [83, 84, 85] indicate that the principal adverse components of air are moisture and oxygen. For example, the fatigue life of copper in dry air is but one-tenth that in a vacuum of 10^{-6} torr [86] (1 torr = 1 mm Hg). The additional presence of moisture further reduces the fatigue life by a factor of three. For lead [87], as with copper, oxygen alone has been found to have an adverse effect. Moisture decreases the fatigue strength of copper, aluminum, magnesium, and iron by 5 to 15%. An example of the influence of moisture content on the fatigue properties of an aluminum alloy is given in Fig. 8.28a. The influence of air on a Fe-0.5% C alloy is shown in Fig. 8.28b.

Investigators have generally agreed that the main effect of a normal atmosphere on fatigue is on the propagation of cracks rather than in their initiation. This seems to be a reasonable interpretation, for the initially present oxide films will not be greatly altered by testing at low pressure. The principal effect of the environment will be in its reaction to the clean metal exposed at the tip of a fatigue crack. Cracks that would have been of a nonpropagating variety in vacuum, propagate in the presence of oxygen or moisture and thereby the fatigue resistance is lessened.

In environments that are more corrosive than air and that substantially reduce the resistance of the surface layers to fatigue crack initiation, both initiation as well as propagation may be affected. An example is shown in Fig. 8.29 [88]. Environments that lead to stress-corrosion cracking may also adversely affect fatigue properties [89].

The influence of atmospheric effects on the rate of crack propagation in notched sheet specimens has been studied. At a pressure of 4×10^{-8} torr the rate of crack growth of an aluminum alloy is an

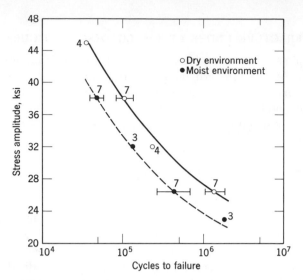

Fig. 8.28. (*a*) *S-N* curves for clean specimens of aluminum alloy 6060-T6 in high and low humidity environments. In all tests the mean tensile stress on the test surface was one-third of the stress amplitude. Numbers adjacent to points indicate the number of replicate tests; vertical bars show the range of ± one standard deviation [84].

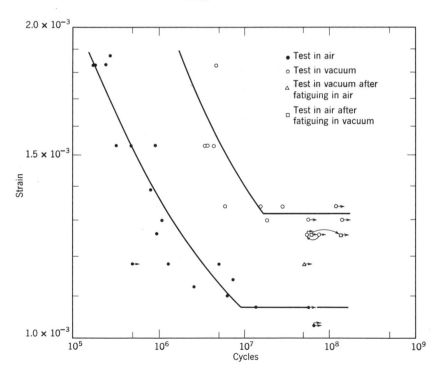

Fig. 8.28. (*b*) Effect of environment on fatigue properties of an Fe-0.5% C alloy [125].

order of magnitude less than in air for low values of the stress intensity factor [91]. At higher values of the stress intensity factor the rate of crack growth is less sensitive to the environment. Therefore the exponent that expresses the power law dependence of the rate of crack growth on the stress intensity factor will depend on the nature of the environment.

If tests are made as a function of air pressure, the results shown in Fig. 8.30 for lead are obtained. Two regions in which the fatigue properties are independent of pressure are separated by a transition region. This transition region for lead occurs at a surprisingly high level of pressure, for even at 10^{-6} torr a surface is covered with gas in about one second if all impinging atoms adhere to the surface. The occurrence of such a high transition pressure in lead has been attributed to a low sticking probability of gas atoms and to the impedance of the crack on the diffusion of air molecules to the crack tip [85]. The existence of the high pressure plateau is thought to be due to the attainment of saturation of the surface by gas atoms for all pressures above the transition region. For other metals these plateaus may not be so well defined as they are for lead. For example, in nickel the fatigue properties continue to improve down to a pressure of but 7×10^{-9} torr [85]. At elevated temperatures the environment can even exert a beneficial influence if oxidation strengthening occurs as in nickel at 816°C for air pressures greater than 10^2 torr [85]. Figure 8.31 indicates that certain coatings can also have a beneficial effect by protecting a specimen surface from the environment.

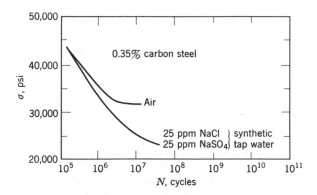

Fig. 8.29. Effect of environment on fatigue properties of a 0.35% carbon steel [88].

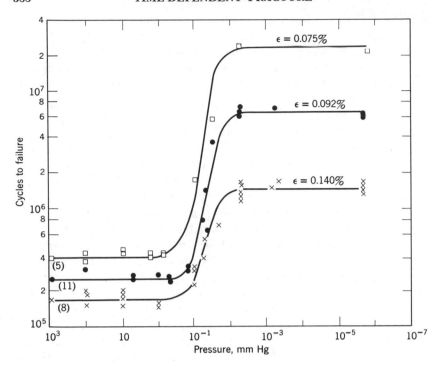

Fig. 8.30. Variation of fatigue life of lead with air pressure [87].

The fracture appearance of fatigued specimens can at times reflect the influence of the environment. A shift from a vacuum to an air environment increased the rate of fatigue crack propagation in an aluminum alloy by a factor of 4 and striations were much more pronounced in air [85]. The Stage II crack growth in Al-Zn-Mg alloys tested in salt-water solutions is more brittle in nature than if tests were conducted in air, and fracture occurs close to {100} planes [90]. The titanium alloy, Ti-7Al-2Cb-1Ta, which normally exhibits only ductile striations, shows evidence of quasi-cleavage and increased crack growth rates when tested in a salt-water environment [92].

    The effect of the environment appears to be the result of a corrosive reaction at the crack tip rather than in the prevention of reweldment of cracked metal during the compression portion of the loading cycle. Evidence for reweldment has been found in stainless steel, copper, and aluminum but not in carbon steels, although the fatigue properties of all were better in vacuum than in air. Reweldment may therefore be a contributing but not a controlling factor [93]. With respect to the nature of the corrosion reaction it has been proposed that hydrogen

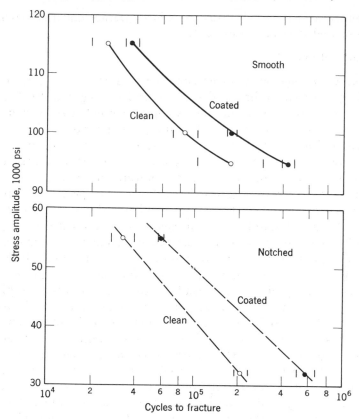

Fig. 8.31. *S-N* curves for smooth and notched specimens of SAE 4340 steel. "Coated" indicates that a polar organic liquid had been applied to the test sections of the specimens [84].

liberated in the breakdown of water molecules may be responsible for the embrittlement of aluminum alloys [94]. Oxide film formation alone under high stress at a fatigue crack tip may promote the rate of advance of the crack, especially at low rates of crack propagation, where the oxide film thickness may be comparable to the amount of crack growth per cycle.

The nature of the environment is important in a more active sense in rolling contact fatigue, where the nature of the lubricant can influence the wear process taking place in the region of contact. Fretting corrosion, which is the result of the relative displacement of materials in contact, is also influenced by the environment. In this instance the presence of water vapor can be beneficial, for it apparently acts as a lubricant to reduce friction and the attendant corrosion and wear processes [95].

Thermal fatigue

Thermal fatigue refers to cracking or fracture due to stresses induced by thermal cycling rather than load cycling [96, 97, 98, 99]. It can be a serious problem; for example, thermal-stress fatigue, the predominant cause of failure of military jet aircraft engines [100], has occurred in a form of low cycle fatigue in steam powerplant components operated on an intermittent basis. The thermal stresses can appear in a number of ways. A temperature change can lead to the development of stress if external constraints are present to prevent free expansion or contraction or if nonlinear thermal gradients develop. A marked anisotropy of thermal expansion coefficients in polycrystalline noncubic metals, such as zinc or cadmium, can give rise to thermal stresses high enough to cause plastic deformation [101]. In an alloy containing a dispersed phase a difference in expansion coefficients between phases can lead to thermal stresses and plastic deformation, as in an austenitic steel containing particles of eutectic niobium carbide [102]. Repeated thermal cycling involving an allotropic phase transformation, such as the ferrite-austenite transformation, can also lead to plastic deformation and fatigue failure [103]. An important factor in thermal fatigue as well as high-temperature fatigue can be the rupture of protective oxide films as the result of localized and reversed grain boundary deformation, which leads to repeated ruptures of the protective oxide film and accelerated oxidation in the region of deformation. The localized oxidation leads to a notching at the surface, a further localization of strain and to the initiation of cracks [104, 105]. Other factors that affect the thermal fatigue resistance are the times at the maximum and minimum temperatures, the stability of the metallurgical structure, and the temperature dependence of the flow stress, expansion coefficient, and modulus. The extent of plastic deformation is important, and it has been observed for a series of cast steels that the resistance to thermal cracking parallels their resistance to creep [106]. The fracture path is also similar to that observed in creep, with transgranular cracking predominant at low values of the maximum temperature and intergranular cracking at high values, where creep is more pronounced (cf. Chapter 9).

To evaluate resistance to thermal fatigue, constrained cylinders, either solid or hollow, are often used in tests. The spring constants of the constraints can significantly affect the magnitude of the thermal stresses developed [107]. Specimens are resistance-heated to the

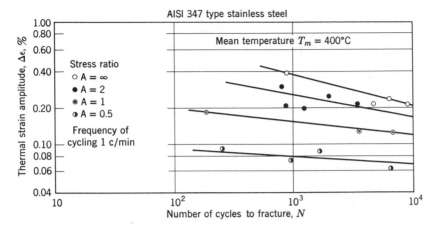

Fig. 8.32. Fracture life of AISI-type 347 stainless steel subjected to cyclic thermal stress superimposed on mechanical stress. A is the ratio of the thermal stress amplitude to the mean tensile stress [108].

desired temperature and then cooled by a compressed gas or simply by conduction. By clamping the specimens at a given point in the thermal cycle we can obtain control of the ratio of the stress amplitude to the mean stress. An example of the influence of mean stress on thermal fatigue lifetime for AISI 347 stainless steel is shown in Fig. 8.32 [108]. As the mean stress level rises, the combined effects of creep and thermal cycling considerably increase the plastic strain per cycle and shorten the total lifetime.

Figure 8.33 shows the results obtained in thermal fatigue tests of a nickel base alloy, Hastelloy N [109], tested under fully reversed strain amplitude. The influence of the temperature range, the maximum temperature, and the time at maximum temperature are evident. If these data, together with those obtained in isothermal fatigue tests, are plotted in terms of the plastic strain range per cycle, a useful correlation, as in low cycle fatigue at ambient temperature, is obtained (see Fig. 8.34). For this alloy the fatigue life under a variety of test conditions can be expressed simply as a function of the total plastic strain amplitude per cycle; that is, a Manson-Coffin type relationship is applicable.

The isothermal fatigue strength generally exhibits the same temperature dependence as the flow stress or the tensile strength at temperatures below that at which creep becomes an important consideration [110]. An example of the fatigue strength as a function of temperature is shown in Fig. 8.35 [111], together with the corresponding tensile strength properties. The fatigue strength of this BCC metal

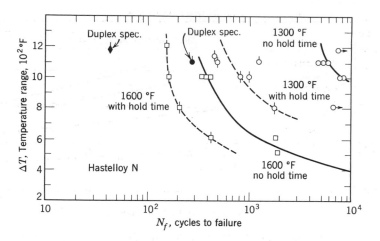

Fig. 8.33. Temperature range versus cycles to failure [109].

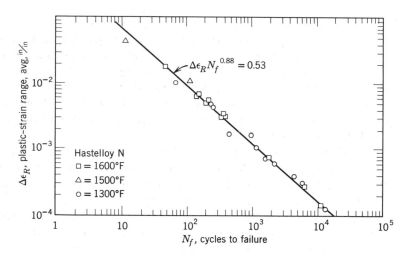

Fig. 8.34. Fatigue lifetime as a function of plastic strain range at elevated temperatures. Rectangular symbols for 1600°F maximum, thermal fatigue only. Circular symbols, 1300°F maximum: open, thermal fatigue; solid, isothermal fatigue. Triangular for 1500°F, isothermal fatigue only [109].

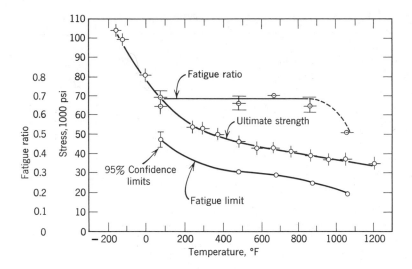

Fig. 8.35. Fatigue and tensile strength of unalloyed arc-cast molybdenum as influenced by temperature [111].

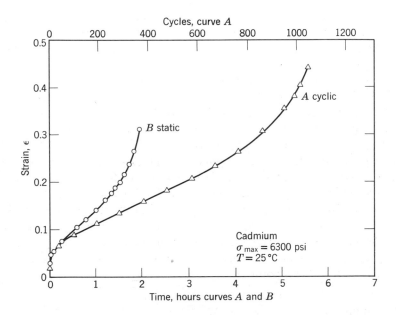

Fig. 8.36. Comparison of the static and cyclic creep behavior of cadmium at 25°C [112].

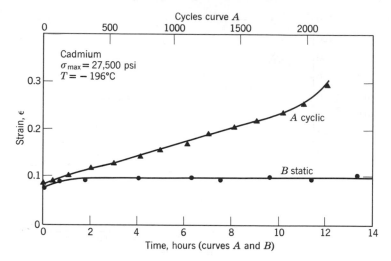

Fig. 8.37. Comparison of the static and cyclic creep behavior of cadmium at —196°C [112].

is remarkably high with respect to tensile strength up to 900°F where the influence of creep becomes manifest. Although the combined effects of creep and fatigue are complex, there is some evidence that the number of cycles to failure can be expressed in terms of the total plastic strain per cycle, at least for fully reversed strain.

A further example of the influence of temperature in combined creep and fatigue is given in Figs. 8.36 and 8.37 [112] for the case of cadmium, which has a low melting point and creeps considerably at room temperature. At room temperature cadmium elongates more rapidly under static load than under a pulsating tensile load to the same maximum stress. On the other hand, at the temperature of liquid nitrogen, time-dependent elongation under a static load is suppressed, but cyclic stress greatly accelerates the creep rate. In general, cycle-dependent deformation processes appear to be dominant when the test temperature is below one-quarter of the melting point, and those that are time-dependent dominate when the test temperature is greater than one-half the melting point.

8.6   Summary

In this chapter the nature of the fatigue process has been discussed in terms of the plastic deformation processes involved in the initiation and propagation of fatigue cracks. The various types of nucleating sites were described and the influence of the nature of the glide process was indicated. Fatigue crack propagation consists of two stages. Stage I pertains to growth along the slip planes involved in crack initiation. Stage II involves propagation more directly under the influence of the tensile stress. A general mechanism for Stage II propagation involves the advance and blunting of a fatigue crack during tensile loading and the resharpening of the crack during unloading or compression. The fraction of total life spent in crack propagation was discussed and means for predicting the total lifetime, presented. Finally, the effects of environment and temperature on fatigue properties were considered. Although the various aspects of the fatigue process are understood in a qualitative fashion, there is much to be done before the subject is understood on a truly quantitative basis.

References

The asterisk indicates that published work is recommended for extensive or broad treatment.

[1] O. H. Basquin, *Proc. ASTM* **10**, 625 (1910).
[2] F. A. McClintock and A. S. Argon, *Mechanical Behavior of Materials,* Addison-Wesley, Reading, Mass. (1966).
[3] M. Kikukawa, K. Ohji, and K. Ogura, *ASME Paper No. 65-Met-4* (1965).
*[4] W. A. Wood, S. M. Cousland, and K. R. Sargant, *Acta Met.,* **11**, 643 (1963).
[5] W. A. Wood and R. L. Segall, *Proc. Roy. Soc.,* **A242**, 180 (1947).
*[6] S. S. Manson, *Thermal Stress and Low Cycle Fatigue*, McGraw-Hill, New York, (1966).
*[7] J. Morrow and F. Tuler, *Trans. ASME, J. Basic Eng.,* 275 (June 1965).
[8] T. H. Alden, *Trans AIME,* **224**, 1287 (1962).
*[9] A. J. McEvily and T. L. Johnston, *Int. Conf. on Fracture*, Sendai, Japan, 1965.
[10] J. E. Pratt, *Univ. Illinois T and A. M. Report No. 652* (1965).
[11] C. E. Feltner and C. Laird, to be published.
[12] E. E. Laufer and W. N. Roberts, to be published.
[13] C. E. Feltner, *Phil. Mag.,* **12**, 1229 (1965).
[14] T. Broom and J. M. Summerton, *Phil. Mag.,* **8**, 1847 (1963).
[15] C. A. Stubbington, *Acta Met.,* **12**, 931 (1964).
[16] J. B. Clark and A. J. McEvily, *Acta Met.,* **12**, 1359 (1964).
[17] N. H. Polakowski, *Proc. ASTM,* **52**, 1086 (1952).
[18] R. J. Warrick, D. J. Schmatz and W. F. Eagleson, Ford Motor Co., unpublished results.
[19] G. J. Moyar and J. Morrow, *Univ. Illinois Eng. Exper. Station Bull. 468* (1964).
*[20] N. Thompson and N. J. Wadsworth, *Adv. Physics,* **7**, 72 (1958).
[21] W. H. Munse, J. E. Stallmeyer, and J. W. Rone, *Univ. Illinois Rept.* (1965).

[22] G. B. Lutz and R. P. Wei, *U.S. Steel Appl. Res. Lab. TR Proj. No. 40112-011(1)* (July, 1961).

*[23] W. J. Harris, *Metallic Fatigue,* Pergamon Press, New York (1961).

[24] P. J. E. Forsyth, *J. Inst. Metals,* **83,** 173 (1954–1955).

[25] C. A. Stubbington and P. J. E. Forsyth, *J. Inst. Metals,* **86,** 90 (1957–1958).

[26] G. M. Sinclair, *Proc. ASTM,* **52,** 743 (1952).

[27] J. C. Levy and S. L. Kanitkar, *J. Iron Steel Inst.,* **199,** 296 (1961).

[28] T. Broom and R. K. Ham, *Proc. Roy. Soc.,* **A251,** 186 (1959).

[29] R. K. Ham and T. Broom, *Phil Mag.,* **7,** 95 (1962).

[30] O. Helgeland, *J. Inst Metals,* **93,** 570 (1964–1965).

[31] Anon, *ASM Metals Handbook,* Metals Park, Ohio (1961).

[32] J. T. McGrath and G. W. J. Waldron, *Phil. Mag.,* **9,** 249 (1964).

*[33] D. H. Avery and W. A. Backofen, *Acta Met.,* **11,** 653 (1963).

[34] A. J. McEvily and R. C. Boettner, *Acta Met.,* **11,** 725 (1963).

[35] M. L. Ebner and W. A. Backofen, *Trans AIME,* **215,** 520 (1959).

[36] D. Hull, *J. Inst. Metals,* **86,** 425 (1958).

[37] A. J. McEvily and E. S. Machlin, in *Fracture,* B. L. Averbach et al. eds., M.I.T., Wiley, New York, p. 450 (1959).

[38] D. M. Fregredo and G. B. Greenough, *J. Inst. Metals,* **87,** 1 (1958).

[39] N. H. Mott, *Acta Met.,* **6,** 195 (1958).

[40] R. L. Segall and P. G. Partridge, *Phil Mag.,* **4,** 912 (1959).

[41] J. C. Grosskreutz and P. Waldow, *Acta Met.,* **11,** 717 (1963).

[42] J. T. McGrath and W. J. Bratina, *Phil. Mag.,* **12,** 1293 (1965).

[43] H. J. Gough, *The Fatigue of Metals,* Benn, London (1926).

[44] A. F. Moore and J. B. Kommers, *The Fatigue of Metals,* Benn, London (1926).

[45] F. C. Lea, *Engineering,* **115,** 217, 252 (1923).

*[46] C. Laird and G. C. Smith, *Phil. Mag.,* **7,** 847 (1962).

[47] J. C. Grosskreutz, *ASTM Conf. on Fatigue Crack Propagation,* Atlantic City (1966).

[48] R. W. Hertzberg and P. C. Paris, *Int. Conf. on Fracture,* Sendai, Japan (1965).

[49] H. W. Liu, *Appl. Mat. Res.,* **3,** 229 (1964).

[50] D. P. Wilhelm, *ASTM Conf. on Fatigue Crack Propagation,* Atlantic City (1966).

[51] W. Weibull, *Acta Met.,* **11,** 745 (1963).

[52] P. C. Paris, M. P. Gomez, and W. E. Anderson, *The Trend in Engineering,* **13,** 9 (1961).

[53] W. Illg and A. J. McEvily, *NASA,* **TN D-52** (1959).

*[54] P. C. Paris and F. Erdogan, *J. Basic Eng., Trans. ASME, Series D,* **85,** 528 (1963).

[55] J. Weertman, *Proc. Int. Conf. on Fracture,* Sendai, Japan (1966).

[56] F. A. McClintock, *Fracture of Solids,* J. J. Gilman and D. C. Drucker, eds., Wiley (1963).

[57] J. R. Rice, *ASTM Conf. on Fatigue Crack Propagation,* Atlantic City (1966).

[58] J. Krafft, *Appl. Mat. Res.,* **3,** 1963.

[59] A. J. McEvily and W. Illg, *NACA,* **7N,** 4394 (1958).

[60] A. J. McEvily, R. C. Boettner, and A. P. Bond, *J. Inst. Metals,* 93, 481 (1964–1965).
[61] A. J. McEvily, W. Illg, and H. Hardrath, *NACA,* TN 3816 (1956).
[62] C. Laird and G. C. Smith, *Phil. Mag.,* 8, 1945 (1963).
[63] S. S. Manson, *NACA,* TN 2933 (1954).
*[64] L. F. Coffin, *Trans. ASME,* 76, 931 (1954).
*[65] R. C. Boettner, C. Laird, and A. J. McEvily, *Trans. AIME,* 233, 379 (1965).
*[66] S. S. Manson, *Exp. Mech.,* 5, 193 (1965).
[67] B. F. Langer, *J. Basic Eng., Trans. ASME,* 84, Series D, 389 (1962).
[68] J. Morrow and F. R. Tuler, *J. Basic Eng. ASME,* 87, 275 (1965).
[69] J. F. Tavernelli and L. F. Coffin, *J. Basic Eng. ASME,* 84 (December 1962).
*[70] R. E. Peterson, *ASTM Edgar Marburg Lecture,* 1962.
[71] G. H. Rowe, in *Fatigue—An Interdisciplinary Approach,* Syracuse Univ. Press, 22 (1964).
[72] G. M. Sinclair and T. J. Dolan, *Trans. ASME,* 75, 867 (July 1953).
[73] A. M. Freudenthal and E. J. Gumbel, *Int. Conf. on Fatigue of Metals,* Inst. Mech. Engineers (1956).
[74] W. Weibull, *Trans. ASME,* 73, 293 (1951).
[75] F. McClintock, in *Metal Fatigue,* McGraw-Hill, New York (1959).
[76] H. F. Hardrath in *Fatigue—An Interdisciplinary Approach,* Syracuse Univ. Press, 345 (1964).
[77] L. Kaechele, Review and Analysis of Cumulative-Fatigue Damage Theories, Rand Report RM-3650-PR, August 1963.
[78] L. Kaechele, Probability and Scatter in Cumulative Fatigue Damage, Rand Report RM-3688-PR, Dec. 1963.
[79] A. M. Freudenthal and R. A. Heller, in *Fatigue in Aircraft Structures,* Academic Press, New York, p. 146 (1956).
[80] H. Grover, in *Fatigue—An Interdisciplinary Approach,* Syracuse Univ. Press, 361 (1964).
[81] P. Kuhn and H. F. Hardrath, *NACA,* TN 2805 (1952).
[82] H. Neuber, *Kerbspannungslehre,* Springer (1958).
[83] P. T. Gilbert, *Metallurgical Rev.,* 1, 379 (1956).
*[84] J. A. Bennett, in *Fatigue—An Interdisciplinary Approach,* Syracuse Univ. Press, 209 (1964).
*[85] M. R. Achter, *ASTM Conf. on Fatigue Crack Propagation,* Atlantic City (1966).
[86] N. J. Wadsworth and J. Hutchings, *Phil. Mag.,* 3, 1154 (1958).
[87] K. U. Snowden, *Acta Met.,* 12, 295 (1964).
[88] H. J. Gough and D. G. Sopwith, *J. Inst. Metals,* 49, 92 (1932).
[89] A. J. McEvily and P. A. Bond, in *Environment Sensitive Mechanical Behavior,* Gordon-Breach, New York (1966).
[90] C. A. Stubbington and P. J. E. Forsyth, *J. Inst. Metals,* 90, 348 (1961–1962).
[91] F. J. Bradshaw and C. Wheeler, Royal Aircraft Establishment, TR 65073, April 1965.
[92] R. W. Judy et al., *NRL Report 6330,* January 1966.
[93] D. E. Martin, *ASME Paper No. 65—Met. 5* (1965).
[94] T. Broom and A. Nicholson, *J. Inst. Metals,* 89, 183 (1961).

[95] H. H. Uhlig, *J. Appl. Mech.,* **21,** 401 (1954).

[96] E. Glenny, J. E. Northwood, S. W. K. Shaw, and T. A. Taylor, *J. Inst. Metals,* **85,** 294 (1958–1959).

[97] E. Glenny and T. A. Taylor, *J. Inst. Metals,* **88,** 449 (1960).

[98] H. Thielsch, *Welding Research Council Bulletin Series No. 10* (1952).

[99] S. S. Manson, *NACA Report, 1170* (1954).

[100] H. C. Cross, *Metals Progr.,* **87,** No. 3, 67 (1965).

[101] W. Boas and R. W. K. Honeycombe, *Proc. Roy. Soc.,* **A188,** 427 (1946–1947).

[102] L. H. Toft and T. Broom, *Joint. Int. Conf. on Creep,* 1963, Inst. Mech. Engineers, London, 3–77 (1963).

*[103] A. J. Kennedy, *Processes of Creep and Fatigue in Metals,* Wiley, New York (1963).

[104] L. F. Coffin, *Trans. ASM,* **56,** 339 (1963).

[105] W. A. Stauffer and A. Keller, *Joint Int. Conf. on Creep 1963,* Inst. Mech. Engineers, London, 3–29 (1963).

[106] W. Gysel, A. Werner, and K. Giet, *Joint Int. Conf. on Creep 1963,* Inst. Mech Engineers, London 3–33 (1963).

[107] E. Krempl and H. Neuber, *Air Force Materials Laboratory Tech. Report No. AFML-TR-65 25* (1965).

[108] S. Taira and M. Ohnami, *Joint Int. Conf. on Creep 1963,* Inst. Mech. Engrs., London, 3–57 (1963).

[109] A. E. Carden, *Trans. ASME,* **87,** Series D 237 (1965).

[110] R. D. McCammon and H. M. Rosenberg, *Proc. Roy. Soc.,* **A242,** 203 (1957).

[111] G. W. Brock and G. M. Sinclair, *Proc. ASTM,* **60** (1960).

[112] C. E. Feltner and G. M. Sinclair, *Joint Int. Conf. on Creep 1963,* Inst. Mech. Engineers, London 3–9 (1963).

[113] A. J. McEvily and R. C. Boettner, in *Fracture of Solids,* J. J. Gilman and D. C. Drucker, eds., Wiley, p. 383 (1963).

[114] F. Borik, W. F. Justusson, and V. F. Zackay, *Trans. ASM,* **56,** 327 (1963).

[115] A. J. McEvily, R. L. Snyder, and J. B. Clark, *Trans. AIME,* **277,** 452 (1963).

[116] R. C. Boettner, A. J. McEvily, and Y. C. Liu, *Phil. Mag.,* **10,** 95 (1964).

[117] R. C. Boettner and A. J. McEvily, *Acta Met.,* **13,** 937 (1965).

[118] P. J. E. Forsyth and D. A. Ryder, *Metallurgia,* **63,** 117–124 (March 1961).

[119] G. Jacoby, *Exp. Mech.* **5,** 65 (1965).

[120] L. F. Coffin and J. F. Tavernelli, *Trans. AIME,* **215,** 794 (1959).

[121] B. J. Lazan and T. Wu, *Proc. ASTM,* **51,** 639 (1951).

*[122] P. J. E. Forsyth, *Proc. Crack Propagation Symposium,* **1,** College of Aeronautics, Cranfield, England, 76 (1962).

[123] P. G. Forrest and A. E. L. Tate, *J. Inst. Metals,* **93,** 438 (1964–1965).

*[124] C. Laird, *ASTM Conf. on Fatigue Crack Propagation,* Atlantic City (1966).

[125] N. J. Wadsworth, *Phil Mag.,* **6,** 397 (1961).

[126] J. Morrow, *ASTM,* **STP 378,** 45 (1965).

[127] E. Orowan, *Proc. Roy. Soc.,* **A171,** 79 (1939).

[128] A. H. Cottrell and D. Hull, *Proc. Roy. Soc.,* **A-242,** 211, (1957).

[129] W. A. Wood, in Fatigue of Aircraft Structures, A. Freudenthal, ed., Academic Press, **1** (1956).

[130] R. W. Smith, M. H. Hirschberg and S. S. Manson, NASA, TN-D-1574 (1963).
[131] S. R. Swanson, F. Cicci and W. Hoppe, *ASTM Conf. on Fatigue Crack Propagation,* Atlantic City (1966).
[132] C. Lipson and R. C. Juvinall, Handbook of Stress and Strength, Macmillan Co., New York, 111 (1963).

Bibliography

1. *Fatigue in Aircraft Structures,* A. M. Freudenthal, ed., Academic, New York, 1956.
2. *Fracture of Solids,* Drucker and Gilman, eds., Interscience, New York, 1963.
3. *Metallic Fatigue,* W. J. Harris, Pergamon, New York, 1961.
4. *Processes of Creep and Fatigue in Metals,* A. J. Kennedy, Wiley, New York, 1963.
5. *Mechanical Behavior of Materials at Elevated Temperatures,* Dorn, ed., McGraw-Hill, 1961.
6. *Fatigue, An Interdisciplinary Approach,* Burke, Reed, and Weiss, eds., Syracuse Univ. Press, 1964.
7. *Current Aeronautical Fatigue Problems,* Schijve, Heath-Smith, and Welbourne, eds., Pergamon, New York, 1965.
8. *Proceedings of the Crack Propagation Symposium,* I and II, College of Aeronautics, Cranfield, England. (1962).
9. *Fatigue of Metals,* P. G. Forrest, Pergamon, New York, 1962.
10. *Manual on Fatigue Testing,* ASTM STP 91, 1949.
11. *The Statistical Treatment of Fatigue Experiments,* L. G. Johnson, Elsevier, New York, 1964.
12. *Fatigue and Fracture of Metals,* Murray, ed., Wiley, New York, 1952.
13. *Fatigue of Metals,* R. Cazaud, translated by A. J. Fenner, Chapman and Hall, 1953.
14. *Thermal Stress and Low Cycle Fatigue,* S. S. Manson, McGraw-Hill, New York, 1966.
15. *Edgar Marburg Lecture,* H. J. Gough, ASTM, 33, 1933.
16. *Internal Stresses and Fatigue in Metals,* Rassweiler and Grube, eds., Elsevier, Amsterdam (1959).
17. *International Conference on Fatigue of Metals,* London and New York, 1956, Institute of Mechanical Engineers and ASME.
18. *Fatigue of Welded Steel Structures,* W. H. Munse, ed., Welding Research Council, 1964.
19. *Residual Stresses and Fatigue in Metals,* J. O. Almen and P. H. Black, McGraw-Hill, New York, 1963.
20. *Symposium on Acoustical Fatigue,* ASTM STP 284, 1961.
21. *Fatigue Resistance of Materials and Metal Structural Parts,* A. Buch, ed., Pergamon, New York, 1964.
22. *Designing Against Fatigue of Metals,* R. B. Heywood, Rheinhold, New York, 1962.
23. *The Fatigue of Metals and Structures,* H. J. Grover, S. A. Gordon, and L. R. Jackson, U.S. Government Printing Office, 1960 (revised).
24. *Symposium on Basic Mechanisms of Fatigue,* ASTM STP 237, 1958.

25. *Fatigue Testing and the Analysis of Results,* W. Weibull, Pergamon, New York, 1961.
26. *A Guide for Fatigue Testing and Statistical Analysis of Fatigue Data,* ASTM STP 91-A, 1963.
27. *Colloquium on Fatigue,* Stockholm, 1955, Weibull and Odgvist, eds., Springer, Berlin, 1956.
28. *Fatigue Resistance,* P. E. Kravchenko, translated by O. M. Blunn, N. L. Day, ed., Pergamon, New York, 1964.
29. *Stress Concentration Design Factors,* R. E. Peterson, Wiley, New York, 1953.
30. *Metal Fatigue,* J. A. Pope, Chapman, Hall, London, 1959.
31. *Metal Fatigue,* G. Sines and J. L. Waisman, eds., McGraw-Hill, New York, 1959.
32. *Fatigue of Aircraft Structures,* Proceedings of Paris Symposium, 1961, W. Barrois and E. L. Ripley, eds., Pergamon, New York, 1963.
33. *Symposium on Fatigue Tests of Aircraft Structures,* ASTM STP 338, 1963.
34. *Full Scale Testing of Aircraft Structures,* Plantema and Schivje, eds., Pergamon, New York, 1961.
35. *The Fatigue of Metals,* H. J. Gough, E. Benn, London, 1926.
36. *The Fatigue of Metals,* H. F. Moore and J. B. Kommers, McGraw-Hill, New York, 1927.
37. *Strength and Structure of Engineering Materials,* N. H. Polakowski and E. J. Ripling, Prentice-Hall, Englewood Cliffs, N.J., (1966).
38. *Mechanisms of Fatigue in Crystalline Solids, Acta Met.,* **11**, No. 7 (July 1963).

Problems

1. Discuss the Orowan [127], Mott [39], and Cottrell-Hull [128] models of fatigue crack initiation in the light of recent experimental evidence concerning fatigue crack initiation.

2. The unnotched fatigue strength of the steel is approximately one-half its tensile strength. Compute the fatigue limit of a wide sheet containing a small hole as a function of hole size for two steels, one of 100,000 psi tensile strength, the other of 300,000 psi. Use Fig. 8.27 in this determination.

3. The rate of fatigue crack propagation can be expressed as

$$\frac{dc}{dN} = A \frac{\sigma^4 c^2}{[(\sigma_y + \sigma_u)/2]\epsilon_u \sigma_u^2 E}$$

Compute an appropriate value for the constant A. Using the relation $\sigma = \sigma_0 \epsilon^n$ and taking the yield stress as the flow stress at 0.2% plastic strain, rewrite the above expression for crack growth in items of A, c, σ_0, and n for an ideal material.

4. Using Eq. 8.5, determine the variation of this number of cycles

to failure as a function of the critical crack length for $\sigma = 10,000$ psi and $\sigma_0 = 50,000$ psi, $n = 0.2$, $E = 10^7$ psi, and $c_0 = 0.1$ in.

5. Plot the expected number of cycles to failure as a function of total strain range for a steel whose tensile strength is 100,000 psi and reduction in area is 80%.

6. Draw a modified Goodman diagram [2] for a steel of 100,000 psi tensile strength and 80,000 psi yield strength and determine the safe working range at a stress of 30,000 psi.

7. The accompanying Fig. a indicates the effect of mean stress on the fatigue limit (R is the ratio of the minimum to maximum stress). Fig. b shows the influence of surface condition on the fatigue limit. Assuming that the fatigue strength of polished specimens at $R = -1$ is one half the tensile strength, construct and compare modified Goodman diagrams for each surface condition for 100- and 200-ks tensile strengths. How would yielding limit the useful design range of the Goodman diagram?

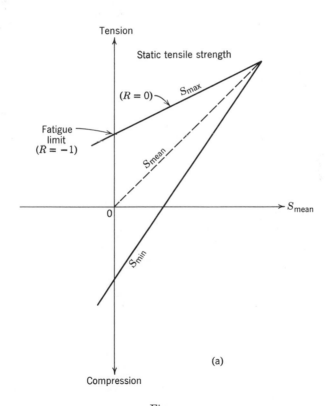

(a)

Fig. a

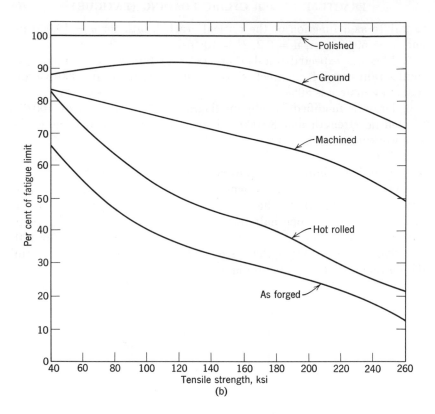

Fig. b

9

Fracture Under Static Loading

9.1 Creep Rupture

At temperatures in excess of $0.5T_M$ the time-dependent deformation and rupture characteristics of materials are primary design considerations. In many applications, for example, the gas turbine, the operating temperature, and hence the efficiency, is limited by the creep characteristics of materials, and therefore there is an incentive to develop economic alloys of superior creep resistance. Designs are usually made on the basis of a maximum permissible amount of creep such as 0.1 or 1% during the expected service life of the particular component. These service lives can vary, from minutes for rocket nozzles to thousands of hours for jet engine turbine buckets between inspections and overhauls, to a hundred thousand hours for high pressure steam lines. However, the amount of creep deformation may not be the only factor governing service lifetime; many new alloys have frequently been found to fracture after very limited deformation, with the result that service life can be governed not by the time to reach some critical strain, but rather by the time to rupture [1].

Creep fractures in most commercial alloys are intergranular and appear as normal and shear ruptures. At high stresses and low temperatures in the creep range these fractures originate at voids at grain boundary triple points. At lower stresses and higher temperatures rupture can result from the formation of a multiplicity of voids along grain boundaries, especially those transverse to an applied tensile stress. This process of void formation is known as *cavitation*. In certain metals (e.g., high-purity aluminum and lead) grain boundary voids are not observed and fracture occurs in a more ductile manner by pure shear [2].

The environment can also increase the tendency for brittle fracture. A form of high-temperature stress corrosion cracking of copper can occur as a result of oxide film rupture [3]. In addition, oxygen absorbed from the atmosphere can facilitate the nucleation of grain

403

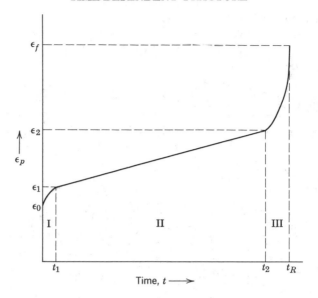

Fig. 9.1. Schematic creep curve.

boundary voids (cavitation) [4], and if oxygen is excluded from the test environment, the extent of cavitation in copper is decreased [5, 6].

The characteristic form of a creep curve with its three stages is shown in Fig. 9.1. Region I, the primary stage, often can be described by Andrade's relation

$$\epsilon_p = \beta t^{1/3} \qquad (\epsilon_0 < \epsilon_p < \epsilon_1) \tag{9.1a}$$

where ϵ_1 is relatively independent of stress and temperature. In the second, or steady state stage, the rate of extension is a minimum and is independent of time:

$$(\epsilon_p - \epsilon_1) = \alpha t \tag{9.1b}$$

$$\dot{\epsilon}_p = \alpha \qquad (\epsilon_1 < \epsilon_p < \epsilon_2)$$

This rate is of particular interest with respect to rupture lifetime because of the existence of an empirical relation [7] between the steady state creep rate $\dot{\epsilon}_p$ and the time to rupture t_R, which for a given material is

$$\dot{\epsilon}_p t_R = \text{const} \tag{9.2}$$

This relationship implies (a) that the amount of strain occurring in the steady state region is a constant for a given material, and (b) that the time spent in the steady state range in developing this critical strain far outweighs the time spent in regions I and III. Therefore those factors which affect the steady state creep rate may also be expected to have an effect on the time to rupture.

The steady state creep rate is thought to be governed by the process of dislocation climb [8, 9] which depends on volume diffusion, whereas the rupture process ordinarily depends on diffusion, along grain boundaries or on grain boundary sliding. Further, the time to rupture can be varied without affecting the value of the minimum creep rate by the imposition of hydrostatic pressure. It seems, therefore, that the underlying reason for the simple interrelation expressed in Eq. 9.2 is not directly evident.

The third stage is characterized by an increased rate of creep and terminates in rupture. The amount of extension in this stage can vary greatly, but generally has little effect on the total time to rupture. This increase in strain rate is due either to an increase in stress owing to the loss of cross-sectional area as internal voids grow and coalesce, or to the onset of necking.

Influence of metallurgical factors on creep strength

In the same metal system it is generally found that the close-packed structures, HCP or FCC, generally have greater creep resistance and greater creep rupture strengths than BCC structures at temperatures above about $0.4T_M$ because of the higher diffusion rates in the more open BCC structure. For this reason austenitic steels, nickel-base alloys, and titanium alloys are used for high-temperature service rather than ferritic alloys. The creep resistance of any alloy is also strongly dependent on its microstructure and its thermal stability at the service temperature during its lifetime. For extended service of ferritic or low alloy steels, normalizing is generally recommended to obtain a fine bainite or pearlite structure. Spherodized structures are less creep-resistant, and tempered martensites lose resistance to plastic deformation rapidly above 1000°F. Elements (e.g., Mo and Cr in ferrite) that retard recovery and recrystallization during creep are most effective in maintaining or increasing creep strength. Prior cold work can also be effective in certain circumstances in improving creep resistance [10].

Table 9.1 contains the nominal compositions of gas turbine materials for higher temperature service [11]. For strength at elevated temperatures, *cobalt alloys* depend primarily on solid solution hardening and on precipitation and suitable dispersion of stable complex metal carbides in the grain boundaries. The sluggishness of the carbide

Table 9.1 [11]
Nominal Composition of Gas Turbine Materials

	C	Cr	Ni	Co	Mo	W	Cb	Ti	Al	Fe	B	Other
British Nimonic Alloys												
80	0.10	20.0	Bal	2.0				2.3	1.3	5.0		
80A	0.10	20.0	Bal	2.0				2.3	1.3	5.0		
90	0.13	20.0	Bal	18.0				2.5	1.5	5.0		
100	0.30	11.0	Bal	20.0	5.0			1.5	5.0	2.0		
105	0.20	15.0	Bal	20.0	5.0			1.2	4.5	1.0		
115	0.20	15.0	Bal	15.0	3.5			4.0	5.0			
Combustion Can Materials												
Type 310	0.25	25.0	20.5							Bal		
Nimonic 75	0.10	20.0	Bal					0.4		5.0		
Inconel 600	0.04	15.8	Bal							7.2		
HS-25	0.10	20.0	10.0	Bal		15.0						
Hastelloy X	0.10	22.0	Bal	1.5	9.0	0.6				18.5		
Turbine Disk Materials												
Timken 16-25-6	0.08	16.0	25.0		6.0					Bal		0.15 N$_2$
A-286	0.05	15.0	26.0		1.2			2.15	0.2	Bal	0.003	0.3 V
Discaloy	0.04	13.5	26.0		2.7			1.7	0.1	Bal		
Greek Ascoloy	0.12	13.0	2.0			3.0				Bal		0.5 V
V57	0.08	15.0	27.0		1.2			3.0	0.2	Bal		
CG-27	0.05	13.0	38.0		5.5		0.6	2.5	1.5	Bal	0.01	
Inconel 901	0.05	13.5	42.7		6.2			2.5	0.2	34.0	0.01	
Inconel 718	0.04	19.0	52.5		3.0		5.2	0.8	0.6	18.0		
René 41	0.09	19.0	Bal	11.0	10.0			3.1	1.5		0.01	

	C	Cr	Ni	Co	Mo	W	Cb	Ti	Al	Fe	B	Other
Cobalt Alloys												
HS-21	0.25	27.0	3.0	Bal	5.0							
S-816	0.38	20.0	20.0	Bal	4.0	4.0	4.0					
HS-31	0.50	25.0	10.0	Bal		7.5						
W1-52	0.45	21.0	1.0	Bal		11.0	2.0					
MAR-M302	0.85	21.5		Bal		10.0					0.005	0.2 Zr, 9.0 Ta
MAR-M322	1.00	21.5		Bal		9.0		0.7				2.0 Zr, 4.5 Ta
Nickel Alloys												
Inconel X-750	0.04	15.0	73.0				0.85	2.5	0.8			
M-252	0.15	20.0	Bal	10.0	10.0			3.0	1.1			
Waspaloy	0.07	19.5	Bal	13.5	4.3			3.0	1.4		0.006	0.09 Zr
Inconel 700	0.12	15.0	46.0	28.5	3.7			2.2	3.0			
Udimet 500	0.08	19.0	Bal	19.5	4.0			2.9	2.9		0.01	
GMR-235D	0.15	15.5	Bal		5.0			2.5	3.5		0.05	
Udimet 700	0.15	15.0	Bal	18.5	5.2			3.5	4.2		0.05	
Inconel 713C	0.12	12.5	Bal		4.2		2.0	0.8	6.1		0.012	0.1 Zr
MAR-M200	0.15	9.0	Bal	10.0		12.5	1.0	2.0	5.0		0.015	0.5 Zr
IN-100	0.15	10.0	Bal	15.0	3.0			4.7	5.5		0.015	0.01 Zr, 1.0 V
TaZ-8	0.12	6.0	Bal		4.0	4.0			6.0			1.0 Zr, 2.5 V, 8.0 Ta
Soviet Turbine Alloys												
EI-437B	0.06	20.0	Bal					2.5	0.7		0.005	
EI-617	0.08	15.0	Bal		3.0	7.0		2.0	2.0		0.008	0.3 V
EI-826	0.08	13.0	Bal		3.0	7.0		2.0	3.0		0.012	0.3 V
ZhS-3	0.16	15.0	Bal		4.0	5.0		2.0	2.0		0.02	
ANV-300	0.10	15.5	Bal	5.0		8.5		1.7	5.0		0.05	
EI-867	0.10	9.5	Bal		10.0	5.0		1.7	4.5		0.10	
EI-929	0.12	10.5	Bal	14.0	5.0	5.5			4.0		0.10	
ZhS-6	0.15	12.5	Bal		4.7	7.0		2.6	5.0		0.01	
ZhS-6K	0.17	11.5	Bal	5.0	4.0	5.0		2.7	5.5		0.01	

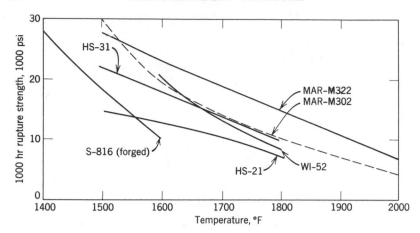

Fig. 9.2. Effect of temperature on the stress at which rupture occurs in 1000 hours for cobalt-base alloys [11].

reactions (diffusion and solution) and the higher melting point of cobalt versus nickel are major factors in strengthening these alloys above 1700°F [11]. Figure 9.2 compares the 1000 hour rupture strengths of the cobalt-base alloys.

Nickel-base alloys are strengthened by two mechanisms: first, by a precipitation-hardening reaction involving an ordered aluminum and titanium compound γ' Ni_3Al (Ti); second, by additions of the high-melting metals molybdenum, tungsten, and columbium which provide solid solution strengthening and produce a matrix in which the diffusion rate and resolution rate of the precipitated γ' are reduced. Cobalt, boron, and zirconium are other important constituents. Cobalt varies the solvus temperature of alloys strengthened by the precipitation of Ni_3Al (Ti) and thus elevates the permissible working temperature. Boron and zirconium added in small quantities improve the stress rupture properties [11]. Figure 9.3 shows how Al + Ti contents influence the 100-hour rupture strength of nickel alloys at 1600°F. A comparison of the 1000-hour rupture strengths of some of these alloys is given in Fig. 9.4.

One of the means to increase thermal stability is to make use of a fine dispersion of insoluble particles within a metal matrix. For example, at a temperature of 930°F, the solid solution range for precipitation-hardened aluminum alloys, aluminum strengthened by a dispersion of fine aluminum oxide particles (SAP), remains stable for prolonged periods [10]. At temperatures above 2000°F, thoria-dispersed nickel (TD nickel) is superior to most other nickel-base alloys [12]. The high melting point refractory metals are also good in this temperature range [13] (Fig. 9.5).

Grain size is another factor which influences the rupture life of a material for two reasons. First, because grain size has a moderate effect on the steady state creep rate [14, 15]. In general, the creep rate decreases with increasing grain size, but in coarse-grained (2d > 0.1 mm) materials the creep rate can actually increase with increasing grain size. The second and more important reason is because grain boundaries transverse to an applied tensile stress tend to be sites for void nucleation. The rupture life will be increased if these boundaries are not present. Recently a directional solidification process has been developed in a nickel-base alloy [16] which allows the production of jet engine turbine blades that contain no transverse boundaries perpendicular to the blade axis. This results in creep ductilities that are about twice those obtained in conventionally processed materials.

A preferred orientation of the grains can also influence creep re-sistance, for high angle boundaries have been found to undergo greater slip mobility [17]. For example, the textured copper always seems to exhibit somewhat higher ductility than the random copper [18]. The steady state creep rate is also a function of boundary misorientation, generally being slightly lower for textured samples.

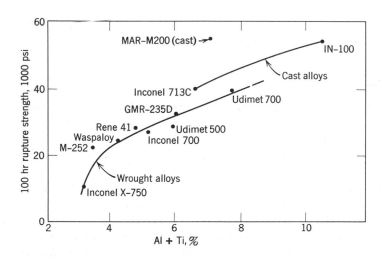

Fig. 9.3. The effect of total Al and Ti content on the 100 hour rupture strength at 1600°F for nickel-base alloys [11].

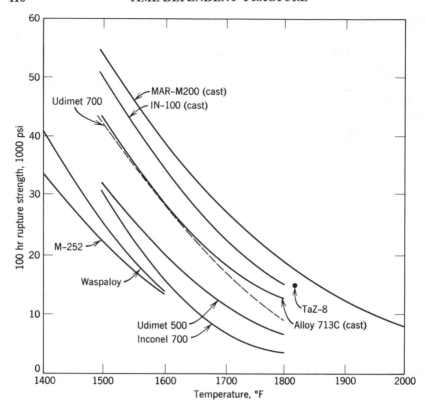

Fig. 9.4. Effect of temperature on the 1000 hour rupture strength of nickel-base alloys [11].

Prediction of service lifetime

Time-temperature parameters are used to correlate the results of stress rupture tests for purposes of comparison of materials and to provide a means for extrapolating the results of short-time tests on a given material to predict the (much longer) service lifetime of a part at a given stress and temperature. These parameters are based on the assumption that a relation between time to rupture, temperature, and stress exists such that a knowledge of the rupture life at one temperature for a given stress allows a prediction of the rupture life at that stress for any temperature in the creep range to be made. These extrapolations should be made bearing in mind that extrapolations from limited results can lead to large errors and that more basic knowledge about creep rupture is needed before such extrapolations can be accomplished with confidence [15]. An example of one of the approaches used to establish a parameter is as follows:

Assume that the steady state creep rate $\dot{\epsilon}_p$ is given by

$$\dot{\epsilon}_p = A e^{-(H-v\sigma)/RT} - A e^{(H+v\sigma)/RT} \tag{9.3}$$

where A is a constant, H is the activation energy v is the activation volume, and σ the applied tensile stress. The stress biases the thermal fluctuations of atoms so that a net extension occurs. For large values of the applied stress, Eq. 9.3 can be written as

$$\dot{\epsilon}_p = 2A e^{-H/RT} e^{v\sigma/RT} \tag{9.4}$$

The rupture lifetime t_R in Eq. 9.2 can be obtained by substituting Eq. 9.3 for $\dot{\epsilon}_p$. Upon rearrangement we obtain

$$T(\ln t_R + C_1) = \frac{H - v\sigma}{R} = f(\sigma) \qquad \text{(high stress)} \tag{9.5}$$

where C_1 is a constant. The quantity on the left-hand side of Eq. 9.5 is of the form of the Larson-Miller parameter [19]. An example of a master curve for rupture life of inconel X based upon this parameter is shown in Fig. 9.6a, which also includes a master

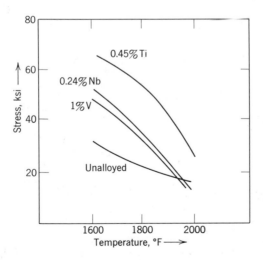

Fig. 9.5. Effect of temperature on the 100 hour rupture strength for molybdenum-base alloys [13].

curve for the minimum creep rate. Because of the interrelation between rupture life and the steady state creep rate it is possible to use a generalized master curve for both quantities, as shown in Fig. 9.6*b*.

Another approach to the determination of the stress and temperature dependence of the creep rate is due to Dorn [21], who found that for polycrystalline materials an expression of the form

$$\dot{\epsilon}_p = \sigma^n e^{-H/RT} \tag{9.6}$$

provided better agreement with data than did Eq. 9.4, at least at moderate and low stress levels. This has been confirmed by others

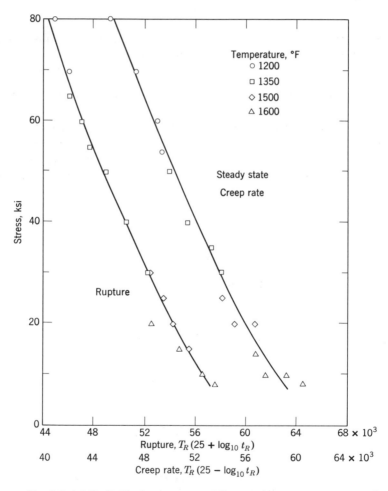

Fig. 9.6. (*a*) Individual master curves for steady state creep rate and creep life of inconel X [34].

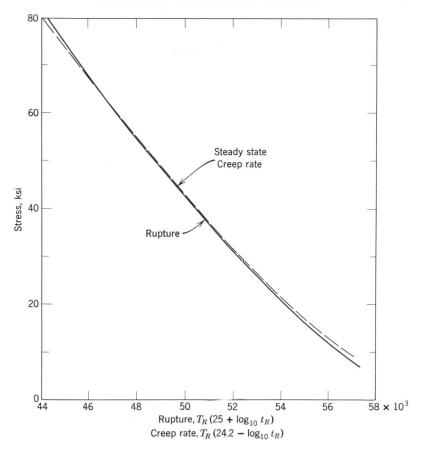

Fig. 9.6. (b) Generalized master curves for steady state creep rate and creep life of inconel X [34].

[22]. The value of the exponent n is approximately 5. Recently Eq. 9.6 has been modified to include the effects of the variation with temperature of the unrelaxed elastic modulus E on the activation energy [22] and also to include the effects of stacking fault energy δ on creep [24]. The resultant expression for $\dot{\epsilon}_p$ is

$$\dot{\epsilon}_p = A'\delta^{3.5}\left(\frac{\sigma}{E}\right)^n e^{-Q/RT} = A''D\delta^{3.5}\left(\frac{\sigma}{E}\right)^n \quad \text{(moderate and}$$

$$\text{low stress)} \quad (9.7)$$

where D is the diffusivity and Q is the activation energy. At low temperatures in the creep range this activation energy may be that

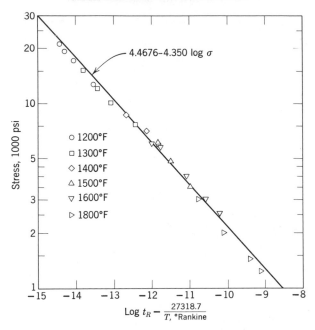

Fig. 9.7. Correlation of data for Timken 35-15 stainless steel, using method of reference 27 [20].

for volume diffusion enhanced by the presence of dislocations, whereas at high temperatures the activation energy is simply that for volume self-diffusion [25]. At very low stress levels, near the melting point, the creep rate is given by the relation

$$\dot{\epsilon}_p = K\sigma \qquad \text{(very low stress)} \qquad (9.8)$$

Combination of Eqs. 9.2 and 9.6 leads to a parametrical relationship of the form [27]

$$[t_R \, e^{-H/RT}] = f(\sigma) \qquad (9.9)$$

As shown in Fig. 9.7, this relationship is in good agreement with data obtained [20] on stainless steel over a wide range of temperature and stress. Excellent agreement, for a wide range of materials, can be obtained if Eqs. 9.7 and 9.2 are combined, accounting for the variation of E with T. Table 9.2 lists a number of parameters which are used to predict service lifetime. C_1, C_2, t_A and T_A are material constants.

Table 9.2

Name	Parameter	Applicability
Larson-Miller [19]	$T(C_1 + \ln t_R) = $ const for given σ	High Stress
Orr-Sherby-Dorn [27]	$\ln t_R - C_2/T = $ const for given σ	Most stress
Barrett-Sherby [24]	$\ln t_R - C_2/T = $ const for given σ/E	levels
Manson-Haferd [28]	$\dfrac{\ln t_R - \ln t_A}{T - T_A} = $ const for given σ	

Macroscopic aspects of creep rupture

The presence of a stress raiser such as a notch can affect the rupture lifetime in one of three ways: the lifetime can be increased, decreased, or remain essentially unchanged if the stress considered for comparison with a smooth specimen is that on the net section. The notch rupture strength of stable materials is generally a constant percentage of the smooth strength over a wide range of rupture time [29]. In most cases the notch rupture strength ratio—defined as the ratio of the stress for rupture of notched specimens to that for smooth specimens at the same lifetime—of a stable alloy is well above unity. However, in a few instances [29, 32], values of unity or slightly below have been observed. Figure 9.8 shows an example of a case where the ratio is less than unity. This type of behavior is often associated with materials which, for certain combinations of heat treatment, prior deformation history, test temperature, and applied stress, may exhibit property changes. This structural instability may greatly affect the notch strength ratio, and both the smooth and notch ductility of the metal [30]. The results of Fig. 9.8 are typical of unstable alloys in general. With increasing rupture time, the notch rupture strength first decreases at the same relative rate as the smooth rupture strength. Following this, a time range exists where the notch rupture strength decreases more rapidly and frequently to values considerably lower than the smooth rupture strength. Another range is frequently observed at lower stresses where the trends are reversed, and such a recovery may lead to the same or even higher rupture strength as that of the smooth specimen.

Structural changes taking place during creep need not be detrimental and can, in fact, be employed deliberately to improve creep resistance. For example, notch strengthening in austenitic steels

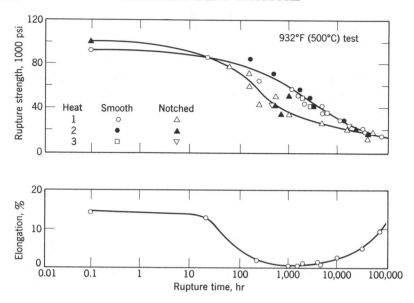

Fig. 9.8. Notch and smooth rupture characteristics for several heats of a ferritic steel 0.11 C, 1.53 Ni, 0.72 Cr, 0.88 Mo. Heat treatment: 1598°F, oil; 1100°F, aid [29].

tested at 600–700°C has been attributed to two factors: first, relaxation of stress concentrations at the base of the notch as the result of plastic yielding, and second, plastic deformation resulting in a warm-working process which accelerates precipitation at the notch and thereby increases the strength of the material beneath the notch [31].

The effects of specimen size on creep rupture behavior are indicated in Fig. 9.9 for a Cr-Mo-V steel tested at 1050°F (566°C). The notch sensitivity of this alloy was varied by prior heat treatment. In the notch-insensitive or ductile condition, a continuous decrease in strength, both smooth and notched, was found with increase in specimen size. In the notch-sensitive or brittle condition, the strength of smooth specimens increased with size, but the converse was true for the notch strength [33]. Such tests indicate that results of tests utilizing small-scale laboratory specimens, as usual, are in general not conservative, and any application of laboratory test results to component design should be done with these facts in mind. More work is needed to develop relations between the macroscopic and microscopic aspects of creep rupture.

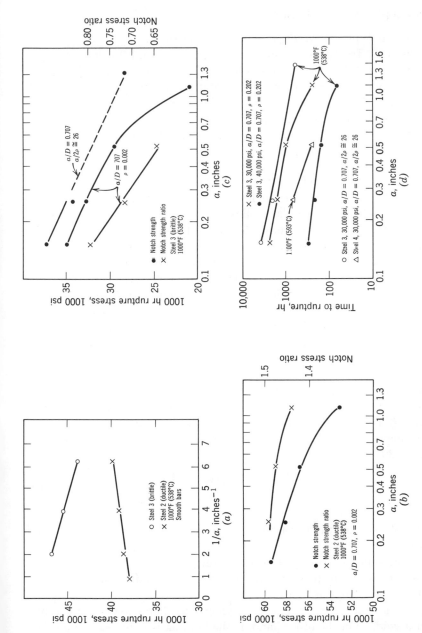

Fig. 9.9. Effect of size (diameter a) on the rupture strength of smooth and notched bars of a tough and a brittle steel [33].

417

Microscopic aspects of fracture

The two types of voids which can form during creep and lead to low ductility fractures are shown in Figs. 9.10a and 9.10b. The type shown in Fig. 9.10a [34] is found to occur at grain boundary junctions, and is known as a wedge or w-type void. The second type, shown in Fig. 9.10b [35], can occur along grain boundaries and is either elliptical or round in shape and is referred to as an r-type void. In general, the w-type is associated with a higher stress and lower temperature than the r-type.

Formation and Growth of w-Type Voids

Voids of the w-type are nucleated at grain boundary junctions as the result of grain boundary sliding, which develops a high strain concentration at a triple point [60, 61, 36], as illustrated in Fig. 9.11 [36]. Recrystallization during creep or grain boundary migration will lessen the strain concentrations at triple points and hence reduce the tendency for void formation to occur; a material brittle at one temperature may be ductile at a higher temperature, owing to increased grain boundary migration. Voids are not observed in lead or high-purity aluminum because of the mobility of the grain boundaries in these materials. In pure aluminum, plastic accommodation or "folding" in the vicinity of a triple point can also prevent void formation.

In order that a wedge-type crack be nucleated, it is necessary that the local stress level exceed the cohesive stress. A sliding grain boundary behaves the same as a planar slip band at low temperature. Consequently, from Eq. 6.3 the shear stress required for crack nucleation is [39]

$$\tau_N = \sqrt{\frac{2G\gamma_m}{d}} \tag{9.10}$$

where γ_m is the work required to make the fracture surface and $\tau_i \cong 0$.

The presence of precipitate or impurity particles at a triple point can serve to facilitate wedge formation if the binding of the particle to its surroundings is weak, that is, γ_m is low. The limiting case is that for a nonwetting particle at which any boundary sliding would open a void. On the other hand, if the binding is high (e.g., aluminum oxide particles in aluminum) voids may not develop. The tendency for void formation can also be strongly stress-dependent as in the case

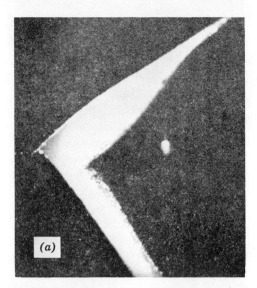

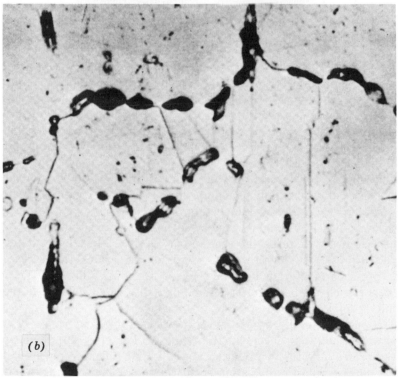

Fig. 9.10. (a) Wedge-type intergranular crack in aluminum-alloy creep specimen; polarized light micrograph. 140×. *Courtesy H. Chang and N. Grant* [36] *and AIME*. (b) r-type voids formed in copper during creep at 480°C under a stress of 4730 psi. Tensile axis vertical. 2500×. *After Kennedy* [35].

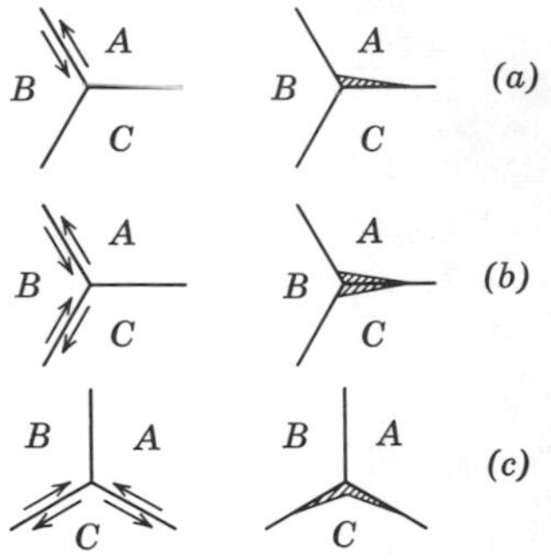

Fig. 9.11. Schematic views show three of the simple basic means of initiating intercrystalline cracks directly owing to grain boundary sliding. The arrows along a grain boundary indicate that this boundary underwent sliding [36].

of an Al-2.6 Mg solid solution alloy containing a fine dispersion of an impurity phase, Mg_2Si [41].

The heterogeneous metallurgical structure of precipitation-hardened alloys can also promote wedge-type crack formation. If zones depleted of precipitate exist along grain boundaries, plastic deformation can take place preferentially within them and develop the localized strains at grain boundary junctions which lead to void formation. The free sliding length of these boundaries can be reduced and wedge-type failures avoided if precipitates form along the grain boundaries [42, 43, 44]. For example, if the alloy nimonic 80A is heat-treated such that appreciable precipitation of chromium carbide occurs at triple points, the ductility is low at 600°–700°C owing to the formation of wedge-type voids. On the other hand, if the alloy is heat-treated to develop a more uniform distribution of carbide particles along the grain boundaries together with zones, the ductility increases [42]. Vacuum melting to prevent the formation of deleterious oxides can also increase the ductility.

The effect of grain size on void formation is not straight forward. The contribution of grain boundary sliding to creep strain is greater, the smaller the grain size [45, 46], and therefore more void formation and lower ductilities might be expected in finer-grained material [47]. This was observed to be the case for an Al-$1\frac{1}{4}$% Mn alloy, but for a 70/30 α-brass, the ductility decreased as the grain size increased. The presence of a grain boundary phase in the Al alloy and the absence of such a phase in the brass may also have influenced these results [47].

The mechanism by which w-type cracks grow and cause brittle fracture is not well documented. Some workers [48, 49] have proposed that a Griffith-type relation is applicable. This seems unlikely, since sufficient plastic relaxation occurs at these temperatures to blunt the crack and keep the local stresses well below the cohesive stress. Furthermore, w-type cracks tend to grow at temperatures at which diffusion processes are too slow to cause appreciable atom migration from the tip to the sides of the wedge. Most likely, growth is plastically induced (Chapter 5) and occurs unstably only when a critical shear displacement is produced near the tip of a growing crack.

Formation and Growth of r-Type Voids

Cavitation, or r-type void formation, is an important cause of embrittlement at high temperatures and low stresses [50]. These cavities are most prevalent along boundaries transverse to an applied tensile stress as well as at triple points. In their earliest stages they tend to be spherical, but as they grow and link up with adjacent cavities they become more elliptical in shape. The linking of the voids results in a serrated fracture surface in contrast to the flat fractures associated with wedge-type failures. Density measurements [51] indicate that these voids are present throughout the creep test; they form continuously and grow in size until final failure [52]. The voids are more numerous near the surface of test specimens, suggesting that growth occurs by a grain boundary diffusion process with the free surface being the ultimate source of vacancies [51]. Voids are not observed in single crystals or in polycrystalline samples produced by recrystallizing a single crystal [51], thus indicating that the formation of voids is related to the presence of particles initially present within the grain boundaries of polycrystalline materials. These particles also serve to prevent grain boundary migration during the void growth stage [61] and thereby reduce the resistance to creep rupture, for if grain boundary migration occurs, the voids left within the grains are relatively innocuous [62].

As in the case of wedge-type fractures, grain boundary sliding is a requisite for cavity formation. Tests on copper bicrystals showed that a shear stress along the boundary produced cavities, but that a tensile stress normal to the boundary did not [6]. Void formation will occur preferentially along those grain boundaries with the greatest amount of sliding, and is favored by increased misorientation across a boundary [17]. Consistent with this observation is the fact that voids are not observed along twin boundaries. Grain boundary serrations can develop during creep and, in conjunction with grain boundary sliding and migration, influence the process of void formation [53].

Various proposals have been advanced to account for the heterogeneous nature of cavity formation. Stress concentrators such as grain boundary ledges, subgrain boundary configurations, or grain boundary jogs created by intersecting slip do not of themselves appear to be effective enough to raise the local stress due to sliding to the

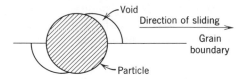

Fig. 9.12. Nucleation of a void at a
particle by grain boundary sliding [4].

cohesive strength level [4]. Most evidence indicates that second-
phase particles are required in the nucleation process, as indicated in
Fig. 9.12. The ease with which a cavity can be nucleated by such a
sliding process will depend on the binding of the particle to the
matrix. As with wedge-type cavities, the void nucleation process
becomes easier as the degree of wetting between particle and matrix
decreases.

The growth of these r-type voids occurs by a flux of vacancies and
a counterflux of atoms from the voids to grain boundary sites (Herring-Nabarro creep). Under an applied tensile stress σ, when an atom
is removed from a void of radius r and deposited in a transverse grain
boundary, the work done in the direction of the stress is $\sigma\Omega\gamma/r$, where
Ω is the atomic volume. For the void to grow simply by vacancy
condensation, the work done must exceed the increase in surface energy $2\Omega\gamma/r$ caused by the removal from a void of radius r and surface
energy γ_s. The condition for void growth is [54, 4, 58]

$$\sigma \geqq \frac{2\gamma_s}{r} \tag{9.11}$$

For values of σ less than the critical value, the void will diappear by
sintering. For values of σ in excess of the critical, the void will grow
in size. It has been pointed out that growth by this mechanism is a
particularly interesting process in that the Griffith criterion need not
be satisfied [38]. For example, voids can grow in magnesium alloys
under stresses of only 100 psi at 450°C [55, 56]. At this stress level,
with $E \simeq 5 \times 10^{11}$ dynes per cm², $\gamma_s \simeq 500$ ergs per cm², and $\sigma \simeq 10^7$
dynes per cm², very long cracks ($c \simeq 2.5$ cm) would be needed to
satisfy the Griffith criterion. The critical importance of this mechanism of cavity growth lies in the fact that if small cavities exist on
transverse boundaries, a creep specimen must ultimately fail at high
temperatures, even under stresses far below the creep strength of the
grains [38].

Evidence for the vacancy diffusion mechanism has been obtained by studying the effects of a hydrostatic pressure p, superimposed upon the tensile stress σ [54]. The shear stress due to σ is unaffected, but the normal stresses are reduced by the amount p. Under these circumstances Eq. 9.11 becomes

$$\sigma - p \geqq \frac{2\gamma_s}{r} \tag{9.12}$$

Therefore, as $\sigma \to p$, voids should not grow but should sinter out because of the much larger critical value of r required. When $\sigma = p$, no voids are observed [54, 57]. An example of the effect of a hydrostatic pressure on formation of cavities is seen in Fig. 9.13. The corresponding creep curves are in Fig. 9.14 [57]. The steady state creep rate, which depends on the shear stress, is unaffected, but the secondary creep stage and total extension are greatly increased by the imposition of hydrostatic pressure.

In addition to the vacancy flux mode of void growth, grain boundary voids can also grow by a continuation of the void nucleation process

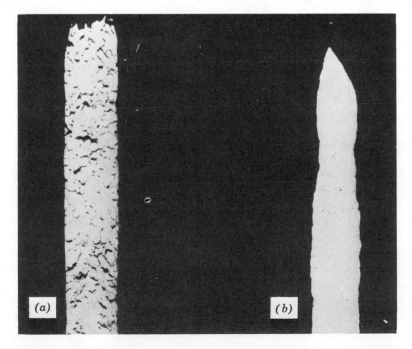

Fig. 9.13. Sections of magnesium specimens after creep at 300°C under an initial stress σ of 700 psi. (a) Without hydrostatic pressure. (b) With hydrostatic pressure $P = \sigma$ [52].

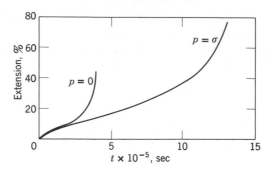

Fig. 9.14. Effect of hydrostatic pressure on creep curves for magnesium at 300°C [52].

of grain boundary sliding as has been observed in the case of an MgO bicrystal [58]. Void formation and growth in copper bicrystals have been found to be a function of the amount of extension and not of the temperature in the range from 600° to 900°C [59], and therefore vacancy condensation cannot be a controlling factor in such experiments. Recent studies of rupture in a Cu = 15.4 at % Al alloy have shown [204] that although reversing from tensile to compressive stress during a test causes voids to disappear compression applied at right angles to the direction of original tension causes them to grow. This observation is consistent only with a grain boundary sliding mechanism of growth.

9.2 Stress Corrosion Cracking

Stress corrosion cracking refers to the embrittlement of a metal or alloy as a result of interaction with a chemical environment other than one consisting of a liquid metal or hydrogen. The embrittlement may be reflected as a loss of ductility in a tensile test [63], as a delayed static fracture, or as surface cracking if stress is applied to a material previously exposed to the environment in the unstressed conditions [64]. The static fatigue of glass and crystalline ceramics and the stress cracking of polymers are analogous processes occurring in nonmetallic materials. A large number of alloys are susceptible to stress corrosion cracking, including those of aluminum, copper, iron, magnesium, and titanium; therefore this type of embrittlement can be widespread and is a matter of considerable concern since it has occurred in critical components of aircraft, chemical installations, and power plants, both conventional and nuclear. Stress corrosion cracking may be of even greater concern when materials of higher strength are used in the future, as there is a general tendency for the susceptibility to environmental effects to increase as the operating stress level is raised. For example, 18 Ni maraging steels of 250 ksi yield strength are more susceptible to stress corrosion cracking in sea water, marine, and industrial environments than are similar steels of 200 ksi yield strength [65].

Not all combinations of environments and alloys lead to embrittlement, of course, for in a given environment a material under stress may be (a) completely passive, that is, free from any aggressive attack, (b) subject to a general, over-all corrosive attack which can eventually lead to failure simply owing to the increase of stress with loss of net section, or (c) attacked locally such that cracks form and grow to result in failure when the more advanced of these cracks reach critical size. Only specific combinations of material and environment result in this last type of behavior, which is known in a generic sense as stress corrosion cracking [66]. Table 9.3 [67] lists a number of alloys and associated environments which have led to embrittlement under stress. In general, these environments are aqueous, but stress corrosion cracking has been reported for pure copper in air at 650°C [68], as well as for pure nickel in molten caustic soda [69]. Titanium alloys in the temperature range 500°–800°F have been embrittled by nothing more than a seemingly innocuous dried perspiration stain. Figures 9.15a and 9.15b show the results of tests to determine the susceptibility of a number of structural titanium alloys to a coating of synthetic sea salt [70]. Certain high strength titanium alloys containing fatigue cracks are also susceptible to cracking in salt water at room temperature [71].

The stress required for stress corrosion cracking is usually tensile, but cracking can also occur owing to a shear stress. There is a wide latitude in the manner of application of the stress which may be either externally applied or a residual. A well-known example of cracking due to residual stresses is that of cartridge brass exposed to moist air containing ammonia vapor. This type of cracking was first noted in India during the rainy season, hence the term "season cracking." Constrained corrosion products themselves can lead to a buildup of stress leading to failure, since most solid corrosion products occupy a larger volume than that of the metal destroyed. An indication of the magnitude of the stresses developed is given by the fact that rivets have been observed to snap owing to the formation of rust between plates, and also by the observation that concrete has shattered because of the corrosion of steel encased by the concrete [72].

Table 9.3

Environments That May Cause Stress Corrosion of Metals and Alloys [67]

Material	Environment
Aluminum alloys	NaCl-H_2O_2 solutions
	NaCl solutions
	Sea water
	Air, water vapor
Copper alloys	Ammonia vapors and solutions
	Amines
	Water, water vapor
Gold alloys	$FeCl_3$ solutions
	Acetic acid–salt solutions
Inconel	Caustic soda solutions
Lead	Lead acetate solutions
Magnesium alloys	NaCl-K_2CrO_4 solutions
	Rural and coastal atmospheres
	Distilled water
Monel	Fused caustic soda
	Hydrofluoric acid
	Hydrofluosilicic acid
Nickel	Fused caustic soda
Ordinary steels	NaOH solutions
	NaOH-Na_2SiO_3 solutions
	Calcium, ammonium, and sodium nitrate solutions
	Mixed acids (H_2SO_4-HNO_3)
	HCN solutions
	Acidic H_2S solutions
	Sea water
	Molten Na-Pb alloys
Stainless steels	Acid chloride solutions such as $MgCl_2$ and $BaCl_2$
	NaCl-H_2O_2 solutions
	Sea water
	H_2S
	NaOH-H_2S solutions
Titanium alloys	Red fuming nitric acid
	Salt solutions

The influence of stress on the time to failure of unnotched speci-
mens of copper-base alloys in an ammonia environment is shown in
Fig. 9.16 [73]. Each of these alloys exhibits an endurance stress
limit below which failure will not occur, and only in the case of the
Cu-30 Zn alloy is this limit less than the yield strength of the alloys.
For all the other alloys shown, plastic deformation is clearly an essen-

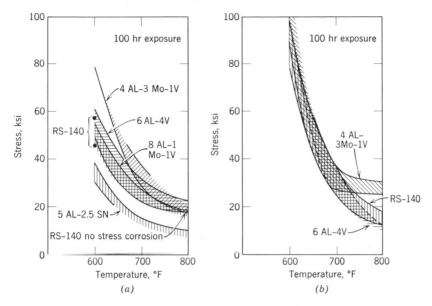

Fig. 9.15. Threshold stress corrosion of titanium alloys verus temperature: (a) Annealed alloys. (b) Solution-treated and aged alloys. Upper edge of band is minimum stress where stress corrosion was observed; lower edge of band is maximum stress where no stress corrosion was observed [70].

tial requirement if failure is to occur. When notched specimens are used, only the region at the notch root may be in the plastic range, as is the case [74], for example, for high strength steel specimens containing preexisting surface flaws tested in an aqueous $3\frac{1}{2}\%$ NaCl solution with results as shown in Fig. 3.39. The time to failure is plotted in terms of the initial stress intensity factor $K_I = \sigma \sqrt{a\pi c}$. For an exposure time in this environment of the order of one hour, the load-carrying capacity of these steels is less than one-third of the initial capacity. As shown in Chapter 3 the fracture process involves the growth of a crack to critical size. The initial flaw propagates slowly when $K_{Iscc} < K_I < K_{Ic}$. Unstable fracture occurs (Fig. 3.38) when the crack is sufficiently long such that $K_I = K_{Ic}$.

Up to this point we have dealt with the principal macroscopic aspects of the stress corrosion process, but have said nothing about the microscopic mechanisms involved. Over the years a number of mechanisms have been proposed, but because of the complexity of this type of environmental cracking, it has been difficult to devise critical checks of these mechanisms; indeed, there may be more than just one correct mechanism [76]. Recently the examination of replicas of fractures

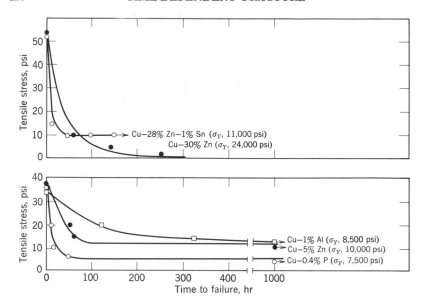

Fig. 9.16. Effect of applied stress on the time for intergranular failure of copper alloys in ammonia [73].

in the electron microscope and transmission electron microscopy have helped to clarify the microscopic processes involved. In the following sections we shall discuss some of the factors which are involved in proposed mechanisms, such as electrochemistry, stress, plastic deformation, surface films, short range order, stacking fault energy, prestrain, the path of cracking (i.e., whether intergranular or transgranular), and the specific effects of the environment.

Fracture of austenitic stainless steels

Austenitic stainless steels in boiling 42% magnesium chloride solutions (154°C) are susceptible to stress corrosion cracking. As indicated in Fig. 9.17, the resistance to cracking increases with nickel content, [75, 76] and since nickel raises the stacking fault energy, it has been proposed that this change is related to the increased resistance. The mode of cracking under these conditions is transgranular, and an increase in stacking fault energy would make the slip process less planar. As shown in Chapter 6, this reduces the stress concentration associated with a blocked slip band and consequently increases the difficulty of rupturing the protective oxide film on the metal surface. Once the protective film is ruptured, localized corrosive attack of the exposed metal can then occur to initiate the embrittlement process [76]. When the behavior of 304 stainless steel (18 Cr-8 Ni) was compared with that of the alloy Incoloy 800 (20 Cr-33 Ni), it was also found that the alloy of lower nickel content was more susceptible [77], and, further, the dislocation arrays at the same value of plastic stain are more planar in the case of the more susceptible alloy [78]. It appears, therefore, that stacking fault energy is an important consideration in determining the resistance to cracking for many alloys. A low stacking fault energy is not, however, an essential for cracking, for cracking has been observed in a 20 Cr-20 Ni-1.5 Mo-0.3 C alloy, despite the fact that the dislocations were in a cellular array [79]. Also the presence of planar dislocation arrays alone is not sufficient to make an alloy susceptible, for example, nichrome (20 Cr-80 Ni) is quite resistant to stress corrosion cracking although the glide is planar [80]. Nevertheless, planar glide and the rupture of a surface film do appear to be an important aspect of the transgranular cracking of many austentic stainless steels.

An important characteristic of the surface films on stainless steel is that they can be quickly reformed, even in the boiling magnesium chloride solutions which have been acidified by the addition of 0.1%

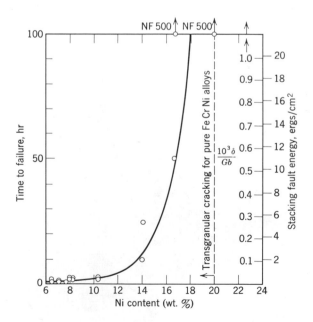

Fig. 9.17. The effect of nickel on the stacking fault energy and time for stress corrosion failure of 18% Cr stainless steel [75, 76].

HCl. This has been demonstrated by plastically deforming a specimen and placing it immediately in a chloride solution which contained platinum in solution to decorate the anodically active regions of the slip steps. This specimen was decorated profusely along all exposed slip steps. A second specimen was immersed in a (0.1% HCl) MgCl solution which did not contain platinum. After one minute it was transferred to the platinum-containing solution, but the degree of decoration of slip steps was greatly reduced, indicating that the protective film had largely reformed during the first minute in the standard magnesium chloride solution. These results indicate that during stress corrosion there is a continual interplay between (a) the tendency for films to reform over exposed metal and thereby protect against further attack, and (b) the counter effects of plastic deformation which tend to break down the protective film to continue the local corrosion process. This means that a more or less continuous form of plastic deformation under load, that is, creep, is required to cause failure.

The corrosion pits which form on the exposed slip steps do not appear to be associated with dislocations as, for example, are the etch pits in LiF. In stainless steels these pits can grow in ⟨111⟩ directions as corrosion tunnels, as shown in Fig. 9.18 [81]. These pits can link up to create a sharp notch, at the tip of which a weak, brittle corrosion product exists. Chloride ions incorporated in the surface film may reduce its mechanical strength. These ions are larger than

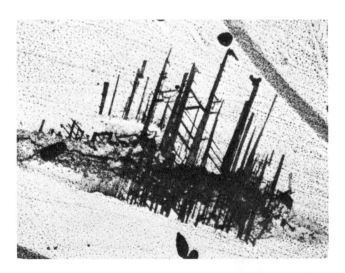

Fig. 9.18. Complex corrosion tunneling in stainless steel [81].

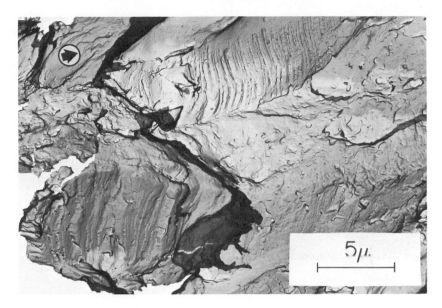

Fig. 9.19. Replicated surface of stress corrosion fracture in a notched specimen of type 321 stainless steel cold-reduced 47%. Specimen exposed in boiling 42% $MgCl_2$ solution. Area shown is $\sim 200\mu$ from origin of fracture. Direction of crack propagation left to right. 4000× [83].

oxygen ions and they diffuse to regions of high strain in the film where they can be more readily incorporated into the film structure [82]. In addition, chloride ions increase the tendency for pitting because they increase the solubility of elements such as nickel and chromium. Once a sufficient number of pits have linked up to create a crack, propagation in bulk specimens may occur by the mechanical rupture of the corrosion product, followed by the advance of the crack by plastic deformation until the tip has been blunted sufficiently to arrest further advance, and the reformation of the film before it again breaks down to result in another increment of advance. Evidence for such a discontinuous process is seen in the striations of Fig. 9.19, which is an electron micrograph of a replica of a fracture surface [83].

Fracture of copper-base alloys

Annealed Cu-30 Zn alloys stressed in an ammoniated copper sulfate solution fail by intergranular cracking. However, transgranular cracking can be made to occur if the specimens are cold-worked prior to testing. It has been proposed that the low stacking fault energy of the dislocations introduced by cold work favors transgranular

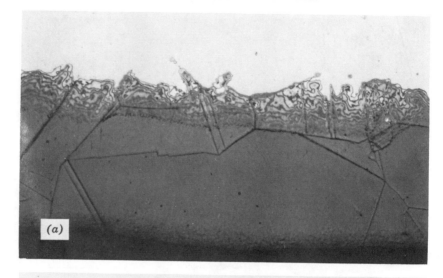

Fig. 9.20. (a) Taper section (taper ratio 12.3:1) through tarnish formed on annealed brass in ammoniacal copper sulfate after 16 hour exposure. 300× [69]. (b) Taper section (taper ratio 9.8:1) through tarnish formed on cold-worked brass in ammoniacal copper sulfate after 9 hour exposure. 250× [69].

cracking, as in the case of the austenitic stainless steels [73, 76, 84]. Another factor affecting the nature of the crack path is the structure of the film formed on α-brass in the solution. As shown in Fig. 9.20 [69], for the annealed condition the initial grain boundary structure is retained within the tarnish which replaces the surface layers of the brass. For the cold-worked condition no structure is evident within the tarnish film. The presence of a grain boundary structure within the tarnish may of itself favor intergranular cracking, for the tarnish

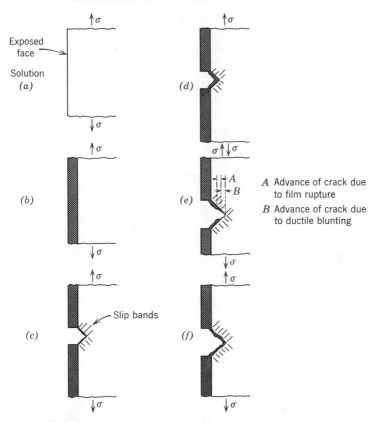

Fig. 9.21. Schematic representation of initiation and growth of stress corrosion cracks by tarnish rupture [85].

is a weak, friable substance, and cracking generally occurs preferentially at grain boundaries in brittle, polycrystalline materials. The absence of a preferred path in the tarnish on cold-worked material would contribute to transgranular cracking.

As indicated in Fig. 9.16, the Cu-30 Zn alloy is susceptible to cracking at stresses less than the macroscopic yield strength. The mechanism of crack initiation at these low stresses is not established, but may involve epitaxial stresses developed as the film forms. Once the film ruptures, a crack can penetrate the substrate metal until arrested by plastic blunting of the crack tip, as shown in Fig. 9.21. A tarnish film reforms to replace the anodic, unprotected metal exposed. This tarnish can then in turn fail for the same reasons as did the initial tarnish, or because of transient plastic deformation (creep) in the plastic zone ahead of the crack tip. The crack continues to advance by the repetition of these two electrochemical and mechanical stages until a critical size is attained. An example of a fracture surface containing striations consistent with this two-stage film rupture mechanism is shown in Fig. 9.22 [84].

The rate-controlling process in this mechanism is the formation of the film at the tip of the crack in each sequence. It is expected that the higher the applied stress the less thick the film will be before it ruptures, because of the higher strain rate within the plastic zone at

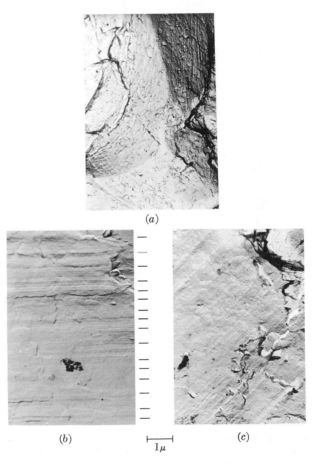

Fig. 9.22. Electron micrographs of replicas of stress corrosion fracture surfaces. (a) Annealed brass at 14,600 psi. (b) Cold-rolled brass at 30,900 psi (main rupture markings indicated in margin). (c) Cold-rolled brass at 67,000 psi. Magnification approximately 7000× [85].

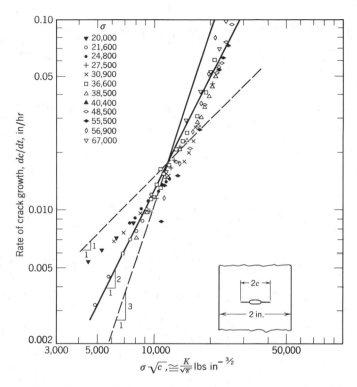

Fig. 9.23. Rate of stress corrosion crack propagation in cold-rolled brass exposed to $0.05M$ $CuSO_4 + 0.48M$ $(NH_4)_2SO_4$; $pH = 7.25$ [85].

the crack tip. If the probability that the film will be ruptured by an avalanche of dislocations is proportional to the plastic zone size, then the rate of crack growth should be proportional to the plastic zone size, and hence to the square of the stress intensity factor $K = \sigma \sqrt{\alpha \pi c}$. This interpretation indicates that the rate of film formation must be large with respect to the rate of release of dislocations, otherwise the film could not reform and the crack tip would continue to become more blunt.

The experimentally determined rate at which cracks grow under constant load in cold-worked α-brass in sheet form is shown in Fig. 9.23. The log of the rate of crack growth, dc/dt, is plotted in terms of the log of a stress intensity factor $\sigma \sqrt{c} \cong K/\sqrt{\pi}$. The data over much of the range investigated plot as a single-valued function of this parameter which is related to the stress intensity, as well as to the size of the plastic zone, at the tip of the crack. The slope of the

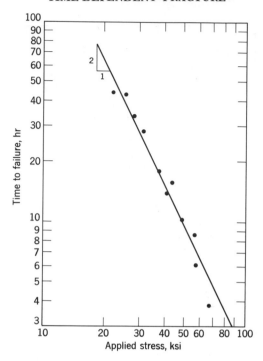

Fig. 9.24. Time to rupture as a function of applied stress for notched, cold-rolled brass specimens exposed to cracking solution [85].

line drawn through the data points indicates that the rate of crack growth can be expressed as

$$\frac{dc}{dt} = k\sigma^2 c \tag{9.13}$$

where k is a proportionality constant. This relation is in accord with the aforementioned probability of a thermally activated slip avalance rupturing the tarnish film. Integration of Eq. 9.13 between the limits of the initial c_0 and final c_F crack sizes yields

$$\ln \frac{c_F}{c_0} = k\sigma^2 t_R \tag{9.14}$$

$$(\sigma \sqrt{\pi c_0} > K_{Iscc})$$

The rupture lifetime t_R is not particularly sensitive to the exact value of c_F as a function of stress since most of the lifetime is spent while the crack is small, and growth is rapid as c_F is approached. Therefore, if c_0 is fixed, the time to rupture t_R should vary inversely as the square of the applied stress level.   The data of Fig. 9.24 are in agreement with this prediction.

Fracture of aluminum base alloys

The role of electrochemical effects in the stress corrosion of aluminum alloys in chloride solutions has been emphasized ever since it was discovered that during aging of Al-Cu alloys the grain boundary regions became anodic with respect to the grain interiors, as shown in Fig. 9.25 [86].   This is an important aspect of the corrosion of these alloys in the absence of stress, but in the case of stress corrosion cracking, the nature of the surface film and the heterogeneous metallurgical structure of these alloys also must be considered.   For example, Fig. 9.26 gives an indication of the metallurgical structure along a grain boundary in an age-hardened Al-5.5 Zn-2.5 Mg alloy. In this alloy, corrosive pitting in the absence of stress takes place preferentially within the grains rather than along the grain boundaries, so that the grain boundaries are cathodic regions rather than anodic as in the case of Al-Cu alloys.   However, in a chloride solution under a tensile stress which can be less than the macroscopic yield stress,

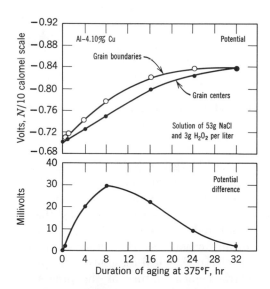

Fig. 9.25. Effect of duration of aging at 375°F on the electrode potentials of grains and grain boundaries of a high-purity aluminum-copper alloy containing 4.10% Cu [86].

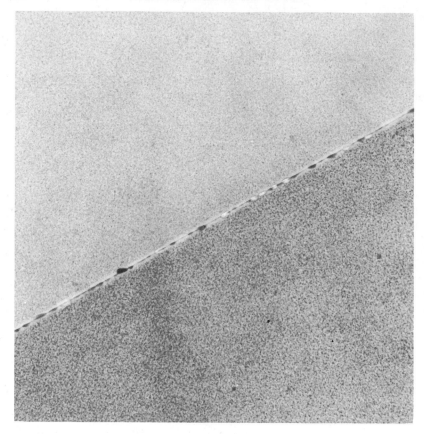

Fig. 9.26. Structure of a grain boundary in Al-Zn-Mg alloy after aging. 20,000×.

cracking occurs along the grain boundary regions. This cracking occurs because plastic deformation has taken place within the narrow solute depleted zones adjacent to the grain boundaries, and the protective film, which may have been weakened by the incorporation of chloride ions, is ruptured. This film-free region of metal is more highly anodic than any other area in contact with the solution [86], with the result that electrochemical dissolution can occur there preferentially. The electrochemical nature of the process can be demonstrated by reducing the anodic potential of the specimen which increases the time to rupture, as shown in Fig. 9.27.

As cracks penetrate the alloy along the grain boundary regions, the competing reactions of anodic dissolution and film formation will

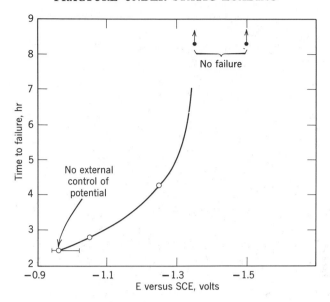

Fig. 9.27. Time to failure for quenched and aged Al-Zn-Mg in deaerated $0.5M$ NaCl at 44 ksi. Potential-controlled by means of a potentiostat.

occur. Striations are observed on the fracture surfaces, indicating that at regular intervals the films do succeed in reforming before being again ruptured, with the crack advancing by mechanical blunting and anodic dissolution. Again creep within the plastic zone at the crack tip may be responsible for the breakdown of the film in each successive stage. If these alloys are cold-worked prior to aging, the weak grain boundary structure can be sufficiently altered so that cracking becomes transgranular and the alloy is much more resistant to cracking. Shot peening has also been found to be useful in improving resistance to stress corrosion cracking of aluminum alloys [87].

The rate of crack growth as a function of the stress intensity parameter $\sigma \sqrt{c} \cong K/\sqrt{\pi}$ is shown in Fig. 9.28, for two conditions of an Al-5.5-Zn-2.5 Mg alloy in two environments. For the conventionally treated material a line of slope 2 has been drawn through the limited data in accord with the more detailed results obtained with α-brass. For the cold-worked prior to aging condition, the material is so resistant to cracking that net section stresses approaching the general yield level of the alloy must be applied. In this case the size of the plastic zone is no longer proportional to $\sigma^2 c$, and this may

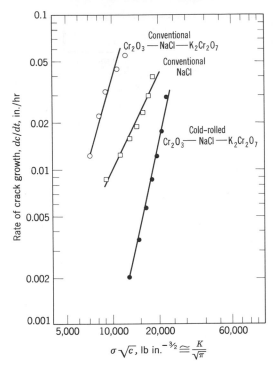

Fig. 9.28. Rate of growth of stress corrosion cracks in Al-Zn-Mg alloy in two environments as a function of the stress intensity factor, $\sigma \sqrt{c} \cong K/\sqrt{\pi}$. Conventional refers to the quenched and aged condition. Cold-rolled refers to material worked before aging.

account for the steeper slope observed for this condition. A similar effect is noted in the growth of fatigue cracks.

9.3   Liquid Metal Embrittlement

The embrittlement of a stressed metal or alloy by a molten metal is known as liquid metal embrittlement (LME). Failures due to LME are not as commonplace as those due to stress corrosion cracking, but nonetheless they are a matter of practical concern (e.g., when liquid metals such as sodium are used as coolants in atomic reactor systems). As in the case of stress corrosion cracking, only certain "couples" (combination of a liquid metal and solid metal) lead to

Table 9.4 [89]

Behavior of Various Liquid Metals on Common Engineering Alloys

Engineering Alloy	Test Temperature (°C)												
	30	50	125	180	210	250	260	300	325	350	380	450	475
Liquid Metal	Hg†	Ga	NA	In	Li	Se	Sn	Bi	Tl	Cd	Pb	Zn	Te
Al alloys	E	E	E	E	N	N	E	N	N	N	N	E	
Mg alloys	N	N	E	N	N	N	N	N	N	N	N	E	
Steel	N	N	N	E	E	N	N	N	N	E	N	E	
Ti alloys	E	N	N	N	N	N	N	N	N	N	N	N	N

† Hg = 3% Zn amalgam.   E = embrittlement.   N = nonembrittlement.

LME.   Table 9.4 contains a summary of findings of the embrittling tendency of a number of liquid metals on alloys of aluminum, magnesium, titanium, and steel [89].   The selective nature of LME is clearly indicated by the fact, for instance, that mercury embrittles titanium alloys but not steel, whereas indium embrittles steel but not titanium alloys.   In addition to the couples listed in Table 9.4, embrittlement can occur for many other couples as well; for example, Cd-Ga, brass-Hg, Zn-Ga, and Zn-Hg.   However, the present state of knowledge is such that an a priori determination of which couples will result in embrittlement is not yet possible.

Characteristics of LME

In order for LME to occur it is necessary that the liquid wet the solid metal.   Since the solid metal is usually covered by an oxide film which interferes with the wetting process, it is necessary to break down the film.   In the case of α-brass the film can be dissolved by immersion in an aqueous solution of mercurous nitrate.   The normal oxide on brass is soluble in the solution and the solute content provides the source of metallic mercury which is deposited on the brass by chemical displacement [89].   As a result, a true interface between brass and mercury is developed.   Other means for removal of the oxide film include mechanical abrasion and chemical dissolution prior to wetting.   When the film has been removed, wetting will occur if the contact angle, Fig. 9.29, is less than 90°.   Wetting can sometimes be accomplished by the addition of alloying elements to the liquid

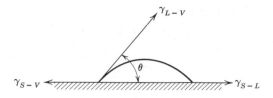

Fig. 9.29. Relation between interfacial energies: $\gamma_{S-V} = \gamma_{L-V} \cos \theta + \gamma_{S-L}$, where θ is the contact angle.

metal. For example, additions of a few per cent zinc or gallium reduce the contact angle of mercury on aluminum oxide. For very small values of the contact angle, penetration of the grain boundaries by the liquid metal can occur, as in the case of steel wetted by molten copper, and of copper by molten bismuth. Such grain boundary penetration, however, is not very common and is not a natural consequence of wetting. Such penetration is a sufficient condition for embrittlement, but it is not a necessary one, and this penetration process is not considered to be a characteristic of LME [89]. Since embrittlement can take place at temperatures as low as the solidus of the liquid metal, without a marked effect of temperature in the brittle range, the embrittling process appears to involve the chemisorption of atoms of the liquid metal and is not a thermally activated process. Diffusion, therefore, is not considered to be a controlling factor.

A number of factors appear to be characteristic of the couples which lead to embrittlement [89].

1. The solubilities of the liquid in the solid and of the solid in the liquid are very small.

2. They should not form intermetallic compounds of high melting point [90].

3. The solid metal and active liquid metal should be of similar electronegativity for maximum embrittlement [90]. In addition, certain mechanical requirements must be met; namely, the material must contain a barrier to dislocation motion at which high localized stresses can develop.

For polycrystalline engineering alloys the path of cracking is usually intercrystalline, although transgranular cleavage has been observed in the case of notched cadmium specimens wet by gallium [91]. Transgranular cracking can result in the case of large-grained specimens if only the interior of the grains is wetted. The use of single crystals can also result in transgranular cracking as in the case of crystals of zinc [92], cadmium [92], tin [93], and iron-3% silicon [94].

The means of investigating the quantitative aspects of the role of stress, strain, and time in LME are similar to those employed in the study of stress corrosion cracking. The effect of a liquid metal environment on the stress-strain properties in constant strain-rate tests, as for the iron-aluminum alloys shown in Fig. 9.30 [95], is one type of test commonly employed. In this instance, as the aluminum content of the alloys increases, embrittlement, as determined by the loss

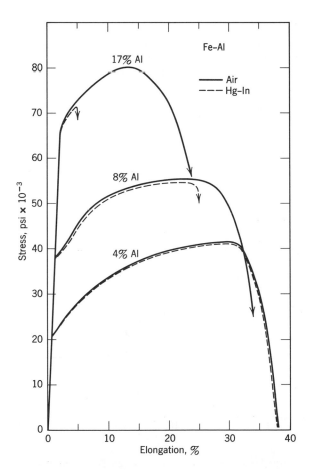

Fig. 9.30. Embrittlement of iron-aluminum alloys by mercury-indium [95].

in elongation, becomes more pronounced. Note that the stress-strain curves of embrittled specimens are almost identical to those of the base material up to the point of fracture, indicating that a certain strain must be achieved to nucleate cracking and that, once nucleated, cracks propagate rapidly to cause failure. The rate at which these brittle cracks propagate appears to be limited only by the speed with which the liquid metal can advance to wet the tip of the crack. If there is an insufficient supply of liquid, the brittle crack growth can be arrested by the blunting of the crack tip [91]. Any process that will speed up the rate at which the liquid can wet the advancing crack tip, such as the imposition of hydrostatic pressure [96], will cause the cracks to grow more rapidly [94]. A small decrease in the rate of crack growth in brass wetted by mercury has been observed as the test temperature is reduced, but this may only reflect the increase in viscosity of the mercury with decreasing temperature [89]. Rates of crack growth have been observed generally to be in the range of 1 to 100 cm per sec [89]. A value of as low as 1 cm per hr has been observed [97], but even this low value is higher than rates of stress corrosion cracking in similar alloys.

Figure 9.31 shows the results obtained in a study of the delayed fracture characteristics of an aluminum alloy wetted by a mercury amalgam. This type of curve contains all the general characteristics

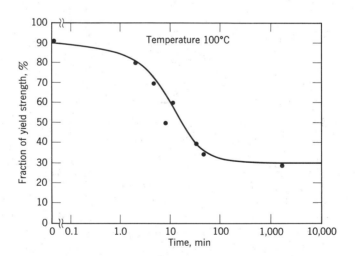

Fig. 9.31. Time to failure of a 2024-T4 aluminum alloy wetted with mercury amalgam at 100°C, as a function of statically applied stress [89].

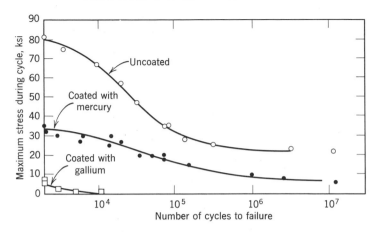

Fig. 9.32. Dynamic fatigue in axial loading of 7075-T6 aluminum alloy, at a stress ratio of 0.89 [89].

of a static fracture curve. The time to failure is extremely sensitive to stress, with fracture times in the range of 1–30 min. This sensitivity to stress provides further evidence that the fracture process is not thermally activated [98]. The influence of a liquid metal environment can also be studied under cyclic load, with results as shown in Fig. 9.32. The effects of gallium on the fatigue resistance of this alloy are particularly striking. At only 10^4 cycles the fatigue strength is almost nonexistent. In static fatigue tests, mercury can cause failure of this alloy in 30 min at but 10% of the 0.2% yield stress [89].

Microscopic aspects of LME

The interfacial energy can be directly determined for very brittle materials by means of the Obriemov [100] -Gilman [101] technique. It has been found that for zinc crystals wetted by mercury the surface energy is about 0.6 that of zinc alone. A decrease in γ_s will also produce a decrease in $\gamma_m \cong (\rho/a_0)\gamma_s$, the work done in microcrack formation, and hence in the stress σ_G (Chapter 6)

$$\sigma_G = \frac{4G\gamma_m}{k_y} d^{-\frac{1}{2}}$$

required for microcrack growth [91, 95, 99, 102]. *Consequently the same metallurgical factors which increase the tendency for cleavage*

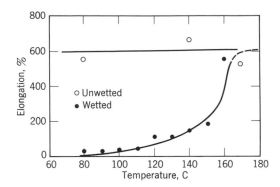

Fig. 9.33. Elongation to fracture of zinc monocrystals as a function of temperature when unwetted and when wetted with mercury. *After Rozhanski et al* [110].

fracture increase the tendency for LME. For example, LME is favored in coarse-grained as compared with fine-grained materials [91, 99].

The influence of temperature of the degrees of embrittlement is shown in Fig. 9.33, where a typical ductile-brittle transition-type of behavior is obtained. The effect is primarily related to an increasing homogeneity of deformation as the temperature increases. This behavior has been considered more explicitly in terms of the nature of the glide process [95]. Planar glide favors brittle behavior (because k_y is raised), whereas wavy glide does not. The effect of an increase in planarity in making alloys more brittle has been observed in ordered alloys as well as for alloys of low stacking fault energy. Figure 9.34 shows the dependence of the degree of embrittlement for a series of copper-base alloys. One of these, the Cu-Al series, is also shown in Fig. 9.30. The degree of embrittlement increases with a decrease in stacking fault energy and a corresponding increase in planarity of the slip process. The importance of planar glide lies in the development of high strain concentrations which cannot be relaxed by cross slip at barriers such as grain boundaries.

The results of studies of LME lead to the conclusion that embrittlement is due to a local decrease of cohesive energy brought about by the chemisorption of atoms of the liquid metal on the surface of the solid metal. In contrast to stress corrosion cracking, thermally activated processes such as diffusion are not considered to be important.

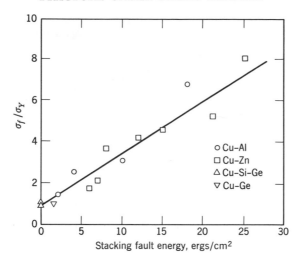

Fig. 9.34. Normalized fracture stress σ_f/σ_Y as a function of stacking fault energy for copper-base alloys tested in mercury. *After Johnston et al* [109].

However, electrochemical effects may play a role, for the fatigue life of polycrystalline zinc or copper specimens immersed in mercury can be increased if the mercury is made cathodic with respect to the specimen [103].

In order for LME to occur, strong barriers to dislocation motion must exist. The importance of this factor has been clearly shown by tests of zinc single crystals [104]. If the crystals are completely wetted by mercury, brittle fracture occurs in the vicinity of the grips where plastic deformation is complex and barriers can develop. The effect of this embrittlement as determined by the specimen elongation is shown in Figs. 9.33 and 9.35*a*. However, if the mercury is applied only to the central portion of the specimens remote from grip constraints, where deformation proceeds by basal glide and no barriers are developed, the crystals are not embrittled, as shown in Fig. 9.35*b*. If tilt boundaries are present in zinc, dislocations passing through them can lead to the development of high strains and embrittlement [105], whereas grain boundaries can act as barriers in polycrystalline zinc [104, 106, 107].

Let us consider the state of strain at a crack tip under stress, bearing in mind that the same considerations apply to the strained bonds at a dislocation pile up. A schematic of the atomic configuration at

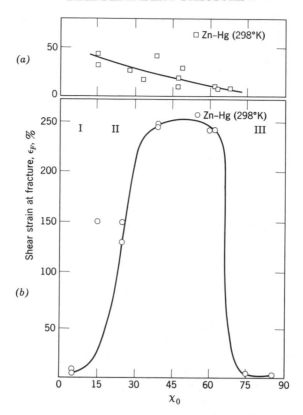

Fig. 9.35. Orientation dependence of shear strain at fracture for: (a) Amalgamated zinc mono-crystals [108]. (b) Uncoated and partially amal-gamated zinc monocrystals [104].

a crack tip is shown in Fig. 9.36 [91]. Adsorption of an atom of the liquid metal is thought to reduce the binding energy between atoms A-A at the crack tip. The crack grows by the successive rupture of these bonds which have potential energy-separation curves of the form $U(r)_0$, Fig. 9.37a. The minimum in the value of $U(r)_0$ at $r = a_0$, where a_0 is the equilibrium distance between the planes, may be regarded as the binding energy. The stress σ required to extend and rupture the bond A-A, Fig. 9.37b, varies as the derivative dU/dr, from a value of $\sigma = 0$ at $r = a_0$ to a maximum value $\sigma = \sigma_c$ at the point of inflection of $U(r)_0$, as shown in Chapter 2.

By assuming that the form of the $\sigma(r)_0$ curve is sinusoidal for values of x between $x = 0$ and $x = \lambda/2$, where λ is a parameter in the direction of x which describes the range of force interaction between the planes as they are separated, the maximum stress to break A-A bonds is given by

$$\sigma_c = \frac{E\lambda}{2\pi a_0} \tag{2.2}$$

where E is Young's modulus of the bonds A-A. By making the additional assumption that the work necessary to break A-A bonds reappears as a surface energy of the created fracture surfaces, it was shown (Eq. 2.4) that

$$\sigma_c = \left(\frac{E\gamma_s}{a_0}\right)^{1/2} \tag{2.4}$$

Adsorption of an atom from the liquid reduces the potential function from $U(r)_0$ to that of $U(r)_g$ with a corresponding reduction in cohesive stress from σ_c to σ'_c. This reduction in bond strength is the cause of LME. The atom-by-atom nature of the crack propagation process is reflected in the absence of detail on the fracture surface of

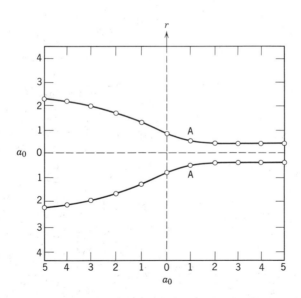

Fig. 9.36. Schematic diagram of atomic configurations at a crack tip; bonds to left of A-A already broken [91].

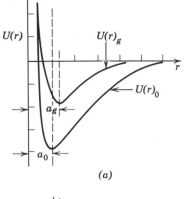

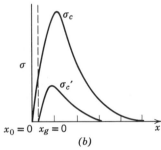

(b)

Fig. 9.37. (*a*) Potential energy functions $U(r)$ and $U(r)_0$ versus r in the presence and absence of liquid metal. (*b*) Stresses σ_c and σ_c' to separate atomic planes in the presence and absence of liquid metal [91].

an alloy embrittled by liquid mercury.

9.4 Hydrogen Embrittlement

Hydrogen embrittlement is one of the most common and serious types of time-dependent fracture. In laboratory testing, the presence of hydrogen results in a decrease in the ductility of unnotched tensile specimens and hence a decrease in the tensile strength of notched specimens. In service, failures can occur without warning, minutes to years after a static load has been applied to a structure containing hydrogen. The most spectacular and expensive failures have occurred in the petroleum industry, particularly in sour gas wells (i.e., containing H_2S) producing natural gas [111]. Fracture in a string of steel casing or tubing has led to the loss of the well and, in some cases, to the entire oil- or gas-bearing formation. Numerous failures of cadmium-plated high strength steel parts, where the hydrogen was introduced during electroplating [112, 113, 114, 115], have plagued the aircraft industry for the past decade. Cracking (flaking) has also been observed in massive steel castings which retained a substantial amount of the hydrogen introduced during solidification [116] and in rapidly cooled weld metal [117, 118, 148]. There are also some indications that the embrittlement of ultra-high strength steels by water vapor results from the presence of dissociated hydrogen [119]. Besides steels having a BCC crystal structure,† titanium and zirconium and their alloys can also be embrittled by hydrogen [120]. The

† Austenitic steels having a FCC crystal structure (e.g., 300 series stainless steel) are usually not susceptible to hydrogen embrittlement.

hydrogen can be introduced into these materials during processing, cleaning (acid pickling), or electroplating operations. In addition, hydrogen can be introduced into zirconium-cladding elements in water-cooled nuclear reactors when the chemical reaction $Zr + 2H_2O \rightarrow ZrO_2 + 2H_2$ takes place, since the oxide coating is relatively permeable to the passage of hydrogen. Finally hydrogen embrittlement can also occur in the refractory metals [121, 122, 123]. This will be discussed in Chapter 11.

Two characteristics of hydrogen embrittlement are now well established. First, it is *not* a form of stress corrosion cracking [111]. In fact, fracture often occurs when the metal has served (or serves) as a cathode, during electroplating or in a cathodic "protection" operation. Second, embrittlement results from hydrogen contents that are greater than the equilibrium solubility limit (about 10^{-3} ppm by weight in iron and 20–33 ppm in Ti and Zr at room temperature and at one atmosphere hydrogen pressure). The "excess" hydrogen which causes embritttlement can be as low as 1 ppm in high strength steel [124, 125], and 35 ppm in Ti and Zr. The mechanism of embrittlement, however, is not the same in all materials, and hence it is necessary to consider the different cases individually.

Embrittlement of iron-base alloys

The Effect of Hydrogen on the Iron Lattice. The equilibrium solubility C_H (in ppm) of hydrogen located in interstitial sites in the iron lattice, varies with temperature T and external hydrogen pressure P_e according to the relation [126, 127]

$$C_H = 42.7 P_e^{1/2} \exp\left(\frac{-6500}{RT}\right) \tag{9.15}$$

Consequently, iron is an endothermic absorber of hydrogen [129] (i.e., the solubility increases with increasing temperature). The diffusivity D_H (cm^2 per sec) of hydrogen also increases exponentially with temperature, according to the relation [124]

$$D_H = 1.4 \times 10^{-3} \exp\left(\frac{-3200}{RT}\right) \qquad (T > 150°C) \tag{9.16a}$$

Measurements of diffusivity by the gas effusion technique have shown that below about 150°C the diffusivity decreases sharply [124] and varies with temperature as

$$D_H = 0.12 \exp\left(\frac{-7820}{RT}\right) \qquad (T < 150°C) \tag{9.16b}$$

This suggests that excess hydrogen is contained in traps and that the low temperature diffusivity, as measured by gas effusion, is dependent on the rate of release of hydrogen from these traps. Recent experimental work has shown that voids are created by the expansion of hydrogen gas that has precipitated out of the iron lattice [129], and it is reasonable to associate these voids with the traps that cause the anomalous diffusion behavior.

Consider an iron specimen heated at a high temperature T_1 in the presence of hydrogen gas (thermally charged) and then rapidly cooled (quenched) to a lower temperature T_2. Immediately after quenching, the hydrogen content of the specimen $C_H(T_1)$ will be greater than the equilibrium content at $T = T_2$, $C_H(T_2)$. Since there is no evidence of any hydride formation in iron-base alloys [120], equilibrium can only be achieved when the excess hydrogen, $C_H(T_1) - C_H(T_2)$, diffuses out of the iron lattice. Hydrogen atoms quenched into regions near the metal's surfaces will be able to diffuse out of the metal in a short period of time, $t = x^2/D$, where x is (approximately) the diffusion distance. Hydrogen atoms quenched into the interior of the crystal will have a longer distance and time of travel to allow equilibrium to be reached. Some of them find it easier to recombine with other hydrogen atoms $(H + H \rightarrow H_2)$ and precipitate internally as molecular hydrogen gas.

The pressure of the hydrogen that has precipitated internally will be determined by the activity of the hydrogen atoms remaining in the lattice near the precipitation site and by the constraints imposed by the mechanical properties of the crystal [125]. Immediately after quenching, the pressure will be very high and, to relieve strain energy, the volume containing the hydrogen gas must expand, creating voids. If the material is ductile (e.g., pure iron at ambient temperature), the voids will expand by plastic deformation and appear as spherical "bubbles" [130]. This deformation causes the X ray diffraction peaks to be broadened after hydrogen charging [131]. When the material has a low resistance to cleavage crack propagation (e.g., Fe-3% Si at ambient temperature) the voids will appear as long, sharp cleavage cracks on {100} planes (Fig. 9.38) [129]. Increasing the difference between the temperatures T_1 and T_2 increases the amount of excess hydrogen $\{C_H(T_1) - C_H(T_2)\}$ and hence the number and length of these cracks. Thus the hydrogen pressure [132, 133] alone is able to nucleate cracks *in the absence of any externally applied stress.*

Similarly, during cathodic charging (e.g., an electroplating operation) the metal serves as a cathode and hydrogen ions are reduced to hydrogen atoms at its surface. Most of these atoms recombine with

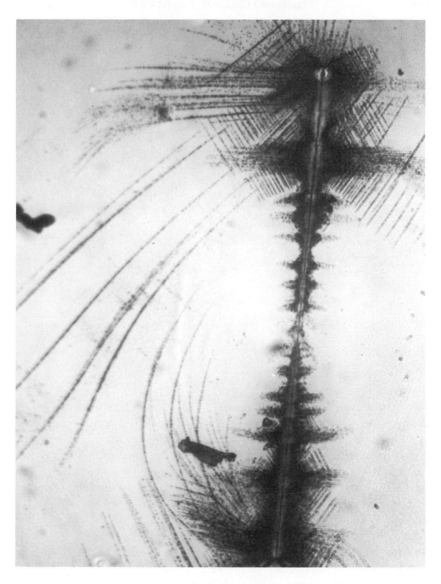

Fig. 9.38. Microcrack produced in Fe-3% Si by cathodic charging of hydrogen. Strain pattern around crack revealed by dislocation etch pitting. 350×. *After A. Tetelman and W. Robertson* [138].

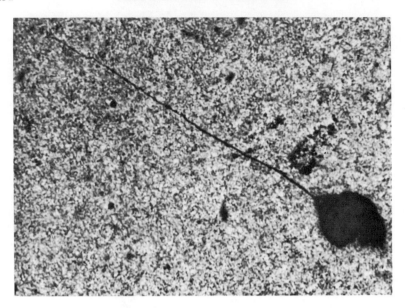

Fig. 9.39. Microcrack produced by cathodic charging in high strength steel. 500×. *Courtesy K. Farrell and A. Quarrell* [134] *and Iron and Steel Inst.*

others and are evolved as H_2 gas molecules; the remainder are driven into the metal by the very high effective hydrogen pressure (fugacity) and precipitate internally in the form of voids or cracks. The amount of hydrogen that is introduced in this fashion, and hence density and size of defects, increases with increasing current density and charging time [129] and varies from one electrolytic bath to another. A crack that was formed by cathodic charging of Fe-3% Si in 4% H_2SO_4 solution is shown in Fig. 5.8a.

The precipitation process that leads to crack nucleation is heterogeneous and requires the presence of internal surfaces on which the recombination of hydrogen atoms can (initially) occur.† Recent investigations have shown [134, 135] (see Fig. 9.39) that interfaces between nonmetallic inclusions and the metal matrix serve as the sites for the initial precipitation and that when the inclusion content is decreased the crack density will decrease as well. Similarly it is reasonable to expect that large carbide particles might also serve as

† Once the crack has been formed, recombination can occur on its surfaces.

precipitation sites for hydrogen gas in iron-carbon alloys, and there is some evidence that this is so [136].

On a microscopic level, cleavage crack growth under internal hydrogen pressure is similar to that which occurs under an externally applied tensile stress. Initially the growth is slow and occurs plastically, by a process of dislocation coalescence similar to that illustrated in Fig. 6.14 and described in Section 6.1, Crystallographic Mechanisms of Crack Nucleation. When the crack has grown to a length† such that the Griffith criterion for elastic propagation is satisfied, then propagation becomes unstable and the crack propagates rapidly [137].

There is, however, one important difference between crack propagation under internal hydrogen pressure and crack propagation under external tensile stress. Because the rate of crack growth is greater than the mobility of hydrogen atoms in the lattice, essentially no hydrogen can enter the crack once it has started to propagate [125]. Consequently the pressure P inside the crack decreases as the crack extends, according to the appropriate form of the ideal gas law for the internal pressures involved. The crack will extend until P decreases to $\sqrt{2E\gamma_P/[\pi c(1 - \nu^2)]}$, at which point it will be forced to stop. Since the crack tip becomes blunted by plastic relaxation when the crack stops, a higher pressure

$$P = \sqrt{\frac{2E\gamma_P{}^*}{\pi c(1 - \nu^2)}} \tag{9.17}$$

will be required to restart it. Because time is required for the necessary diffusion (and recombination) of hydrogen atoms to the crack surfaces, the growth process is discontinuous [138], during both the plastic (early) and elastic (later) stages of growth.

The single crystal studies described here provide a basis for understanding, at least qualitatively, the complicated nature of hydrogen embrittlement that occurs under tensile loading. They indicate that the excess hydrogen provides an internal pressure which allows crack growth to occur at a lower value of the externally applied tensile stress σ. For example, when hydrogen is present inside microcracks formed by plastic deformation, the criterion for initial microcrack growth (Eq. 6.19) can be written [139, 140]

$$(\sigma + P_n)nb = 2\gamma_m \tag{9.18}$$

† Typically, about 0.02–0.04 cm in Fe-3% Si single crystals charged at room temperature.

so that σ decreases with increasing pressure P_n. Because P_n decreases as the crack extends, it is necessary that

$$(\sigma + P_n) > \left(\frac{2E\gamma_m}{\pi c(1 - \nu^2)}\right)^{\frac{1}{2}} \qquad (9.19)$$

at all stages of crack growth if a completely unstable cleavage fracture is to develop. This implies that $\sigma \approx 2P$ for microcrack lengths of the order of one-tenth the grain size in polycrystalline iron [125]. Hydrogen can therefore assist in the initial growth of a crack, but a sufficient amount of external stress must be present to guarantee a complete failure.

Hydrogen Embrittlement of Steel

The effect of hydrogen on the tensile ductility of low strength (1020) steel is shown in Fig. 9.40. Three pertinent conclusions can be drawn from these curves [141]. First, the degree of hydrogen embrittlement† $\{[\epsilon_f - \epsilon_f(H)]/\epsilon_f\}$ increases with increasing hydrogen content. Second, there is no hydrogen embrittlement at very low $(T < T_A)$ or relatively high temperatures, and the degree of hydrogen embrittlement reaches a maximum at some intermediate temperature T_H where standard specimens fracture by a mixture of fibrous tearing and cleavage or fibrous tearing alone. Third, although increasing loading rates increase the transition temperatures of standard samples (see Chapters 6 and 7), the degree of hydrogen embrittlement *decreases* with increasing loading rates, for a given hydrogen content.

The absence of hydrogen embrittlement under high rates of loading indicates that Charpy impact tests are of little value in predicting whether a given material is particularly susceptible to this type of fracture. In fact, the static loading test gives the most reliable estimate of a material's susceptibility to hydrogen embrittlement, although very slow tensile tests ($\dot{\epsilon} = 0.05$ per min) are also able to predict this as well.

The absence of any deleterious effects of hydrogen at very low temperatures and under high rates of loading, and the return of ductility following relatively low (100–300°C) aging treatments to remove hydrogen by outgassing ("baking") [142, 143], suggest that hydrogen must be present and must be mobile to cause embrittlement. The exact nature of the embrittlement process has been the subject

---

† ϵ_f is the tensile ductility of uncharged specimens and $\epsilon_f(H)$ is the tensile ductility of specimens containing hydrogen.

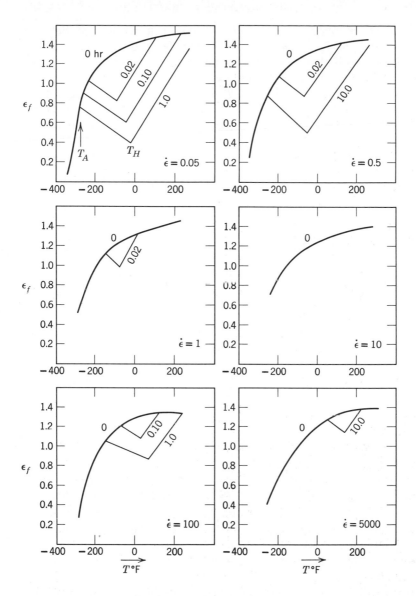

Fig. 9.40. Tensile ductility of 1020 steel at various temperatures and strain rates $\dot{\epsilon}$. The numbers adjacent to the curves indicate the time of cathodic charging prior to tensile testing. T_A is the lowest temperature at which ductility is decreased by a given hydrogen content; T_H is the temperature at which the degree of hydrogen embrittlement is a maximum. *After T. Toh and W. Baldwin* [141].

457

of numerous investigations. These have been reviewed in detail elsewhere and need not be repeated here [118, 120, 124, 125, 144, 145]. Instead, we shall present what appears to be the most reasonable explanation of the problem at this writing.

In the transition region $T > T_s$ unnotched, uncharged tensile specimens fracture by a mixture of fibrous tearing and cleavage. The fibrous cracks advance discontinuously, by the formation of microcracks or the cracking of particles in the highly strained region ahead of their tips and by the subsequent tearing of material between these defects and the tips. Since a fibrous crack is essentially an internal notch, the fracture process is similar to that described in detail in Chapter 7 for notched specimens of dirty materials fractured at temperatures $T > T_{s(N)}$. It was shown there that the toughness associated with this type of crack propagation is related to the density of microcracks and voids formed ahead of the advancing crack and that a large microcrack or void density (i.e., low temperature) implied smaller displacements (and hence a lower toughness) during plastic tearing. The tensile ductility ϵ'_f is a measure of the nominal strain required to achieve a critical displacement for instability, $V^*(c)$, at the tip of an advancing crack. The fracture stress is the stress level required to produce this strain, owing to work hardening.

Consider a crack that contains a fixed amount of hydrogen gas. When microcrack or microvoid nuclei (dislocation pile ups or cracked inclusions or carbides) are formed in the region of high plastic-stress and strain intensification ahead of the crack, a pressure gradient will exist between it and these nuclei. Consequently hydrogen will diffuse from the crack into the nuclei, building up a pressure P_n inside them (Fig. 9.41). Since the nuclei are small, only a small amount of hydrogen† needs to diffuse into them before the pressure in the nuclei P_n is a relatively large fraction of the pressure inside the crack P. This diffusion raises the local stress available for microcrack growth or void formation, from $\sigma_{yy} = K_{\sigma(p)} \{\sigma_Y + \Delta\sigma\}$ to $\sigma_{yy} + P_n$. Consequently, for any distribution‡ of values of σ_G, an increasing density of microcracks and/or voids will be formed ahead of the advancing crack when hydrogen is present (Fig. 9.42). This causes a decrease in the tip displacement required for crack propagation and hence a decrease in the tensile ductility ϵ'_f and the nominal fracture stress.

---

† When the nuclei are much smaller than the crack, the decrease in pressure in the crack due to diffusion to the nuclei is negligible.

‡ σ_G varies from grain to grain, because of the distribution of grain orientations and the distribution of carbide and inclusion shapes and sizes in the grains (Chapter 6).

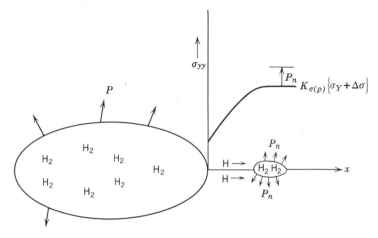

Fig. 9.41. The diffusion of hydrogen from an advancing crack to a micro-crack nucleus and the build up of pressure P_n in the nucleus.

At a fixed rate of specimen elongation, the amount of pressure that can build up in the nuclei before these are elongated and joined to the crack tip by tearing, and hence the probability of microcrack and/or microvoid formation, depends (1) on the pressure gradient between the nuclei and the crack tip, and (2) on the diffusivity of hydrogen. Since the pressure in the crack decreases as the crack extends, a fixed amount of hydrogen will produce its maximum embrittling effect when the crack is short rather than when it is long [125]. Increasing the amount of hydrogen in the crack (i.e., increasing the hydrogen content of the material) increases the value of P and P_n for all crack lengths and increases the microcrack density, thereby lowering ϵ'_f at a given temperature.

Below a temperature $T = T_A$, hydrogen diffusivity is too low to produce any significant lowering of ϵ'_f. Above T_A the ductility decreases with increasing temperature up to T_H (Fig. 9.40). This indicates that in this temperature range the increased rate of pressure buildup, owing to an increased hydrogen diffusivity, more than compensates for the decrease in σ_{yy} (owing to a decreasing σ_Y) with increasing temperature. Consequently the microcrack and/or microvoid density increases and ϵ'_f decreases with increasing temperature. Above T_H this is not the case, probably because the maximum potential pressure in the nuclei P_n is limited by that pressure which exists in the crack P, and because, as shown above, microcrack growth will not occur over any appreciable distance when P_n is a large fraction of the total stress $(\sigma_{yy} + P_n)$ acting across the nucleus. Any increase

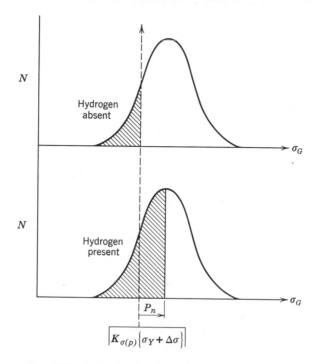

Fig. 9.42. N is the fraction of grains that crack or voids that form at a particular tensile stress level σ_G; shaded area is the total fraction of grains that will crack or voids that form when $\sigma = \sigma_G$. The presence of a hydrogen pressure P_n increases the density of fractured grains and/or voids ahead of an advancing crack, by increasing the local tensile stress level.

in σ_G with increasing temperature would also contribute to the increasing ductility in this temperature range. At any temperature an increasing rate of straining decreases the amount of pressure buildup and the probability of microcrack formation before tearing and hence decreases the degree of hydrogen embrittlement. With increasing applied strain rate T_A and T_H also increase.

These considerations indicate that since hydrogen causes embrittlement by producing an increase in the local stress field around an advancing crack, the same *microstructural variables* (e.g., fine-grain size, spheroidal carbides rather than plates, low inclusion content) that increase ductility in the absence of hydrogen will cause a smaller degree of embrittlement when hydrogen is present [146, 147]. Thus austenitic stainless steel (FCC crystal structure) is usually not embrittled by hydrogen, and soft BCC steels (Rockwell C hardness less than 25) are not embrittled to any significant degree.

Alloy composition does not directly affect a particular steel's susceptibility to hydrogen embrittlement, in the sense that there is no one alloying element that strongly increases or decreases the susceptibility. However, the *strength level* of the steel is an important factor and, as in the case of notch brittleness (Chapter 7), susceptibility increases with increasing strength level. Since changes in strength level can be achieved by changes in composition as well as by variations in heat treatment, alloy composition can have an indirect effect on the degree of hydrogen embrittlement.

The effect of hydrogen content on the room temperature ductility of some quenched and tempered steels, heat-treated to the indicated tensile strength levels, is shown in Fig. 9.43. The uncharged steels, or those that have been charged and degassed, exhibit about 45% RA at fracture [134]. The fracture surface contains a large number of relatively deep dimples, indicative of a relatively tough tear-type fracture. In charged (or charged and partially degassed) specimens, regions of cleavage fracture, 5–150 μ in diameter, exist in the vicinity of inclusions (Fig. 9.44) in the high strength (120 tsi) material. These areas of cleavage were probably formed by a process similar to that described above for the low strength alloys, although the

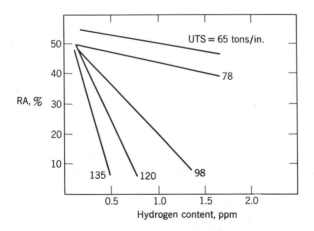

Fig. 9.43. The effect of hydrogen content and strength level on the tensile ductility of a quenched and tempered alloy steel. *After K. Farrell and A. Quarrell* [134].

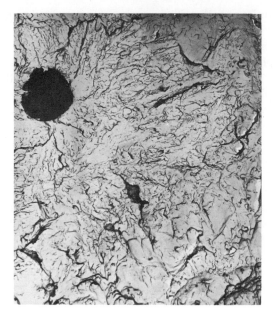

Fig. 9.44. Fracture surface of a hydrogen embrittled high strength steel. *Courtesy K. Farrall and A. Quarrell* [134] *and J. Iron Steel Inst.*

fine-grained structure in the tempered steel severely limits the extent of growth of individual microcracks. Increasing the strength level increases the probability of microcrack formation, thereby decreasing the ductility to a large degree when hydrogen is present (Fig. 9.43).

There is also evidence of quasicleavage and a shallow dimple structure on other regions of the fracture surface. These may result from the high degree of stress intensification set up ahead of the sharp cleavage crack, which causes the tip displacement required for tear fracture to decrease (Chapter 7). They may also result from the cracking, by hydrogen pressure buildup, of numerous particle-matrix interfaces which normally do not fracture (in uncharged materials) at the tip of an advancing crack at room temperature.

Hydrogen embrittlement also occurs under *static loading* at temperatures that are too low for creep to be a factor [111, 129, 142]. Tne effect of applied (nominal) stress on the time to failure is shown in Fig. 9.45. There is usually an incubation period during which the fracture stress is essentially independent of time, followed by a period where the time to fail increases with decreasing applied stress. Below a certain lower critical stress, which is analogous to the endurance limit observed in fatigue tests, embrittlement does not occur [149, 150]. For a given applied stress and hydrogen content the time to fail decreases with increasing strength level of the material. Within a certain strength range, notched specimens of relatively low-yield strength materials actually have greater load-carrying capacity than higher strength materials when exposed to hydrogen.

The embrittlement under static loading is reversible with respect to hydrogen content and if the hydrogen is removed by baking [142, 143], the properties become similar to those measured on uncharged materials. This effect is shown in Fig. 9.45. Since the diffusivity of hydrogen increases with increasing temperature, shorter baking times are required to improve strength and ductility when baking is performed at higher temperatures.

Electrical-resistance measurements made during the static loading of notched steel specimens containing hydrogen indicate that fracture occurs discontinuously, in the form of short bursts of crack growth, before the specimen actually fails. This slow growth process is probably similar to that which occurs under tensile loading, with internal hydrogen pressure causing the formation of voids and microcracks in

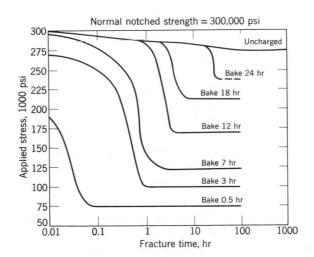

Fig. 9.45. Effect of baking at 300°F on the failure time, at a given level of applied stress, of notched specimens of 4340 steel heat-treated to a 230 psi strength level. *After J. O. Morlett et al* [142].

the region of high stress intensification ahead of the notch. Decreasing the nominal stress increases the length of the critical-sized crack which can propagate unstably and also the rate of slow crack growth (Section 3.4). Consequently the time to fail increases with decreasing nominal stress. The lower critical stress for static failure is probably related to the maximum length and sharpness of the crack that can form in a material of fixed hydrogen content. As shown in Fig. 9.45, it increases with decreasing hydrogen content. It also decreases with decreasing test temperature and notch root radius [150].

Embrittlement of titanium and zirconium

Both pure titanium and zirconium are group IV metals which have a HCP crystal structure (α) at room temperature. These metals exhibit a similar form of embrittlement when they contain hydrogen and, for convenience, will be discussed together. The equilibrium solubility limit C_H of hydrogen in the α-phase increases rapidly with increasing temperature. For example, at room temperature $C_H = 0.1$ atomic per cent (20 ppm by weight) in titanium [120], but at 300°C C_H is about 8 atomic per cent. In zirconium $C_H = 0.3$ atomic per cent (33 ppm) at room temperature and 1 atomic per cent at 300°C.

In contrast to the case of iron, hydride formation can occur in these metals when hydrogen concentrations greater than C_H are introduced (e.g., by cooling to room temperature after thermal charging at high temperature). Increasing hydrogen contents above C_H cause increasing amounts of hydride formation. The number, size, and orientation of the hydride platelets play a large role in the determining of the degree of embrittlement.

In the absence of applied stress, hydride precipitation tends to occur on a few preferred planes such as {1010} and {1011} in titanium [151] and {1012} in zirconium [120]. The former are principal slip planes in Ti, the latter is a principal twinning plane in Zr. The application of hydrostatic stress during precipitation can cause the hydride platelets to be aligned parallel to a compressive stress and perpendicular to a tensile stress [152, 153]. The extent of this effect is controlled by the texture, and a strong basal plane texture in the working direction lowers the stress required for platelet orientation. This effect, called "stress orientation," has been attributed to the fact that the density of the hydride is much lower than that of either Ti or Zr, so that precipitation occurs in an orientation that will relieve the greatest amount of tensile strain.

The size of the hydrides is related to the kinetics of the precipitation process, hence to the time and temperature in which the precipitation

Fig. 9.46. Fracture of hydride platelets in hydrogenated α-zirconium. 1000×. *Courtesy D. Westlake* [157] *and American Society for Metals.*

occurs. Slow cooling after thermal charging leads to relatively large (2×10^{-3} cm long) hydrides [154]. Rapid cooling or lower excess (above C_H) hydrogen contents causes the hydride precipitation to be suppressed (i.e., a supersaturated solid solution exists at room temperature) or to occur on a much finer scale. However, plastic strain at room temperature can cause the nucleation and growth of hydrides in quenched specimens [155], probably because of the enhanced diffusivity that can take place in a plastically deformed metal.

Fracture in hydrogenated α-Ti and Zr is associated with the cracking of the hydride platelets [156, 157] (Fig. 9.46) or the interface between the hydride and the matrix [153]. This cracking is initiated most easily when twins impinge on the platelets, but slip-nucleated cracking can also occur [157]. There is some evidence that the cleavage resistance of the hydrides increases as their size decreases [154].

At ambient temperature and low rates of straining, pure Zr and Ti matrices are extremely ductile so that cleavage fracture cannot develop. The hydrogenated specimens fail by the ductile rupture process described in Section 5.2. Decreasing the volume fraction of hydride which will crack (by decreasing the hydrogen content or by quenching to prevent large-scale precipitation) increases the ductility (Fig. 9.47) since the void density will be decreased. Similarly, when the hydrides are oriented, ductility will be decreased when specimens are strained perpendicular to the planes on which the hydrides are aligned. This effect tends to lower the strength and ductility of zircalloy-2 tubing [158, 159], since the basal plane texture in the

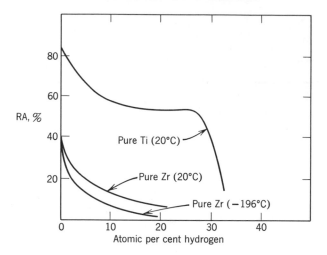

Fig. 9.47. Effect of hydrogen content on the tensile ductility of α-zirconium [44] and α-titanium [166] at the indicated temperatures.

radial and longitudinal (tube) directions causes the hydrides to be aligned perpendicular to the circumferential direction. Consequently there is a large circumferential (hoop) stress acting across the hydrides when internal pressure (e.g., owing to swelling of fuel elements) is set up inside the tube which favors hydride cracking.

The hydrogen embrittlement of α-titanium and zirconium is more pronounced at low temperatures [154, 160] and under high rates of strain [161] and there is no real evidence for delayed failure under static loading. The decrease in notch toughness with decreasing temperature and increasing hydrogen content (Fig. 9.48) results from the fact that these changes decrease tensile ductility (Fig. 9.47) and, as shown in Chapter 7, a decrease in tensile ductility leads to a decrease in notch toughness. The reason for the decreased tensile ductility at lower temperature appears to be related to the fact that the yield strength and strain-hardening rate both increase with decreasing temperature and increasing strain rate [154]. This raises the stress level in the plastic zone ahead of a cracked hydride and favors cleavage crack propagation or brittle intergranular propagation through the matrix.

In addition to the single-phase α-alloys, the αβ-alloys can also be embrittled by hydrogen [120, 161, 162]. The β-phase has the BCC structure which exists above 900°C in pure titanium. The addition of alloying elements such as molybdenum and manganese can allow this phase to exist at room temperature. During cooling from high temperature, the amount of β-phase present in a given alloy decreases. Since the hydrogen is more soluble in the β- than in the α-phase, a supersaturated solution of hydrogen in either or both the remaining amount of β-phase and the α-phase will be formed. There is no evidence for a hydride phase in the $\alpha\beta$-alloys [161]. However, there is autoradiographic evidence that the excess hydrogen segregates to $\alpha\beta$-phase boundaries upon thermal aging above room temperature [163].

In contrast to the embrittlement in the α-alloys, hydrogen embrittlement in the $\alpha\beta$-alloys is favored by slow strain rates and moderate temperatures. This suggests that, as in the case of the embrittlement of steel, the diffusivity of the hydrogen is an important factor in the embrittlement process. At low temperature the diffusivity is too low to allow appreciable embrittlement; at higher temperatures the solubility of hydrogen in the β-phase is increased and the alloys are inherently more ductile [164]. Consequently there is a certain temperature at which embrittlement is a maximum, for a given amount of hydrogen and rate of straining. At slow rates of tensile testing, maximum embrittlement is attained at room temperature [120]. These $\alpha\beta$-alloys are also susceptible to delayed failure

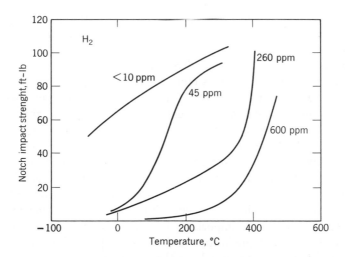

Fig. 9.48. Effect of hydrogen content and temperature on the notch impact strength of α-zirconium. *After G. Muhlenkamp and A. Schwarpe* [160].

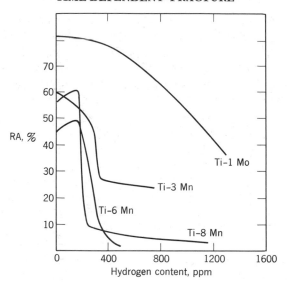

Fig. 9.49. Effect of hydrogen content and alloy content on the tensile ductility of $\alpha\beta$-titanium alloys. *After P. Cotterill* [120].

under static tensile loading [120, 164, 165], the time to fail increasing with decreasing nominal stress.

Crack propagation in $\alpha\beta$-alloys occurs intergranularly [120], along the $\alpha\beta$-phase boundary. Since hydrogen can segregate there, it is reasonable to associate the embrittlement with the presence of a large supersaturation of hydrogen. To date, there are only suggestions [120] as to the nature of this embrittlement. In view of the absence of the hydride phase and the fact that embrittlement requires a minimum concenration of hydrogen (Fig. 9.49) for a given rate of elongation, it seems most reasonable to attribute the embrittlement either to the buildup of hydrogen pressure at the $\alpha\beta$-phase boundary or to the lowering of the surface energy of the boundary, in a fashion similar to that produced by oxygen in decarburized iron. Either of these effects would lower the work done in crack propagation, for reasons discussed earlier, and hence lower the tensile ductility. The hydrogen diffusivity is important because it determines the rate at which hydrogen will precipitate out of the supersaturated solid solution and either enter the crack or the highly strained region ahead of it [142].

9.5 Neutron Irradiation Embrittlement

The recent expansion in the use of nuclear power plants has introduced a new set of engineering problems that are related to the embrittlement of materials during or after exposure to neutron irradiation. Neutron irradiation produces a variety of lattice defects in materials. From the point of view of fracture the most important of these are the atoms displaced from their normal lattice sites (interstitials), the vacant lattice sites left behind (vacancies), and clusters of these point defects which form small dislocation loops. These defects interfere with the motion of slip or twinning dislocations, thereby raising the yield strength σ_Y. This in turn can lead to the embrittlement of metals used in reactors.

In addition to producing the lattice defects mentioned above, neutron irradiation can cause nuclear transmutations in fuels, control rods, cladding, and structural components, which lead to the production of inert gas atoms such as xenon, krypton, and helium. These atoms are able to cause swelling of the fuels and other materials inside the reactor, a process which severely limits the operating lifetime of the reactor materials. They can also produce embrittlement in fuel element canning materials used in high-temperature reactors.

The extent of embrittlement by either of these processes depends on numerous reactor variables such as irradiation temperature, neutron flux ϕ, and time of irradiation t, as well as on material variables such as grain size, alloy content, and so on. Consequently the embrittlement problems are quite complex and many of them are still unresolved, except at a very simplified level.

Embrittlement of iron and steel

Many of the details of this phenomenon have been reviewed elsewhere [167, 168, 169, 199] and only the more pertinent aspects will be discussed here. Figure 9.50 shows the effect of neutron irradiation on the yield stength σ_Y, fracture stress σ_f, and ductility of unnotched tensile specimens of polycrystalline, high-purity (0.003% carbon) iron [170]. Several pertinent facts may be observed on this figure. First, below a test temperature of $-140°C$ a fast neutron dose[†] of 2×10^{18}

---

† Fast or fission neutrons, which are the ones that produce appreciable hardening, are those having energy greater than, say, 1 MEV. The fast neutron dose may be measured using radioactivation techniques. Typically, the ratio of slow (thermal) neutrons to fast ones is about (1–10)/1. All doses discussed here are of the "fast" type.

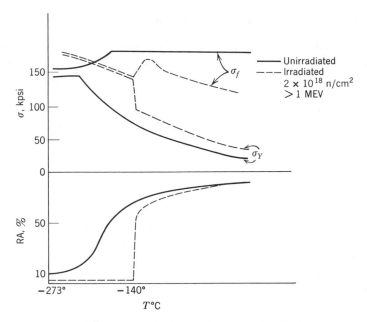

Fig. 9.50. Effect of temperature and neutron irradiation on the tensile yield strength σ_Y, fracture stress σ_f, and ductility of pure iron. *After J. Chow et al* [170].

neutrons per cm² (n per cm²) produces a substantial increase in σ_Y. Above $-140°$C the yield strength of the irradiated material decreases sharply and then parallels that of the unirradiated material as the test temperature is increased. Second, the *cleavage fracture stress* σ_f is essentially unaltered by irradiation when cleavage occurs at plastic strains less than about 10%. This has also been noted for low carbon steels [170]. Since σ_Y is increased and σ_f is unaffected, irradiation raises the brittleness-transition temperature T_D (where $\sigma_Y = \sigma_f$). The sharp drop in σ_Y of the irradiated material at $-140°$C accounts for the sharp rise in tensile ductility at that temperature. Third, the strain-hardening rate $d\sigma/d\epsilon$ and the uniform strain ϵ_u of the irradiated material are considerably less than that of the unirradiated material. This behavior is typical of that observed by other investigators [171–175] on a variety of low carbon and low alloy steels. The ductile fracture strain ϵ'_f for a full shear fracture is unaffected by neutron doses less than about 10^{19} n per cm², but it decreases with increasing doses above this value in low alloy steels [172, 173]. Similar behavior (increases in σ_Y, UTS, decreases in $d\sigma/d\epsilon$, ϵ_u, and ϵ'_f) has also been observed [176] in irradiated austenitic stainless steels. Since these steels have the FCC cystal structure, they do not fracture by cleavage, even after irradiation.

At test temperatures between −50° and 150°C the strengthening of steels produced by neutron irradiation $\Delta\sigma_Y$ is essentially independent of test temperature and usually obeys a relation of the form [169, 175]

$$\Delta\sigma_Y = A(\phi t)^{\frac{1}{2}} \qquad (9.20)$$

where ϕt is the integrated fission neutron dose in units of 10^{18}n per cm^2 (Fig. 9.51). High fluxes introduced over a short time period lead to the same hardening as low fluxes introduced over longer time periods, provided that (ϕt) remains the same [167, 200]. This behavior is in agreement with theoretical predictions [177] based on the interaction of moving dislocations and radiation-produced obstacles, the number of which increase with increasing neutron dose. At high doses ($\approx 10^{19}$n per cm^2) the number of obstacles approach a saturation value [177] and σ_Y increases more slowly with increasing dose [169, 175, 177].

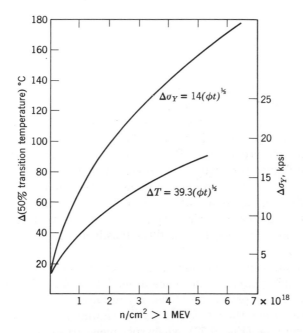

Fig. 9.51. Effect of integrated neutron dose on the change in 50% transition temperature and yield strength of silicon-killed mild steel. *After R. Nichols and D. Harries* [169].

Numerous investigators [169, 170, 171, 175, 178, 179, 180] have measured the effect of neutron irradiation on the parameters σ_i and k_y in the Petch equation $\sigma_Y = \sigma_i + k_y d^{-\frac{1}{2}}$. All investigators reported that σ_i was increased by irradiation. In low cabon steel [169, 171, 179, 180] and silicon iron [178] k_y is unchanged by irradiation. Some workers have noted that in pure iron k_y is unchanged [175] by irradiation; others have noted [170, 179] that k_y is decreased by irradiation. A decrease in k_y seems surprising since σ_f, which is extremely sensitive to changes in k_y in a clean material (see Eq. 6.27), is unaffected by irradiation [170]. One possible explanation of this effect is that σ_i is changed more by irradiation in coarse-grained rather than fine-grained material, (owing perhaps to differences in substructure or defect annealing at grain boundaries in a fine-grained material). This would cause the slope of a σ_Y versus $d^{-\frac{1}{2}}$ plot to decrease and *give the impression* that k_y was lowered by irradiation. Although the situation in pure iron† remains ambiguous at the moment, there is little doubt that in low carbon and low alloy steels all changes in σ_Y upon irradiation are due to changes in σ_i. Consequently

$$\Delta\sigma_Y = \Delta\sigma_i = A(\phi t)^{\frac{1}{2}} \qquad (9.21)$$

The value of A is dependent on irradiation temperature and the type of steel being irradiated. Typically A varies between 10 and 16 when irradiation is performed at 100°C, with $\Delta\sigma_i$ in units of ksi and ϕt in units of 10^{18} fission neutrons per cm² [200].

A is smaller in aluminum-killed than silicon-killed mild steels. For a given steel, A decreases as the irradiation temperature increases and irradiations performed above 350°C produce little change in tensile and impact properties. The reason is that annealing of defects occurs simultaneously with the production of new ones. Consequently the net defect content, and hence the change in properties after irradiation, decreases as the rate of annealing (i.e., annealing temperature) increases. Similarly, postirradiation heat treatment causes the annealing of defects and the tensile and impact properties to recover [169].

In aluminum-killed and low alloy steels there is some indication [169] that aging effects take place around 100°–130°C which lead to an increase in the parameter A when irradiation is performed at this temperature rather than at 50°C. In a similar fashion $\Delta\sigma_i = \Delta\sigma_Y$ is found to increase (Fig. 9.52) when specimens irradiated at 20°C are annealed in this temperature range. These effects are analogous to

† And pure molybdenum [181] as well, where a decrease in k_y has been noted after irradiation.

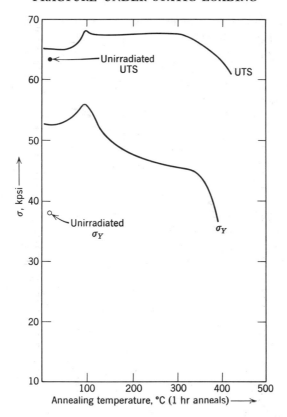

Fig. 9.52. Effect of postirradiation one-hour annealing treatments at the indicated temperatures on the yield and ultimate strengths of aluminum-killed steel. (Prior irradiation treatment was 3×10^{18} n/cm^2 > 1 MEV). *After R. Nichols and D. Harries* [169].

those found in precipitation-hardening alloys where low-temperature annealing allows the clusters to reach a critical size and spacing such that they can impart maximum resistance to dislocation motion; at high temperatures large clusters grow at the expense of small ones and the dispersion strengthening effect is reduced. The kinetics of these processes are quite complicated and are influenced by the alloy content of the steel.

The effect of neutron irradiation on the impact properties of steels used in reactor pressure vessels has received considerable attention [168, 169, 172, 173, 174, 182, 183, 184]. Figure 9.53 shows the

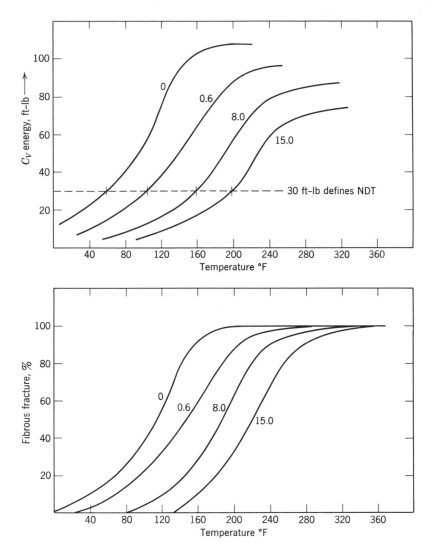

Fig. 9.53. Effect of indicated irradiation dose (units of 10^{18} n/cm² > 1 MEV) on the Charpy energy and per cent fibrosity at various temperatures, for an A302B steel. *After F. Brandt and A. Alexander* [184].

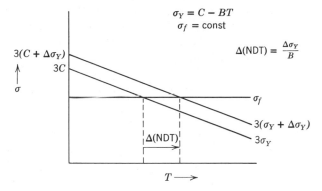

$$\sigma_Y = C - BT$$
$$\sigma_f = \text{const}$$

$$\Delta(NDT) = \frac{\Delta\sigma_Y}{B}$$

Fig. 9.54. Effect of a change in yield strength $\Delta\sigma_Y$ on the NDT when the cleavage strength σ_f is independent of temperature and the yield strength decreases linearly with temperature.

behavior of Charpy V notch specimens of an A 302 B steel after doses ranging from 0 to 1.5×10^{19} n per cm². It is noted (1) that the NDT temperature† increases with increasing irradiation exposure, (2) that the maximum energy absorbed C_V(max) decreases after irradiation, particularly at high doses, and (3) that the slope of the Charpy curves decreases (i.e., the transition region spreads out) particularly at high doses.

The NDT is *approximately* the temperature where plastic constraint and strain hardening have raised the tensile stress level σ_{yy} ahead of the V notch or an advancing tear fracture from σ_Y up to $3\sigma_Y = \sigma_f$. For those cases where the cleavage fracture stress σ_f is independent of temperature and σ_Y decreases linearly with temperature (i.e., $\sigma_Y = C - BT$), then an increase in σ_Y, $\Delta\sigma_Y$, will cause the NDT to increase by $\Delta\sigma_Y/B$ (Fig. 9.54). Physically this means that a given increase in $\Delta\sigma_Y$ will produce a larger increase in the NDT when the yield strength varies slowly with temperature (i.e., small B, due, for example, to high alloy content or high loading rate) than when it varies more sharply. Fracture toughness measurements show [189] that K_{Ic} for unirradiated steel specimens tested at low temperature is about the same (25 $k$psi in.½) as for irradiated specimens tested at a higher temperature, where σ_Y of the irradiated material is the same as that of the unirradiated material at the lower temperature. Since K_{Ic}

† Which corresponds to the 30 ft-lb temperature for this steel as well as for the A 212 B and other reactor steels [168].

depends on the *ratio* (σ_f/σ_Y) (see Chapter 7), this would be expected if σ_f varied only slightly with temperature.

For a typical value of B for mild steel there is about a 2°C increase in the NDT per 1000 psi increase in σ_Y [185]. Hence, from Eq. 9.21 we have

$$\Delta(\text{NDT}) \cong 2A(\phi t)^{\frac{1}{2}}$$
$$\cong 28(\phi t)^{\frac{1}{2}} \tag{9.22}$$

for $A = 14.0$. Recent experimental work has shown that the 50% transition temperature does increase linearly [169] with $(\phi t)^{\frac{1}{2}}$ (Fig. 9.51). Figure 9.55 shows that for a wide variety of steels, the change in NDT upon irradiation [168] varies between 28 and $39(\phi t)^{\frac{1}{2}}$.

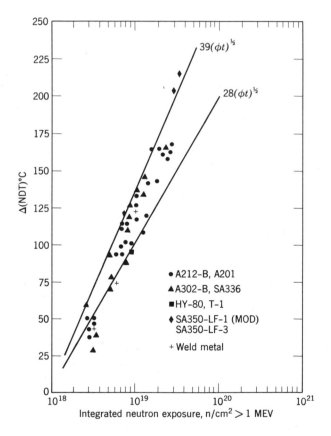

Fig. 9.55. $\Delta(\text{NDT})$ as a function of fast neutron dose (>1 MEV) for a series of steels. *After W. Pellini et al* [168].

These differences probably arise because of differences in the temperature sensitivity of σ_Y and the constant A for a given steel composition and grain size.

The analysis presented here predicts that $\Delta(\text{NDT})$ is independent of grain size when A and B are assumed independent of grain size.[†] However, some workers have noted [169, 186] that in low carbon steel $\Delta(\text{NDT})$ is increased less, for a given dose ϕt, in fine-grained as compared with coarse-grained steels; other data indicate [187] that $\Delta(\text{NDT})$ is increased more in fine-grained specimens of nickel ferrite than in coarse-grained specimens. These anomalies may result from the fact that $\Delta\sigma_i$ (i.e., A) is itself a function of grain size.

The decrease in $C_V(\text{max})$ and the lowering of the Charpy slope after irradiation result from the decrease in tensile ductility ϵ'_f and strain-hardening rate $d\sigma/d\epsilon$, for reasons outlined in Chapter 7. This effect is most noticeable at high doses where large decreases in tensile ductility are produced [172, 173]. In the 212B and 302 reactor steels $C_V(\text{max})$ still remains a sufficiently high value after irradiation so that low-energy tear is not a problem. However, in higher strength steels [182] such as HY-80, a high dose can cause $C_V(\text{max})$ to decrease to a relatively low value (40 ft-lb) such that low-energy tear fracture might occur in service. The fact that the reactor is being operated well above its NDT temperature is then no longer a guarantee that unstable fracture will not develop (see Chapter 3).

The effect of irradiation temperature on $\Delta(\text{NDT})$ is similar to its effect on $\Delta\sigma_i$ discussed above. Increasing the irradiation temperature lowers $\Delta\sigma_i$ and hence lowers $\Delta(\text{NDT})$ [188], except when aging effects take place. This behavior is shown in Fig. 9.56. Figure 9.57 is a composite diagram showing $\Delta(\text{NDT})$ as a function of irradiation temperature as well as the estimated dosage for the indicated years of service. Since the NDT temperature for these steels is about 0°F, $\Delta(\text{NDT})$ is also equal to the NDT temperature of the steel after a given amount of exposure. Embrittlement is a potential problem [168] when the operating temperature is less than about 60°F greater than the NDT temperature of the steel; this would be the case after a few years of service for high power, compact reactors operated at the lower temperature range, but after 20 years for the high power, large reactors operating at 550°F.

The decrease in $\Delta\sigma_i$ following annealing (Fig. 9.52) [169] would indicate that apart from the technological difficulties of performing the operation, annealing of a pressure vessel after exposure to ir-

† The *absolute* value of the NDT is, of course, lower in fine-grained steels.

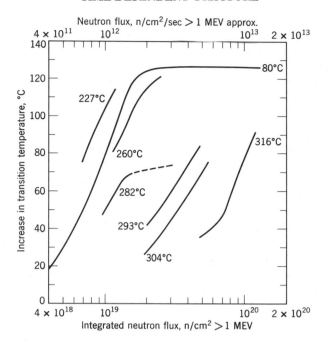

Fig. 9.56. Increase in impact-transition temperature of irradiated A 212 B steel, as a function of the temperature of irradiation and irradiation dose. *After R. Berggren* [188].

radiation might also produce a decrease in $\Delta(\text{NDT})$. Figure 9.58*a* indicates that postirradiation annealing can produce a decrease in the NDT. It is noteworthy that, for a given heat treatment, the degree of recovery depends on irradiation temperature. In some instances, larger amounts of recovery are attainable in steels that were irradiated at lower temperature [168]. Consequently the relative positions of the two Charpy curves are reversed after annealing. This results from the fact that, for reasons which are not yet clear, $\Delta\sigma_Y = \Delta\sigma_i$ decreases more slowly when the material is irradiated at a higher temperature [201], as shown schematically in Fig. 9.58*b*.

Transmutation effects on reactor materials

The *swelling* which occurs in reactor materials imposes severe limits on the operating lifetime of fuels and canning materials. The distortions or cracking produced by swelling can decrease the effective thermal conductivity of a fuel element, which in turn could lead to the melting of the element.   A second potential problem is that the swelling can lead to the cracking of the cladding material which surrounds the fuel; this would allow the release of fission products into the reactor coolant.   Only a brief outline of this problem will be presented here.   The interested reader is referred to reference [190] for a detailed treatment.

For every four atoms of uranium that are fissioned, about one atom of krypton or xenon is produced.   The solubility of these atoms in uranium is extremely small [191] and, like the excess hydrogen atoms in a charged steel, these inert gas atoms will form stable nuclei and precipitate (either heterogeneously or homogeneously) out of the uranium (or UO_2) lattice.   Since the inert gas atoms are relatively large, their mobility will be low and strongly influenced by temperature.   At low temperatures fine, closely spaced bubbles are formed;

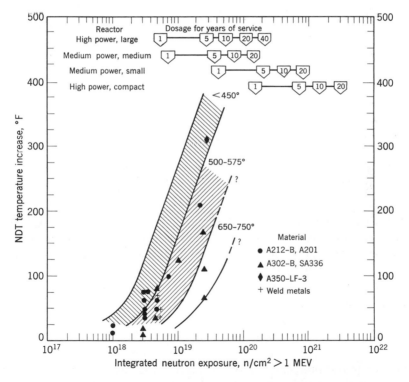

Fig. 9.57. Increase in NDT as a function of irradiation dose and temperature of irradiation. Estimated dosage for years of service in particular reactors are indicated. *After W. Pellini et al* [168].

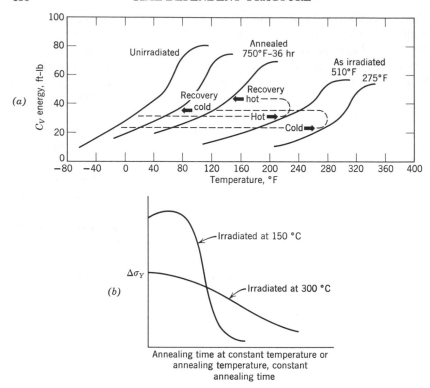

Fig. 9.58. (a) Effect of irradiation temperature and annealing temperature on the Charpy curves of a reactor-grade steel. Neutron dose about 2.15×10^{19} n/cm^2 > 1 MEV. (b) Annealing of irradiation damage as a function of irradiation temperature in mild steel. *After Wechsler and Berggren* [201].

at higher temperatures the bubbles, like precipitates in an age-hardened alloy, can coalesce and grow rapidly at the expense of one another [192]. The number of bubbles then decreases while their size and spacing increases [193].

A spherical bubble of volume $V = \frac{4}{3}\pi r^3$ and radius r, containing m atoms of an ideal gas at a temperature T, is in equilibrium when [194]

$$\frac{4}{3}\pi r^3 \left(P_e + \sigma + \frac{2\gamma}{r} \right) = mkt \qquad (9.23)$$

where σ is the pressure due to the strength of the material and P_e is the externally applied pressure (if any); $2\gamma/r$ is the pressure corresponding to the surface tension γ of the material. At low temperatures where diffusion is slow and the bubbles are small, $2\gamma/r$ is the predominant term [194]. For a given temperature and gas content, the number of bubbles is inversely proportional to their radius r [195]. The amount of swelling is small.

At higher temperatures (above 400°C in uranium, for example) [192, 196], the bubbles are larger (Fig. 9.59) and their growth rate is controlled by the creep strength† of the material σ. The amount of swelling which occurs in this temperature range is considerably greater than at low temperatures since the material itself is actually deforming. Calculations of the magnitude of this effect based on Eq. 9.23 have been reported [197].

Various methods have been suggested [192] for minimizing the amount of swelling ΔV by maximizing one or more of the three pressure terms in Eq. 9.23.

1. Increasing P_e by pressuring the fuel cans or by using stronger cladding to constrain the fuel.

2. Increasing the creep strength σ by techniques such as alloying and heat treatment.

† In very brittle materials it might be expected that pressure buildup would lead to the formation of cleavage cracks, as in the case of hydrogen-charged silicon iron. This has been observed in UO_2 [198]. Then σ would refer to the crack-propagation stress rather than the creep stress.

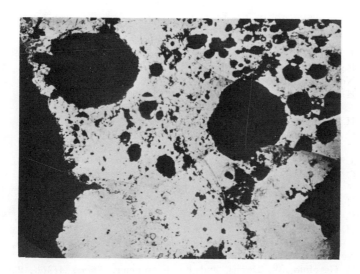

Fig. 9.59. Voids observed in uranium after 0.3% burnup at 600°C in five months, after thermal cycling. Volume increase equals 141%. *Courtesy A. Churchman* [196] *and AIME.*

3. Increasing the surface tension force $(2\gamma/r)$ by increasing the number of bubbles, thereby decreasing their radius. This might be accomplished by increasing the number of sites for heterogeneous precipitation (e.g., by adding a finely dispersed second-phase or inert particles) or by preirradiating at low temperature to provide sites for fine homogeneous precipitation at higher temperatures.

4. Reducing the mobility of the inert gas atoms in the fuel, thereby preventing agglomeration. This might be accomplished by the addition of alloying elements to the fuel.

In addition to swelling, transmutation effects can also produce embrittlement in austenitic stainless steels and nickel-base alloys that are used as fuel element canning materials in high-temperature reactors [199, 200, 203]. Thermal (slow) neutrons are able to transmute the B^{10} (a boron isotope) impurity atoms in these materials to lithium and helium. The helium appears to agglomerate into gas bubbles at grain boundaries and cause intergranular fracture, as described in Section 9.1. The embrittlement is pronounced above 600°C and is observed as a decreased tensile ductility or decreased lifetime in a stress rupture test.

References

The asterisk indicates that published work is recommended for extensive or broad treatment.

*[1] J. R. Low, Jr., *Fracture of Solids,* Interscience, New York, (1963), p. 197.
[2] R. D. Gifkins, in *Fracture,* B. L. Averbach et al. eds., M.I.T., Wiley, New York (1959), p. 579.
[3] H. H. Bleakney, *Can. Met. Quart.,* 4, 13 (1965).
[4] J. E. Harris, *Trans. AIME,* 233, 1509 (1965).
[5] H. H. Bleakney, *Can. J. Tech.,* 30, 340 (1952).
[6] C. W. Chen and E. S. Machlin, *Trans. AIME,* 209, 829 (1957).
[7] F. C. Monkman and N. J. Grant, *Proc. ASTM,* 56, 593 (1956).
*[8] J. Weertman, *J. Appl. Phys.,* 26, 1213 (1955).
[9] C. R. Barrett and W. D. Nix, *Acta Met.,* 13, 1247 (1965).
[10] F. J. Clauss, *Mat. Design Eng.,* 53, 139 (January 1961).
[11] H. C. Cross, *Metals Progr.,* 87, No. 3, 67 (1965).
[12] F. J. Anders, G. B. Alexander, and S. Warkel, *Metal Progr.,* 82, 88 (December 1962).
[13] J. J. Harwood and M. Semchysen, *High Temperature Materials,* Wiley, New York (1959), p. 243.
[14] F. Garofalo, W. Domis, and F. Gemminger, *Trans. AIME,* 230, 1460 (1964).
*[15] F. Garofalo, *Fundamentals of Creep and Creep Rupture,* Macmillan, New York (1965).

[16] B. G. Piearcey and F. L. Ver Snyder, *Pratt and Whitney Report No. 65-007* (April 1965).

[17] F. N. Rhines, W. E. Bond, and M. A. Kissel, *Trans. ASM,* **48,** 919 (1956).

[18] C. R. Barrett, Stanford University, unpublished results.

[19] F. R. Larson and J. Miller, *Trans. ASME,* **74,** 765 (1952).

[20] F. J. Clauss, *Proc. ASTM,* **60,** 905 (1960).

*[21] J. E. Dorn, *Creep and Recovery,* ASM, Metals Park, Ohio (1957).

[22] D. McLean and K. F. Hale, *Structural Processes in Creep,* Iron and Steel Inst., London (1961), p. 19.

[23] C. R. Barrett, A. Ardell, and O. Sherby, *Trans. AIME,* **230,** 200 (1964).

*[24] C. R. Barrett and O. Sherby, *Trans. AIME,* **233,** 1116 (1965).

[25] C. R. Barrett and O. Sherby, *Trans. AIME,* **230,** 1322 (1964).

[26] F. Garofalo, *Trans. AIME,* **227,** 351 (1963).

[27] R. L. Orr, O. D. Sherby, and J. E. Dorn, *Trans. ASM,* **46,** 113 (1954).

[28] S. S. Manson and A. M. Haferd, *NACA,* TN 2890 (1953).

[29] A. Thum and K. Richard, *Mitt. Ver. Grosskesselbesitzer,* No. 85, 198, (1941).

[30] W. F. Brown et al., *Influence of Stress Concentrations at Elevated Temperature,* ASTM STP 260 (1959).

[31] N. T. Williams and G. Willoughby, *Joint Int. Conf. on Creep 1963,* Inst. Mech. Engrs., London, 4–11 (1963).

[32] H. R. Voorhees and J. W. Freeman, WADC TR No. 54-175, Part 3 (1956).

[33] R. M. Goldhoff, *Joint Int. Conf. on Creep 1963,* Inst. Mech. Engrs. London, 4–19 (1963).

[34] G. J. Heimerl and A. J. McEvily, *NACA,* TN 4112 (1957).

*[35] A. J. Kennedy, *Processes of Creep and Fatigue in Metals,* Wiley, New York (1963).

[36] H. C. Chang and N. J. Grant, *Trans. AIME,* **206,** 544 (1956).

[37] I. S. Servi and N. J. Grant, *Trans. AIME,* **191,** 909 (1951).

*[38] A. H. Cottrell, *Structural Processes in Creep,* Iron and Steel Inst. (1961), p. 1.

[39] A. N. Stroh, *Proc. Roy. Soc.,* **A223,** 404 (1954).

[40] D. McLean, *J. Inst. Metals,* **85,** 468 (1956–57).

[41] A. S. Nemy and F. N. Rhines, *Trans. AIME,* **215,** 992 (1959).

[42] J. Heslop, *J. Inst. Metals,* **91,** 28 (1963).

[43] C. M. Weaver, *Acta Met.,* **8,** 343 (1960).

[44] W. Betteridge and A. W. Franklin, *J. Inst. Metals,* **85,** 475 (1956–57).

[45] D. McLean, *J. Inst. Metals,* **81,** 293 (1952–53).

[46] R. C. Gifkins, *Acta Met.,* **4,** 98 (1956).

[47] D. M. R. Taplin and U. N. Whittaker, *J. Inst. Metals,* **92,** 426 (1963–64).

*[48] C. Crussard and J. Friedel, *Proc. NPL Symposium on Creep and Fracture of Metals at High Temperatures,* Nat. Physical Lab., London (1956), p. 243.

[49] D. McLean, *J. Inst. Metals,* **85,** 468 (1956–57).

[50] J. N. Greenwood, D. R. Miller and J. W. Suiter, *Acta Met.,* **2,** 250 (1954).

[51] R. C. Boettner and W. D. Robertson, *Trans. AIME,* **221,** 613 (1961).

[52] R. T. Ratcliffe and G. W. Greenwood, *Phil. Mag.,* **12,** 59 (1965).

[53] A. W. Mullendore and N. J. Grant, *Deformation and Fracture at Elevated Temperatures,* M.I.T. Press, Cambridge (1965), p. 165.

[54] D. Hull and D. Rimmer, *Phil. Mag.,* **4,** 673 (1959).

[55] J. C. E. Olds and G. M. Michie, *J. Inst. Metals,* **88,** 493 (1959–60).

[56] P. E. Brookes, *J. Inst. Metals,* **88,** 500 (1959–60).

[57] M. A. Adams and G. T. Murray, Materials Research Corp., *Res. Report AT (30-1)-2178* (1960).

[58] J. Intrater and C. S. Machlin, *Acta Met.,* **7,** 140 (1959).

[59] C. Zener, *Fracturing of Metals,* ASM, Cleveland (1948), p. 1.

[60] R. Eborall, *Proc. NPL Symposium on Creep and Fracture of Metals at High Temperatures,* Nat. Physical Lab., London (1956), p. 229.

[61] N. B. W. Thompson and R. L. Bell, *J. Inst. Metals,* **94,** 116 (1966).

[62] P. W. Davies, et al., summary in *J. Inst. Metals,* **93,** 320 (1964–65).

[63] E. G. Coleman, D. Weinstein and W. Rostoker, *Acat Met.,* **9,** 491 (1961).

[64] A. J. Forty and P. Humble, *Phil. Mag.,* **8,** 247 (1963).

[65] S. W. Dean and H. R. Copson, *Corrosion,* **21,** 95 (1965).

*[66] E. N. Pugh, *Environment-Sensitive Mechanical Behavior,* Gordon and Breach, New York (1966).

[67] M. G. Fontana, *Indust. Eng. Chem.,* **46,** 99A (May 1954).

[68] H. H. Bleakney, *Can. Met. Quart.,* **4,** 13 (1965).

[69] A. J. McEvily and A. P. Bond, *Environment-Sensitive Mechanical Behavior* Gordon and Breach, New York (1966).

[70] R. Newcomer, H. C. Tourkakis and H. L. Turner, *Corrosion,* **21,** 307 (1965).

[71] B. F. Brown et al., Marine Corrosion Studies, Third Interim Report of Progress, *NRL Memo Report 1634* (July 1965).

[72] U. R. Evans, *An Introduction to Metallic Corrosion,* Arnold, London (1948), p. 106.

*[73] W. D. Robertson and A. S. Tetelman, *Strengthening Mechanisms in Solids,* ASM, Metals Park, Ohio (1962), p. 217.

[74] B. F. Brown and C. D. Beachem, *Corrosion Science,* **5,** 245 (1965).

[75] C. Edeleanu, *J. Iron Steel Inst.,* **175,** 390 (1953).

*[76] P. R. Swann and J. D. Embury, *High Strength Materials* Wiley, New York (1965), p. 327.

[77] K. C. Thomas, R. Stickler and R. J. Allio, *Corrosion Science,* **5,** 71 (1965).

[78] D. J. Burr, *Corrosion Science,* **5,** 733 (1965).

[79] M. N. Saxena and R. A. Dodd, *Environment-Sensitive Mechanical Behavior,* Gordon and Breach, New York (1966).

[80] D. L. Douglas, G. Thomas and W. R. Roser, *Corrosion,* **20,** 15 (1964).

[81] N. A. Nielsen, *Corrosion,* **20,** 104 (1964).

[82] C. R. Bergen, *Corrosion,* **20,** 269 (1964).

[83] H. L. Logan, M. J. McBee and D. J. Kahan, *Corrosion Science,* **5,** 729 (1965).

[84] D. Tromans and J. Nutting, *Fracture of Solids,* Interscience, New York (1962), p. 637.

*[85] A. J. McEvily and A. P. Bond, *J. Electrochem. Soc.,* **112,** 131 (1965).

[86] *ASM Metals Handbook,* ASM, **1,** 918 (1961).

[87] H. R. Copson, *Corrosion Handbook* Wiley, New York (1948), p. 569.

[88] E. N. Pugh, W. G. Montague and A. R. C. Westwood, *Trans. ASM,* **58,** 665 (1965).

*[89] W. Rostoker, J. M. McCaughey and H. Markus, *Embrittlement by Liquid Metals,* Reinhold, New York (1960).

[90] M. H. Kamdar and A. R. C. Westwood, to be published.

[91] N. S. Stoloff and T. L. Johnston, *Acta Met.,* **11,** 251 (1963).

[92] Y. V. Goryunov, N. U. Pertsov and P. A. Rehbinder, *Dokl. Acad. Nauk U.S.S.R.,* **127,** 784 (1959).

[93] J. W. Obreimov, *Proc. Roy. Soc.,* **A127,** 290 (1930).

[94] H. Nichols and W. Rostoker, *Acta Met.,* **9,** 504 (1961).

*[95] N. S. Stoloff, R. G. Davies and T. L. Johnston, *Environment-Sensitive Mechanical Behavior,* Gordon and Breach, New York (1966).

[96] F. N. Rhines, J. A. Alexander and W. F. Barclay, *Trans. ASM,* **55,** 22 (1962).

*[97] A. J. McEvily and A. P. Bond, *Environment-Sensitive Mechanical Behavior,* Gordon and Breach, New York (1966).

[98] L. S. Brgukhanova, I. A. Andreeva and V. I. Lichtman, *Soviet Physics (Solid State)* **3,** 2025 (1962).

[99] E. G. Coleman, D. Weinstein and W. Rostoker, *Acta Met.,* **9,** 491 (1961).

[100] J. W. Obreimov, *Proc. Roy. Soc.,* **A127,** 290 (1930).

[101] J. J. Gilman, in *Fracture,* B. L. Averbach et al. eds., M.I.T., Wiley, New York (1959), p. 193.

[102] A. R. C. Westwood and M. H. Kamdar, *Phil. Mag.,* **8,** 787 (1963).

[103] N. A. Tiner, *Trans. AIME,* **221,** 261 (1961).

[104] M. H. Kamdar and A. R. C. Westwood, *Environment-Sensitive Mechanical Behavior,* Gordon and Breach, New York 1966.

[105] A. N. Stroh, *Phil. Mag.,* **3,** 597 (1958).

[106] R. Bullough, *Phil. Mag.,* **9,** 917 (1964).

[107] J. J. Gilman, *Trans. AIME,* **212,** 783 (1958).

[108] E. D. Shchukin, N. V. Pertsov and U. V. Goryunov, *Soviet Physics (Crystallography),* **4,** 840 (1959).

[109] T. L. Johnston, R. G. Davies and N. S. Stoloff, *Phil. Mag.,* **12,** 305 (1965).

[110] V. N. Rozhanski et al, *Doklady Akad. Nauk U.S.S.R.,* **116,** 769 (1957).

*[111] A. E. Scheutz and W. D. Roberston, *Corrosion,* **13,** 437 (1957).

[112] G. Sachs, *WADC Report TR 53-254* (1954).

[113] P. N. Vlannes, S. W. Strauss and B. F. Brown, *NRL Report 4906* (1957).

[114] N. M. Geyer, G. W. Lawless and B. Cohen, *Hydrogen Embrittlement in Metal Finishing,* Reinhold, New York (1961) p. 109.

[115] A. H. Sully and W. A. Bell, *J. Iron Steel Inst.,* **178,** 15 (1954).

[116] J. H. Andrew et al, *J. Iron Steel Inst.,* **153,** 67 (1946).

[117] W. D. Biggs, *Brittle Fracture of Steel,* Macdonald and Evans, London (1960), p. 333.

[118] M. Smialowski, *Hydrogen in Steel,* Addison-Wesley, Reading, Mass. (1962).

[119] G. L. Hanna, A. R. Troiano and E. A. Steigerwald, *Trans. ASM,* **57,** 658 (1964).

*[120] P. P. Cotterill, *Prog. in Mat. Sci.,* **9,** No. 4, 201 (1961).

[121] B. W. Roberts and H. C. Rogers, *J. Metals,* **8,** 1213 (1956).

[122] W. M. Baldwin, Jr., *The Metal Molybdenum,* ASM, Cleveland (1957) p. 279.

[123] A. L. Eustice and O. N. Carlson, *Trans. ASM,* **53,** 501 (1961).

[124] M. L. Hill, *Hydrogen Embrittlement in Metal Finishing,* Reinhold, New York (1961), p. 46.

*[125] A. S. Tetelman, *Fracture of Solids,* Interscience, New York (1963), p. 671.

[126] W. Geller and T. Sun, *Arch. Eisenhuttenw.,* **21,** 437 (1950).

[127] M. L. Hill and E. W. Johnson, *Trans. AIME,* **215,** 717 (1959).

[128] D. P. Smith, *Hydrogen in Metals,* Univ. Chicago Press, Chicago (1948).

[129] A. S. Tetelman and W. D. Robertson, *Trans. AIME,* **224,** 775 (1962).

[130] A. S. Tetelman, Ph.D. thesis, Yale University (1961).

[131] A. S. Tetelman, C. N. J. Wagner and W. D. Robertson, *Acta Met.,* **9,** 205 (1961).

[132] C. Zapffe and C. Sims, *Trans. AIME,* **145,** 225 (1941).

[133] F. J. de Kazinsky, *J. Iron Steel Inst.,* **177,** 85 (1954).

*[134] K. Farrell and A. G. Quarrell, *J. Iron Steel Inst.,* **202,** 1002 (1964).

[135] M. Gell and W. D. Robertson, to be published.

[136] D. DeSante and A. S. Tetelman, to be published.

[137] A. S. Tetelman and T. L. Johnston, *Phil. Mag.,* **11,** 389 (1965).

[138] A. S. Tetelman and W. D. Robertson, *Acta Met.,* **11,** 415 (1963).

[139] F. Garafolo, Y. Chow and V. Ambegaokar, *Acta Met.,* **8,** 504 (1960).

[140] B. A. Bilby and J. Hewitt, *Acta Met.,* **10,** 587 (1962).

[141] T. Toh and W. M. Baldwin, Jr., *Stress Corrosion Cracking and Embrittlement,* Wiley, New York (1956), p. 176.

[142] J. O. Morlett, H. Johnson and A. Troiano, *J. Iron Steel Inst.,* **189,** 37 (1958).

[143] H. C. Rogers, *Trans. AIME,* **215,** 666 (1959).

*[144] A. Troiano, *Trans. ASM,* **52,** 54 (1960).

*[145] A. R. Elsea and E. E. Fletcher, *DMIC Report 196* (1964).

[146] N. J. Petch, *Phil. Mag.,* **1,** 331 (1956).

[147] P. Bastien, *Physical Metallurgy of Stress Corrosion Fracture,* Interscience, New York (1959) p. 311.

[148] R. G. Baker and F. Watkinson, *Hydrogen in Steel, BISRA Report 73* (1962), p. 123.

[149] H. H. Johnson, J. G. Morlet and A. Troiano, *Trans. AIME,* **212,** 528 (1958).

[150] E. A. Steigerwald, F. W. Schaller and A. Troiano, *Trans. AIME,* **218,** 832 (1960).

[151] T. S. Liu and M. A. Steinberg, *Trans. ASM,* **50,** 455 (1958).

[152] M. R. Louthan, Jr., *Trans. AIME,* **227,** 1166 (1963).

[153] M. R. Louthan, Jr., *Trans. ASM,* **57,** 1004 (1964).

[154] C. J. Beevers, *Trans. AIME,* **233,** 780 (1965).

[155] F. Forscher, *Trans. AIME,* **206,** 536 (1956).

[156] A. P. Young and C. M. Schwartz, *Trans. AIME,* **212,** 309 (1958).

[157] D. G. Westlake, *Trans. ASM,* **56,** 1 (1963).

[158] D. L. Douglass, *Corrosion,* **17,** 589 (1961).

[159] H. H. Klepfer and D. L. Douglass, ASTM, Philadelphia, STP 368 (1964), p. 118.

[160] G. T. Muhlenkamp and A. D. Schwarpe, U.S. A.E.C. Report BMI 845.

[161] O. Z. Rylski, *Dept. Mines and Tech. Surveys,* Canada P. M. 203 (CAN).

[162] C. M. Craighead, G. A. Lenning and R. I. Jafee, *J. Metals,* **8,** 923 (1956).

[163] O. J. Huber et al, *J. Metals,* **9,** 918 (1957).

[164] H. M. Burte, *WADC Report TR-54-616* (1954).

[165] D. N. Williams, *Battelle Mem. Inst., Titanium Lab Report* 100 (1958).

[166] R. I. Jaffe, *J. Metals,* **3,** 247 (1955).

*[167] M. S. Wechsler, *Interaction of Radiation with Solids,* North Holland, Amsterdam (1964).

*[168] W. S. Pellini, L. E. Steele and J. R. Hawthorne, *Welding J.* (October 1962).

*[169] R. W. Nichols and D. R. Harries, *Rad Effects on Metals and Neutron Dosimetry,* ASTM, Philadelphia, STP 341 (1963) p. 162.

[170] J. G. Y. Chow, S. B. McRickard and D. H. Gurinsky, *Rad Effects on Metals and Neutron Dosimetry,* source cited in [169], p. 46.

[171] A. T. Churchman, I. Mogford and A. H. Cottrell, *Phil. Mag.,* **2,** 1271 (1957).

[172] R. G. Berggren and J. C. Wilson, *ORNL Report CF-56-11-1* (1957).

[173] J. C. Wilson and R. G. Berggren, *Proc. ASTM,* **55,** 689 (1955).

[174] D. R. Harries, *J. Iron Steel Inst.,* **194,** 289 (1960).

[175] I. L. Mogford and D. Hull, *J. Iron Steel Inst.,* **201,** 55 (1963).

[176] S. H. Bush and J. C. Tobin, *ASTM Symposium on Stainless Steels,* Atlantic City (1963).

[177] A. D. Whapham and M. J. Makin, *Phil. Mag.,* **5,** 237 (1960).

[178] D. Hull, *Acta Met.,* **9,** 191 (1961).

[179] J. D. Campbell and J. Harding, *Response of Metals to High Velocity Deformation,* Interscience, New York (1961), p. 51.

[180] D. Hull and I. L. Mogford, *Phil. Mag.,* **3,** 1213 (1958).

[181] A. A. Johnson, *Phil. Mag.,* **5,** 413 (1960).

*[182] L. E. Steele and J. R. Hawthorne, *NRL Report 5984* (1963).

[183] A. H. Cottrell, *Steels for Reactor Pressure Circuits, BISRA Report 69* (1961), p. 281.

[184] F. A. Brandt and A. J. Alexander, *Rad Effects on Metals and Neutron Dosimetry,* source cited [169], p. 212.

[185] N. J. Petch, in *Fracture,* B. L. Averbach et al. eds., M.I.T., Wiley, New York (1959) p. 54.

[186] I. L. Mogford, A. T. Churchman and D. Hull, *AERE (Harwell) Report M/R, 2485* (1958).

[187] L. P. Trudeau, *Steels for Reactor Pressure Circuit,* source cited [183], p. 382.

[188] R. G. Berggren, *Steels for Reactor Pressure Circuits,* source cited [183], p. 370.

[189] A. M. Sullivan, *Int. Conf. on Fracture,* Sendai, Japan (1965) E, p. 95.

[190] *Nuclear Metallurgy,* AIME, **6** (1959).

[191] D. E. Rimmer and A. H. Cottrell, *Phil. Mag.,* **2,** 1345 (1957).

[192] J. A. Brinkman, *Nuclear Metallurgy,* AIME, **6,** 1 (1959).

[193] A. T. Churchman, R. S. Barnes and A. H. Cottrell, *AERE (Harwell) Report M/R 2510* (1958).

[194] A. T. Churchman, R. S. Barnes and A. H. Cottrell, *J. Nuc. Eng.,* **2,** 88 (1958).

[195] R. S. Barnes, *Nuclear Metallurgy,* AIME, **6,** 21 (1959).

[196] A. T. Churchman, *Nuclear Metallurgy*, AIME, **6**, 13 (1959).

[197] R. S. Barnes et al, U. N. Conf. on Peaceful Uses of Atomic Energy A/Conf. 15/(1958), p. 81.

[198] J. A. L. Robertson et al, *Nuclear Metallurgy*, AIME, **6**, 45 (1959).

[199] *Flow and Fracture of Materials in Nuclear Environments*, ASTM, Philadelphia, STP 380 (1966).

[200] P. J. Barton, D. R. Harries and I. L. Mogford, *J. Iron Steel Inst.*, **203**, 507 (1965).

[201] R. G. Berggren and M. Wechsler, private communication.

[202] J. C. Tobin, M. S. Wechsler and A. D. Rosin, *Proc. Third U.N. Conf. on Peaceful Uses of Atomic Energy*, Paper **242** (1962).

[203] G. H. Broomfield, D. R. Harries and A. C. Roberts, *J. Iron Steel Inst.*, **203**, 502 (1965).

[204] P. W. Davies and R. Dutton, *Acta Met.*, **14**, 1138 (1966).

Problems

1. Calculate the best average value of the activation energy H (in kcal per mole) in the stress-rupture parameter $t_R\ e^{-H/RT}$, and the best average value of the constant C_1 in the Larson-Miller parameter, if given the following information on Timken 25-20 stainless steel:

$$(H = 80 \text{ kcal/mole } C_1 = 13.5)$$

Stress, psi	Temperature °F	Rupture Time, hr
10,000	1800	0.6
	1600	4.0
	1500	25
	1400	250
	1300	1950
5,000	1800	13
	1600	120
	1500	800
	1400	6000 (extrapolated)

2. Calculate the values of T_A and $\log t_A$ in the Manson-Haferd parameter from the information given in Problem 1 above.

$$(T_A = 550°K; \log t_A = 9.5)$$

3. Suppose that high temperature rupture occurs when significant pore growth takes place in the vicinity of a notch. $\sigma_F = 10$ ksi when pores 10^{-4} in. exist near a notch, whose length is 2.0 in., which lies in a thin plate. What is the maximum notch depth that can exist without causing unstable fracture initiation at 10 ksi if the initial pore size is raised to 3×10^{-4} in. Assume that $V^*(c)$ is unchanged. (.67 in.)

4. An investigation was made of the rate of crack growth in cold-rolled brass exposed to ammonium sulfate under a static stress σ. Some of the results are as follows:

Rate	(dc/dt) in. per year	Nominal Stress, psi	c, in.
0.01		500	2×10^{-2}
0.02		500	4×10^{-2}
0.04		1000	2×10^{-2}

G_{Ic} for unstable fracture is 300 in. lb per in.^{2} in this environment and $E = 12 \times 10^6$ psi. It is proposed to use this brass in a piping system in an ammonium sulfate plant. Previous experience has shown that the pipes must sustain a tensile stress of 10,000 psi and that machine marks 10^{-3} in. deep are formed in the pipes and collars. How long would you estimate that a pipe would last without fracturing in a catastrophic manner, once the sulfate started to flow through it?

(4.71 × 10^{-2} years)

5. The tensile flow stress of a single crystal of a certain alloy is 20 ksi at $-30°C$. The fracture stress σ_f, measured at the transition grain size of 0.02 in., is 30 ksi. What is the transition grain size of this alloy after being exposed to a certain liquid which lowers its surface energy by 33.3%? (5 × 10^{-3} in.)

6. The elastic strain energy U, stored in each half of the split cantilever beam shown in the accompanying figure, is given by $U = 2P^2l^3/E\omega t^3$

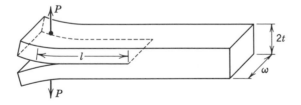

(a) In terms of the surface energy γ, compute the force P required to propagate an atomically sharp crack of length l.

(b) How will an increase of tip radius due to blunting affect force for propagation?

7. The potential energy U between two atoms can be expressed as a function of separation distance r in terms of the Morse potential function

$$U = L(e^{-2a(r-r_0)} - 2e^{-a(r-r_0)})$$

where L and a are constants, and r_0 is the equilibrium separation. Derive an expression for the maximum stress required for separation in terms of the initial modulus and L.

8. Suppose that a crack of initial length 0.0215 cm containing hydrogen gas spreads isothermally to a length 0.032 cm, at which point the pressure is insufficient to maintain elastic propagation. What is the ratio of the work done in starting the crack to the work done in spreading it under isothermal conditions, assuming that a Griffith-type relation can be used to describe the starting of the crack. HINT: the volume of a crack is proportional to $(h \cdot c)$.   (2.22)

Fracture of Specific. Materials

10

Fracture of Steel

There are literally thousands of steels (i.e., iron-base alloys) available today, each one characterized by a particular trade name or alloy composition. Although a quantitative value of fracture toughness parameters (e.g., NDT temperature and K_{Ic}) for each grade would greatly facilitate the selection of a material for a particular application, these parameters are available for only a very few of the steels (or other materials as well). There are primarily two reasons for this. First, because a wide range of microstructures can be obtained in a steel of given alloy composition, simply by variations in thermomechanical treatment. Secondly, because the concentration of fabrication defects (i.e., blow holes, inclusions, and so on) is extremely sensitive to mill practice and can vary between heats of steel of the same composition or even in different parts of the same billet. Since it is microstructure and defect concentration that primarily determine toughness, rather than composition per se, a large variation in toughness can be produced in a given steel simply by varying the thermomechanical treatment and fabrication practice.

A detailed understanding of the fracture of steel therefore requires an understanding of both the physical metallurgical aspects of the material (e.g., what microstructure will result from a given heat treatment) as well as an understanding of how this particular microstructure affects the toughness of a structure of given geometry. The physical metallurgy of steel is well documented in several excellent texts [1, 2, 3, 4] and need not be reviewed here. It is assumed that the reader is familiar with the elementary principles of heat treatment as discussed in an introductory course in metallurgy or engineering materials.

The general effects of microstructural parameters on strength and toughness were discussed in detail in Chapters 6 and 7. Briefly, it was shown that when *cleavage* is the primary mode of unstable fracture, the tensile ductility ϵ_f (Chapter 6) and hence the notch toughness

(Chapter 7) are related to the *difference* between the tensile stress level σ_f at which unstable microcrack propagation can occur and the tensile yield stress σ_Y, according to the (approximate) relation

$$\epsilon_f = \frac{\sigma_f - \sigma_Y}{d\sigma/d\epsilon} = \frac{\sigma_Y}{(d\sigma/d\epsilon)}\left(\frac{\sigma_f}{\sigma_Y} - 1\right) \tag{10.1}$$

The yield stress is given by

$$\sigma_Y = \sigma_i + k_y d^{-\frac{1}{2}} \tag{10.2}$$

When the first microcrack that forms is able to spread unstably,

$$\sigma_f = \frac{4G\gamma_m}{k_y} d^{-\frac{1}{2}} \qquad \text{(clean material)} \tag{10.3}$$

or

$$\sigma_f = \left[\frac{2E\gamma_B}{\pi\alpha(1 - \nu^2)}\right]^{\frac{1}{2}} d^{-\frac{1}{2}} \qquad \text{(dirty material)} \tag{10.4}$$

Although expressions for σ_f are not available for the case of cleavage (in dirty materials) initiated at the tip of a stopped microcrack or a group of them linked together by tearing, similar variations with these microstructural variables are expected (Section 6.3, p. 273).

The individual effects of microstructure on toughness may thus be summarized as shown in Table 10.1. At the present time there is only a small amount of quantitative information about these individual relationships. For example, as discussed in Section 9.4, a 1 ksi (1000 psi) increase in $\Delta\sigma_i$ will cause the NDT to increase by about 2°C (Fig. 9.5), for those cases where σ_f is temperature-independent. Similarly it has been shown [5, 6] (Fig. 10.1a) that the NDT will decrease by 2.3°C per increase in $d^{-\frac{1}{2}}$, when d is measured in inches. This amounts to about a 14°C decrease in NDT per increase in ASTM grain-size number [7] (Fig. 10.1b).

It must be borne in mind that these relationships only apply when *individual parameters alone* are being varied. For example, if grain size is varied by changing the alloy content, then σ_i, γ_m, etc., will be changed as well as d, and the *net change* in NDT will be different from that measured when grain size alone is varied (e.g., by varying the time of recrystallization of ferrite at a constant temperature below the A_1 temperature). Similarly, when unstable fracture occurs by *low-energy tear* there are four primary factors (Section 7.3) that influence the fracture toughness. These are also listed in the table. At present there are no quantitative relationships that can be used to

Table 10.1
Effect of Microstructural Changes on Toughness

(a) CLEAVAGE FRACTURE

Parameter	Tensile Ductility ϵ_f	Notch Toughness G_{Ic}	NDT	Propagation Transition, $T_{B(N)}$
(1) $d\uparrow$	$\rightarrow$ $\downarrow$	$\downarrow$	$\uparrow(2.3°C/\Delta d^{-½})$ (in.)$^{-½}$	$\uparrow$
(2) $\sigma_i \uparrow$	$\rightarrow$ $\downarrow$	$\downarrow$	$\uparrow(2°C/ksi)$	$\uparrow$
(3) $\gamma_m \downarrow$	$\rightarrow$ $\downarrow$	$\downarrow$	$\uparrow$	$\uparrow$
(4) $\gamma_B \downarrow$	$\rightarrow$ $\downarrow$	$\downarrow$	$\uparrow$	$\uparrow$
(5) $k_y \uparrow$	$\rightarrow$ $\downarrow$	$\downarrow$	$\uparrow$	$\uparrow$
(6) $d\sigma/d\epsilon \uparrow$	$\rightarrow$ $\downarrow$	$\downarrow$	$\uparrow$	$\rightarrow$
(7) $\rho \downarrow$	$\rightarrow$ $\downarrow(\epsilon'_f)$	$\downarrow$	—	$\uparrow$

(effective root radius of advancing tear fracture) determined by shape and content of dispersed second-phase particles

(b) LOW-ENERGY TEAR FRACTURE

Parameter	Tensile Ductility, ϵ'_f	G_{Ic}	DWTT Energy	$C_V(max)$
(8) Volume fraction or size of second-phase particles $V_f \uparrow$	$\rightarrow$ $\downarrow$	$\downarrow$	$\downarrow$	$\downarrow$
(9) $\sigma_Y \uparrow$	$\rightarrow$ $\downarrow$	$\downarrow$	$\downarrow$	$\downarrow$
(10) $d\sigma/d\epsilon \downarrow$	$\rightarrow$ $\downarrow$	$\downarrow$	$\downarrow$	$\downarrow$
(11a) Fiber content perpendicular to notch plane or parallel to tensile axis $\uparrow$	$\rightarrow$ $\uparrow$	$\uparrow$	$\uparrow$	$\uparrow$
(11b) Fiber content parallel to notch plane or perpendicular to tensile axis $\uparrow$	$\rightarrow$ $\downarrow$	$\downarrow$	$\downarrow$	$\downarrow$

$\downarrow$ indicates decrease.
$\uparrow$ indicates increase.

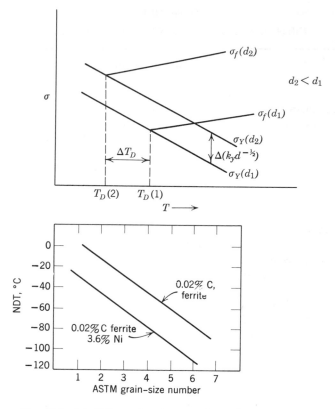

Fig. 10.1. (*a*) Effect of a decrease in grain size, from $2d_1$ to $2d_2$, on σ_f, σ_Y, and T_D for a low carbon steel. (*b*) Effect of grain size on the NDT for ferrite and nickel ferrite. *After J. Hodge et al* [7].

treat these types of fractures and discussion is limited to qualitative predictions of behavior.

Most of the variations of both cleavage and low-energy tear fracture toughnesses, because of variations in alloy content, heat treatment, and so on, can be explained in terms of these 11 parameters. To facilitate discussion it is necessary to assign the many types of steels to certain groups that can be considered as a unit. The most convenient grouping is in terms of microstructure and this will be used here. In Section 10.1 we begin with a discussion of ferritic-pearlitic steels. In Sections 10.2 and 10.3 we consider the bainitic and martensitic steels, and in Section 10.4 we discuss some of the fracture problems associated with the use of austenitic stainless steels. Finally, in Section 10.5 we consider briefly some of the fracture problems that are associated with welding.

10.1   The Fracture of Ferritic-Pearlitic Steels

Ferritic-pearlitic steels account for most of the steel tonnage produced today. They are iron-carbon alloys that generally contain 0.05–0.20% carbon† and a few per cent of other alloying elements that are added to increase yield strength and toughness. In these steels the microstructure consists of BCC iron (*ferrite*), containing about 0.01% carbon and soluble alloying elements, and Fe_3C (*cementite*). In very low carbon steels the cementite particles (carbides) lie in the ferrite grain boundaries and grains, but when the carbon content is greater than about 0.02%, most of the Fe_3C forms a lamellar structure with some of the ferrite. This lamellar structure is called *pearlite* and it tends to exist as "grains" or nodules, dispersed in the ferrite matrix (Fig. 10.2). In low carbon (0.10–0.20%) steel (i.e., mild steel) the pearlite accounts for between 10–25% of the microstructure.

For carbon contents less than about 0.2%, σ_Y is independent of the amount of pearlite in alloys having the same ferrite grain size [6].

† High carbon pearlitic steels will be discussed separately at the end of the section.

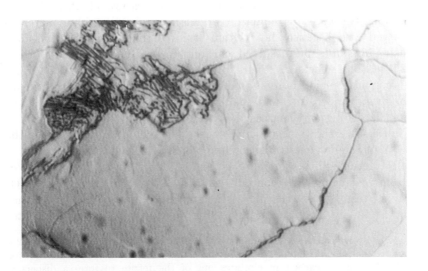

Fig. 10.2. Typical microstructure of a low (0.1–0.2%) carbon steel. 1600×. *Courtesy R. Wullaert.*

Although the pearlite grains are very hard, they are so widely dispersed that the ferrite matrix can deform around them with little difficulty. It should be noted, however, that the ferrite grain size generally decreases with increasing pearlite content [8] because the formation of pearlite nodules during the transformation interferes with ferrite grain growth. Consequently the pearlite can indirectly raise σ_Y by raising $d^{-\frac{1}{2}}$.

After plastic flow has begun and the cross-sectional area of the tensile specimen is reduced, the pearlite nodules are closer together and they can then exert a significant plastic constraint upon further deformation of the ferrite. This increases the strain-hardening rate [8]; thus the UTS of low-carbon steel is increased by pearlite additions to a much greater extent than the yield strength [6, 8].

The effect of carbon content

From the point of view of fracture analysis, two ranges of carbon content are of most interest in the low carbon steels: (1) steels containing less than 0.03% carbon where the presence of pearlite nodules has little effect on toughness, and (2) steels containing higher carbon contents where the pearlite does have a direct effect on toughness and the shape of the Charpy curve.

Figure 10.3 indicates that carbon additions increase the yield stress (0.05% proof stress) of ferrite [9]. This occurs because carbon lowers the austenite-ferrite transformation temperature [1, 2, 3, 4] and consequently causes a reduction in ferrite grain size—this increases $d^{-\frac{1}{2}}$—and also because the friction stress σ_i increases [10] with increasing amounts of dissolved carbon, up to the solubility limit of 0.02%. Both effects are more pronounced as the cooling rate from the austenitic region is increased because the carbon then precipitates at a lower temperature and the grain refining action and the precipitation hardening† are increased.

Figure 10.4 shows that small carbon additions also increase the 50% V notch impact-transition temperature of high-purity iron. A small fraction of this increase is the result of slight increases in σ_i with increasing carbon content up to 0.02%. The principal effect, however, results from the increasing size and number of carbides that form at ferrite grain boundaries, for a given rate of cooling after

---

† 0.02% is the solubility limit at the transformation temperature and 0.002% carbon is the solubility limit at room temperature. During the transformation 0.02–0.002 = 0.018% carbon precipitates out of the ferrite to form a dispersion of Fe_3C which strengthens the ferrite by raising σ_i. Thus all carbon up to 0.02% increases σ_i.

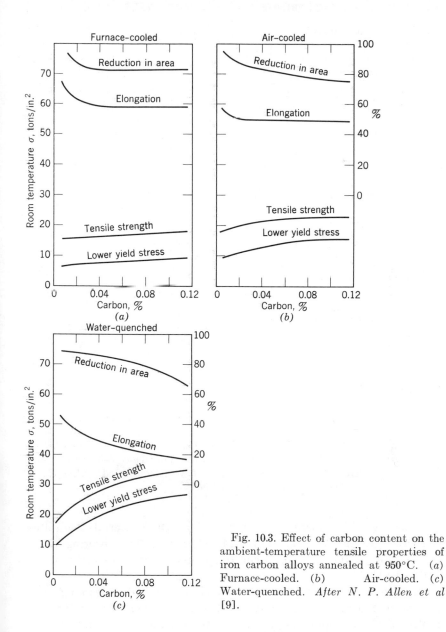

Fig. 10.3. Effect of carbon content on the ambient-temperature tensile properties of iron carbon alloys annealed at 950°C. (a) Furnace-cooled. (b) Air-cooled. (c) Water-quenched. *After N. P. Allen et al* [9].

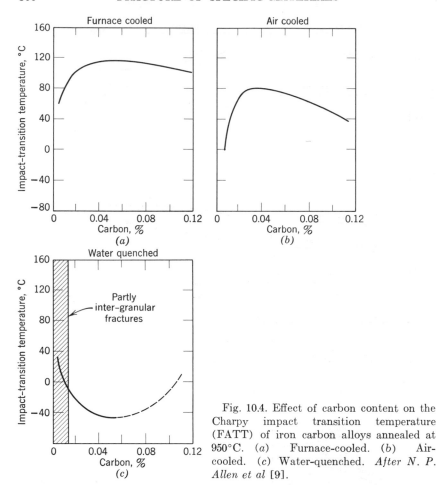

Fig. 10.4. Effect of carbon content on the Charpy impact transition temperature (FATT) of iron carbon alloys annealed at 950°C. (a) Furnace-cooled. (b) Air-cooled. (c) Water-quenched. *After N. P. Allen et al* [9].

austenitizing. The increased "dirtiness" of the material facilitates microcrack formation [11, 12, 13, 14] (Fig. 6.17) by lowering γ_m so that ϵ_f (Fig. 6.21) and the other toughness parameters are reduced (Table 10.1).

The degree of embrittlement is strongly dependent on cooling rate. In furnace-cooled steels essentially all of the carbon has time to precipitate at the boundaries upon cooling so that the transition temperature increases continuously with increasing carbon content, γ_m being the only parameter that is changed to any degree. Carbides do not form in air-cooled steels containing less than 0.015% carbon [9]. The transition temperature is then unaffected by carbon additions up to this value since increases in σ_i are balanced by an increasingly finer grain size. Grain boundary cementite is able to form in the air-cooled steels at carbon contents greater than 0.015%, and the transition temperature rises with increasing carbon content above this value. The finer grain size and carbide size keep the transition temperature of the air-cooled steels somewhat below that of the furnace-cooled steels for any given carbon content. Increasing carbon contents above 0.03% up to 0.12% produce no further increase in transition temperature and may actually cause the transition temperature to decrease slightly, if the grain size is refined. In this range of carbon contents the amount of grain boundary carbide remains constant or decreases, most of the carbide existing in the form of pearlite nodules which do not especially facilitate cleavage crack growth in primary ferrite.

Figure 10.5 shows the effect of carbon (or pearlite) content on the Charpy V notch impact strength of normalized iron-carbon alloys. Increasing carbon (pearlite) content causes the slopes of the Charpy curves to be decreased [8, 15], $T_{B(N)}$ to be raised, and $C_V(\text{max})$ to be lowered, so that the 50% transition temperature is increased. In low carbon steels this results from the increased rate of strain hardening due to pearlite [16] (No 6, Table 10.1) and the decreasing radius ρ of an advancing fibrous crack (No 7, Table 10.1) in materials containing larger volume fractions of hard, second phase.†

† This increases the degree of stress intensification ahead of the advancing tear (Fig. 7.12) and facilitates cleavage.

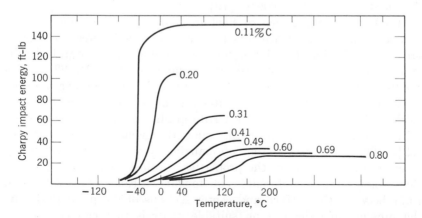

Fig. 10.5. Effect of carbon content on the Charpy impact energy curves for normalized iron carbon alloys. *After K. Burns and F. Pickering* [8].

Grain boundary carbides are not formed in pure iron-carbon alloys quenched into water from 950°C unless the carbon content is greater than about 0.05% [9]. Since fast cooling rates and higher carbon contents lead to increasingly finer grain sizes, σ_Y increases (Fig. 10.3) and the transition temperature decreases (Fig. 10.4) as the carbon content increases. However, at higher (0.05%) carbon contents, even the fast cooling rate cannot suppress grain boundary carbide formation and the 50% transition temperature begins to rise as the carbon content is increased above this value.

Intergranular fractures are observed in quenched samples containing very low carbon contents [17, 18, 37], as shown in Fig. 10.4. The embrittlement appears to be related to the presence of oxygen in ferrite grain boundaries that do not also contain a minimum concentration of carbon. The role of the carbon in preventing intergranular fracture is unclear. It may be that the carbon segregates preferentially at the boundaries in place of oxygen, thereby preventing the oxygen embrittlement [18].

The effect of processing variables

It has been pointed out (Fig. 10.4) that the impact properties of water-quenched steels are superior to those of annealed or normalized steels because the fast cooling rate prevents the formation of grain boundary cementite and causes a refinement of ferrite grain size. The grain size of normalized steels can also be refined by lowering the normalizing temperature and Fig. 10.6 shows that the transition temperature can be lowered by 70°C if the normalizing temperature is decreased from 1200° to 900°C [19].

Many commercial grades of steel are sold in the "hot-rolled" condition and the rolling treatments have a considerable effect on impact properties. Rolling to a lower finishing temperature (controlled rolling) lowers the impact-transition temperature [20, 21, 22]. This results from the increased cooling rate and corresponding reduced ferrite grain size. Since thick plates cool more slowly than thin ones, thick plates will have a larger ferrite grain size and hence are more brittle than thin ones after the same thermomechanical treatment [23]. Therefore, post rolling normalizing treatments are frequently given in order to improve the properties of rolled plate [24]. In very thick plates, large differences in toughness can even exist (Fig. 10.7) between the surface and center sections of the plate [25]. At the present time there is no suitable analysis that can be used to develop fracture-safe design criteria for "composite" structures such as these.

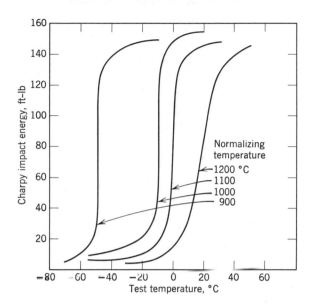

Fig. 10.6. Effect of normalizing temperature on the Charpy impact energy curves for a low carbon steel. *After F. Pickering and T. Gladman* [6].

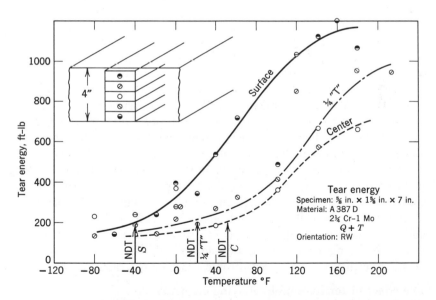

Fig. 10.7. Variations in toughness across the thickness of a 4 in. thick plate of A387D steel, as determined by making subsize DWTT on material cut from the indicated locations of the plate. *After W. Pellini et al* [25].

503

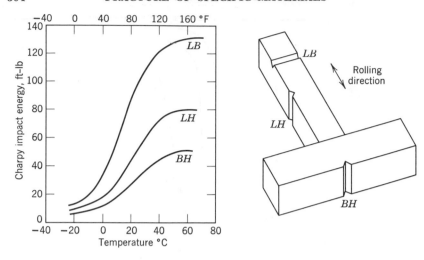

Fig. 10.8. Effect of directionality on the Charpy impact energy curves for mild steel. *After P. Puzak et al* [26].

Hot rolling also produces an anisotropic or directional toughness [22, 23, 24, 26, 27] owing to combinations of texturing, pearlite banding, and the alignment of inclusions and grain boundaries in the rolling direction. Texturing is not considered to be important in most low carbon steels [28]. Pearlite bands (due to phosphorous segregation during casting) and elongated inclusions are dispersed on too coarse a scale to have an appreciable effect on notch toughness at the low temperature (15 ft-lb) end of the Charpy transition temperature range [22, 29]. The small differences that do arise there (Fig. 10.8) appear to result from finely dispersed inclusions which lead to splitting (Fig. 7.24) and hence produce a reduction in triaxiality of stress ahead of the notch [22]. When fracture occurs by fibrous tearing, less energy is required for propagation through pearlite bands or along inclusion-matrix interfaces. Similarly transverse splitting on a coarser scale can interfere with the slowly propagating fibrous crack. Consequently there is a marked effect of directionality on $C_v(\text{max})$ and hence on the Charpy slope (Fig. 10.8), with transverse specimens (BH) absorbing less energy than longitudinal ones (LH or LB). The tensile ductility ϵ'_f is also anisotropic [23] for the same reasons.

The effect of ferrite-soluble alloying elements

Most alloying elements that are added to low carbon steel produce some solid solution hardening at ambient temperature and thereby raise the lattice friction stress σ_i. At the same time, they may cause the austenite-ferrite transformation to occur at lower temperature, which produces a refinement in ferrite grain size. Both of these effects raise the yield stress according to Eq. 10.2.

An extensive investigation based on quantitative metallography techniques and a multiple regression analysis has recently allowed these effects to be separated. For low carbon steels (0.02–0.2% carbon) it has been shown [6] that at ambient temperature the tensile yield stress is given (in units of ksi) by the relation

$$\sigma_Y = 15.1 + 0.504(2d)^{-\frac{1}{2}} + \sum_{i=1}^{N} \alpha_i \qquad (10.5)$$

where $2d$ is the ferrite grain size in inches and α_i is the strength increase in ksi per wt per cent of alloy addition *dissolved in the ferrite;* no quantitative account of precipitation-hardening effects has been reported at this time. The summation is taken over all N alloying elements that are dissolved in the ferrite. Table 10.2 lists the values of α_i for various alloying additions. These values only apply at ambient temperature, and at the present time no values of α_i have been reported for other test temperatures. There are strong indications [96], however, that ferrite-soluble alloying elements such as nickel and manganese lower the temperature dependence of the yield stress of iron (i.e., α_i decreases with decreasing temperature) and, in some cases [143, 144, 145] can even lower σ_Y at very low temperatures (i.e., α becomes negative).

Table 10.2

Effect of Alloying Additions on Ambient Temperature Flow Stress of Ferrite [6]

Element	$\Delta\sigma_i$ (1000 psi) per 1 wt % addition (measured at 18°C)
C, N (up to 0.02%)	805
P	99
Sn	18
Si	12.2
Cu	5.6
Mn	4.7
Mo	1.5
Ni	0
Nb(up to 0.06%, as rolled steels)	224

It is important to appreciate that this equation *cannot* be used to predict the lower yield stress *unless* the resultant grain size is known. This, of course, depends on factors such as normalizing temperature and cooling rate. The importance of this type of approach is that it allows prediction of the extent that individual alloying elements will decrease toughness by increasing σ_i, since NDT increases by about 2°C per ksi increase in σ_i (Table 10.1).

In addition to changing σ_i, alloying additions can affect the various impact-transition temperatures by affecting $d^{-\frac{1}{2}}$ and the other parameters listed in Table 10.1. These changes are much more difficult to evaluate because of the large number of effects that can be produced by a given alloy addition, some of which cancel out in the regression analysis, others of which are not single-valued (i.e., they depend on the concentration of other alloying additions, precipitate size, and so on). For silicon or aluminum-killed steels containing approximately 0.1% carbon and about 0.008% nitrogen, the 50% fibrous transition temperature in a Charpy V notch impact test is given by the relation [6]

$$T_{N(50\%)}°C = 63 + 44.1(\% \text{ Si}) + 2.2(\% \text{ pearlite}) - 258(\% \text{ Al})$$
$$- 2.3(2d)^{-\frac{1}{2}} \quad (10.6)$$

Regression analyses for NDT temperatures or other Charpy transition temperatures have not been reported at this time and it is only possible to discuss the effects of the individual alloying additions on a qualitative basis.

Manganese

Most commercial steels contain about 0.5% manganese to serve as a deoxidizer and to tie up sulfur as manganese sulfide, thereby preventing the occurrence of hot-cracking [24]. Table 10.2 indicates that manganese additions produce a small increase in σ_i and hence a small tendency towards brittleness. In low carbon steels this effect is outweighed by the ability of manganese: (1) to decrease the tendency for the formation of films of grain boundary cementite in air-cooled or furnace-cooled specimens containing 0.05% carbon [9], thereby lowering the value of γ_m; (2) to cause a slight reduction in ferrite grain size; (3) to produce a much finer pearlite structure [9, 30]. The first two of these effects account for the lowering of the NDT temperature (approximately 20 ft-lb, (Fig. 10.9) with increasing Mn additions; the third effect as well as the first cause the Charpy curves to become sharper, as shown on the figure. In steels containing

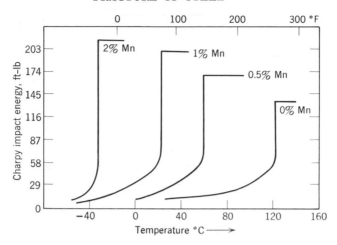

Fig. 10.9. Effect of manganese on the Charpy impact energy curves of furnace-cooled iron–0.05% carbon alloys. *After N. P. Allen et al* [9].

higher carbon contents the effect of manganese on the 50% transition temperature is less pronounced (Fig. 10.10), probably because the amount of pearlite rather than the distribution of grain boundary cementite is the most important factor in determining this transition temperature when the pearlite content is high. It should also be noted that if the carbon content is relatively high (greater than 0.15%) a high manganese content may have a detrimental effect on the impact properties of normalized steels [24] because the high hardenability of the steel causes the austenite to transform to the brittle upper bainite structure rather than ferrite or pearlite.

Nickel

Nickel, like manganese, is able to improve the toughness of iron carbon alloys [7, 15, 31–34]. The magnitude of the effect is dependent on carbon content and heat treatment. In very low (about 0.02%) carbon steels, nickel additions up to 2% are able to prevent the formation of grain boundary cementite in hot-rolled and normalized alloys and cause a substantial decrease in the initiation-transition temperature $T_{s(N)}$ (Fig. 10.11), and a sharpening of the Charpy curves [34]. Further additions of nickel produce substantially smaller improvements in impact properties; these are the result of grain-size refinement and a lowering of the k_y parameter (Table 10.1) [34]. In alloys containing carbon contents lower than this,

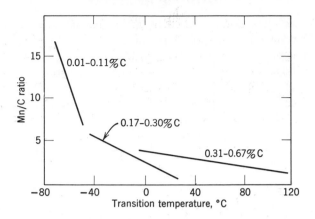

Fig. 10.10. Effect of manganese to carbon ratio on the transition temperature of ferritic steels. *After J. Rinebolt and W. Harris* [15].

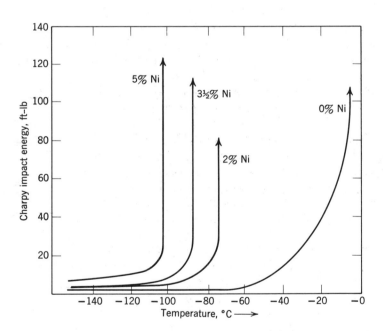

Fig. 10.11. Effect of nickel content on the Charpy impact energy curves for hot-rolled, 0.02% carbon ferrite. *After R. Wullaert and A. Tetelman* [34].

such that carbides are not present after normalizing, nickel has a smaller effect on the transition temperature (Fig. 10.11) (about 10°C/ per cent nickel) [7]; this is the result of the lowering of d and k_y. The principal beneficial effect of nickel additions to commercial steels containing about 0.1% carbon results from the substantial grain-size refinement and reduction of free nitrogen content after normalizing [34]. The reasons for this behavior are not clear at present; it may be related to the fact that nickel is an austenite stabilizer and consequently lowers the temperature at which the austenite decomposition will take place.

Phosphorous

In pure iron-phosphorus alloys, intergranular embrittlement can occur from the segregation of phosphorous at ferrite grain boundaries [32], which lowers the value of γ_m (Table 10.1). Also, phosphorus additions produce a significant increase in σ_i (Table 10.2) and a coarsening of ferrite grain size since phosphorus is a ferrite stabilizer. These effects combine to make phosphorus an extremely effective embrittling agent, even when fracture occurs transgranularly. Increases in the average energy transition of 72°C per 0.1% phosphorus additions have been reported [15].

Silicon

Silicon is added to some commercial steels to deoxidize or "kill" them, and in this respect the silicon produces beneficial effects on impact properties [15]. When manganese and aluminum are present, a large fraction of the silicon is dissolved in the ferrite and this raises σ_i by solid solution hardening (Table 10.2). This effect, coupled with the fact that silicon additions raise k_y, causes the 50% transition temperature to increase by about 44°C per wt per cent silicon in iron-carbon alloys of constant grain size [6] (Eq. 10.6). In addition, silicon, like phosphorus, is a ferrite stabilizer and hence promotes ferrite grain growth. The net effect of silicon additions in normalized alloys is to raise the average energy-transition temperature by about 60°C per wt per cent silicon added [15, 31].

Nitrogen

In iron carbon-nitrogen alloys the primary effect of nitrogen additions is to raise the value of σ_i by about 8 ksi psi per 0.01% free nitrogen added, up to the solubility limit of 0.02%. According to Table 10.1 the transition temperatures should thus rise by about 16°C per 0.01% added. Actually the observed increase is about three times as great as this [35]. This fact may be due to the strong dislocation locking produced by the nitrogen (increase in k_y) [5] and the corresponding decrease in γ_m which would also occur.† In cold-worked steels, nitrogen also produces strain aging and strain-age embrittlement. These effects will be discussed separately in the following section.

Aluminum

The effect of alloying or killing a steel with aluminum is twofold [24]: First, the aluminum combines with some of the nitrogen in solution to form AlN. The removal of this free nitrogen leads to a decrease in transition temperature because σ_i is decreased and γ_m/k_y is increased [5], as described above. Second, the AlN particles that form interfere with ferrite grain growth and consequently refine the ferrite grain size. These combined effects cause the transition temperature to decrease—about 40°C per 0.1% aluminum added [36]. However, additions of aluminum greater than that required to tie up the nitrogen have little effect.

Oxygen

Oxygen additions promote intergranular fracture in iron alloys [9, 17, 18, 37]. Figure 10.12 shows that very small additions can produce large increases in the transition temperature of high-purity iron.

† See Eqs. 5.12 and 5.13.

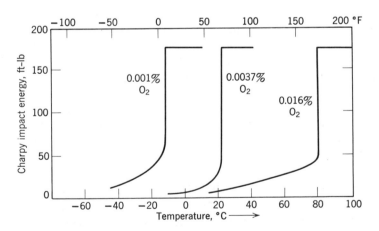

Fig. 10.12. Effect of oxygen content on the Charpy impact energy curves of ferrite. *After W. Biggs* [24].

These fractures are thought to result from the segregation of oxygen to ferrite grain boundaries. This segregation lowers the values of the true surface energy γ_s and hence of $\gamma_m = (\rho/a_0)\gamma_s$ so that the transition temperature increases. In alloys that contain a high oxygen content (greater than 0.01%), fracture occurs along the continuous path provided by the embrittled grain boundary. In alloys of lower oxygen content, cracks are nucleated at the grain boundary and then propagate transgranularly. The problem of oxygen embrittlement can be solved by the addition of deoxidizing elements such as carbon, manganese, silicon, aluminum, and zirconium, which react with the oxygen to form oxide particles, thereby removing the oxygen from the boundary region. These oxide particles are beneficial in their own right because they retard the growth of the ferrite grains, thereby increasing $d^{-1/2}$.

The effect of cold working and aging

Plastic deformation raises σ_i by strain hardening so that the impact-transition temperature is raised. The effect on the NDT is most pronounced at small prestrains [38, 39]. An increase of about 2.5°C in the 25% fibrous transition is observed [5] per 1 ksi increase in σ_i after strains up to 2%, in agreement with Table 10.1. At 10% prestrain the increase in transition temperature per ksi increase in σ_i is about half of this, probably because γ_m/k_Y is increased when the material is homogeneously deformed. Heavy cold work (80% by rolling) can introduce a large amount of directionality into a plate, and effects similar to those described in Section 10.1, the effect of processing variables, have been noted [40].

In addition to the effect of strain, subsequent aging can produce additional increases in σ_i, the magnitude of which depends on alloy content, aging time, and temperature. These changes are believed to result from dispersion-hardening effects produced by the precipitation of iron nitride from ferrite solid solution [15]. The return of the upper yield point results from the pinning of dislocations by free nitrogen atoms. The increases in σ_i lead to strain-age embrittlement [41, 42, 43], with the NDT increasing about 2°C per 1 ksi increase in σ_Y after aging, consistent with Table 10.1 [5]. Alloying elements such as manganese, aluminum, and vanadium lower the tendency towards strain-age embrittlement, the former by retarding the precipitation of nitrides, the latter two by "gettering" the nitrogen in the form of AlN and VN during normalizing or hot rolling. Silicon is also beneficial in this respect because it serves as a deoxidizer and leaves the aluminum free to getter the nitrogen.

Strain aging can be a problem in steel plates that are roll-straightened even though the strains that are introduced are only of the order of a few per cent. In parts that are fabricated, the problem can be more serious because the magnitude of the strain, and hence the amount of strain hardening, is increased. This is especially true if the parts are galvanized after fabrication [44], for this process is carried out at temperatures between 200 and 300°C where aging occurs quite rapidly. In order to remove the tendency for strain-age embrittlement, it is necessary to lower the value of σ_i. This may be accomplished by recrystallizing [43] below A_1, which removes the plastic strain and hence the strain-hardening contribution to σ_i (as well as causing the carbides and nitrides to grow) or by normalizing. This latter treatment causes the dissolution of the carbides and nitrides as well as the removal of the cold work, and the structure that results after the austenite transformation takes place will show increased ductility.

Quenching and aging treatments can also lead to embrittlement in mild steels, by a process similar to that which occurs during strain aging. Since *both* carbide and nitride precipitation can occur after quenching,† the change in σ_i and hence the increase in transition temperature can be much greater. This effect, like strain aging, is dependent upon both time and temperature since the precipitation process is diffusion controlled. Figure 10.13 shows the increase in transition temperature observed [45] after aging at room temperature in a 0.17% carbon, 0.75% manganese steel that had been quenched from 690°C.

Closely related to the problems of quench-age and strain-age embrittlement is *blue brittleness*, which is the embrittlement encountered in low carbon steels that have been strained at temperatures around 250–300°C. At these temperatures carbon and nitrogen atoms are sufficiently mobile to move along with the dislocations and exert a dragging force on them. This causes an increase in σ_i which results in an increased yield strength and a decrease in impact energy [46] (Fig. 10.14). Similarly low-impact energy values are obtained from specimens that have been strained in the blue brittle range and fractured at temperatures below it, such as room temperature. Small ($\approx 0.05\%$) additions of titanium, aluminum, and probably niobium and vanadium as well, can eliminate blue brittleness in low carbon steel [47]. These elements interact with carbon and nitrogen and

---

† Carbide precipitation usually does not occur in normalized or furnace-cooled steels since all of the carbon has already precipitated in the form of carbides.

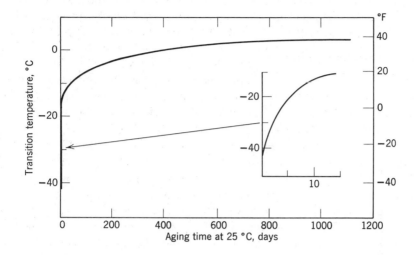

Fig. 10.13. Effect of aging after quenching from 690°C on the impact-transition temperature of a low carbon steel. The short-time behavior is shown at higher magnification. *After J. Low* [45].

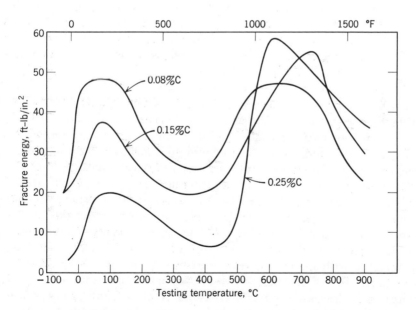

Fig. 10.14. Impact energy curves for low carbon steels in which blue brittleness occurs around 300°C. *After O. Reinhold* [46].

decrease their diffusivities, or perhaps form carbides and nitrides with them (see below), so that their interaction with the moving dislocations is decreased.

Effect of niobium additions

A recent innovation in the steel industry is the development of high strength, low alloy ferritic steels ($\sigma_Y = 50,000-60,000$ psi) which also exhibit good impact properties (50% transition below $-10°C$). Many of these steels contain small amounts of Nb (or V) which precipitates during transformation as NbC. These ferrite precipitates produce appreciable dispersion strengthening in the ferrite which raises σ_i [6, 48]. In addition, fine grain sizes are produced in conventionally rolled steels [6, 35, 49]. These result from the presence of a fine austenite grain size prior to transformation. The latter is due either to Nb in solution, which retards the recrystallization of austenite during working, or to the presence of larger, undissolved carbides, formed by precipitation during hot rolling or the coarsening of fine precipitates during normalizing, which hinder austenite grain growth [49].

The magnitude of the strengthening effects due to increases in σ_i and $d^{-\frac{1}{2}}$ varies with thermal mechanical treatment because these determine: (a) the *amount* of precipitate, by controlling the amount of carbide that dissolves during austenization and hence the amount of niobium available for fine-scale precipitation, (b) the precipitate *size*, and (c) the grain-refining effects in the austenite. In addition, the amount of pearlite present is a function of niobium content and, if carbide precipitation occurs in ferrite grain boundaries, γ_m may be affected as well. For these reasons it is difficult to make even qualitative predictions of the effect of niobium content on impact properties.

Figure 10.15a indicates that niobium additions lower the transition temperature of hot-rolled steels containing about 0.15% carbon and 0.85% manganese. For steels containing more than about 0.15% niobium, the transition temperature decreases with increasing amounts of niobium. This occurs because in this composition range the pearlite content decreases with increasing niobium content and the ferrite grain size is slightly reduced (for Nb $> 0.35\%$). Since σ_i does not change appreciably with niobium additions greater than 0.05%, the slopes of the Charpy curves are raised with increasing Nb content and the 50% transition is decreased. The decrease in NDT is small. A comparison of Figs. 10.15a and 10.15b indicates that at all compositions in this range (Nb $> 0.1\%$), normalizing after rolling decreases the transition temperature as compared to the hot-rolled steels. This is so

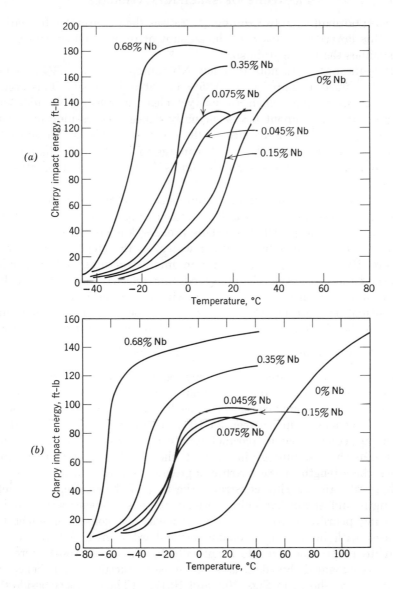

Fig. 10.15. Effect of niobium content on the Charpy impact energy curves for (a) hot-rolled and (b) normalized (950°C) low carbon steel. *After R. Phillips et al* [49].

because normalizing after rolling (a) causes the precipitates to coarsen and this lowers σ_i, (b) lowers the amount of pearlite still further, and (c) refines the ferrite grain size.

It is interesting to note that low Nb contents (0.04–0.08%) reduce $T_{D(N)}$ as much as additions of 0.35% in hot-rolled steel. This occurs because, for reasons which are not yet clear, increasing Nb additions between 0.04 and about 0.35% actually cause a coarsening of ferrite grain size. Since σ_Y is essentially the same for 0.68% $\geq$ Nb $\geq$ 0.05% there does not appear to be any advantage to adding more than 0.05% Nb to mild steel unless high toughness at very low temperature is desired.

Another method for refining grain size and reducing the pearlite content in hot-rolled steels is by rapid cooling through the transformation temperature [50, 51]. This process, known as "control cooling," is accomplished by spraying water on the steel before coiling. Figure 10.16 shows the difference in microstructures and properties between the conventionally processed (air-cooled) and control-cooled steel [35]. Further increases in σ_i, with about the usual increase in transition temperature (2°C per ksi), can occur if these alloys are aged in the vicinity of 600°C to produce NbC precipitation hardening.

The effect of carbon additions between 0.3 and 0.8%

In hypoeutectoid steels containing between 0.3 and 0.8% carbon, proeutectoid ferrite is the continuous phase and forms primarily at austenite grain boundaries. The pearlite forms inside the austenite grains and makes up between 35–100% of the microstructure. More than one colony (set of parallel ferrite and cementite plates) forms within each austenite grain so that the pearlite is polycrystalline. Since the strength of the pearlite is greater than that of the proeutectoid ferrite, the pearlite constrains the flow of the ferrite. The yield strength and strain-hardening rate of these steels increase with increasing pearlite (carbon) content because the constraint effect increases with increasing amounts of the hard aggregate and because pearlite refines the size of the proeutectoid grains. These increases are accompanied by decreases in tensile ductility and hence in $C_V(\max)$, as shown in Figs. 10.5 and 10.17. (The numbers beside the data points in Fig. 10.17 refer to the 0.2% offset yield stress values at 18°C in ksi and to the carbon contents.)

A recent investigation [52] of the tensile behavior of hypoeutectoid steels has shown that the ductility-transition temperature is determined solely by the grain size and σ_i value of the proeutectoid ferrite and that pearlite has no direct effect on the fracture behavior in this

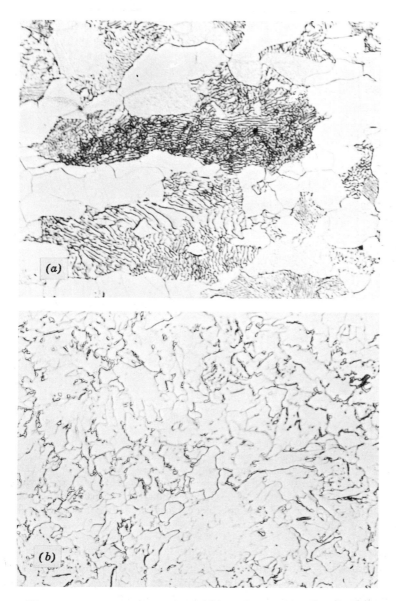

Fig. 10.16. Variation in microstructure and properties between conventionally cooled and control cooled 0.12% carbon, niobium-bearing hot-rolled steel. 1000×. *Courtesy J. Bucher and D. Grozier* [35] *and American Society for Metals.* (a) Conventionally cooled, $\sigma_Y = 49.5$ ksi, 50% transition temperature in one-half size Charpy test = +10°F. (b) Control-cooled, $\sigma_Y = 59.5$ ksi, 50% transition temperature in one-half size Charpy = −90°F.

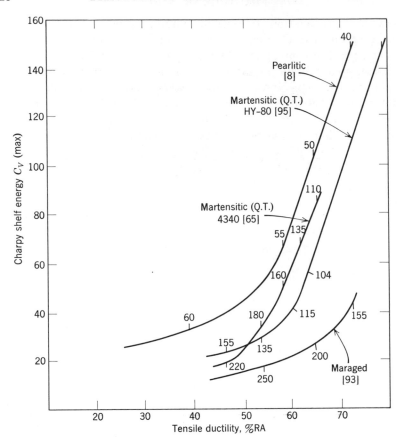

Fig. 10.17. Relation between Charpy shelf energy C_V(max) and tensile ductility for various steels. The numbers adjacent to the curves are the 0.2% offset yield strengths in ksi. These variations in strength, ductility, and tear toughness were obtained by changing the carbon content in the pearlitic steels, the tempering temperature in the martensitic steels, and the molybdenum content in the maraged steels.

range (i.e., cleavage does not initiate in the pearlite colonies). The constraint effect which increases σ_i is balanced by increased grain-size refinement so that the ductility transition temperature T_D is essentially unchanged by carbon additions from 0.2 up to 0.6%.

Above the ductility transition of the "clean" ferrite, pearlite plays a direct role in the fracture process. Microcracks form by shear in the pearlite colonies [52, 53] (Fig. 10.18) and these link up to form fibrous cracks which can develop into fast running cleavage cracks below the propagation transition. Similar processes occur in the impact specimens; $C_V(\text{max})$ decreases and $T_{B(N)}$ increases with increasing pearlite content [8, 15] for reasons stated previously.

In steels containing large-volume fractions of pearlite, deformation in the pearlite can initiate microcleavage crack formation at low temperatures and/or high strain rates. Deformation of individual colonies occurs by kinking. Microcracks can form in the colony that kinks most sharply, but are observed to extend over several colonies [54, 55]. This indicates that the effective grain size for fracture is *not* the colony size. Electron microscopy techniques have shown [56, 57] that the effective grain size is, in fact, the prior austenite grain size.

Since the fracture path is primarily along the cleavage plane in the ferrite plates (although there is some intercolony fracture), this indicates that there is some preferred orientation between ferrite plates in adjacent colonies within a prior austenite grain. However, there is also evidence that pearlite grains are able to arrest microcracks in hypoeutectoid alloys [8, 24], so that it cannot be stated unequivocally that cleavage crack propagation will always occur with little difficulty in pearlite grains. In some cases (Fig. 10.19) crack propagation can occur along the ferrite-cementite interface.

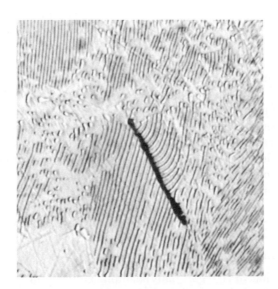

Fig. 10.18. Shear cracking of pearlite, 1800×.
Courtesy D. Hodgson [52].

Fig. 10.19. Ferrite cementite interface cracking in pearlitic steel. 1800×.
Courtesy D. Hodgson.

These results indicate that in eutectoid steels the impact-transition temperature can be lowered when the austenite grain size is reduced and when the pearlite spacing is increased; the former change being to increase (generally) the resistance to the crack propagation, the latter to prevent sharp kink formation and hence to increase the resistance to cleavage crack initiation. In hypoeutectoid steels the most effective means of lowering the transition temperature are refinement of ferrite grain size [8, 52, 54] (the dominant effect) and increased pearlite spacing.

The toughness of pearlitic steel can be increased by spherodization [58]. This removes the danger of crack initiation in the pearlite by kinking and lowers the tendency for initial growth of microcracks in the proeutectoid ferrite by lowering the constraint and hence σ_i. In addition, the spheroidal particles promote fibrous rather than cleavage cracking,† if they are dispersed on a fine enough scale (i.e., γ_m is raised) ; this also causes the NDT temperature to be lowered.

10.2 The Fracture of Bainitic Steels

The addition of 0.05% molybdenum and boron to low carbon (0.1%) steels is able to suppress the austenite-ferrite transformation,

† See Fig. 6.10a.

which normally occurs between 700° and 850°C, without affecting the kinetics of the austenite-bainite transformation which then takes place between 675° and 450°C. Bainite formed between 675° and about 525°C is called "upper bainite" and bainite formed between 525° and 450°C is called "lower bainite." Both structures consist of acicular ferrite and dispersed carbides. The tensile strength of these untempered bainites increases [59, 60] from 85,000 to 170,000 psi as the transformation temperature drops from 675° to 450°C. Since the transformation temperature is determined by the amount of alloying elements (e.g., Mn and Cr) that are present, these elements exert an indirect effect on the yield and tensile strengths. For example, the UTS of a 0.75% Mn steel (upper bainite) is 100,000 psi, while steels containing 1.5% Mn and 1.5% Cr have tensile strengths of the order of 170,000 psi. The high strengths obtained in these steels is the result of two effects: (1) the progressive refinement of the bainitic ferrite plate size as the transformation temperature is lowered, and (2) the fine carbide dispersion which occurs within the grains of the lower bainite. In upper bainite the carbides form along the bainitic and prior austenitic grain boundaries and have little effect on strength [60].

Figure 10.20 indicates that the fracture characteristics of these steels is strongly dependent on the tensile strength and hence on the transformation temperature. Two effects should be noted. First, at a given tensile strength level the impact properties of tempered lower bainite are far superior to that of untempered upper bainite. The reason for this behavior is that in upper bainite, as in pearlite, the cleavage facets traverse several bainite grains and the "effective grain size" for fracture is the prior austenite grain size rather than the ferrite grain size [56, 57, 60]. This implies alignment of cleavage planes within ferrite plates formed from a single austenite grain. In lower bainite the cleavage planes in the acicular ferrite are not aligned so that the effective grain size for quasicleavage fracture is the ferrite needle size. Since this is one to two orders of magnitude smaller than the prior austenite grain size, the transition temperature of the lower bainite is much below that of upper bainite, at the same strength level. A second feature that is important is the distribution of the carbides. In upper bainite these lie along grain boundaries and may promote brittleness by lowering γ_m as described previously in connection with furnace-cooled ferritic steels. In tempered lower bainite the carbides are more uniformly distributed in the ferrite and raise γ_m by interfering with cleavage cracks and promoting tearing [60] (quasicleavage) as in the case of spherodized pearlites just described. These differences may be observed on the fractographs shown

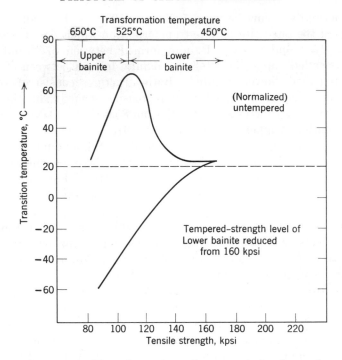

Fig. 10.20. The effect of tensile strength level on the Charpy impact-transition temperature of bainitic steels. Tensile strength was varied by varying the austenite-bainite transformation temperature or the tempering temperature for tempered lower bainite. *After K. Irvine and F. Pickering* [60].

in Fig. 10.21. Thus, at the same strength level, the relatively higher γ_m and smaller effective d in the lower bainite cause it to have a much lower transition temperature than upper bainite.

A second effect that should be noted is the variation of transition temperature with tensile strength in the untempered alloys. In the upper bainite a decrease in transformation temperature produces a refinement of ferrite needle size and this raises σ_Y. Since σ_f is unaffected by these changes (it depends primarily on the prior austenite grain size), the transition temperature is increased. Tensile strength levels of 120,000 psi or greater are obtained in lower bainite and the transition temperature decreases with increasing tensile strength. This is consistent with Table 10.1 since the increased strength is primarily obtained by a refinement of ferrite grain size, which, in the

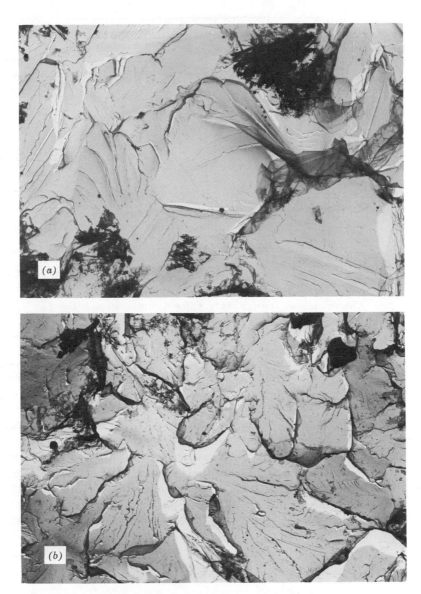

Fig. 10.21. Cleavage fracture surfaces of upper bainite (a) and lower bainite (b). 3000×. *Courtesy K. Irvine and F. Pickering* [60] *and JISI.*

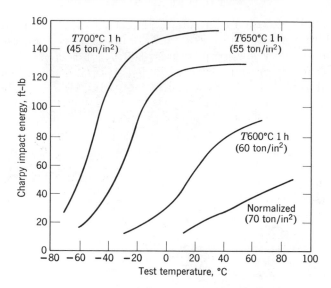

Fig. 10.22. Effect of tempering temperature T on the Charpy impact energy curves for bainitic steels. The tensile strength values are as indicated. *After K. Irvine and F. Pickering* [60].

lower bainites, controls fracture stress as well as yield and tensile stress.

Because the fracture stress of the upper bainite is dependent on austenite grain size, and since the carbide particles are already large, tempering has little effect on tensile and impact properties. In lower bainite, tempering causes particle coarsening; this lowers σ_i and hence lowers the value of the 50% transition temperature by about 2°C per 1 ksi decrease in σ_Y (Fig. 10.22). At the same time $C_V(\max)$ and the Charpy slopes are increased. This is consistent with the increase in tensile ductility and decrease in σ_Y (No. 9, Table 10.1) that occurs upon tempering.

10.3   The Fracture of Martensitic Steels

General characteristics

The addition of carbon and other alloying elements to steel retards the transformation of austenite to either ferrite and pearlite, or bainite, and if the cooling rate after austenitizing is sufficiently rapid, the austenite will transform to martensite by a shear process that requires no measurable diffusion of atoms. The features that are pertinent to the fracture of martensite are as follows: (a) Because the transformation occurs at very low temperatures (200°C or lower) the size of the tetragonal ferrite or martensite needle is very small, at least in two of its three dimensions. (b) Because the transformation occurs by shear, the carbon atoms do not have time to diffuse out of their lattice position in the austenite and hence the ferrite is supersaturated with carbon; this causes the martensite to have an elongated (body-centered tetragonal) crystal structure and leads to lattice expansion. (c) The martensitic transformation occurs over a range of temperatures because the formation of the first martensite plates increases the difficulty of transforming the remaining austenite. Thus transformation structures can be mixtures of martensite and retained austenite.

The distortion set up because of the transformation and thermal contraction produces internal or "residual" stresses which may lead to cracking in the absence of applied stress or at very low applied stresses [61]. This is particularly true if large amounts of hydrogen are present (i.e., after welding or hydrogen annealing). The magnitude and distribution of these stresses increase with increasing cooling rate and size of the plate or fabricated part [2]. In many structures the residual stresses are tensile at the surface and compressive in the interior. This is so because the outer surface cools first and hard martensite is formed there. When the interior region cools and transforms, it must expand against an already hardened layer and this sets up tensile stresses in the outer layer.

The high supersaturation of carbon (σ_i effect)† coupled with the very fine needle size ($d^{-\frac{1}{2}}$) causes untempered martensite to have a high yield strength and hardness, both of which increase with carbon content. However, the presence of residual stress, the high σ_i value, and the retained austenite which has a large effective grain size for fracture, all cause untempered martensite to be extremely brittle. Furthermore, the transformation of retained austenite to "isothermal" martensite after aging at room temperature can produce additional inhomogeneous strain in an already strained matrix, thereby increasing the chances of crack formation [62, 67].

To produce a stable steel that can be satisfactorily used in engineering applications, it is necessary to temper it. Three stages of

† At least as far as the measured σ_Y is concerned. There is evidence that the large strengthening effect of carbon additions is not due to solid solution hardening but rather to an increase in the work-hardening rate in the microstrain region [64].

tempering occur in high (greater than 0.3%) carbon martensites, tempered for about one hour in various ranges [1, 2, 3, 4].

1. At temperatures up to about 100°C some of the supersaturated carbon precipitates out of the martensite to form very fine particles of epsilon (hexagonal) carbide which are dispersed in a martensite that consequently has a decreased carbon content. This process can actually cause σ_Y to increase because of the dispersion strengthening effect of the carbides.

2. Between 100° and 300°C any retained austenite is able to transform to bainite and epsilon carbide. The residual stresses are also reduced by low-temperature recovery processes.

3. In the third stage of tempering, which begins about 200°C, depending on carbon content and alloy composition, the epsilon carbides dissolve and the low carbon martensite loses both its tetragonality and its carbon; the microstructure is then composed of ferrite $+ Fe_3C$ or $(Fe, M)_3C$, where M represents atoms of alloying elements that have a strong affinity for carbon. As the temperature of tempering increases up to the eutectoid temperatures (723°C), the carbide precipitates coarsen and σ_Y (i.e., σ_i) decreases. Tempering just below the eutectoid temperature causes the cementite particles to assume a relatively large (1–10 μ) spheroidal shape, similar to that obtained by annealing a pearlitic structure for long periods of time. Since these processes are diffusion controlled, increasing the tempering temperature decreases the time required to produce a given microstructure. In alloys containing strong carbide formers (titanium, vanadium, molybdenum, tungsten, and chromium), complex $(Fe, M)_yC$ carbides can form in the vicinity of 550°–650°C.

Fracture of medium strength steels (90 ksi $< \sigma_Y <$ 180 ksi)

In addition to the removal of residual stress there are two effects associated with tempering that increase notch toughness. The first is the transformation of retained austenite for reasons discussed above. The austenite should be transformed at low temperatures (around 300°C) to the tough, acicular lower bainite. If it is transformed by tempering at a higher temperature, say 600°C, the brittle pearlite structure will form. Consequently steels that are to be tempered at 550°–600°C are first tempered at around 300°C to avoid this problem. This procedure is called "double tempering."

Secondly, there is the decrease in yield strength (σ_i decreases) and the increase in dispersed carbide content (γ_m increases), both of which cause the impact-transition tempering range to be lowered as the

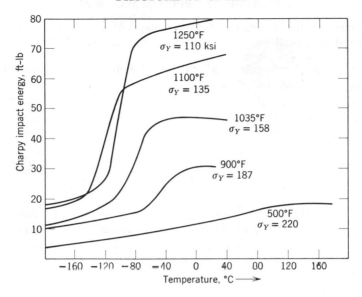

Fig. 10.23. Effect of tempering $1\frac{1}{2}$ hours at the indicated temperatures on the Charpy impact energy curves of 4340 steel. The yield strength values are as indicated. *After J. Nunes and F. Larson* [65].

tempering temperature is increased (Fig. 10.23). In addition, the decreases in σ_Y causes C_V (max) to increase and the Charpy curves to sharpen. This reduces the tendency for low-energy tear fracture. Figure 10.17 indicates that for martensitic steels, C_V(max) decreases with decreasing tensile ductility as in the case of ferritic-pearlitic steels. It is noteworthy that the relation between these two parameters is different for the different means employed to produce a given strength level. In general, tensile ductility and C_V (max) increase, at the same strength level, as the microstructure is refined.

The effect of tempering on the fracture surface topography is summarized in Fig. 10.24 [25]. At *high strength levels* (low-tempering temperatures) the dimples associated with fracture above the transition temperature range are shallow. In the transition region, fracture occurs by mixtures of dimple formation, quasicleavage and intergranular separation along prior austenite grain boundaries.† At very low temperatures the latter two modes are dominant. At *low strength*

† When this mode is dominant the transition-temperature range can be lowered by the refinement of the prior austenite grain size [66].

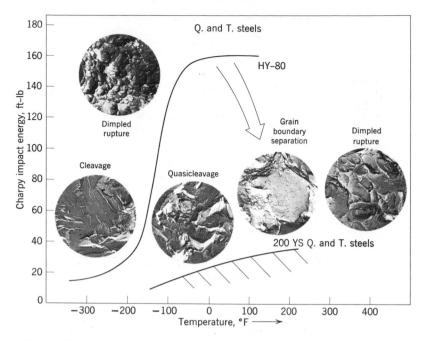

Fig. 10.24. Relation between fracture appearance, as observed by replica electron microscopy, and Charpy impact energy curves for **HY-80** and 4340 steel. 2000×. *After W. Pellini et al* [25].

levels ($\sigma_Y < 90$ ksi) the fracture transition is from dimpled rupture to cleavage and is similar to that observed in ferritic steels. The transition temperature range is sharp.

In the *medium strength range* the behavior falls in-between these two extremes, as seen from the Charpy slopes shown in Fig. 10.23. Because the yield strength of steel increases with decreasing temperature, medium strength steels have yield strengths that fall in the high strength range (i.e., $\sigma_Y > 180$ ksi) when they are tested at low temperature. Since the tear toughness decreases with increasing yield strength (Fig. 3.23), part of the decrease in prebrittle, plane-strain fracture energy with decreasing test temperature is a result of the fact that the energy associated with dimple formation is decreasing. The other part, of course, results from the increasing amounts of quasi-cleavage and intergranular separation that make up the radial zone as the temperature is lowered. The total decrease in plane-strain toughness for rapid propagation and the increase in σ_Y imply that decreasing amounts of slow fibrous fracture and shear lip formation will occur before and after† the onset of rapid fracture (Chapter 3). The decrease in prebrittle and postbrittle fracture energies with decreasing temperature produces a decrease in total impact energy through the transition region.

As shown in Fig. 3.23, heat-treated medium strength steels can sustain large gross plastic strains without fracturing by unstable tear in an explosion bulge test in the presence of 2 in. flaws. Consequently they possess both good impact and strength properties and are used extensively in engineering applications where these combinations are required. Fully martensitic structures have the best impact properties [67]. When mixed microstructures are present, the transition temperature is raised and the tear energy is reduced, at an equivalent strength level (Fig. 10.25).

One of the most serious fracture problems that arises in medium strength (and some high strength) quenched and tempered steels is that due to *temper embrittlement.* This type of fracture tends to occur in certain steels that have been tempered in the 750°–1050°F (400°–600°C) range and *slowly cooled* to room temperature. It also

† Postbrittle fracture energy.

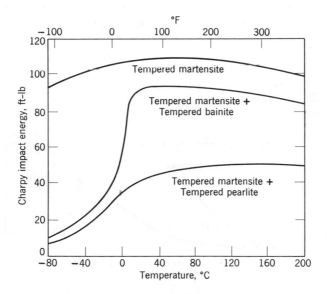

Fig. 10.25. Effect of mixed microstructure on the Charpy impact energy curves for quenched and tempered steels. UTS = 125 ksi in all three cases. *After J. Holloman et al* [67].

occurs after prolonged heating during tempering, regardless of cooling rate. At high temperature the effect begins to disappear after prolonged annealing. It is not observed in materials that are quenched after tempering, but this is not necessarily a good method of avoiding the problem because of the residual stresses and distortion that may be produced by the quench. The embrittlement is reversible and may be removed by heating above 1050°F followed by a rapid cool through the dangerous temperature region. This type of fracture has been the subject of numerous investigations and review articles [65–71] and the reader is referred to these for aspects related to particular steels.

Thermal treatments that produce temperature brittleness do not effect the tensile properties or hardness when these are measured at room temperature. At low temperatures (i.e., −196°C) the ductility and fracture stress of unnotched tensile specimens are lowered by these treatments and the yield stress is unaffected [72].

The most common measure of temper embrittlement is the shift of an impact-transition temperature (Fig. 10.26) such as the NDT or the propagation transition $T_{B(N)}$. The change in transition temperature increases linearly with $t^{1/2}$, where t is the time of tempering at a given temperature in the embrittlement range [73]. Increasing the tempering temperature also increases Δ(NDT) for a fixed tempering time (Fig. 10.27) [74]. The activation energy for the process varies between 40 and 60 kcal per mole, which suggests that the embrittling species is not carbon since the activation energy for its diffusion in iron is considerably less than this [73].

Temper embrittlement is reversible. If the tempering temperature is raised above the critical range, the transition temperature is low-

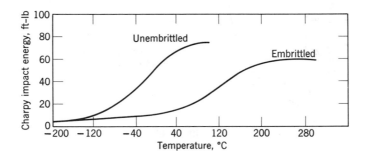

Fig. 10.26. Effect of a temper-embrittlement treatment (48 hours at 932°F) on the Charpy impact energy curves for Q.T. 3140 steel having a coarse prior austenite grain size. *After F. Carr et al* [72].

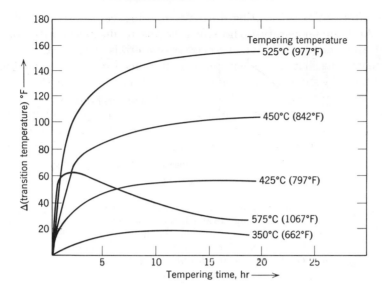

Fig. 10.27. Effect of tempering time at the indicated temperatures on the increase in transition temperature of a Q.T. steel. *After G. Vidal* [74].

ered, but it can be raised back again if the material is reheat treated in the critical range (Fig. 10.28). The presence of trace elements appears to be responsible for the embrittlement [75, 76]. The most important of these are antimony, phosphorus, tin, and arsenic, with manganese and silicon having a small effect. Molybdenum reduces temper brittleness when other alloying elements are present. Nickel and chromium appear to have little effect.

The decrease in σ_f at low temperatures and the fact that σ_Y is unaffected by the critical thermal treatments suggest that embrittlement is caused by the segregation of the trace impurities to prior austenite grain boundaries, where they lower the boundary surface energy γ_s and hence γ_m [69, 77]. In accord with this view are the observations that fracture due to temper brittleness occurs along prior austenite grain boundaries [56] and that susceptibility increases with the prior austenite grain size. It should also be noted that precipitates have never been observed [56] on the fracture surfaces by either the electron microscope or the microprobe. This would indicate that the segregation required for embrittlement occurs on an extremely fine scale. At present these effects can only be explained in terms of a lowering of grain boundary energy by a reversible form of

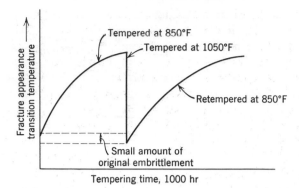

Fig. 10.28. Effect of tempering in the temper-embrittlement range (850°F), subsequently tempering above it (1050°F) and then retempering at 850°F on the fracture appearance transition temperature of a Q.T. steel. *After G. Gould* [73].

segregation that occurs under certain conditions. The only means of avoiding the problem are to remove the trace impurities that cause it, add molybdenum to the steel to prevent it or avoid the critical temperature range during tempering.

Fracture of high strength steels ($\sigma_Y > 180$ ksi)

High strength steels are produced by basically one of three processes; quenching and tempering, deforming the austenite before quenching and tempering (ausforming), or annealing and aging to produce precipitation hardening (e.g., maraging). In addition, further increases in strength can be achieved by straining and retempering or by straining during tempering. The high strength level of these steels makes them extremely brittle (Chapter 7), especially when particular environments such as water vapor or hydrogen are present (Chapter 9). The toughness parameters of these steels are compiled in the aerospace materials handbook [78] and need not be repeated here. Some additional data are given in [102] and [103].

From the microstructural point of view, the most interesting and serious form of embrittlement in the *quenched and tempered steels* is that which occurs after tempering at relatively low temperatures [63, 79, 80, 81, 82] such as 450°–650°F in 4340 steel (this is known as 500°F embrittlement) or 900°F in a 300 M steel. The latter is a modified version of 4340, which contains about 1.5% silicon (compared with about 0.25% Si in **4340**) to retard the embrittlement to higher temperatures and allow high yield strength levels (**260** ksi) to be achieved by tempering at 500°F without reduction of toughness.

Figure 10.29 shows the effect of tempering temperature (and indicated yield strength level) on the Charpy V notch impact energy absorbed at 18 and −196°C. Also indicated on the figure are the relative amounts of the various types of fracture that are observed by fractography in the flat (radial) portion of the broken charpy bars.†
The amount of dimple (tear) fracture and dimple depth [82, 84] increases and the amount of quasicleavage decreases with decreasing strength level (i.e., increasing tempering temperature or test temperature) as discussed previously. It is also obvious from this figure that the 500°F embrittlement is the result of increasing amounts of fracture along prior austenite grain boundaries. The nature of the grain

† At −196°C the amount of shear lip and slow fibrous tearing before rapid propagation is negligible. At room temperature, shear lips make up a larger fracture of the fracture surface (about one-third after tempering at 800°F) and there is more microscopic slow growth before the onset of rapid propagation.

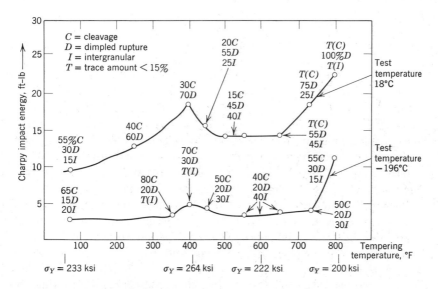

Fig. 10.29. Estimated percentage of various fracture modes in the radial fracture region. These were determined by replica electron microscopy on Charpy specimens of 4340 steel, broken at 18° and −196°C, as a function of tempering for one hour at the indicated temperatures to the indicated yield strength level σ_Y. *After J. Bucher et al [82].*

boundary embrittlement reaction is uncertain at present. The fact that it occurs at tempering temperatures at which epsilon carbide is transforming to cementite and that this temperature is much below that at which temper embrittlement occurs in medium strength steels suggests that it may be due to the precipitation of fine films of Fe_3C or alloy carbide along the prior austenite grain boundary [80, 81]. Another possibility [82, 85] is that it results from the segregation of trace elements (P, As, etc.) to epsilon carbide and the deposition of these elements in the prior austenite grain boundary when the epsilon carbide transfers to cementite.

Ausformed steels are typically processed [86, 87] by austenitizing at about 1000°C, air cooling to 570°C, and then rolling (50–90% reduction) at this temperature prior to cooling to room temperature and tempering. The heavy working of the austenite produces numerous sites (dislocations) for the precipitation of alloy (Mo, Cr, V) carbides during the ausforming processes [88]. These carbides remain after the austenite transforms to martensite and they dispersion strengthen the martensite [88, 89]. Typical ausformed steels contain 3% chromium, to develop a "bay" between the ferrite and bainite transformation temperatures so that the austenite can be deformed by multipass rolling without transforming to one of these products, as well as 1–2% Ni, Si, Mn, Mo, and V. The carbon content varies between 0.3 and 0.6% and it is found that the strength increases with increasing amounts of deformation and carbon content. Typical room temperature yield strengths vary between 250,000 and 350,000 psi, depending on these variables.

Large amounts of deformation by rolling (e.g., 84% thickness reduction) cause the austenite grains to elongate in the rolling direction. Some carbide precipitation occurs in them during rolling. When these steels are tempered in the secondary hardening range, the carbides are able to grow and produce planes of weakness that lie along the rolling direction [90]. As shown in Fig. 7.26, the transverse tensile stresses σ_{xx} below the notch root are able to produce splitting along these boundaries and relax hydrostatic stresses, thereby improving the notch toughness when the notch plane is perpendicular to the rolling direction. The effect of prior deformation and test temperature on the Charpy V notch impact strength of a 3% Ni, 3% Mo, 0.2% carbon die steel tempered at 565°C is shown in Fig. 10.30a. Increasing the amount of ausform deformation from 0 to 55% increases the yield strength of the tempered martensite from 220 to 250 ksi and causes the transition temperature to be increased and $C_V(\max)$ to be decreased, for reasons discussed earlier. However, when the prior de-

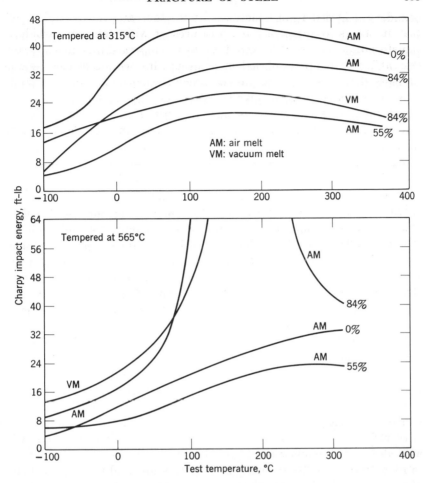

Fig. 10.30. Effect of tempering temperature and per cent ausform deformation on the Charpy impact curves of an ausformed steel. *After A. McEvily and R. Bush* [90].

formation is increased to 84%, the transition temperature is lowered and the tear energy can be greatly increased. This indicates that large amounts of hot working are required in order to orient sufficiently the embrittled austenite grain boundaries perpendicular to the rolling direction such that they can effectively relax the hydrostatic stress state ahead of the notch. It should be noted that this effect is most beneficial below about 300°C. Above this temperature the yield strength of the steel is so low that the transverse stresses in the plastic zone ahead of the notch are unable to crack the prior austenite grain boundaries and the tear energy decreases rapidly with increasing temperature (decreasing σ_Y). Similarly if tempering is performed below the secondary hardening range (e.g., at 315°C), an insufficient amount of carbide formation and growth will occur in the prior austenite grain boundaries to weaken them seriously and longitudinal splitting along them will not occur to any significant degree [90]. Consequently the toughness of the material deformed 84% is only slightly greater than that of material deformed 55% and less than the conventionally processed material having the smaller yield strength (Fig. 10.30b).

Maraging steels contain between 15 and 25% nickel (generally 18%), a low carbon content (0.03%), various percentages of cobalt (7–9%), molybdenum (3–5%), and small amounts of titanium and aluminum. The steels are slowly cooled from an austenitizing temperature of 1500°F to produce a low carbon martensite. This structure is then aged for five hours at 850° to 950°F and cooled back to room temperature [91]. The high strengths of the steels result from the precipitation of very fine alloy carbides in the martensite plates during the aging treatment [92, 93].

Figure 10.31 shows the effect of temperature on the Charpy V notch impact strength of commercial 250 grade (i.e., $\sigma_Y \approx 250$ ksi at room temperature) maraging steel [94]. The low and broad transition temperature range is the result of the facts that: (1) the grain size of the tempered martensite plate is very fine and is the effective grain size for quasicleavage since any austenite retained after initial cooling is transformed during aging; (2) the yield strength increases slowly with decreasing temperature, probably because the nickel content is so high [96]; and (3) the carbon content is low and no large carbides which might initiate fracture are present. Similarly, low carbon martensites deform by slip rather than twinning [97] and this too would make cleavage initiation more difficult. From −100°C to about 450°C the Charpy impact strength remains constant at about 20 ft-lb as fracture occurs by void formation and coalescence (dimpled rupture). Above this temperature the alloys begin to soften (owing to particle growth) and the impact strength increases.

As shown in Figs. 3.26 and 10.17, maraged steels are tougher than other steels when evaluated at the same strength level. The reason for this behavior is not completely clear at present. It may be related to the fact that the alloy carbides which cause strengthening are more finely distributed than in conventional Q.T. steels or ausformed steels and consequently are less liable to fracture and cause

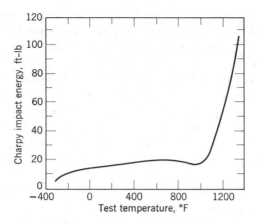

Fig. 10.31. Charpy impact energy curve for a typical 250 grade maraging steel. The impact strength remains low until about 1000°F at which point the alloy softens and the tear toughness increases. *After E. Kula and C. Hickey* [94].

void formation, for reasons discussed in Chapter 7. Another factor may be the relatively high (5%) molybdenum content, which prevents grain boundary segregation during aging and fracture along prior austenite grain boundaries, in a manner similar to the way in which it reduces temper brittleness in medium strength steels. As shown in Fig. 10.32, the toughness of these steels is seriously decreased when molybdenum is absent and fracture occurs by intergranular separation [93].

Effect of casting procedures

It was pointed out in Chapter 3 that many service failures result from poor casting and welding procedures, which lead to the formation of large, elongated inclusions or stringers in wrought metals. These defects crack easily and subsequently behave as internal notches which act as strain concentrators and lower the nominal stresses and strains at which unstable fracture can occur. In view of the low fracture toughness of high strength steels, defects such as these are particularly dangerous and must be detected by adequate inspection techniques in order to guarantee reliability. In addition, it was shown in Chapter 7 that inclusions which are too small to behave as microcracks and initiate unstable fracture are still able to lower the toughness of the material in the presence of defects formed by other processes (e.g., welding and corrosion). Inclusions are also known to reduce the fatigue strength, particularly at high strength levels [98] (Chapter 8).

The inclusion content of high strength steels is related to the total amount of phosphorus plus sulfur that is present, the sulfur appearing in the form of sulfides, the phosphorus in the form of a brittle eutectic. It may be reduced by consumable electrode vacuum remelting, vacuum induction melting, or vacuum degassing. These processes cause the Charpy shelf energy to increase [98, 99, 100] and the impact energy anisotropy ratio† to decrease [100] in medium and high strength steels having yield strengths below 250 ksi. In ultrahigh strength steels the Charpy test is not able to delineate the influence of inclusions unambiguously [100]. However, it is obvious from Fig. 10.33 that decreasing sulfur contents produce a higher notch

† The ratio of the Charpy impact energy in the longitudinal and transverse directions, at a given test temperature.

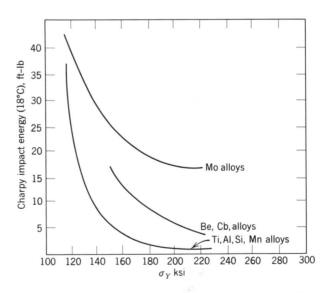

Fig. 10.32. Effect of yield strength level and alloy content on the ambient-temperature Charpy impact strength of maraged steel. *After S. Floreen and G. Speich* [93].

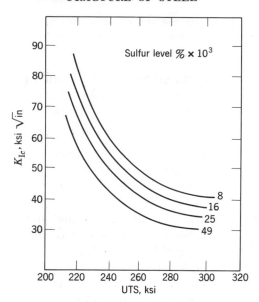

Fig. 10.33. Effect of tensile strength level and sulfur content on K_{Ic} for Q.T. 4335 steel. *After R. Wei* [101].

toughness when all other composition and strength variables are held constant.

10.4 The Fracture of Stainless Steels

Stainless steels are basically iron-chromium and iron-chromium-nickel alloys to which small amounts of other elements have been added to improve mechanical properties and corrosion resistance. Their resistance to corrosion arises from the formation of an impervious layer of chromium oxide on the metal surface which, in turn, prevents any further oxidation of the metal. Consequently these steels are corrosion-resistant in oxidizing atmospheres which strengthen this layer, but are susceptible to corrosion in a reducing environment which breaks down the layer. The corrosion resistance (in oxidizing environment) increases with increasing chromium content and also with increasing nickel content. The latter element increases the overall passivity of the iron. Carbon is also added to improve mechanical properties (yield and tensile strength) and to stabilize the austenitic stainless steel.   Generally speaking, the stainless steels can be classified by their microstructures:

Martensitic

These are iron-chromium alloys that can be austenitized and subsequently heat-treated to form martensite. They normally contain about 12% chromium and 0.15% carbon (type 410).

Ferritic

These alloys contain about 14–18% chromium and 0.12% carbon (type 430) and are completely ferritic since chromium is a ferrite stabilizer; the austenite phase is completely suppressed in alloys containing more than 13% Cr.

The fracture characteristics of ferritic and martensitic stainless steels are similar to those of other ferritic or martensitic steels at the same strength level, grain size, and so on, and need not be repeated here. Specific information related to particular grades, particularly the high strength martensitic stainless steels formed by precipitation, or cold working and precipitation, may be found in the literature [78, 102, 103, 104, 105].

Austenitic

Nickel is a strong austenite stabilizer and alloys containing 8% nickel and 18% chromium (type 300) are austenitic at room temperature and below, as well as at high temperatures. These steels, like the ferritic grade, cannot be hardened by martensitic transformation.

Austenitic stainless steels have a FCC structure and consequently do not fracture by cleavage, even at cryogenic temperatures. After heavy cold rolling (80%), 310 type steels have an extremely high yield strength [106] ($\sigma_Y = 260$ ksi) combined with a notch sensitivity ratio of 1.0 at temperatures as low as $-253°C$ and consequently are used in missile systems for storage tanks for liquid hydrogen. Similarly 301 type stainless can be used down to $-183°C$ (e.g., for liquid oxygen storage tanks), but below this temperature the austenite is unstable and deforms to brittle, untempered martensite if any plastic deformation occurs at the low temperatures [107, 108].

Most austenitic stainless steels are used in corrosive environments. When they are heated in the temperature range $500°–900°C$ (e.g., during welding), chromium carbide precipitates at austenite grain boundaries, resulting in a depletion of chromium from the region near to the boundaries. This depleted layer is very susceptible to corrosive attack (particularly in hot chloride environments), and localized corrosion, in the presence of applied stress, leads to inter-granular brittle fracture. To alleviate this problem, small quantities of elements which are stronger carbide formers than chromium, such as titanium (type 321), or niobium (type 347), are commonly added. These elements combine with the carbon to form alloy carbides, which prevents chromium depletion and subsequent susceptibility to stress corrosion cracking. This process is called "stabilizing."

The stability of the austenite phase depends on the presence of nickel and carbon. If carbon is removed from the austenite to a large degree (by reactions with titanium or niobium or if the alloy is initially carbon-free), the austenite will decompose to a mixture of austenite and ferrite, the proportions of which depend on composition and heat treatment [70]. If this structure is heated to temperature between 600° and 950°C, some of the ferrite will, in high chrome alloys (>20%), decompose to form a mixture of ferrite and another phase which is labeled sigma. The rate of this decomposition in-creases with increasing chromium content and increasing amounts of prior cold work [109, 110].

The sigma phase has a β-uranium crystal structure and is ex-tremely hard and brittle, probably because it is essentially an ordered structure composed of iron and chromium. When it precipitates in the ferrite phase it increases the hardness, probably by a dispersion-strengthening mechanism. Although they are not very detrimental to high-temperature strength [110], these hard precipitates can in-duce brittleness at temperatures below about 500°F [110, 111] (Fig. 10.34), as would be expected for reasons discussed in Chapters 5 and 7. In order to prevent the formation of the sigma phase in austenite (or ferritic) stainless steels, it is necessary to reheat treat them at temperatures between 1050° and 1150°C and rapidly cool them back to ambient temperature.

Austenitic stainless steels are used extensively in high-temperature applications (e.g., pressure vessels) where both corrosion resistance and creep resistance are required. Some of these steels (e.g., type 347) are susceptible to cracking in the heat-affected zone near welds during postwelding heat treatments and/or elevated temperature service [112]. The cracking is the result of precipitation of niobium or titanium carbides in the grains and grain boundaries when the weld is reheated [110, 113]. Plastic strain produced by thermal stress during welding refines the size of the precipitate and enhances the dispersion strengthening of the grains, relative to the depleted zone adjacent to the grain boundaries [113]. Consequently high-tempera-ture creep occurs primarily by grain boundary sliding and, as shown

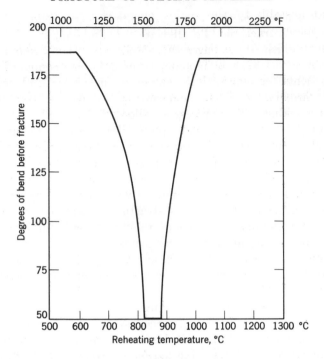

Fig. 10.34. Effect of reheating temperature on the room temperature bend ductility of austenitic stainless steel. *After H. Kirkby and J. Morely* [111].

in Chapter 9, this can lead to grain boundary cracking around triple points. This problem can be alleviated through the use of steels containing higher ratios of carbon or nitrogen to niobium [114] such as types 316 or 321.

10.5   The Relation Between Welding and Brittle Fracture

It has long been recognized that the use of welding as a fabrication technique can pose some extremely difficult problems for the engineer who is designing for a high degree of reliability [141]. This has become even more true in recent years since welding has been used to join high strength steel parts that are subject to high stresses and have low fracture toughness. Whereas relatively large flaws may have little deterimental effect in low strength materials operated at ambient temperatures (and anyway are easy to detect), small flaws can cause unstable fracture at low operating temperatures or in high strength materials [115, 116] and these flaws are difficult to detect. Furthermore, subcritical-size flaws formed during welding can grow to critical size (for a given nominal stress) if residual stresses or macroscopic stress concentrators (e.g., sharp corners) are present, if the weld or base metal has been embrittled (e.g., by quench aging), or in the presence of corrosive environments or alternating stress [115] (Chapter 3). Consequently a greater degree of control of weld perfection is required as the stringency of the operating conditions increases. The safest way to ensure this is to have experienced and well-qualified personnel working on the more difficult jobs and to design such that welding operation can be performed with a minimum amount of difficulty (i.e., that the welder has accessibility to deposit a weld bead that will penetrate completely and fuse adequately) [24].

In addition to flaws formed by incomplete fusion, shrinkage cavities, porosity, or arc strikes, welding can lead to cracking in the weld metal itself, along the fusion line, or in the heat-affected zone in the base metal adjacent to the fusion line. Hot cracks are those which form at high temperatures when the weld starts to cool; cold cracks are those formed below 400°F. Either of these types of cracks can lead to fracture under service conditions, as can embrittlement of the weld or base metal by microstructural effects such as quench aging or the formation of mixed microstructures in the HAZ.

Hot cracking (hot shortness)

At the present time there is no universally accepted theory of this phenomenon although numerous ones have been proposed [115, 117, 118, 119]. It is generally accepted that thin liquid films of a low melting point phase (e.g., FeS, NiS) are present for a certain period of time during solidification of the weld metal. The strains (resulting from thermal gradients) set up during solidification are concentrated in these liquid films and cracking occurs when they reach a critical value. Increasing strain rate or time of film life (because of solute segregation of a phase which has a low solidus) increases the probability of achieving the critical strain and favors cracking. The ratio of the solid-liquid interface energy to the grain boundary energy is also important [118]. If the liquid completely wets the boundary (dihedral angle $\theta = 0°$) it can flow into an incipient crack and heal it. If not all of the grain boundaries are wet, then this may not be possible and stable cracks can be initiated. However, if there is little wetting between phases, then all second-phase segregation will be confined to grain corners and long, sharp cracks do not tend to form. Similar considerations indicate that hot cracking can form in the base metal owing to the segregation and melting of low melting point phases [119].

Geometrical parameters such as the shape and size of the parts being welded and the type of weld are important [120] because they determine the magnitude of the thermal strains which are produced during solidification. Thermal factors [121] such as the coefficient of thermal expansion, the thermal conductivity, and welding speed are also significant and the probability of cracking increases as these parameters increase. Grain size is also a factor. Since the grain boundary area per unit volume decreases with increasing grain size, the grain boundary concentration of a given amount of impurities, and hence the tendency for hot cracking, increases with increasing grain size [122]. Finally the alloy composition [115, 123, 124] plays an important role by determining the impurity content. The presence of sulfur is extremely detrimental to welds because of the formation of the low melting FeS or NiS. One of the functions of manganese in steels is to react with the sulfur and form the higher melting compound MnS to prevent hot cracking. The presence of phosphorus is also detrimental because of the possibility of iron phosphorus eutectic formation at grain boundaries.

Cold cracking

Chemical analyses performed on welded structures indicate that large amounts of hydrogen exist in the *weld metal* [24, 125, 126, 127]. This hydrogen, which comes from both the moisture in the atmosphere and from the components in the flux coating, is very soluble in the weld metal when the weld is molten. The equilibrium solubility of hydrogen in solid iron, particularly BCC ferrite, is four or five orders of magnitude lower at room temperature. Consequently, it is to be expected that this excess hydrogen would embrittle the weld, in accordance with the discussion of hydrogen embrittlement presented in the preceding chapter. There it was shown that hydrogen embrittles iron by precipitating out of the iron lattice as molecular gas and exerting an internal pressure sufficient to cause microcrack formation and propagation. When the weld metal is rapidly cooled, the excess hydrogen is entrapped and produces cracks. In welds that are cooled slowly a large fraction of the excess hydrogen is able to leave the metal. The small amount that is left is not able to produce large fissures; these are not formed until an external stress is applied. In either case, fracture is a form of hydrogen embrittlement and the same factors that were discussed in the preceding chapter (e.g., strength level [126, 127]) are applicable here.

Another characteristic of weld metal that gives it different properties from the base metal is its high oxygen content [24]. The high rate of cooling of the weld prevents the removal of many of the impurities into the slag and these, in turn, appear as very fine inclusions in the weld metal. This is actually not too bad from the point of view of fracture resistance, provided the inclusions are small and finely dispersed, because they inhibit ferrite grain growth and produce a fine-grained structure in the weld metal.

The *base metal* adjacent to the weld (HAZ) will be heated to very high temperatures, which reach the melting point at the plate surface and then taper off as the distance from the weld metal increases. Thus a whole series of different microstructures can be observed at increasing distances from the plate surface within the heat-affected zone, depending on the cooling rate. This, in turn, is a function of geometrical variables and plate temperature [128, 129]. In steels of moderate hardenability the rapid cooling rate will cause the formation of martensite and bainite, as well as proeutectoid ferrite and pearlite, in the HAZ. Since the martensite will exist in the untempered condition it will be extremely brittle. When excess hydrogen from the weld metal diffuses into the heat-affected zone, and regions inside this zone transform to the brittle martensite phase, "underbead cracks" will form [142]. Similarly since the solubility of hydrogen is much higher in austenite than in the BCC phases, cracking can also occur when retained austenite transforms isothermally to martensite and the excess hydrogen is precipitated in the highly strained region where the transformation takes place [63, 129]. Thus the material in the base metal is subject to problems of hydrogen embrittlement similar to those encountered by the weld metal.

One way of alleviating some of these problems is to preheat the base parts prior to welding them together [24, 130, 131]. This, in turn, results in a slower cooling rate after welding and allows the hydrogen time to diffuse out of the weld. Furthermore, the transformation structures (in a mild steel-base metal) will be a mixture of ferrite and pearlite rather than the undesirable untempered martensite. A second advantage of either a preheat or a postheat treatment is that the magnitude of any shrinkage stresses around the weld will be reduced. Care must be taken to avoid heating for too long a time because this will result in an increase in ferrite grain size and also the possible solution of carbon, which may reprecipitate as the dangerous grain boundary cementite upon cooling. Postheating has the additional advantage of tempering any of the martensite that may have formed during the rapid cool.

Almost all engineering materials are polycrystalline and anisotropic on a microscopic scale. In order for continuity to be maintained at grain boundaries when a load is applied, the amount of stress and strain in each grain of the aggregate will be different. When the load is removed, *residual stresses* and strains will be left in each grain [132]. They are set up in the welding process because the contraction of the weld metal during cooling is constrained by the large masses adjacent to it. It has been shown [133, 134] that the stresses left in the weld metal are of the order of the tensile yield stress within about three fused widths on either side of the weld. If the weld metal is subsequently loaded to produce a nominal stress sufficient to cause plastic deformation, the effect of residual stress can be canceled.

As stated earlier, residual tensile stresses can be important in fracture if they are able to cause the crack to extend to such a length that it can propagate unstably under a low nominal stress. This effect has been noted by numerous investigators [135–139]. These residual stresses may be relieved by postwelding heat treatment [24, 135, 140].

10.6 Summary

1. The NDT of mild steel increases by about $2°C$ per 1 ksi increase in σ_i and decreases by $2.3°C$ per increase in $d^{-\frac{1}{2}}$, when d is measured in inches. This amounts to about a $14°C$ decrease in NDT per increase in ASTM grain size number.

2. The effect of alloying additions on the toughness of mild steel (and other materials as well) must be analyzed in terms of the effect of these additions on the microstructure. The microstructure, in turn, determines the various parameters σ_i, d, etc., in the fracture equations and hence the transition temperature and ductility.

3. Most alloying additions refine the ferrite grain size (which is beneficial for ductility), and raise σ_i (which is harmful). Carbon additions are particularly harmful; small amounts tend to segregate at grain boundaries (as Fe_3C) and lower the work required for microcrack initiation. Larger amounts lead to the formation of pearlite, which increases the tendency for fibrous cracking and lowers the energy for tear fracture. Nitrogen, silicon, and phosphorus are also harmful because they increase σ_i and cause plastic deformation to be more inhomogeneous, thereby raising k_y.

4. Manganese and nickel are beneficial additions because they refine the grain size (especially Ni) and also lower k_y. In addition, manganese prevents the formation of the deleterious grain boundary films of Fe_3C and refines the pearlite nodules. Aluminum is particularly beneficial because it removes nitrogen from solution and because the AlN particles also refine the ferrite grain size. Niobium additions also improve toughness because they produce a large grain-refining action and lower the amount of pearlite.

5. Processing conditions and cold working also affect the toughness of mild steel. Low finishing temperatures (controlled rolling), rapid cooling rates after rolling (control cooling), and normalizing all tend to refine the grain size. Cold working, especially when followed by aging in steels containing free nitrogen (rimmed or semikilled steels), raises σ_i and decreases ductility.

6. At a given strength level, the toughness of tempered lower bainite is far superior to that of upper bainite. The reasons are: (1) the effective grain size for fracture is the large prior austenite grain size in upper bainite, but in the lower bainite it is the ferrite needle size, and (2) carbides are homogeneously distributed in lower bainite but lie along grain boundaries in upper bainite (thereby lowering γ_m).

7. As-quenched martensite is very hard and brittle. The hardness results from the high supersaturation of carbon (σ_i is large) and fine needle size (d is small). The brittleness results from the presence of residual (transformation) stresses, retained austenite, and the large value of σ_i. Tempering improves the toughness by removing residual stress, transforming the retained austenite and lowering σ_i (by carbide precipitation). Increasing the tempering time or temperature causes σ_Y to decrease, the tear energy to increase, and the Charpy V transition curves to sharpen. In the absence of temper embrittlement, quenched and tempered martensitic steels are tougher than ferritic or bainitic steels (when evaluated at the same strength level), by virtue of their fine effective grain size for fracture (i.e., the needle size).

8. Temper embrittlement can be a serious problem in medium strength steels tempered for long times between 750°–1050°F or slowly cooled to ambient temperature after tempering in this range. The embrittlement is reversible and may be removed by annealing above 1050°F and rapidly cooling through the dangerous temperature region. The embrittlement (observed as an increase in NDT) appears to result from the segregation of impurities (antimony, phosphorus, tin, and arsenic) to prior austenite grain boundaries, where they lower the work done in microcrack initiation γ_m. The only means of avoiding the problem are to remove the trace impurities that cause it, add molybdenum to prevent it, or avoid the critical tempering range during tempering.

9. High strength *quenched* and *tempered steels* are also embrittled when tempered in the range 450°–650°F. This embrittlement is thought to result from the precipitation of Fe_3C or alloy carbides along prior austenite grain boundaries. The toughness of *ausformed steels* is anisotropic. It is very high (at a given strength level) when evaluated perpendicular to the working direction because of splitting effects which relax constraint ahead of a crack. At a given strength level in the high strength, range-maraged *steels* are tougher than any of the other types of high strength steels. The reason for this behavior is still unclear, but it may result from a low content of finely dispersed precipitates and the high molybdenum content which prevents segregation at prior austenite grain boundaries. High sulfur contents, which lead to high inclusion contents, are detrimental to all types of high strength steels.

10. The fracture characteristics of ferritic and martensitic stainless steels are similar to those of other ferritic and martensitic steels when evaluated at the same strength level and grain size. Austenitic (FCC) stainless steels do not fracture by cleavage and hence have good toughness at cryogenic temperatures. Embrittlement can occur in hot chloride environments if chromium carbide precipitation depletes the chromium content near grain boundaries. This may be alleviated by niobium additions which preferentially react with carbon and leave the chromium in solid solution.

11. Welding can produce embrittlement by numerous means. The more important of these are (1) by forming flaws, owing to incomplete fusion, shrinkage cavities, arc strikes, hot cracking, and cold cracking (hydrogen embrittlement), (2) by setting up residual tensile stresses in the vicinity of the weld, and (3) by embrittling the base metal in the heat affected zone, owing to quench aging or the formation of mixed microstructures (e.g., ferrite and untempered martensite). These effects may be minimized by preheating the parts before welding and by postwelding heat treatment.

References

The asterisk indicates that published work is recommended for extensive or broad treatment.

*[1] R. M. Brick, R. B. Gordon and A. Phillips, *Structure and Properties of Alloys,* 3rd ed., McGraw-Hill, New York (1965).

 [2] R. Reed-Hill, *Physical Metallurgy Principles,* Van Nostrand, Princeton (1964).

[3] W. Crafts and J. L. Lamont, *Hardenability and Steel Selection,* Pitman, New York (1949).

[4] E. C. Bain and H. W. Paxton, *Alloying Elements in Steel,* 2nd ed., ASM, Cleveland (1961).

*[5] N. J. Petch, in *Fracture,* B. L. Averbach et al. eds., M.I.T., Wiley, New York (1959), p. 54.

*[6] F. B. Pickering and T. Gladman, *Metallurgical Developments in Carbon Steels,* BISRA Report 81 (1963), p. 9.

[7] J. M. Hodge, H. D. Manning and H. M. Reichhold, *Trans. AIME,* **185,** 233 (1949).

[8] K. W. Burns and F. B. Pickering, *J. Iron Steel Inst.,* **202,** 899 (1964).

*[9] N. P. Allen et al, *J. Iron Steel Inst.,* **174,** 108 (1953).

[10] J. Heslop and N. J. Petch, *Phil. Mag.,* **1,** 866 (1956).

[11] W. H. Bruckner, *Weld. J. Res. Supp.,* **29,** 467s (1950).

[12] J. C. Danko and R. D. Stout, *Trans. ASM,* **49,** 189 (1957).

*[13] G. T. Hahn et al, in *Fracture,* B. L. Averbach et al. eds., M.I.T., Wiley, New York (1959), p. 91.

[14] C. J. McMahon, Jr., and M. Cohen, *Acta Met.,* **13,** 591 (1965).

*[15] J. A. Rinebolt and W. J. Harris, Jr., *Trans. ASM,* **44,** 225 (1952).

[16] N. J. Petch, *Progr. in Metal Phys.,* **5,** 1 (1954).

[17] J. R. Low, Jr., and R. C. Feustel, *Acta Met.,* **1,** 185 (1953).

[18] C. J. McMahon, Jr., and M. Cohen, *Int. Conf. on Fracture,* Sendai, Japan (September 1965) **B-II,** 37.

[19] K. J. Irvine and F. B. Pickering, *J. Iron Steel Inst.,* **201,** 944 (1963).

[20] J. K. Macdonald, *Ship Structure Committee Report,* SSC-73 (November 1953).

[21] H. M. Banta, R. H. Frazier and C. H. Lorig, *Weld. J. Res. Supp.,* **30,** 79s (1951).

[22] F. de Kazinczy and W. A. Backofen, *Trans. ASM,* **53,** 55 (1961).

[23] I. M. Mackenzie, *J. West. Scot. Iron Steel Inst.,* **60,** 224 (1953).

*[24] W. D. Biggs, *Brittle Fracture of Steel,* Macdonald and Evans, London (1960).

*[25] W. S. Pellini et al, *NRL Report 6300* (June 1965).

[26] P. P. Puzak, E. W. Eschbacher and W. S. Pellini, *Weld. J. Res. Supp.,* **31,** 561s (1952.

[27] P. Matton-Sjoberg, *J. West Scot. Iron Steel Inst.,* **60,** 180 (1953).

[28] W. Barr and C. F. Tipper, *J. Iron Steel Inst.,* **157,** 223 (1947).

[29] W. S. Owen, M. Cohen and B. L. Averbach, *Welding J.,* **23,** 368s (1958).

[30] G. T. Hahn, M. Cohen and B. L. Averbach, *J. Iron Steel Inst.,* **200,** 634 (1962).

[31] W. P. Rees, B. E. Hopkins and H. R. Tipler, *J. Iron Steel Inst.,* **177,** 93 (1954).

[32] B. E. Hopkins and H. R. Tipler, *J. Iron Steel Inst.,* **188,** 218 (1958).

[33] M. Gensamer, *Trans. AIME,* **215,** 2 (1959).

[34] R. Wullaert and A. S. Tetelman, to be published.

[35] J. H. Bucher and J. D. Grozier, *Metals Eng. Quart.,* **5,** No. 1 (1965).

[36] W. Crafts and C. M. Offenhauer, *NBS Circular 520* (1952), p. 48.

[37] W. P. Rees and B. E. Hopkins, *J. Iron Steel Inst.,* **172,** 403 (1952).

[38] J. Heslop and N. J. Petch, *Phil. Mag.,* **3,** 1128 (1958).

[39] E. P. Klier, F. C. Wagner and M. Gensamer, *Weld J. Res. Supp.*, **27**, 71s (1948).

[40] N. P. Allen et al, *J. Iron Steel Inst.*, **202**, 808 (1964).

[41] J. R. Low, Jr., and M. Gensamer, *Trans. AIME*, **158**, 207 (1944).

[42] C. F. Tipper, *J. Iron Steel Inst.*, **172**, 173 (1952).

[43] F. Garafolo and G. V. Smith, *Trans. ASM*, **47**, 957 (1955).

[44] H. Thielsch, *Weld. J. Res. Supp.*, **36**, 401s (1957).

[45] J. R. Low, Jr., *Weld. J. Res. Supp.*, **31**, 253s (1952).

[46] O. Reinhold, *Ferrum*, **13**, 97 (1916).

[47] R. L. Kenyon and R. S. Burns, *Proc. ASTM*, **34**, 48 (1934).

[48] W. C. Leslie, *NPL Conference on Structure and Mechanical Properties of Metals,* Teddington, Eng. (1963), p. 333.

[49] R. Phillips, W. E. Duckworth and F. E. L. Copley, *J. Iron Steel Inst.*, **202**, 593 (1964).

[50] E. R. Morgan, T. E. Dancy and M. Korchynsky, presented at AISI meeting New York (May 1965).

[51] H. Harding and M. Korchynsky, to be published.

[52] D. E. Hodgson, M.S. thesis, Stanford University (1965).

[53] W. H. Bruckner, *Weld. J. Res. Supp.*, **30**, 459s (1951).

[54] J. H. Gross and R. D. Stout, *Weld. J. Res. Supp.*, **30**, 481s (1951).

[55] J. C. Danko and R. D. Stout, *Weld. J. Res. Supp.*, **34**, 113s (1955).

*[56] J. R. Low, Jr., *Fracture*, Wiley, New York (1959), p. 68.

[57] A. M. Turkalo, *Trans. AIME*, **218**, 24 (1960).

[58] J. H. Gross and R. D. Stout, *Weld. J. Res. Supp.*, **34**, 117s (1955).

[59] K. J. Irvine and F. B. Pickering, *J. Iron Steel Inst.*, **187**, 292 (1957).

*[60] K. J. Irvine and F. B. Pickering, *J. Iron Steel Inst.*, **201**, 518 (1963).

*[61] E. P. Polushkin, *Defects and Failures of Metals,* Elsevier, Amsterdam (1956).

[62] B. L. Averbach and M. Cohen, *Trans. ASM*, **41**, 1024 (1949).

[63] L. S. Castleman, B. L. Averbach and M. Cohen, *Trans. ASM*, **44**, 240 (1952).

[64] A. J. McEvily, T. L. Johnston and A. S. Tetelman, *High Strength Materials,* Wiley, New York (1964), p. 360.

[65] F. R. Larson and J. Nunes, *Proc. ASTM*, **62**, 1192 (1962).

[66] L. D. Jaffe, F. C. Holden and H. R. Ogden, *Trans. AIME*, **200**, 652 (1954).

[67] J. H. Holloman et al, *Trans. ASM*, **38**, 807 (1947).

[68] J. H. Holloman, *Trans. ASM*, **36**, 473 (1946).

[69] B. C. Woodfine, *J. Iron Steel Inst.*, **173**, 229 (1953).

[70] M. Szczepanski, *The Brittleness of Steel,* Wiley, New York (1963).

[71] J. R. Low, Jr., *Fracture of Engineering Materials,* ASM, Cleveland (1964).

[72] F. L. Carr, J. Nunes and F. R. Larson, to be published.

[73] G. Gould, presented at Texas Chapter of ASM Symposium on Failure Analysis (November 1965).

[74] G. Vidal, *Revue de Metallurgie*, **42**, 149 (1945).

[75] W. Steven and K. Balajiva, *J. Iron Steel Inst.*, **193**, 141 (1959).

[76] J. R. Low, Jr., and D. F. Stein, to be published.

[77] D. McLean and L. Northcott, *J. Iron Steel Inst.*, **158**, 169 (1953).

*[78] *Aerospace Materials Handbook,* Ferrous Metals, Vol. I, Syracuse U. Press (1964).

[79] B. S. Lement, B. L. Averbach and M. Cohen, *Trans. ASM,* **46**, 851 (1954).

[80] L. J. Klinger et al, *Trans. ASM,* **46,** 1557 (1954).

[81] J. M. Capus, *J. Iron Steel Inst.,* **199,** 53 (1963).

[82] J. H. Bucher et al, AFML Dept. TR-65-60 (1965).

[83] C. H. Shih, B. L. Averbach and M. Cohen, *Trans. ASM,* **45,** 498 (1953).

[84] A. J. Edwards, *NRL Report* (November 1963).

[85] B. G. Reisdorf, *Trans. AIME,* **227,** 1334 (1963).

*[86] J. Shyne, V. Zackay and D. Schmatz, *Trans. ASM,* **52,** 346 (1960).

[87] D. Schmatz, F. Schaller and V. Zackay, *NPL Conference on Structure and Mechanical Properties of Metals,* Teddington, Eng. (1963), p. 613.

[88] A. J. McEvily et al, *Trans. ASM,* **56,** 753 (1963).

[89] G. Thomas, D. Schmatz and W. Gerberich, *High Strength Materials,* Wiley, New York (1964), p. 251.

*[90] A. J. McEvily and R. Bush, *Trans. ASM,* **55,** 654 (1962).

[91] R. F. Decker, J. T. Eash and A. J. Goldman, *Trans. ASM,* **55,** 52 (1962).

[92] A. J. Baker and P. R. Swann, *Trans. ASM,* **57,** 1008 (1964).

[93] S. Floreen and G. R. Speich, *Trans. ASM,* **57,** 714 (1964).

[94] E. B. Kula and C. F. Hickey, Jr., *Watertown Arsenal Report ASD-TDR-63-262* (1963), p. 236.

[95] A. R. Willner and M. L. Salive, *Navy Dept. Report 1605,* D. Taylor Model Basin (1965).

[96] J. Nunes and F. R. Larson, *Trans. AIME,* **227,** 1369 (1963).

[97] R. H. Richman, *Trans. AIME,* **227,** 159 (1963).

[98] D. C. Ludwigson and F. R. Morral, *DMIC Report 128* (1960).

[99] J. M. Hodge, R. H. Frazier and F. W. Boulger, *Trans. AIME,* **215,** 747 (1959).

[100] G. E. Gazza and F. R. Larson, *Trans. ASM,* **58,** 183 (1965).

[101] R. P. Wei, *Fracture Toughness Testing,* ASTM, Philadelphia, STP 381 (1965), p. 279.

*[102] R. T. Ault, *Air Force Materials Symposium,* AFML Report TR-65-29 (1965), p. 333.

*[103] *Problems in the Load Carrying Application of High Strength Steels,* DMIC Report 210 (1964).

[104] S. Floreen and G. W. Tuffnell, *Trans. ASM,* **57,** 301 (1964).

[105] B. R. Banerjee, J. J. Hauser and J. M. Capenos, *Trans. ASM,* **57,** 856 (1964).

[106] J. L. Christian, J. D. Gruner and L. D. Girton, *Trans. ASM,* **57,** 199 (1964).

[107] S. A. Kulen and M. Cohen, *J. Metal,* **188** (September 1950).

[108] J. L. Christian, *WAFB Report ASD-TDR-62-258* (1962).

[109] H. Thielsch, *Weld. J. Res. Supp.,* **29,** 577s (1950).

[110] H. Thielsch, *Defects and Failures in Pressure Vessels and Piping,* Reinhold, New York (1965).

[111] H. W. Kirkby and J. I. Morely, *J. Iron Steel Inst.,* **158,** 289 (1948).

[112] P. P. Puzak, W. R. Apblett and W. S. Pellini, *Weld. J. Res. Supp.,* **34,** 9s (1956).

[113] R. N. Younger and R. G. Baker, *Brit. Weld. J.,* **8,** 579 (1961).

[114] T. M. Cullen and J. W. Freeman, *Trans. ASME,* **85A,** 151 (1963).

*[115] P. A. Kammer, K. Masubuchi and R. E. Monroe, *DMIC Report* 197 (1964).

[116] M. D. Randall, R. E. Monroe and P. J. Rieppel, *Weld. J. Res. Supp.,* **41,** 193s (1962).

[117] W. S. Pellini, *Foundry,* **80,** 124 (1952).

[118] J. C. Borland, *Weld. J.,* **8,** 526 (1961).

[119] W. R. Apblett and W. S. Pellini, *Weld. J. Res. Supp.,* **33,** 83s (1954).

[120] J. T. Berry and R. C. Allan, *Weld. and Met. Fab.,* **30,** 271 (1962).

[121] C. A. Terry and W. T. Tyler, *Weld. and Met. Fab.,* **26,** 103 (1958).

[122] L. N. Cordea, R. M. Evans and D. C. Martin, *BMI Report AF33(616)-7702* (1962).

[123] A. Hoerl and T. J. Moore, *Weld. J. Res. Supp.,* **36,** 442s (1957).

[124] R. P. Sopher, *Weld. J. Res. Supp.,* **37,** 481s (1958).

[125] M. Smialowski, *Hydrogen in Steel,* Addison-Wesley, Reading, Mass. (1962).

[126] R. G. Baker and F. Watkinson, *Hydrogen in Steel,* BISRA Report 73 (1962), p. 123.

[127] J. T. Berry and R. C. Allan, *Weld. J. Res. Supp.,* **39,** 105s (1960).

[128] C. R. Felmley, C. E. Hartbower and W. S. Pellini, *Weld. J. Res. Supp.,* **30,** 451s (1951).

[129] M. W. Mallett and P. J. Rieppel, *Weld. J. Res. Supp.,* **29,** 343s (1950).

[130] R. D. Stout and K. J. McGeady, *Weld. J. Res. Supp.,* **26,** 683s (1947).

[131] *Welding Handbook,* Am. Weld. Soc., New York (1962).

[132] W. R. Osgood, *Residual Stresses in Metals and Metal Construction,* Reinhold, New York (1954).

[133] G. P. Degarmo et al, *Weld. J. Res. Supp.,* **25,** 451s (1946).

[134] A. A. Wells, *Weld. Res.,* **17,** 263 (1952).

[135] T. W. Greene, *Weld. J. Res. Supp.,* **28,** 193s (1949).

[136] A. A. Wells, *Trans. Inst. Nav. Arch.,* **58,** 296 (1956).

[137] H. Kihara and K. Masubuchi, *Weld. J. Res. Supp.,* **38,** 159s (1959).

[138] W. J. Hall, W. J. Nordell and W. H. Munse, *Weld. J. Res. Supp.,* **41,** 505s (1962).

[139] L. J. McGeady, *Weld. J. Res. Supp.,* **41,** 335s (1962).

[140] E. R. Parker, *Weld. J. Res. Supp.,* **36,** 433s (1957).

[141] E. R. Parker, *Brittle Behavior of Engineering Structures,* Wiley, New York (1957).

[142] P. D. Blake, *The Welder,* **25,** 35 (1956).

[143] H. H. Kranzlein, M. S. Burton and G. V. Smith, *Trans. AIME,* **233,** 64 (1965).

[144] N. S. Stoloff, R. G. Davies and R. C. Ku, *Trans. AIME,* **233,** 1500 (1965).

[145] N. S. Stoloff, *Symposium on Fundamental Phenomena in Materials Sciences,* Boston (1966).

11 _____

Fracture of Brittle
Nonferrous Metals

The high temperature and/or high strength-to-weight requirements of aerospace structures, advanced propulsion systems, high-speed aircraft and deep submergence vessels have stimulated the development of certain nonferrous metals that can be used in these applications. In addition to the failures that arise because of extreme environmental conditions (e.g., high operating temperatures and corrosive environments, Chapter 9) or fatigue (Chapter 8), there are also brittle failures that occur in some of these metals because of their inherent low ductility and toughness. These failures can pose problems in fabrication and in certain service applications (e.g., at the low temperatures that exist in outer space. From the point of view of catastrophic fracture, the most important and interesting of the nonferrous metals are the BCC refractories, high strength aluminum alloys† and the HCP metals, magnesium, beryllium, and titanium. Although low strength aluminum, copper- and nickel-base alloys are used extensively as structural materials, these FCC metals are quite ductile and tough in the absence of extreme environments and need not be discussed here.

Cleavage is the only mode of unstable fracture in the BCC refractory and low strength HCP metals, so that the effect of composition and microstructure on toughness and ductility can be described by variations in impact and tensile-transition temperatures; low-energy tear is the primary mode of unstable fracture in the high strength alloys so that their variations in toughness appear as variations in G_{Ic}, DWTT energy, C_V(max), and tensile ductility (Chapter 3). These parameters are summarized for many commercial alloys in numerous DMIC reports [1–7] and the aerospace materials handbook [8] and need not be repeated here.

† Already discussed in Chapter 5.

As in the case of steels, alloy content and processing conditions affect the toughness of nonferrous metals by affecting their microstructures which, in turn, determine their toughness. According to the principles set forth in Chapters 6 and 7, toughness is related to microstructure through the eleven parameters given in Table 10.1. We shall now use these parameters to describe qualitatively† the behavior of the brittle nonferrous alloys in the same manner as was done for ferrous alloys in the preceding chapter.

11.1   The Fracture of BCC Refractory Metals

The fundamental aspects of fracture of BCC refractory metals have received a great deal of attention [9, 10, 11, 12] since brittle fracture during fabrication poses one of the most serious drawbacks to the development of the refractories for use as structural materials. All of the refractory metals show a cleavage-shear transition at a temperature that increases with increasing strain rate and the presence of a notch. Because most of the alloys are relatively (to their modulus) low strength, the shear failures are tough, and low-energy tear does not appear to be a problem at the present time; consequently the cleavage-shear transition is also a brittle-ductile (tough) transition. It has also been noted [9–13] that commercially available group Va refractory metals (V, Nb, Ta) are much tougher than commercially available group VIa metals (Cr, Mo, W). This may be seen (Table 11.1) from a comparison of typical values of their 50% tensile-transition temperatures, either in absolute terms or relative to their absolute melting temperatures.

Table 11.1 also indicates that interstitial solutes such as H, C, N, and O are much more soluble in the VA metals than in the VIa metals. In view of the interstitial contents that normally exist in the commercial metals, it has been estimated [11] that the Va metals are dilute solid solutions with respect to interstitials. On the other hand, the VIa metals are usually two-phase alloys consisting of a saturated or supersaturated solid solution and small amounts of second-phase oxides, nitrides, and carbides segregated at grain boundaries. *It is the presence of impurity segregation or second-phase particles at grain boundaries that causes the VIa metals to be so much more "brittle" than the Va metals* [11]. When extensive purification techniques are carried out to reduce the interstitial content below the solubility

---

† The *quantitative* relation between ΔT_D, $\Delta \sigma_i$, and $\Delta d^{-\frac{1}{2}}$, as given in Table 10.1, only apply for the ferritic steels.

Table 11.1

Estimated Ambient
Temperature Solubility of
Interstitials (PPM) [11]

Group	Metal	T_M°C	H	C	N	O	Transition Tempera-ture° C [13]	Transition Temperature (T/T_M)
Va	V	1900	1000	1000	5000	3000	−100	0.08
	Nb	2415	9000	100	300	1000	−230	0.015
	Ta	2996	4000	70	1000	200	< −250	≈0.003
VIa	Cr	1900	0.1–1	0.1–1	0.1	0.1	350	0.29
	Mo	2610	0.1	0.1–1	1	1	15	0.10
	W	3410	—	≪0.1	≪0.1	1	300	0.15

limit, or when the metals exist in the form of single crystals, then the
transition temperatures of VIa metals can be drastically reduced
(e.g., to temperatures below 4°K in the case of high-purity molybdenum [14]. Alternatively the transition temperature of the VIa
metals increases with increasing interstitial content [11, 15] above
the solubility limit (Fig. 11.1). In the Va metals the transition temperature also increases with increasing interstitial content, both

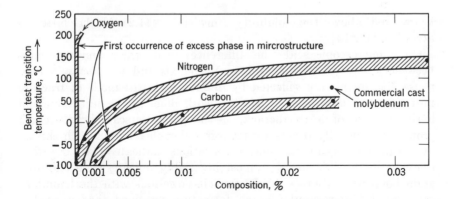

Fig. 11.1. Effect of impurity content on the bend-transition temperature of
cast molybdenum. *After L. Olds and G. Rengstorff [15].*

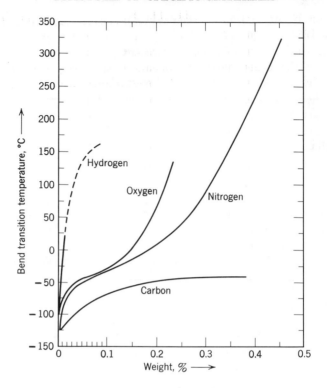

Fig. 11.2. Effect of interstitial content on the bend-transition temperature of vanadium. *After B. Loomis and O. Carlson* [19].

below and above the solubility limit (Fig. 11.2) if the latter is exceeded [11, 13].

On a microscopic scale the differences in ductility and toughness between the normally single-phase Va metals and the normally two-phase VIa metals are reflected by differences in the mode of fracture initiation and the value of γ_m, the work done in the initiation and initial growth of a microcrack. Although large-scale brittle crack propagation usually occurs by transgranular cleavage in both the Va and VIa metals, microscopic observations indicate [11, 16] that cleavage in the VIa metals almost invariably initiates at a ruptured grain boundary. Both slip and twin intersections with the boundary can lead to intergranular crack formation provided that a weakly bound particle is present at the site of the intersection, if the intrinsic surface energy of the boundary has been weakened by impurity segre-

gation, or if dislocation locking by impurities prevents plastic relaxation [38]. Figure 11.3 shows [17] a typical intergranular facet
at the initiation site of a brittle fracture in chromium. In the Va
metals, on the other hand, it appears that brittle fracture usually
initiates by slip or twin-induced transgranular cleavage [10, 11], by
one of the dislocation mechanisms discussed in Chapter 6. In the
terminology of that chapter, the fracture characteristics of the Va
metals are usually those of clean, single-phase materials, whereas
those of the VIa metals are usually those of dirty materials.

It was pointed out in Chapter 6 that the presence of brittle particles or impurity segregation at grain boundaries can cause γ_m to be
decreased. This is apparent from the values of γ_m (obtained by
means of the Cottrell-Petch analysis (Eq. 628)) that are listed in
Table 11.2; except for the case of an unusually clean chromium metal,
γ_m is one to two orders of magnitude lower for the VIa metals than
for the Va metals. This accounts for the lower values of σ_f and ϵ_f
(at constant grain size and the higher transition temperatures that are
observed in VIa as compared with Va metals.

It is difficult to assess the relative embrittling tendencies of *individual, interstitial solute additions* in VIa metals because of the

Fig. 11.3. Intergranular initiation site of cleavage fracture in chromium. 3500×. *Courtesy A. Gilbert et al* [17] *and Inst. of Metals.*

Table 11.2

Tensile Transition Data

(as compiled by Owen and Hull [10])

Metal	Transition Parameters		σ_f at Transition, ksi	γ_m, in. lb per in.^{2}
	$d^{-\frac{1}{2}}$ (cm$^{-\frac{1}{2}}$)	$T_D°$K		
Vanadium	6.3	77	68.5	0.63
	6.8	77	73.0	0.74
Chromium	17.9	77	78.5	0.29
	19.5	373	7.3	0.02
	19.5	373	8.5	0.006
Tungsten†	27.8	433	46.0	0.05
Molybdenum	24.4	200	42.4	0.05

† Estimated values.

experimental problems involved in controlling their content. Consequently, although correlations of γ_m values with impurity content would be extremely helpful in analyzing fracture data, these correlations have not been reported. It is known, however, that oxygen is particularly harmful in recrystallized molybdenum [9, 15] (Fig. 11.1) and tungsten, and that nitrogen and sulfur are particularly harmful in wrought chromium [18] (Fig. 11.4).

Hydrogen is the most effective embrittling agent for Va metals [11, 13, 19], with large increases in transition temperature occurring at contents as low as 100 ppm in tantalum, 50 ppm in niobium, and 10 ppm in vanadium. Nitrogen and oxygen are much less effective.† About 500 ppm of oxygen is required to embrittle niobium [13] and 1500 ppm to embrittle vanadium [19] seriously. Similarly the nitrogen content must reach about 2000 ppm before pure vanadium is appreciably embrittled [19]. Carbon is the least effective embrittling agent for Va metals [11]; the transition temperature is not raised appreciably until the carbon content is well in excess of the solubility limit.

† The well-known embrittlement of tantalum by high oxygen contents also results from impurity levels in excess of the solubility limit [13] and the fracture mechanism is then similar to that described above for VIa metals.

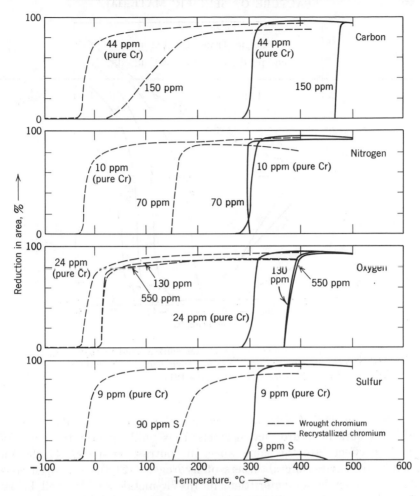

Fig. 11.4. Effect of impurity content and temperature on the tensile ductility of wrought (----) and recrystallized (——) chromium. *After B. C. Allen et al* [18].

The mechanism of interstitial embrittlement in the Va metals is not well understood. It is known that interstitial solutes raise the hardness [20] and flow stress [13, 21, 22], and the embrittlement may result from increases in σ_i or k_y or both of these parameters. The role of hydrogen is particularly confusing since the ductility actually decreases with increasing temperature and decreasing strain rate over a narrow temperature range in vanadium [21, 23, 24] (Fig. 11.5) and

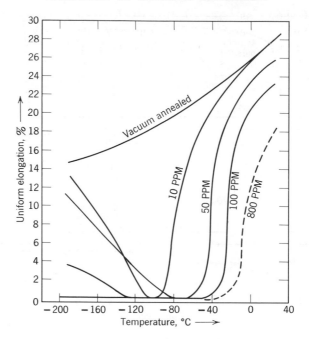

Fig. 11.5. Effect of hydrogen content and temperature on the uniform tensile elongation of vanadium. *After A. Eustice and O. Carlson* [21].

in tantalum [13]. Although this behavior appears similar to that which occurs in hydrogenated steels, it is unlikely that the cause of the two effects are the same since, in contrast to steels, (1) the Va metals are exothermic absorbers of hydrogen; (2) the hydrogen contents responsible for embrittlement in Va metals can be well below the solubility limit; (3) cracking is not observed in the absence of applied stress. Instead it appears that the embrittlement results from dislocation locking [21] and the elevation of the flow stress after hydrogenation. In niobium, for example, it has been shown that the introduction of hydrogen causes k_y to be raised by a factor of 4 at room temperature [22]. The ductility minimum (Fig. 11.5) in vanadium results from the fact that σ_Y is raised more, by a given amount of hydrogen, over this particular low temperature range [21]. The reasons for this particular temperature dependence of solid solution strengthening are obscure at the present time.

After impurity content, *grain size* is the most important variable affecting the ductility and toughness of the refractory metals [12, 13].

Figure 11.6 shows, as expected, that the transition temperature of W, Mo, and Nb decreases with decreasing grain size. In VIa metals the effect is most pronounced in wrought as compared with recrystallized metals [38]. In the case of chromium the results are conflicting because both increases [17] and decreases [25] in transition tempera-

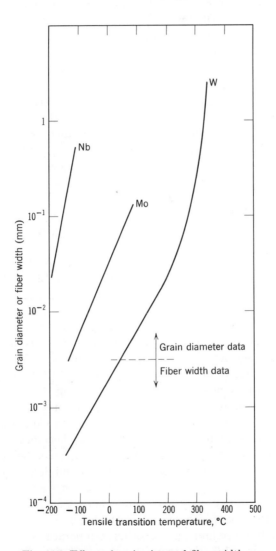

Fig. 11.6. Effect of grain size and fiber width on the tensile transition temperature of Nb, Mo, and W. *After L. Siegle and C. Dickinson* [12].

ture have been observed after grain refinement. As pointed out in Chapter 6, this may reflect the fact that γ_m is a function of grain size since, for a given impurity content, the probability of having an embrittling inclusion in a grain boundary oriented perpendicular to the tensile axis decreases as the grain size increases. Similarly it has been noted that in all of the VIa metals the transition temperature of single crystals is markedly (100°–200°C) *below* that of the finest grained polycrystalline materials [16, 26, 27, 28]. This is not in contradiction to the concepts of Chapter 6 and the data of Fig. 11.6; it simply reflects the fact that γ_m is considerably larger when fracture cannot initiate by grain boundary rupture. Small amounts of prestrain applied above the ductility-transition temperature are also able to nullify the effects of impurities in polycrystalline VIa metals and cause the initiation mode of fracture to change from intergranular to transgranular [16, 29]. Consequently the ductility transition is lowered upon subsequent retesting at lower temperatures.

From a practical point of view, *cold or warm working* below the recrystallization temperature and *alloying* provide the most efficient means of increasing the ductility and toughness of Va and VIa metals, as well as of increasing the yield and tensile strengths. Figure 11.4 illustrates, for example, that the tensile-transition temperature of wrought chromium is about 300°C below that of the recrystallized material [18]. Similar effects have been noted [2] in tungsten and molybdenum (Fig. 11.7) and in Va metals when the interstitial impurity content is above the solubility limit. In general, the transition temperature decreases with increasing amounts of working (Fig. 11.7). The beneficial effects of large amounts of working are thought to result from the production of a fibrous microstructure, with grain boundaries elongated in the working direction [2]. As pointed out in Chapter 7, this tends to favor splitting along the grain boundaries and reduces the stress concentration factor of a crack advancing perpendicular to the boundaries (Fig. 7.24). Consequently crack propagation perpendicular to the boundaries becomes more difficult. The ductility and toughness are thus anisotropic, being high when measured parallel to the working direction and low when measured perpendicular to it. In addition to producing an elongated grain structure, working increases the homogeneity of subsequent deformation and reduces the width of the grains. Consequently the effective grain size for fracture and the transition temperature are reduced [12]. As shown in Fig. 11.6, the transition temperature of tungsten varies with the fiber width in the same manner as it does with grain size in recrystallized material. Large amounts of working also introduce a

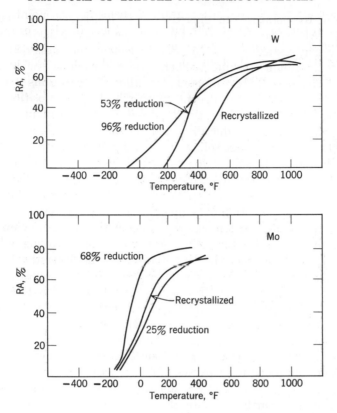

Fig. 11.7. Effect of various amounts of cold work on the tensile ductility of W and Mo at various temperatures [2].

preferred orientation of atomic planes within the elongated grains [2, 12]. In the case of VIa metals, the texture is ⟨110⟩, {100}† so that {100}-type cleavage planes are perpendicular to the plane of the sheet and are inclined at 45° to the working direction. This leads to a reduced ductility for rolled sheets loaded at 45° to the rolling direction and accounts for the fact that cracking in the sheet often occurs at 45° to the rolling direction (45° embrittlement).

The working temperature has little effect on ductility and toughness provided that it is below the recrystallization temperature [2, 12]. Hot working (i.e., working at a temperature such that recrystallization occurs) causes the ductility and toughness to decrease and the

† That is, the ⟨110⟩ direction is parallel to the working direction and lies in the plane of the sheet, which is parallel to a {100} plane.

transition temperature to increase and, consequently, should be avoided. Alternatively, if the working temperature is too low, the tendency for texturing and 45° embrittlement increases [2]. Thus there are optimum working temperatures for each of the refractory metals. These will be affected by alloying additions since re-crystallization kinetics are dependent on the alloy composition. *Sub-stitutional solute additions* and *dispersed particles* can produce sig-nificant changes in the ductility and toughness of refractory metal alloys [11, 31–37]. Most solutes raise σ_i by solid solution hardening, and while this is beneficial for creep resistance, it tends to increase the transition temperature. In some alloys (e.g., W in Ta) the re-crystallization temperature is raised by substitutional solutes [2]; this favors the production of a fine-grained microstructure and tends to lower the transition temperature. From the point of view of re-sistance to brittle fracture, the most important alloying additions are those which change the distribution of the interstitial solutes. Rhenium additions are able to lower the transition temperature of wrought and recrystallized VIa metals [11, 31, 32] (Fig. 11.8). This is thought to result from a decreased interstitial solubility at grain boundaries when rhenium is present [33]. Similarly the addition of reactive elements such as aluminum, titanium, zirconium, and hafnium are able to decrease the transition temperature by reacting with the interstitial solutes to form oxides, nitrides, and carbides and prevent interstitial segregation at grain boundaries [11]. In addition, these finely dispersed particles cause the recrystallization temperature to increase. This tends to lower the transition temperature because of

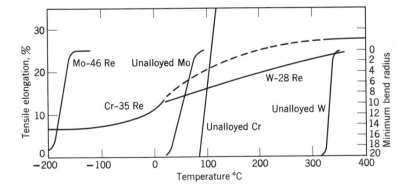

Fig. 11.8. Effect of rhenium additions on the ductility of VIa metals at various temperatures. *After G. Hahn et al* [11].

grain refinement. The addition of *stable dispersed particles* [34, 35] such as thoria or zirconia to tungsten is able to lower significantly the transition temperature of commercial tungsten-alloy sheet loaded in the rolling direction. This results from both the grain-refining effect of the particles and the increased tendency for splitting along the rolling direction which, in turn, helps arrest cracks that propagate perpendicular to the rolling direction.

11.2   The Fracture of HCP Metals

A number of HCP metals, specifically beryllium, magnesium, and titanium, possess physical or mechanical properties, or combinations thereof, which make them attractive as structural materials. For example, the density of either beryllium or magnesium, Table 11.3, is less than that of aluminum (0.098 lb in.$^{-3}$) and that of titanium is less than that of iron (0.284 lb in.$^{-3}$). The modulus-density ratio of beryllium is the highest of any metal, an important consideration where both light weight and elastic stability are desired. However, the fracture behavior of these metals gives rise to limitations. Each exhibits a ductile-brittle transition as a function of temperature, and, in the case of beryllium, brittle behavior even at room temperature and above severely restricts its usage.

The experimentally observed deformation and cleavage modes of a number of HCP metals are listed in Table 11.4, and the slip planes are indicated in Fig. 11.9. It is difficult to make an a priori determination of these preferred slip and cleavage modes in HCP metals. For example, it has been suggested that the tendency for non-basal slip at low temperatures should increase with decreasing c/a ratio, and that the behavior of beryllium is anomalous in this respect. However, nonbasal glide is also not observed in ϵ-AgZn which has an axial ratio of 1.551, which is even less than that of beryllium. With respect to the plane of cleavage, application of the Pauling theory of the metallic bond leads to the conclusion that perfect crystals of zinc, cadmium, magnesium, and beryllium should cleave on $\{10\bar{1}2\}$ planes [70]. The fact that cleavage, when observed, occurs instead on basal planes is ascribed to the influence of dislocations. Indeed, of the HCP metals only zinc and beryllium show clear evidence of cleavage fracture, that is, tensile fracture on a crystallographic plane of low index. The interesting aspect of cleavage fracture in these metals is that it occurs along planes which are also primary slip planes. Fracture can result owing to the splitting of a tilt boundary [43] as already discussed (Chapter 6) or as the result of the blocking of

Table 11.3
Physical Properties of HCP Metals

Metal	Lattice [39] Parameter A	c/a [39]	M.P., °C [39]	Density [39], lb in.$^{-3}$	Poisson's Ratio	Bulk Modulus, psi × 10^{-6}	Shear Modulus [40], psi × 10^{-6}	Basal Plane Stacking Fault Energy [41], ergs/cm^2
Cd	2.9787	1.886	320.9	0.313	0.29	7.1	3.6	15–30
Zn	2.6649	1.856	419.5	0.258	0.27	8.5	5.6	15–30
Mg	3.2088	1.624	650	0.063	0.28	4.8	2.5	∼60
Co	2.5071	1.624	1495	0.322	0.31	2.6	11.0	—
Zr	3.2312	1.590	1852	0.234	0.33	15.1	5.0	—
Ti	2.9503	1.587	1668	0.163	0.36	18.0	5.4	—
Be	2.2858	1.567	1277	0.067	0.08	16.7	19.7	≥190

Table 11.4
Deformation and Cleavage Modes in HCP Metals [41]

Metal	Primary Slip Plane	Secondary Slip Plane	Twinning Planes	Cleavage Plane
Cd	(0001)	$(11\bar{2}2), (10\bar{1}1), (10\bar{1}0)$	$(10\bar{1}2)$	—
Zn	(0001)	$(11\bar{2}2), (10\bar{1}1), (10\bar{1}0)$	$(10\bar{1}2)$	(0001)
Mg	(0001)	$(10\bar{1}1), (10\bar{1}0), (11\bar{2}2)$	$(101n)$	—
Co	(0001)	$(10\bar{1}1)$	$(10\bar{1}2), (11\bar{2}1)$	—
Zr	$(10\bar{1}0)$	$(0001), (10\bar{1}1), (11\bar{2}2)$	$(10\bar{1}2), (11\bar{2}n)$	—
Ti	$(10\bar{1}0)$	$(0001), (10\bar{1}1)$	$(10\bar{1}2), (11\bar{2}n)$	—
Be	(0001)	$(10\bar{1}0), (11\bar{2}2)$	$(10\bar{1}2), (11\bar{2}n)$	(0001)

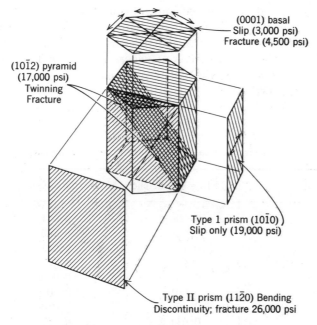

Stress values shown represent minimum critical
applied stresses

Fig. 11.9. Deformation modes in beryllium crystals of
nominal purity. The stress values shown represent minimum critical applied stress [42].

567

dislocations at an obstacle [44, 45, 46]. There is, in addition, some evidence of basal cleavage in magnesium [47] and titanium alloys [48] in conjunction with parting at twin-matrix interfaces, and basal cleavage in Mg_3Cd is known to be favored by planar glide owing to ordering [71]. Cadmium can be cleaved if wetted by liquid gallium [49] and quasicleavage, that is, fracture which closely resembles cleavage but is not known to follow a crystallographic plane [50], has been observed in titanium alloys in salt water solutions [51, 52].

In the next sections we will consider some of the important factors which influence the fracture behavior of HCP metals.

The fracture of beryllium

Beryllium can be melted and cast but, as the castings are brittle and difficult to machine [53], practically all the beryllium used in space, nuclear, and other applications is made from beryllium powders and undergoes powder metallurgy processing [54]. The powder is usually fabricated by the mechanical commutation of vacuum-cast ingots of magnesium-reduced beryllium pebbles or electrolytically produced beryllium flakes. A comparison of the impurity content of extruded ingots of pebble or flake powders is given in Table 11.5 [55]. Each of the powder particles is enveloped by BeO which tends to hinder grain growth during processing.

Table 11.5 [55]
Chemical Analysis of Beryllium Extrusions

	Pebble Ingot Extrusion	Flake Ingot Extrusion
Be, wt/o	0.106	0.041
Al	0.056	0.065
Si	0.119	0.061
Mg	0.001	0.022
Mn	0.004	0.0008
Cu	0.001	0.005
Ni	0.022	0.018
Cr	0.011	0.007
BeO	0.030	0.014
C-combined	0.040	0.004
C-free	0.009	0.011
N	—	0.004
Total	3640 ppm	2600 ppm

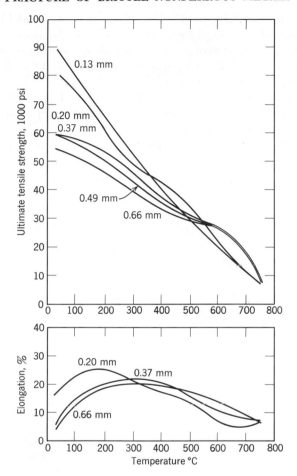

Fig. 11.10. Effect of grain size on elevated-temperature tensile properties of hot-extruded beryllium [56].

In accordance with the principles discussed in Chapter 6, both the strength and ductility increase with decreasing grain size (Fig. 11.10). The impurity content is also important. Hydrogen, which can embrittle titanium, is soluble to a negligible extent in beryllium and has no effect on the transition from ductile to brittle behavior [57]. Oxygen is able to embrittle beryllium, but there is no quantitative measure of the effect. There is some uncertainty as to the analysis of oxygen content. It has been stated that there is no certain proof that polycrystalline beryllium of less than 1000 ppm of oxygen has been prepared [58]. Single crystals have been prepared of 99.83% purity [59], and 8-pass zone refining under special controlled conditions yielded crystals of the lowest oxygen content [60], perhaps less than 10 ppm.

The properties are sensitive to fabrication method. The ductility of pressed block at room temperature is low ($\frac{1}{2}$–3% elongation). However, when some means of working the pressed block is employed that develops a preferred orientation, both the strength and ductility can be increased significantly. Extrusion and rolling tend to align the crystals with basal planes parallel to the axis of the rod or the surface of the sheet, which is ideal for tensile ductility. Cross rolling of sheet develops ductility in a second direction, and specially rolled sheet has in fact been developed, with up to 50% elongation in the plane of the sheet [61]. However, ductility in the third or transverse direction remains essentially zero, as evidenced by the lack of necking in a tensile test.

The effect of temperature on the tensile elongation of extruded pebble beryllium of Table 11.5 is shown in Fig. 11.11 [62]. Also shown are the various modes of fracture observed as a function of tempera-

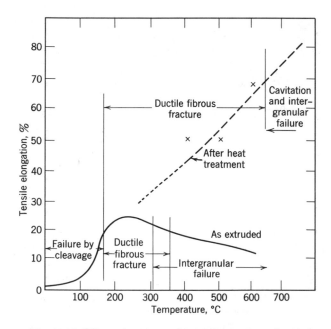

Fig. 11.11. Effect of optimum heat treatment on elevated-temperature ductility of extruded ingot beryllium [62].

ture. The purer flake material exhibited greater ductility and lower strength at all temperatures. The improvement in ductility after a heat treatment at 780°C for 120 hours was attributed to a precipitation reaction [56]. After heat treatment, specimens tested at 600°C failed by necking rather than by intergranular fracture.

Much study has been devoted to the determination of the factors responsible for the brittleness of beryllium. An important factor is that beryllium at low temperatures deforms almost exclusively by basal slip, so that there are but two independent slip systems, hardly enough to impart ductility to a polycrystalline aggregate (Chapter 6). Studies of single crystals show that the stress required to move dislocations on the prismatic planes is much higher than that for basal slip, and increasing the purity of the crystals does not alter this behavior very much at low temperatures, as shown in Fig. 11.12 [59]. It appears that a significant increase in ductility will not occur if only substitutional impurity atoms are removed, but the effect of interstitial atoms is not clearly established [41]. Interest in the role of interstitial atoms in ease of slip in HCP metals is due in part to the fact that, in titanium, interstitials are known to exert an important influence [63]. At a combined oxygen plus nitrogen level of 1000 ppm, the critical stress for basal and prismatic slip in titanium is about equal, whereas at a combined level of 100 ppm the stress for prismatic slip is several times lower than for basal slip. However, in the purest beryllium no such effect has been noted, although the stress for dislocation motion on basal planes is reduced. Carbon can increase resistance to slip, as has been demonstrated by the recontamination of purified beryllium [64]. There is thus no obvious evidence, however, that impurities are responsible for brittleness [65], which would mean that this is an intrinsic characteristic of pure beryllium. At present, data are insufficient to determine whether inability to cross slip or the overcoming of the Peierls-Nabarro stress is responsible for the high resistance to slip on the prismatic planes. The low ductility of beryllium (and zinc) as compared with other HCP metals appears to be due more to the fact that they cleave more easily [41], for other HCP metals also exhibit a wide variation in ease of slip on the basal and prism planes. Ease of twinning may be another factor that imparts greater ductility to HCP metals other than zinc and beryllium.

Alloying additions can improve the properties somewhat. For example, the addition of 5 w/% Cu to zone-refined beryllium resulted in an increase in the shear stress for basal slip which was about three times that for prism slip [64]. At elevated temperatures the creep

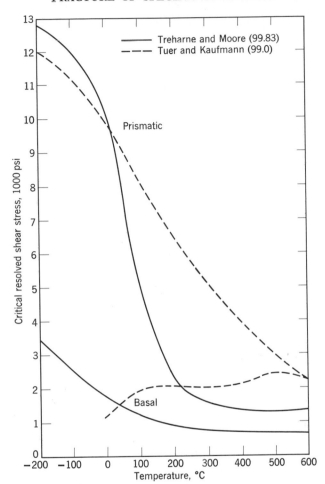

Fig. 11.12. Variation of shear stresses for slip on basal and prismatic planes with temperature. *After Treharne and Moore* [59].

rupture behavior can be improved by the addition of up to 3 w/% BeO. The stress for a 1000 hour rupture life at 1000°F increased from 4,300 psi to 11,000 psi as the BeO content increased from 0.90 w/% to 3.0 w/% [67]. It was also found that the material was notch-insensitive.

A recent approach to maintaining the high modulus and low density of Be but with greater overall ductility is the development of beryllium-aluminum alloys [68]. These alloys consist of a dispersion of beryllium particles in an aluminum matrix. Sheet specimens of these alloys can be bent through an angle of 90° without fracturing, whereas cross-rolled beryllium sheet fails at a bend angle of 26°. The total elongation in the plane of the sheet is largely unaffected by alloying, but Young's modulus decreases from 42×10^6 psi for Be to 26×10^6 psi for Be-43% Al. The corresponding density change is from 0.066 lb in.$^{-3}$ to 0.077 lb in.$^{-3}$.

The fracture of magnesium

The effect of test temperature and grain size on the fracture strengths of high-purity magnesium and of a Mg-2w/% Al alloy are shown in Fig. 11.13 [72]. These data reveal that each material exhibits two characteristically different types of fracture behavior. Over the higher range of temperatures the fracture stress decreases in a manner that parallels the flow-stress temperature relationship, and the fracture is of a ductile type. Over the lower range of temperatures the true fracture stress is independent of temperature but highly dependent upon grain size. As the grain size is decreased the fracture stress increases, and the temperature of transition to the low temperature fracture behavior range is reduced. The addition of aluminum to magnesium has the important effect of increasing the fracture stress at a given grain size as well as reducing the transition temperature.

As shown in Fig. 11.14, a linear relation exists between the fracture stress and the reciprocal of the square root of the grain size. Since this type of fracture dependence on grain size is observed for brittle materials such as steel, the term *brittle* is used in this case also, although up to 7% plastic strain may precede fracture. A possible factor responsible for this sometimes large plastic deformation is that fracture is related to the occurrence of twinning, which occurs over the temperature range from ambient down to 78°K. The stress for prismatic slip which results in plastic deformation is close to that for the twinning process which results in fracture [73].

In polycrystalline pure magnesium and in commercial magnesium alloys a significant number of microcracks and voids are initiated at a strain which is about equal to one-half of the total strain to fracture. A significant amount of the total plastic deformation therefore occurs as the microcracks grow and link up in the final stages of fracture [85]. The ductility of these alloys increases abruptly with increasing testing temperature at temperatures slightly above ambient. This ductility rise results from a microstructural instability, that is, recovery and recrystallization during straining, of the metals at the testing temperature [85, 86].

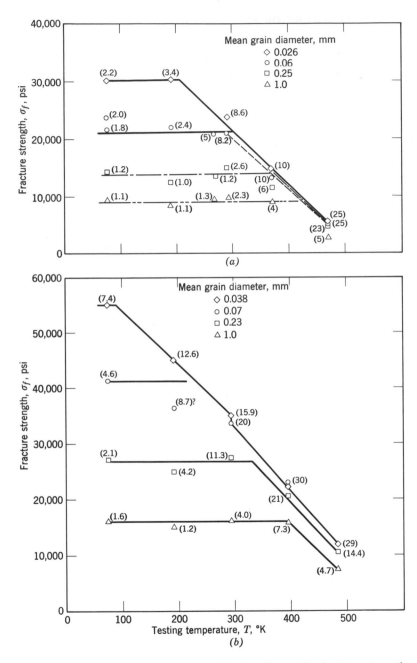

Fig. 11.13. Effect of temperature and grain size on the fracture strength of (a) magnesium and (b) Mg-2% Al alloy. Numbers in brackets are the percentage strains to fracture [72].

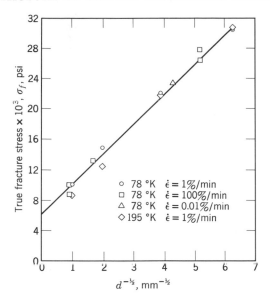

Fig. 11.14. Relation between fracture strength
and grain diameter for pure magnesium [72].

At room temperature, failure of polycrystalline high-purity magnesium occurs on various crystallographic planes of high index as
well as by an intergranular mechanism [74]. Fracture at low temperatures is also intergranular along with complex transcrystalline
modes [75]. Grain boundary shearing, while quite prevalent at
298°K, is rare at 78°K. Basal slip is the major deformation mechanism from 298°K to 78°K, but at the lower temperatures the inci-
dence of prismatic slip increases as grain boundary shearing decreases.
As shown in Fig. 11.15 [41], the stress for prismatic slip in magnesium
can be quite high, even relative to other HCP metals. The occurrence of prismatic slip in polycrystalline magnesium indicates that
high stresses are developed which cannot be relieved by grain boundary sliding. When both basal and prismatic slip are operative, wavy
glide is observed as cross slip from one system to the other occurs.

The addition of lithium to magnesium to produce a low density
alloy has an interesting effect on low-temperature fracture behavior
[76]. The stress for prismatic slip is reduced relative to that for
basal slip, and thereby the low-temperature ductility is increased.
The greater facility for slip is also reflected in a decrease in strain-
hardening behavior as barriers to plastic deformation can be more

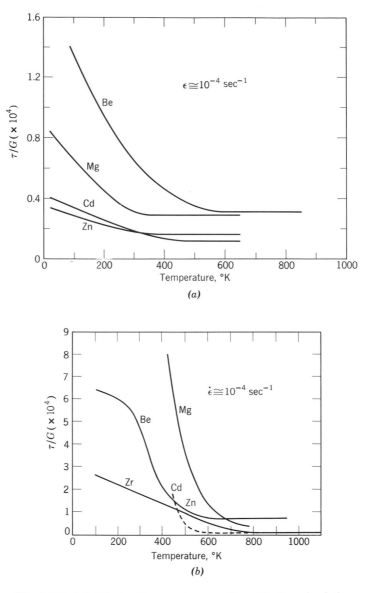

Fig. 11.15. (a) Effect of temperature on the critical resolved shear stress for basal slip in some HCP metals [41]. (b) Effect of temperature on the critical resolved shear stress for $\{10\bar{1}0\}$ prism slip in some HCP metals [41].

easily circumvented. In magnesium alloys of high lithium content (14.5%), fracture even at 78°K occurs by necking, and there is no leveling off of the fracture stress as in the case of pure magnesium. For this particular alloy system, the c/a ratio decreases with increasing lithium content and this may correlate with the greater ease of prismatic slip [76], although this is not true for other systems [71].

The fracture of titanium

Titanium alloys can be grouped into three categories according to the predominant phase or phases in their microstructure. In pure titanium the α-phase, an HCP structure, is stable to 1625°F, and the BCC β-phase is stable from 1625°F to the melting point of 3130°F. Certain alloying additions stabilize the α-phase. Among these are aluminum and the interstial elements carbon, hydrogen, nitrogen, and oxygen. Most alloying elements, such as chromium, columbium, copper, iron, manganese, molybdenum, tantalum, and vanadium stabilize the β-phase to the extent that a mixed $\alpha\beta$-alloy or an entirely β-phase alloy can persist down to room temperature. Some elements, notably tin and zirconium, behave as neutral solutes in titanium and have little effect on the transformation temperature, acting as strengtheners of the α-phase. The single-phase α-alloys are not heat-treatable, but both the $\alpha\beta$-alloys and the β-alloys can be strengthened by heat treatment. The yield strengths of these alloys range from 30,000 psi for commercially pure titanium up to 250,000 psi for certain $\alpha\beta$-alloys.

One group of elements that affects the low-temperature fracture behavior, especially of notched specimens, are the interstitials. Figure 11.16 shows the effect of interstial oxygen content on the strength of mildly notched specimens of an $\alpha\beta$-alloy tested at low temperatures [77]. Because of the deleterious effect of interstials, extra-low interstials grades (ELI) are available for use where structural reliability at low temperatures is of concern. A comparison of the interstial content of a standard α-alloy (Ti–5 Al–7.5 Sn) and its ELI counterpart is as follows:

	Interstial Content, w/% [8]	
	Standard Alloy	ELI Grade
C	0.15	0.05
H	0.02	0.015
N	0.07	0.04
O	0.2	0.12

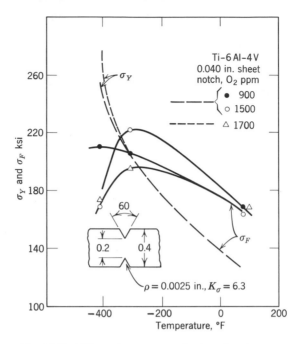

Fig. 11.16. Effect of oxygen content on low-temperature mild notch properties of annealed titanium alloy
sheet [8].

The low-temperature properties of α-alloys can also be adversely
affected by the presence of iron, as shown in Fig. 11.17 [78], and for
this reason the iron content is usually maintained at less than 0.15
w/% in α-alloys intended for application below $-320°F$, whereas the
iron content may ordinarily be as high as 0.5 w/%.   Another aspect
of low temperature fracture of titanium and its alloys is that fresh
fracture surfaces can be extremely reactive in liquid or gaseous oxygen environments.   Under certain conditions, explosive reactions and
the burning of fracture surfaces have been observed [79].   At elevated
temperatures there are no unusual aspects of fracture associated with
titanium alloys per se, with the exception of their susceptibility to
stress corrosion cracking when coated with solid salt (Chapter 9).
The fracture toughness of sheet material as expressed in terms of a
Kc evaluation is shown in Fig. 11.18 [80].   In this case the specimens were tested at $-423°F$, and Kc was determined as a function of
sheet thickness and cooling rate from the annealing temperature.   The
directionality of tensile properties of sheet and bar material is shown

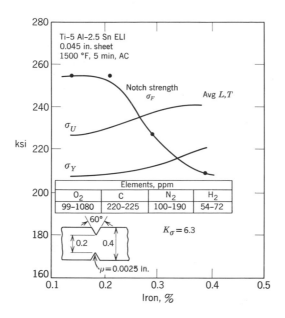

Fig. 11.17. Effect of iron content on −423°F
tensile and mild notch properties of ELI titanium
alloy sheet [8, 78].

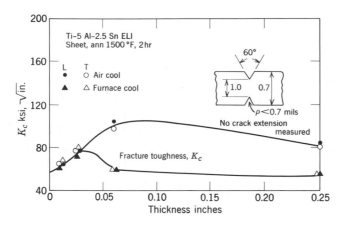

Fig. 11.18. Fracture toughness at −423°F as function of
thickness for ELI titanium alloy sheet cooled slowly and
rapidly from annealing temperature [8, 80].

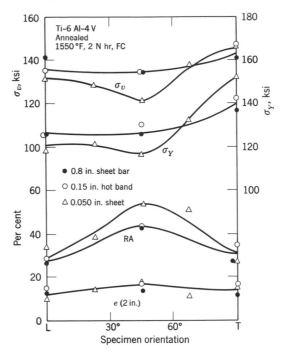

Fig. 11.19. Directionality of tensile properties for titanium alloy strip product in annealed condition [8, 81].

in Fig. 11.19 [81]. This directionality in properties in hexagonal metals as a result of preferred orientation developed during processing is made use of to obtain texture hardening in biaxial stress fields. α-Ti alloys tend to develop a preferred orientation with the basal plane parallel to the sheet surface, which inhibits strain in the thickness direction for a tensile stress parallel to the plane of the sheet [82]. The result of thickness strain inhibition is to raise the yield strength in a biaxial stress field beyond that of an isotropic material. The yield strength elevation can be calculated as follows [83]:

$$\frac{\sigma_1}{\sigma_Y} = \left[1 + \frac{\sigma_2{}^2}{\sigma_1{}^2} - \frac{\sigma_2}{\sigma_1} - \left(\frac{2R}{1+R}\right)\right]^{-\frac{1}{2}}$$

where, σ_Y is the unidirectional yield strength, σ_1 and σ_2 are the principal stresses, and R is the ratio of the true strain in the width direction to that in the thickness direction in a tensile test. For the Ti–5

Al–2 Sn α-alloy the value of R can be as high as 4.0 [82], which can give rise to considerable biaxial strengthening.

Some of the notch fracture-toughness characteristics of titanium alloy plate have also been investigated in order to provide alloy selection design and specification criteria for the use of titanium alloys as hull materials and for other structural applications [84]. Charpy V notch tests, and drop weight tear tests (DWTT) are used in this evaluation (see Fig. 3.25). The dependence of the Charpy V notch energy on temperature for different ranges of yield strength for high strength titanium alloy plate material is shown in Fig. 11.20. No sharp transition in impact properties over a narrow range of temperature is observed. Over the extremes of fracture toughness, temperature, and interstitial levels, the mode of fracture is dimpled, normal rupture. The decrease in impact strength with decreasing temperature results from the fact that the yield strength increases with decreasing temperature, as discussed in Chapter 7 for low-energy tear fracture in medium and high strength materials.

The relation between the energy absorbed in the Charpy test and that absorbed in the DWTT is shown in Fig. 11.21. Above 2500 ft-lb

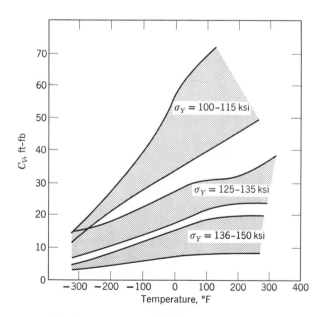

Fig. 11.20. Relation of Charpy V notch energy to temperature for different ranges of yield strength for high strength titanium alloy plate material [84].

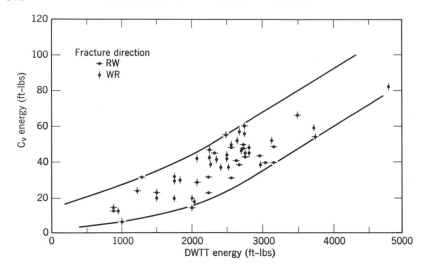

Fig. 11.21. Relation of Charpy V notch energy and drop weight tear energy for 1 inch thick titanium alloy plate [84].

DWTT energy there is a reasonably direct correlation between the two tests. Below the 2500 ft-lb DWTT energy value, the Charpy test shows a decreased sensitivity in fracture toughness as compared to the DWTT. Unfortunately this is also the region where most of the high strength titanium alloys above 110 ksi yield strength lie [84].

11.3   Summary

1. Commercially available group Va refractory metals (V, Nb, Ta) are considerably more ductile and have considerably lower transition temperatures than commercially available group VIa refractory metals (Cr, Mo, W). This results from the fact that the solubility of impurities in the Va metals is much higher than in the VIa metals. Consequently, at the impurity contents that normally exist in commercial materials, impurity atoms and/or second phase particles are segregated at grain boundaries in VIa metals. These lower the plastic work expended in microcrack formation γ_m, and hence lower the toughness. In Va metals the impurities are usually dissolved in the lattice.

2. Oxygen is a particularly strong embrittling agent for molybdenum and tungsten; nitrogen and sulfur are particularly harmful in wrought chromium. Hydrogen is the most effective embrittling agent for Va metals. The hydrogen embrittlement appears to result from increase in dislocation locking strength (k_y) and σ_i after hydrogenation.

3. After impurity content, grain size is the most important variable affecting the ductility and toughness of Va and VIa refractory metals. In general, ductility and toughness are increased by grain refinement.

4. Cold or warm working and alloying provide the most efficient means of increasing the ductility and toughness of refractory metals. The beneficial effects of working are thought to result from the production of a fibrous microstructure which permits splitting and crack-tip blunting when toughness is evaluated in the working direction. Alloying can improve ductility and toughness in VIa metals by lowering the tendency for impurity segregation at grain boundaries or by gettering out the impurities.

5. The fracture behavior of the HCP metals beryllium, magnesium, and titanium, together with that of some of their alloys, has been reviewed. Brittle fracture at low temperatures is common to these HCP metals and, in both titanium and beryllium, interstitial elements such as oxygen influence this behavior.

6. In each of these metals the decrease in the number of deformation modes with decreasing temperature is a common feature involved in low ductility fractures.

References

The asterisk indicates that published work is recommended for extensive or broad treatment.

[1] "The Status and Properties of Titanium Alloys for Thick Plate," *DMIC Report 185* (1963).

[2] "The Effect of Fabrication History anad Microstructure on the Mechanical Properties of Refractory Metals and Alloys," *DMIC Report 186* (1963).

[3] "The Engineering Properties of Columbum and Columbum Alloys," *DMIC Report 188* (1963).

[4] "The Engineering Properties of Tantalum and Tantalum Alloys," *DMIC Report 189* (1963).

[5] "The Engineering Properties of Molybdenum and Molybdenum Alloys," *DMIC Report 190* (1963).

[6] "The Engineering Properties of Tungsten and Tungsten Alloys," *DMIC Report 191* (1963).

[7] "Current Methods of Fracture Toughness Testing of High Strength Alloys with Emphasis on Plane Strain," *DMIC Report 207* (1964).

*[8] *Aeropace Materials Handbook, Non Ferrous Metals,* Vol. II, Syracuse Univ. Press. (1965).

*[9] J. H. Bechtold, E. T. Wessel and L. L. France, *Refractory Metals and Alloys* Vol. I, Interscience, New York (1961), p. 25.

*[10] W. S. Owen and D. Hull, *Refractory Metals and Alloys,* Vol. II, Interscience, New York (1963), p. 1.

*[11] G. T. Hahn, A. Gilbert and R. I. Jaffee, *Refractory Metals and Alloys,* Vol. II, Interscience, New York (1963), p. 23.

*[12] L. L. Siegle and C. D. Dickinson, *Refractory Metals and Alloys,* Vol. II, Interscience, New York (1963), p. 65.

*[13] A. G. Imgram, E. S. Bartlett and H. R. Ogden, *Trans. AIME,* **227,** 131 (1963).

[14] A. J. Lawley, J. Van Den Sype and R. Maddin, *J. Inst. Metals,* **91,** 23 (1962).

[15] L. E. Olds and G. W. P. Rengstorff, *Trans. AIME,* **206,** 150 (1956).

[16] A. Gilbert et al, *Acta Met.,* **12,** 754 (1964).

[17] A. Gilbert, C. N. Reid and G. T. Hahn, *J. Inst. Metals,* **92,** 351 (1964).

[18] B. C. Allen, D. J. Mayhuth and R. I. Jaffee, NASA TN D-837 (1961).

[19] B. A. Loomis and O. N. Carlson, *Reactive Metals,* Interscience, New York (1959), p. 227.

[20] R. W. Thompson and O. N. Carlson, *J. Less Common Metals,* **9,** 354 (1965).

[21] A. L. Eustice and O. N. Carlson, *Trans. AIME,* **221,** 238 (1961).

[22] B. A. Wilcox and R. A. Huggins, *J. Less Common Metals,* **2,** 292 (1960).

[23] B. W. Roberts and H. C. Rogers, *Trans. AIME,* **206,** 1213 (1956).

[24] A. W. Magnusson and W. M. Baldwin, Jr., *J. Mech. Phys. Solids,* **5,** 172 (1957).

[25] H. L. Wain et al, *J. Inst. Metals,* **86,** 281 (1957).

[26] J. A. Belk, *J. Less Common Metals,* **1,** 50 (1959).

[27] J. Orehotsky and R. Steinitz, *Trans. AIME,* **224,** 556 (1962).

[28] P. Beardmore and D. Hull, *J. Less Common Metals,* **9,** 168 (1965).

[29] B. S. Lement and K. Kreder, to be published.

[30] R. J. Stokes and C. H. Li, *Trans. AIME,* **230,** 1104 (1964).

[31] B. C. Allen, C. N. Reid and R. I. Jaffee, *BMI Summary Report to NASA* (1962).

[32] W. D. Klopp, F. C. Holden and R. I. Jaffee, *BMI TR Nonr. 1512* (1960).

[33] R. I. Jaffee, D. J. Mayhuth and R. W. Douglass, *Refractory Metals and Alloys,* Vol. I, Interscience, New York (1961), p. 383.

[34] J. L. Ratliff, D. J. Mayhuth and H. R. Ogden, *Trans. AIME,* **230,** 490 (1964).

[35] A. Gilbert, J. L. Ratliff and W. R. Warke, *Trans. ASM,* **58,** 142 (1965).

[36] R. T. Begley and J. H. Bechtold, *J. Less Common Metals,* **3,** 1 (1961).

[37] L. L. Sherwood, F. A. Schmidt and O. N. Carlson, *Trans. ASM,* **58,** 403 (1965).

[38] K. Farrell, A. C. Schafhauser and J. O. Stiegler, to be published.

[39] *Metals Handbook,* ASM, Metals Park, Ohio (1961).

[40] S. F. Pugh, *Phil. Mag.,* **45,** 823 (1954).

*[41] H. Conrad and I. Perlmutter, *Tech. Report AFML-TR-65-310* (presented at the Grenoble Conference on Beryllium Metallurgy, November 1965).

[42] G. L. Tuer and A. R. Kaufmann, *The Metal Beryllium,* ASM, Metals Park, Ohio (1955), p. 372.

[43] A. N. Stroh, *Phil. Mag.,* **3,** 597 (1958).

[44] J. J. Gilman, *Trans. AIME,* **212,** 783 (1958).

[45] M. Kamdar and A. R. C. Westwood, *Environment-Sensitive Mechanical Behavior,* Gordon and Breach, New York (1966).

[46] R. Bullough, *Phil. Mag.*, **9**, 917 (1964).

[47] J. G. Byrne, *Deformation Twinning*, Gordon and Breach, New York (1964), p. 397.

[48] H. I. Burrier, M. F. Amakau and E. A. Stiegerwald, *Tech. Report AFML-TR-65-239* (July 1965).

[49] N. S. Stoloff and T. L. Johnston, *Acta Met.*, **11**, 251 (1963).

[50] C. D. Beachem, B. F. Brown and A. J. Edwards, *NRL Memo Report 1432* (June 1963).

[51] A. J. McEvily and A. P. Bond, *Environment-Sensitive Mechanical Behavior*, Gordon and Breach, New York (1966).

[52] R. W. Judy et al, *NRL Report 6330* (1966).

[53] P. Corzine and A. R. Kaufman, *The Metal Beryllium*, ASM, Metals Park, Ohio (1955), p. 136.

[54] H. H. Hausner, *Beryllium: Its Metallurgy and Properties*, Univ. Cal. Press (1965), p. 33.

[55] A. B. Brown, F. Morrow and A. J. Martin, *J. Less Common Metals*, **3**, 62 (1961).

[56] W. W. Beaver and K. G. Wikle, *Trans. AIME*, **200**, 559 (1954).

*[57] P. Cotterill, R. E. Goosey and A. J. Martin, *The Metallurgy of Beryllium*, Inst. of Metals, 1961 London Conf. (1963), p. 220.

[58] G. E. Darwin and J. H. Buddery, *Berylium*, Butterworth, London (1960).

[59] P. I. Theharne and A. Moore, *J. Less Common Metals*, **4**, 275 (1962).

[60] M. Herman and G. E. Spangler, *The Metallurgy of Beryllium*, Inst. of Metals, 1961 London Conf. (1963), p. 75.

[61] D. W. Lillie, *The Metal Beryllium*, ASM, Metals Park, Ohio (1955), p. 304.

*[62] A. J. Martin and G. C. Ellis, *The Metallurgy of Beryllium*, Inst. of Metals, 1961 London Conf. (1963), p. 3.

[63] A. T. Churchman, *Proc. Roy. Soc.*, **A226**, 216 (1954).

[64] C. E. R. Tristem and A. Moore, Second Int. Conf. on Beryllium, Philadelphia (October 1964).

[65] A. R. Kaufman, *The Metal Beryllium*, ASM, Metals Park, Ohio (1955), p. 367.

[66] G. E. Spangler et al, *Franklin Inst. Reports Q-B1993-5 and Q-B2089-1* (1963).

[67] R. G. O'Rourke et al, U.S. A.E.C., Physics Section, *Report COO-312* (1956).

[68] R. W. Fenn et al, *J. Spacecraft*, **2**, 87 (1955).

[69] N. S. Stoloff and R. G. Davies, *J. Inst. Met.*, **93**, 127 (1964–65).

[70] M. H. Richman, ARPA SD-86, *Report E28*, Brown Univ., Providence, R.I. (February 1966).

[71] N. S. Stoloff and R. G. Davies, *Trans. ASM*, **57**, 247 (1964).

*[72] F. E. Hauser, P. R. Landon and J. E. Dorn, *Trans. AIME*, **206**, 589 (1956).

[73] R. E. Reed-Hill and W. D. Robertson, *Acta Met.*, **5**, 728 (1957).

[74] F. E. Hauser et al, *Trans. ASM*, **47**, 102 (1955).

[75] F. E. Hauser, P. R. Landon and J. E. Dorn, *Trans. ASM*, **48**, 986 (1956).

[76] F. E. Hauser, P. R. Landon and J. E. Dorn, *Trans. ASM*, **50**, 856 (1958).

[77] Titanium Metals Corporation of America, Tech. Service Dept. (1962).

[78] A. J. Hatch and C. W. Field, *TMCA Tech. Report,* TMCA, Henderson, Nevada (June 27, 1963).

[79] J. D. Jackson, W. K. Boyd and P. D. Miller, Battelle Memorial Inst., *DMIC Memo 163* (Jan. 15, 1963).

[80] J. L. Shannon and W. F. Brown, *Proc. ASTM,* **63,** 809 (1963).

[81] A. E. Leach, Final Report, ASD 62-7-675, Crucible Steel Co., Air Force Contract AF 33(600)-37938 (September 1961).

[82] W. A. Backofen, W. S. Hosford and J. J. Burke, *Trans. ASM,* **55,** 264 (1962).

[83] R. Hill, *Proc. Roy. Soc.,* **A193,** 287 (1948).

[84] R. W. Huber and R. J. Goode, *NRL Report 6228* (April 1965).

[85] N. W. Toaz and E. J. Ripling, *Trans. AIME,* **206,** 936 (1956).

[86] C. S. Roberts, *Magnesium and Its Alloys,* Wiley, New York (1960), p. 103.

12

Fracture of Nonmetallic Materials

Nonmetallic materials constitute a class of structural materials of increasing technological importance. One reason for the greater interest in these materials stems from the high ratios of strength to density obtainable, especially in fiber or whisker form, as indicated in Table 12.1 [1, 2, 3]. However, perhaps more so than in the case of metals, fracture behavior often constitutes a severe limitation to an even more widespread usage of these materials. In this chapter the influence of structure on the fracture characteristics of three types of nonmetallics—glasses, crystalline ceramics, and polymers—will be considered.

12.1 The Fracture of Glasses

Glass is a hard substance, usually brittle and transparent, and commonly composed of silicates and alkali metals fused at a high temperature. Innumerable kinds may be produced but, structurally, the simplest silicate glass is pure fused silica which, on the average, is a random arrangement of silica tetrahedra, with regions of more or less ordered arrangements present [4]. In this structure, oxygen atoms occupy the corners of the tetrahedra and silicon atoms occupy the centers [5]. The bonds between the silicon and oxygen atoms are partially covalent, and on cooling from an elevated temperature there is a tendency to build up a chain network of tetrahedra, which, given sufficient time, would lead to the formation of crystal structure. However, the usual cooling rates are much too fast for this close packing of tetrahedra to result, and as a consequence an open amorphous structure is frozen in below the fictive, or glass transition, temperature. The structure of glasses can be clearly distinguished from the structure of liquids since below the fictive temperature, Fig. 12.1, the glass structure is independent of temperature [4]. On cooling the liquid there is a discontinuous change in volume at the melting point if crystallization occurs. However, if no crystallization occurs, the

Table 12.1 [1, 2, 3]
Strength of Fibers

Material	UTS, psi	E, psi $\times 10^{-6}$	UTS/ E	Specific Gravity	UTS/ SG, psi	E/SG psi $\times 10^{-6}$
Nylon fiber	72,000	0.7	0.102	1.07	67,000	0.65
Nylon molded	9,000	0.3	0.030	1.1	8,800	0.27
Cellulose-hemp	130,000	8.25	0.016	1.5	87,000	5.5
Wood fiber	130,000	10.5	0.012	1.5	87,000	7.0
Asbestos fiber	216,000	26.5	0.008	2.4	90,000	10.1
Quartz fiber	3,500,000	10.0	0.350	2.65	1,329,000	3.28
Glass fiber	500,000	10.8	0.046	2.4	208,000	4.5
Al_2O_3 whisker	2,200,000	72	0.031	4.0	550,000	18.0
Steel piano wire	350,000	30.0	0.012	7.8	45,000	3.85
7075-TG Al alloy	80,000	10.5	0.008	2.8	28,500	3.75
Magnesium alloy	43,000	6.5	0.006	1.8	23,800	3.6
Iron whisker	1,900,000	43	0.044	7.8	244,000	5.5
WC	—	104	—	15.63	—	6.7
Diamond	—	170	—	3.51	—	48.5
NaCl	160,000	6.3	0.025	2.165	—	2.9
Silicon whisker	940,000	24	0.039	2.3	408,000	10.4
Silicon (bulk)	750,000	24	0.031	2.3	326,000	10.4
Boron	350,000	51	0.007	2.3	152,000	22.0
Ausformed steel	450,000	30	0.015	7.8	57,800	3.8
TiC	800,000	70	0.011	4.9	163,000	14.3
Methyl-meth- acrylate	7,000	0.4	0.018	1.2	5,800	0.33
Polystyrene	8,500	0.5	0.017	1.05	8,100	0.48
BeO whisker	2,800,000	49	0.057	3.01	—	—
Beryllium (fine grain)	95,000	45	0.002	1.8	53,000	25
Epoxy-filament wound	250,000	7.0	0.036	2.2	113,000	3.2
Phenolic-glass fabric	58,000	3.4	0.008	1.8	32,000	1.9

volume of the liquid decreases at about the same rate as above the melting point until there is a decrease in the expansion coefficient at a range of temperatures called the transformation range. Below this temperature range the glass structure does not relax at the cooling rate employed, and the expansion coefficient for the glassy state is very nearly equal to that found for the crystalline solid [4]. The appearance of a two-dimensional array of atoms in a crystalline and

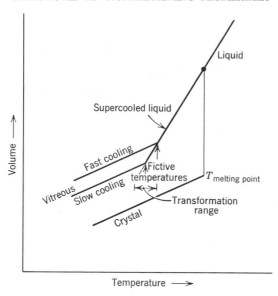

Fig. 12.1. Relation between liquid, crystalline, and vitreous states [4].

in a glassy state as envisioned by Zachariasen and Warren [6] is shown in Fig. 12.2. Because of the open nature of the glassy state, the imposition of a sufficiently high hydrostatic pressure can lead to its collapse into a more densely packed arrangement [7].

This silicate structure is characterized by a high hardness and a high modulus of elasticity. Both the structure and its properties can be altered by the addition of bulky alkali metals and alkali earth ions to form the type of structure shown in Fig. 12.3. The binding of these bulky ions is of an ionic nature, which is weaker than the co-valent Si-O bond. This weaker binding of the bulky ions allows them to diffuse more readily through the structure, a fact that can be important in reactions with the environment [5]. In the network structure shown in Fig. 12.2b the oxygen ions are referred to as bridging ions, and the alkali-oxygen assemblies of Fig. 12.3 are called terminal structures. These terminal structures relax the rigidity of the silicate structure as indicated by the decrease of Young's modulus, Fig. 12.4, as the number of terminal structures increase [5]. The decrease of modulus with increase of sodium leads to a corresponding reduction of the theoretical strength of these soda glasses, which are referred to as soft.

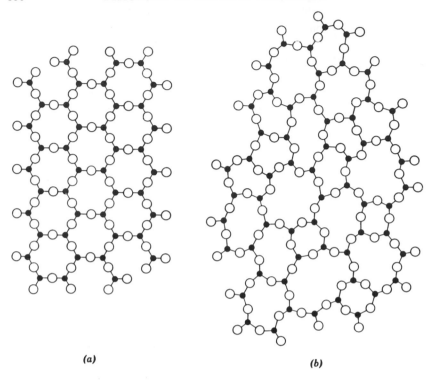

(a)   *(b)*

Fig. 12.2. (*a*) Regular crystalline lattice. (*b*) Corresponding irregular glassy network. *After W. H. Zachariasen and B. E. Warren* [6].

Certain compositions of glass can be converted into fine-grained ($< 1\mu$) crystalline structures with the glassy phase a minor component. These glass-ceramics (e.g., Pyroceram) have interesting properties; for example, the thermal expansion coefficients are only $\pm 1 \times 10^{-7}$ per °C in the range from 0°–750°C, an important characteristic for thermal shock resistance. In addition, flexural strengths range up to 50,000 psi [8].

Observations of glass strength

When Griffith conducted his now classic investigations of the fracture behavior of brittle materials [9] he used glass in the experimental portion of the work. His observations of the breaking strength of glass fibers, Fig. 12.5, led him to conclude that the strength decreased with increasing fiber diameter because of an intrinsic size effect. This view was widely held until recently, when it was found that variables involved in the processing of glass fibers (e.g., the speed and temperature of draw) have a marked effect on properties, and that Griffith probably had had to vary the drawing conditions to obtain a range of fiber sizes [10]. It has been shown [10, 11] that if fibers of different diameter are produced under nearly identical conditions, the breaking strengths are identical within experimental limits and there is no measurable effect of diameter as such; near-theoretical strengths have been achieved with glass rods as much as $\frac{1}{4}$ in. in diameter [12]. It is now recognized that the strength of glass is controlled by *surface defects*. For example, the removal of surface layers by etching leads to a strength which is independent of glass diameter. The fact that glasses fail below their theoretical strength is due to the presence of surface flaws introduced in processing or handling rather than a statistical distribution of flaws throughout the bulk. The importance of surface flaws as stress raisers in brittle

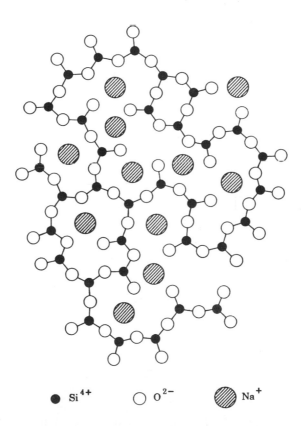

$\bullet$ Si^{4+} $\bigcirc$ O^{2-} ◐ Na$^+$

Fig. 12.3. Schematic representation of a sodium silicate glass. *After B. E. Warren* [6].

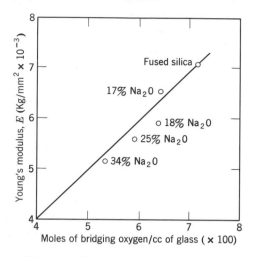

Fig. 12.4. Young's moduli of $SiO_2 \cdot Na_2O$ glasses as a function of bridging oxygen atom density [5].

materials can be seen from Fig. 12.6 [13], which is based on the analysis of Elliot [14]. For stresses at the tip of a flaw up to the order of $E/100$, a level which begins to approach estimates of the theoretical strength, the change in tip radius is small enough that the stresses and displacements can be considered to be linear functions of the applied

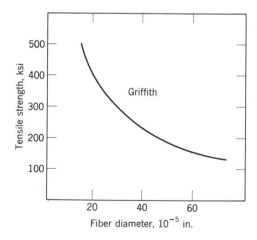

Fig. 12.5. Variation of tensile strength with fiber diameter, according to Griffith [9].

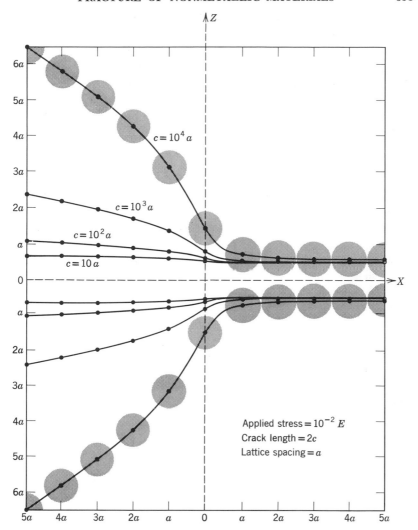

Fig. 12.6. Atomic displacements near the tips of cracks of various lengths [13].

stress. In more precise calculations small correction factors have been included to incorporate the effects of change in tip geometry for large deformations [15]. In the presence of an atomically sharp flaw of depth of the order of 10^4 atoms, the tip stress can easily exceed the applied stress by a factor of 100. Once the strength of the first bond at the tip of a flaw in a truly brittle material is exceeded, the crack must advance catastrophically; there is no subcritical growth (in the absence of environmental effects) as in the case of materials capable of plastic deformation. A great virtue of plastic deformation is its ability to blunt the tip radius and thereby reduce the magnitude of the stress concentration (Chapter 5). It is the inability to relax stress concentrations which makes high strength elastic materials so suscepti-ble to surface flaws. However, as will be discussed, there is evidence that even glasses are capable of some stress relaxation through local-ized viscoelastic deformation.

Static fatigue

When glass is subjected to sustained tensile loading, fracture which is known as *static fatigue* can occur. The nature of the stress and time dependence of this process is illustrated in Fig. 12.7 [15]. This behavior is the result of interaction with the environment, and the effects have been recognized in times as short as 0.01 sec [16], there-fore, even the determination of the tensile strength may be affected. Below a certain stress, failure does not occur even for long times under load, and there exists then a static fatigue limit. Orowan [17] attrib-uted the loss in strength under load to a reduction in surface energy as the result of gaseous adsorption on flaw surfaces. Studies have shown that water vapor (perhaps modified by CO_2) is primarily re-sponsible for the weakening [16], and that the sensitivity increases with alkali content of glass. Cyclic and static fatigue experiments of soda-lime glass [18] indicate that the fatigue limits for steady loads and cyclically applied loads were essentially the same. When higher loads were applied, the time for failure under cyclic loading became

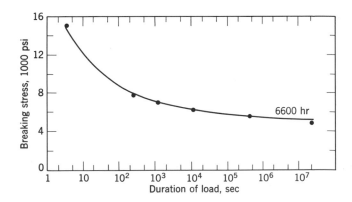

Fig. 12.7. Stress-time characteristics of annealed soda-lime glass rods, ¼ in. in diameter, in bending [15].

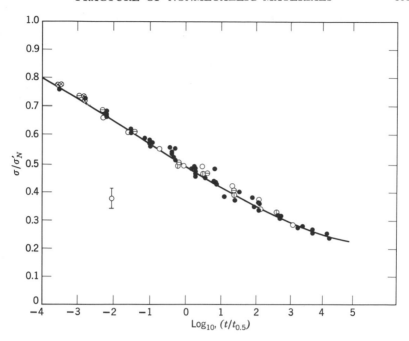

Fig. 12.8. Universal fatigue curve for various abrasions [21].

very much less than under steady loading. This characteristic suggests that below the fatigue limit no structural changes (inelastic
effects) occur at the crack tip, but that as the stresses are increased
this is no longer true [19]. This behavior is to be expected in a material capable of but limited plastic deformation, if any at all.

The susceptibility of soda-lime glass to static fatigue as a function
of temperature is a maximum at 200°C [20]. For temperatures below
200°C, the rate of the reaction, which is thermally activated [5], decreases with temperature and is essentially nil at the temperature of
liquid nitrogen. At temperatures above 200°C there is an increase in
fatigue resistance which may be due to pronounced blunting of the
crack tip and a consequent reduction in the associated stress concentration factor. If the surface of glass specimens are abraded in controlled fashion to introduce surface defects, it has been found that a
universal fatigue curve can be drawn, as shown in Fig. 12.8 [21].
The abscissa of this plot is the strength of the glass at room temperature relative to its strength at liquid nitrogen. The ordinate is the
logarithm of a reduced time to failure, $\ln(t/t_{0.5})$, where $t_{0.5}$ is the
time to failure of a specimen loaded to one-half of its strength at the
temperature of liquid nitrogen.

In developing a mechanism by which a flaw grows to critical size as a result of interaction with environmental constituents, it has been considered that the reaction of terminal structures with water vapor can lead to the advance of the crack by a dissolution process or the formation of a weak corrosion product [5]. Chemical reaction is favored at the tip of a crack because of the higher diffusivity of sodium ions in a more open region of high tensile stress. Diffusion of sodium exposes an oxygen atom and allows a sequence of reactions to be triggered, resulting in the dissolution of the initial glass structure. From the temperature dependence of the time to break, an activation energy of 20 kcal per mol can be computed which is close to that for diffusion of sodium.

A more generalized thermodynamic approach [22] has considered further a mechanism based on stress-enhanced reactivity at the tip of a crack. Figure 12.9 indicates three shapes of the crack tip which may develop during static fatigue. In this figure K_σ is equal to $2(c/\rho)^{1/2}$, the elastic-stress concentration factor at the tip of a sharp flaw. At high stresses, preferential attack at the tip occurs, dK_σ/dt is positive, and the flaw grows in severity until the stress at the tip reaches the bond strength, whereupon brittle fracture occurs. The fatigue limit corresponds to dissolution in a manner to maintain the ratio c/ρ, and hence the stress concentration at a constant value. Below this level the stress is insufficient to cause preferential attack at the tip, and the stress-raising effects of the flaw are reduced (Joffe effect) [23, 24].

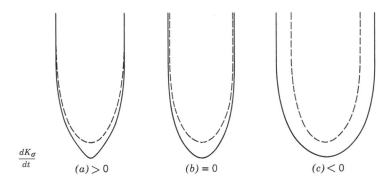

$$\frac{dK_\sigma}{dt}$$ $(a) > 0$ $(b) = 0$ $(c) < 0$

Fig. 12.9. Hypothetical changes in flaw geometry due to corrosion or dissolution. (a) Flaw sharpening as a result of stress corrosion. (b) Flaw growth such that the rounding of the tip by stress corrosion balances the lengthening of the flaw. (c) Simple rounding by corrosion or dissolution (Joffe effect) [22].

Table 12.2 [28]

Fracture Characteristics of Glasses†

Glass	Breaking Stress, psi	Fracture Energy, ergs/cm^2	Elastic Modulus, psi $\times 10^{-6}$	Surface Energy, ergs/cm^2
Lead alkali	6,800	1160	8.8	245
Soda-lime	9,000	1700	10.5	336
96% silica	11,000–12,000	3070	9.6	—
Borosilicate	12,000	3480	9.2	348
Aluminosilicate	13,800	3340	12.7	353

† Initial crack depth 0.002 in.; fracture time, 1 sec.

In this mechanism of static fatigue, glass is treated as a brittle, elastic solid. However, there is growing evidence that glass is capable of plastic deformation, especially at the high stresses which exist at the tip of a crack. This capacity for plastic deformation increases as the number of terminal structures increase, and it is these glasses which are most susceptible to static fatigue. Evidence that glass is capable of undergoing plastic deformation is given by the following observations [25]:

1. Hardness impressions can be made in soda glass which appear similar to those observed in hard metals [26]. An estimated flow stress based on these observations is 500,000 psi. Since glass lacks any capacity for strain hardening, the strength of glass would equal this flow stress, with the glass failing as the result of plastic instability [25].

2. Residual stresses are observed at the roots of cracks when load is removed [27], and the cracks do not close completely. A surface depression, as in cyclic fatigue of metals, is created during unloading [28].

3. The value of the fracture energy of glass computed by means of the Griffith equation is many times greater than the true surface energy, indicating, as for metals, that plastic deformation is involved in the fracture process [19], Table 12.2 [28].

4. Soft glasses fail to reach the predicted terminal velocities for brittle solids [29].

5. Glasses are able to flow under high hydrostatic pressure [6].

It therefore appears to be quite possible that the mechanism of static fatigue in glass is similar to that envisioned for the stress corrosion cracking of metals such as brass in a tarnishing solution. (See Fig. 9.21, Chapter 9.) Environmental constituents attack the structure at the tip of a flaw, creating a weak corrosion product which ruptures, but the crack, being subcritical in length, is blunted out at the corrosion-product-glass interface. The process is repeated until the Griffith length is reached and catastrophic failure occurs.

In order to avoid static fatigue it is necessary to protect the surface from environmental attack or to introduce compressive residual stresses into the surface layers. By coating the polished surface of a glass with a protective lacquer, tensile stresses of 150,000 psi have been sustained for 28 years [30]. Three general classes of prestressing materials to produce a surface compressive stress are recognized: physical chill tempering, high temperature lamination, and ion exchange. In *chill tempering*, the most widely used method, a glass having a relatively high thermal expansion coefficient is chilled rapidly from a high temperature unstressed state. The surface freezes in a low-density state characteristic of the high temperature, while the slower cooling interior shrinks before it freezes. The magnitude of the surface compressive stress developed by this method is usually 10,000–15,000 psi [30]. Fracture of chill-tempered glass often involves a characteristic "dicing" or fragmenting of the entire glass surface [31].

The *high temperature lamination method* involves the encasing of a glass by another glass of lower coefficient of thermal expansion, so that, on cooling, the desired surface compressive stress will develop in the outer layer. Calculations indicate that stresses as high as 500,-000 psi can be induced by this technique [30]. In the *ion-exchange method* large alkali metal ions are diffused into the surface of glass and replace smaller alkali metal ions in the surface layers, at temperatures and times too low for the resulting stress to be relaxed by the silica network. The induced compression stress varies with the size differential of the alkali metal ions and with the ion concentration. Calculated compressive stresses induced by this method also approach 500,000 psi. It is also possible for diffusion treatments to have an adverse effect. Cracks produced by sodium vapor treatment on the surface of glass are fractures due to a superficial tensile stress developed on cooling in a surface layer of higher thermal expansion [32].

12.2 The Fracture of Crystalline Ceramics

Ceramics are glasses or crystalline solids, usually brittle, which are composed in large part of inorganic, nonmetallic materials. If it were not for their brittle nature and attendant sensitivity to surface flaws, crystalline ceramics would be attractive structural materials. For example, MgO has a low density, 0.13 lb in.$^{-3}$, a high modulus, 35×10^6 psi, and a high melting point, 5072°F. It is possible that once the fracture characteristics and limitations of crystalline ceramics are better understood, they may find a wider range of applicability [33]. Ionic crystals such as LiF and MgO have, of course, already proved to be quite valuable in studies of plastic flow and fracture because of the relative simplicity of these processes in ionic crystals, and because individual dislocations can be revealed by suitable etchants [34]. Also the fact that these crystals are transparent and become optically birefringent when stressed allows the stress fields associated with dislocation pile ups to be revealed by photoelastic techniques.

Slip and ductility

In any consideration of the ductility of crystalline ceramics, three factors are most important. First is the stress level required for dislocation motion relative to that for fracture, second is the density of mobile dislocation sources [69], and third is the number of independent deformation modes available. These include slip, twinning, grain boundary sliding, and, at sufficiently high temperatures climb and the Herring-Nabarro creep mechanism, which can enable otherwise brittle polycrystalline materials such as Al_2O_3 and BeO to deform [35, 36]. An important factor affecting the inherent ease of dislocation motion is the nature of the atomic binding forces within the lattice. If this binding force is of a strongly covalent-ionic nature as in Al_2O_3, then dislocation motion below $T/T_M = 0.6$ is extremely difficult owing to the high Peierls force associated with covalent binding. Such materials, lacking the ability to relax stress concentrations by plastic deformation, are extremely sensitive to surface damage. For Al_2O_3 only at elevated temperatures can this lattice resistance be overcome to allow bulk slip to occur [36–40]. If the binding is of an ionic nature as in MgO, dislocations can move more readily, although the lattice resistance does increase rapidly at low temperatures. If the binding is of a mixed ionic-metallic nature as in AgCl [41], then dislocation motion occurs readily.

For a crystal to be able to undergo a general homogeneous strain by slip, five independent slip systems are necessary [42]. The six independent components of the general strain tensor are reduced to these five since plastic flow usually occurs without a change in volume. Five slip systems are independent if the operation of any one produces a change in shape of a crystal which cannot be produced by a suitable combination of amounts of slip on the other four systems [43].

Ionic crystals with the sodium chloride structure possess the family of slip systems $\{110\}$ $\langle 1\bar{1}0 \rangle$, and there are six physically distinct systems. However, only two of these are independent since the change in shape produced by any one of the six can be produced by an appropriate amount of slip on two others. Because the remaining three slip systems correspond to the interchange of slip plane and slip direction with the first three, there are in all only two independent systems. A general deformation is thus not possible by slip [43]. A crystal slipping on the $\{110\}$ $\langle 1\bar{1}0 \rangle$ system cannot be twisted about a $\langle 001 \rangle$ axis, nor extended along $\langle 111 \rangle$. Even crystals extended 10% or more in a $\langle 100 \rangle$ direction are brittle in the sense that they fail by cleavage without necking, since the complex deformation involved in the necking process cannot be developed by just two independent slip systems. Polycrystalline specimens of ionics such as MgO are incapable of extensive plastic extension because of their limited ability to develop compatible deformations across grain boundaries.

An increase in temperature can improve ductility as the stress for dislocation motion on other slip systems approaches that for motion on the primary slip system. In fact, necking of high-purity NaCl crystals is observed at $300°F$ [44], and at room temperature 0.50% plastic strain in polycrystalline NaCl has been obtained [45], where secondary slip systems may have been active. One of the aims of research with ionic (and also HCP) materials is to learn why slip on these secondary systems is so difficult at low temperatures. Hopefully, through control of impurities or by alloying, these secondary systems may be made to operate more freely to provide the necessary deformation modes at about the same critical resolved shear stress. If this can be achieved, then it may be possible to employ the usual techniques of dispersion strengthening to obtain strength as well as ductility [46]. Thus far, however, solid solution alloying additions of KBr added to KCl [47] and NaCl added to AgCl [48] have only *increased* the planarity of glide and have thereby increased the ductile-brittle transition temperature. The addition of discrete particles of Al_2O_3 to AgCl, on the other hand, has resulted in an improvement in that it has made crack propagation more difficult and thereby reduced the ductile-brittle transition temperature [49].

Slip of course is important in the propagation as well as the initiation stages of fracture. For slow rates of crack growth, plastic deformation occurs at the crack tip and tends to blunt the tip shape and generally retard propagation (Chapter 5). As the stress is increased, the crack grows in length and velocity, and less energy is required to create new fracture surface since the stress field of the crack tip moves at too high a velocity to activate plastic deformation. As shown in Chapter 5, this results from the low density of mobile dislocation sources that exist in these crystals [69]. For LiF it has been found that for crack velocities in excess of only $\frac{1}{100}$ of the terminal velocity of crack propagation (6×10^5 cm per sec) no dislocation loops were produced as the crack propagated [50, 51]. Under these conditions the tip radius of the crack must approach atomic dimensions. At a smaller average velocity, $\frac{1}{200}$ of the terminal velocity, crack propagation was found to be unstable. The velocity oscillated from a high value to a low value and, in regions of low velocity, dislocations were observed to have formed.

The fracture surface energy in ionic crystals can be estimated largely on the basis of Coulombic interaction forces, and the following table gives a comparison of theoretical and experimentally determined surface energies. Agreement between the theory and experiment is reasonably good, except perhaps in the case of the NaCl crystals. For these, the greater ease of plastic deformation is reflected in a higher value of the experimental surface energy.

| Crystal | Surface Energy γ_s (ergs per cm^2) [50] | |
	Theoretical Value	Experimental Value
MgO	1362	1200
LiF	700, 169	340
NaCl	77, 155, 130, 188	356–405, 276, 300, 366
KCl	56, 134, 195	

Fracture of ionic single crystals and polycrystals

The appearance of slip lines and dislocation etch pits on the {100} surface of a freshly cleaved and bent MgO crystal is shown in Fig. 12.10. In such a crystal there are present microcracks associated with the tear lines, which under the influence of stress can grow to critical size and in this process emit dislocations from the crack tip, as indicated in Fig. 5.8. However, if the crystal is polished to remove surface damage, then the slip-band interactions themselves can give rise to the nucleation of cracks. Figure 12.11 is a schematic diagram showing the formation of cracks at the intersection of {110} glide bands [52]. In MgO loaded at low strain rates, microcracks have

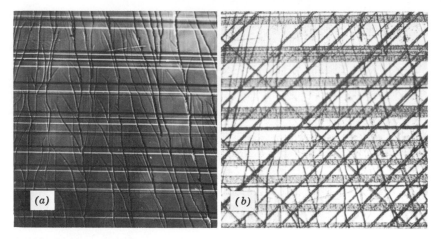

Fig. 12.10. (a) Cleavage surface of MgO crystal after bending. (b) Same surface as in (a), after etching. 100×.

been observed to grow slowly upon increase of load after they have been initiated at glide bands [53]. However, even in MgO it is difficult to establish the details of the fracture process because of the fine scale of the initiating event and the fact that fracture at slip band intersections is a heterogeneous process. That is, although there are many such intersections, only certain intersections are critical. There is evidence from thin film microscopy that individual dislocations on orthogonal systems do not interact strongly, whereas those on non-orthogonal systems do [54]. However, on a more macroscopic level

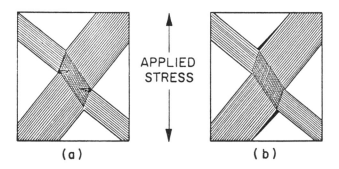

Fig. 12.11. Schematic diagram showing the formation of (a) (100) and (b) (110) cracks respectively at the intersection of {110} glide bands [52].

where two orthogonal slip bands intersect, there must be a degree of interaction giving rise to a rotation or kink accommodation in the region of intersection [55, 56], as shown in Fig. 12.11. In the process of deformation, if this kink accommodation does not keep pace with the widening of the slip bands, a macroscopic stress will be developed which will result in cracking [55].

Another view which also emphasizes the dynamic character of the fracture process, but which stresses the microscopic nature of the initiating process at the dislocation level, is based on the following observations [57, 58]. If the process of plastic deformation is viewed in MgO crystals under polarized light, the stress fields of the dislocations can be observed. This technique reveals that groups of dislocations move along the edges of existing bands not in a continuous fashion but in bursts. It is proposed that most of these dislocations on orthogonal planes pass by each other without interaction, but occasionally some come close enough to interact and cause a crack to be nucleated. Once nucleated the crack can grow by the trapping of more dislocations until the critical Griffith size is reached, with this growth stage will be governed by the macroscopic stresses existing in the vicinity of the slip-band intersection. An important aspect of this process is that the nucleating site may be associated with the presence of impurity particles [52], and there is direct evidence that this is often the case.

At higher temperatures where large strains, that is, $> 10\%$ prior to fracture, can be obtained, a form of kinking can affect fracture behavior. Failure, however, is still brittle in the sense that it occurs without localized necking preceding the actual rupture. The kinking type of failure occurs during bending or in compression as a consequence of a constraint preventing lateral deformation. In compression, this restriction occurs at the interface between the specimen and the loading-head interface, whereas in bending it arises because the lateral Poisson contraction of the surface layers is restricted by the less highly strained portion of the specimen beneath the surface, giving rise to anticlastic curvature of the crystal, Fig. 12.12. In the tension surface, secondary tensile stresses are thereby developed which are at right angles to the main tension stress owing to bending. This secondary tensile stress is a maximum at the center of the surface and diminishes to zero at the edges. It does not interact with the {110} planes which have the bending axis as their zone axis, but it does influence the stress on the complementary planes causing the shear stress on these planes to decrease from the edge to the center. When the specimen is bent past about 10% elongation, slip begins on these

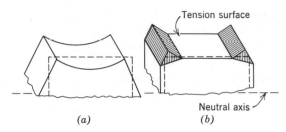

Fig. 12.12. Change in cross-sectional shape of a bent beam. (a) Anticlastic curvature of an elastic solid. (b) Anticlastic kinking of a crystalline solid [59].

complementary systems, and these dislocations interact with the dislocations on the primary bend planes to harden the corner regions. Material in the center of the specimen continues to slip so as to thin down the specimen, with the result that a kink develops [59, 60]. Fracture associated with kinking usually initiates at a subsurface site, but the precise mechanism involved has not yet been established.

Fracture in polycrystalline ceramics such as MgO occurs when a slip band is blocked at a grain boundary (Chapter 6), and Fig. 12.13 gives an indication of the high stresses developed at such sites [61]. Fracture will occur when these stresses exceed the cohesive strength of the boundary. In polycrystals made from powders, residual porosity is a principal variable affecting strength of the boundaries in short-time tests at low temperatures. In addition, the boundaries may contain impurities which also serve to lower the cohesive strength of the boundaries, although not all particles are in this category [62]. In addition to this mode of grain boundary crack nucleation, grain boundary sliding can also result in cracking at elevated temperatures [63]. Of course, the basic cause of low ductility is the lack of sufficient slip systems for generalized deformation.

Thermal shock

A sudden change of temperature, referred to as a thermal shock, can cause the cracking or spalling of brittle materials such as ceramics if thermal stresses induced by the temperature change are of sufficient magnitude. These thermal stresses can arise owing to the development of nonlinear thermal gradients or as a result of constraints which prevent free thermal contraction or expansion [64]. For a brittle material, fracture under thermal shock conditions is to be expected when the maximum thermally induced stress σ_m, reaches the level of the fracture stress σ_f, over a volume sufficient to allow the unstable growth of a microcrack. For the relatively simple case of radial heat flow in a quenched cylinder or thin circular disk insulated on its flat faces, the following expression gives the magnitude of σ_m in terms of Youngs modulus E, Poisson's ratio ν, the thermal conductivity k, the surface-heat transfer coefficient h, the radius of the cylinder or disk a, the linear coefficient of expansion α, and the magnitude of the temperature change ΔT [65]:

$$\sigma_m = \frac{E\alpha\,\Delta T/(1 - \mu)}{2.0 + 4.3k/ah - 0.5\,\exp\,(-16k/ah)} \tag{12.1}$$

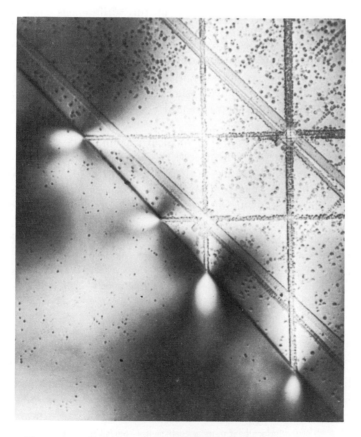

Fig. 12.13. Polarized transmitted-light photomicrograph of MgO bicrystal showing stress concentrations at ends of slip bands. 78× [61].

The magnitude of ΔT sufficient to cause fracture is obtained by setting $\sigma_m = \sigma_f$:

$$\Delta T_f = \frac{\sigma_f (1 - \mu)}{E \alpha \, \Delta T} \left[2.0 + \frac{4.3k}{ah} - 0.5 \exp \left(-\frac{16k}{ah} \right) \right] \quad (12.2)$$

This expression indicates that a low value of α and a high value of k serve to increase resistance to thermal shock, as does a low value of the heat transfer coefficient. Excellent agreement between predicted and experimentally observed values of ΔT_f as a function of the heat transfer coefficient h is shown in Fig. 12.14 for steatite [66].

If the ratio of h/k is either very small or very large, two limiting conditions for thermal shock resistance can be determined. Under these conditions the maximum temperature differences (between the body and the medium into which it is immersed) that can be withstood before the surface fractures are proportional respectively to $k\sigma_f/E\alpha$ and $\sigma_f/E\alpha$ [66]. Therefore, because the importance of the value of the thermal conductivity depends on the heat transfer conditions, the order of the thermal shock resistance of brittle materials may vary depending on the particular characteristics of the quenching media [64].

For materials which do not behave in a completely elastic manner, the onset of plastic flow will reduce the magnitude of σ_m and result in a sharp increase in resistance to thermal shock [64]. The total strain developed will still be given approximately as σ_m/E, so that if the yield stress σ_Y is known, the amount of plastic strain can be deter-

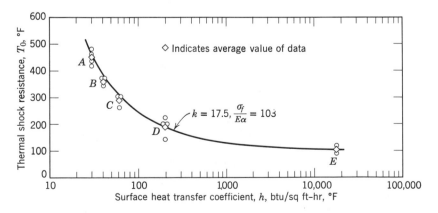

Fig. 12.14. Comparison of experimental thermal-shock resistance for steatite with predicted curve based on direct measurements of physical properties [66].

mined. Since fracture in such materials is the result of plastic deformation, a parameter S, which is proportional to the plastic strain, is used to characterize the magnitude of thermal shock [67]. This parameter is defined as

$$S = \frac{E}{\sigma_Y} \epsilon_p \qquad (12.3)$$

and since

$$\epsilon_p = \frac{\sigma_m}{E} - \frac{\sigma_Y}{E}$$

$$S = \frac{\sigma_m}{\sigma_Y} - 1 \qquad (12.4)$$

with σ_m computed from Eq. 12.1. It must be appreciated that σ_Y can be extremely large if the shock occurs rapidly, since the yielding behavior of ceramics is extremely strain-rate sensitive.

A summary of results relating to the thermal shock resistance of single crystals and bicrystals of MgO as a function of surface condition and the magnitude of the parameter S is as follows [67]:

S	Type and Condition of Crystal	Effect
< -0.7 ($\Delta T = 100°C$)	Any	None
-0.6	Any	Slip just detected
0.15–0.75	Unpolished or imperfect surface	Cracking produced including shattering in some cases
0.06–0.4	Bicrystal, "perfect" surfaces	Single crack
> 0.8	Bicrystal, "perfect" surface	Shatters
> 16 ($\Delta T \approx 1000°C$)	Single crystal, "perfect" surface	No cracking, heavy slip

The results clearly show that the presence of surface imperfections or crystal boundaries, both of which lower σ_f, greatly impairs the thermal shock resistance of MgO. The nature of the fracture processes as the result of a single thermal shock is similar to that observed under tensile stress although the manner of applying the stress is of course quite different. In the thermal shock tests of MgO crystals, no evidence for the presence of microcracks stabilized by plastic deformation was found. Microcracks once formed appeared to propagate readily in spite of the presence of fine slip, and this suggests that the effect of such slip is to inhibit the formation of microcracks rather than to stabilize them at a very small size.

The resistance to cracking of perfect single crystals is also retained if the thermal shock is repeated, and in one case a crystal withstood 1250 shocks at an S value of 0.95 per cycle without evidence of cracking, although considerable fine slip was observed. MgO crystals subjected to thermal fatigue therefore behave as they do when cyclically loaded at room temperature [68]. In both cases a high resistance to the development of cracks is manifested as a result of the limited nature of cross slip and the spread of fine slip across the surface of the crystals.

12.3   The Fracture of Polymers

Long chain molecules are known as polymers, and because there is a wide variety of specific molecular types within this category there is a correspondingly wide range of properties and fracture characteristics. Many polymers such as cotton occur in nature whereas others such as nylon are synthesized. These synthetic polymers are particularly interesting in that their molecular structure is subject to control to yield desired mechanical properties. These properties depend strongly upon a repeating monomer unit along the chain or backbone, as well as upon the properties of strongly attached side groups of molecules and upon interactions with neighboring molecular chains. Carbon atoms are generally important constituents of the main chain, but to improve elevated temperature properties, silicon and oxygen atoms are often substituted. The binding along the chain and to the side groups is of a covalent nature, with van der Waals forces which are one-tenth as strong as the primary bonds existing between adjacent chains. In order to provide structural integrity, occasional strong cross links may exist between the chains as in a linear polymer network, or the chains may be arranged in a three-dimensional branched and entangled network. Mechanical processing can impart a high degree of orientation to the chains, which influence mechanical properties.

The number of monomer units in a chain reflects the degree of polymerization and the molecular weight of the polymer. Polymeric materials in which the chains are arranged in random array are known as amorphous polymers. If the side groups are regularly

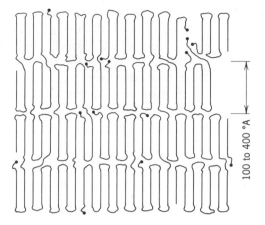

Fig. 12.15. Morphology of crystalline polymers. Schematic diagram of a polymer crystal-
lite made up of folded chains [81].

arranged so that the chains can be folded upon themselves without
steric hindrance, a folded chain structure can develop, Fig. 12.15 [81]
which, through the operation of a screw dislocation, can lead to the
formation of a three-dimensional crystalline array. X ray diffraction
patterns of crystalline polymers are more diffuse than those of annealed metals, but the regularity of structure is indeed evident. A
crystalline monolayer of polyethylene is shown in Fig. 12.16 [70].
Deformation modes characteristic of crystalline materials such as slip
and twinning have also been observed in certain cases in single
crystals as well as in polycrystalline aggregates of polymeric materials, and the role of dislocations in such structures has been discussed
[71, 72, 73, 74].

The side groups which influence the ability to form crystalline
arrangements fall into three categories. If the side groups are regularly spaced on one side of the chain, the polymer is referred to as
isotatic; if they alternate regularly from one side of the chain to the
other, the polymer is syndiotactic, and if they are arranged at random, the polymer is atactic. When the side groups are large only the
first two can form crystalline arrays. Although atactic polymers
tend to remain amorphous on cooling, even they can be strongly
oriented by stretching. Polymers that do crystallize in bulk do not
do so completely, and so both amorphous and crystalline regions co-
exist. The relative proportions of each phase will determine the
properties, for the amorphous phase is usually brittle at low tempera-

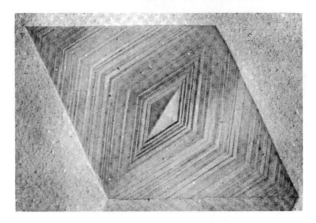

Fig. 12.16. Monolayer crystal of polyethylene. 4000×. The corrugated structure was preserved by a special preparation technique involving sedimentation of the crystal on a liquid glycerine surface and strengthening it with metal and carbon while on the liquid surface. *After Bassett et al* [70].

tures, whereas the crystalline regions tend to be both stronger and tougher. For crystalline polymers the stiffness and yield strength depend more on the degree crystallinity than on the molecular weight. The size of crystalline regions affects fracture also. Figure 12.17 is an example of fibrils, which are bundles of oriented molecules, drawn across breaks in polycrystalline polyethylene [75]. As in metals, a decrease in crystalline size results in an increase in impact strength.

The properties of polymeric materials depend strongly on the test temperature relative to two important reference temperatures. For amorphous polymers this temperature is the glass-transition temperature T_g. Below this temperature there is a decrease in expansion coefficient and the molecular structure is effectively fixed. This temperature is usually more sharply defined for polymers than for glass. The dependence of the modulus on time and temperature for an amorphous polymer is shown in Fig. 12.18 [76, 85]. For those polymers which crystallize on cooling the important reference temperature is the melting point T_M, at which an abrupt change in density occurs. Figure 12.19 [77] indicates the effects of the degree of crystallinity on the mechanical properties. For partially crystalline polymers, empirical relations between T_g and T_M exist which depend on the symmetry of the polymer chains. For example, if the chains are symmetrical, $T_g/T_M = \frac{2}{3}$, and if nonsymmetrical, $T_g/T_M = \frac{1}{2}$ [78, 79].

Typical stress-strain behavior for amorphous and crystalline polymers together with the corresponding monomer units are shown in Fig. 12.20 [80]. Both polycarbonate and polyethylene extend plastically by a process referred to as cold drawing which is phenomenologically similar to the spread of a Luder's band in mild steel but which in polymers is the result of an orientation of the molecular network structure. The types of deformation as affected by temperature and structure are as follows [83]:

Viscous flow. The irreversible bulk deformation of polymeric material associated with the irreversible slippage of chains past each other.

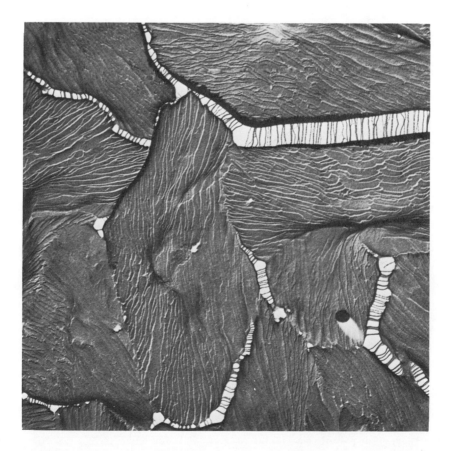

Fig. 12.17. Electron micrograph of a polyethylene film, cast from hot solution onto a hot slide. ×5750. The polygonal areas which separate by pulling threads are single crystals. *Keller* [75].

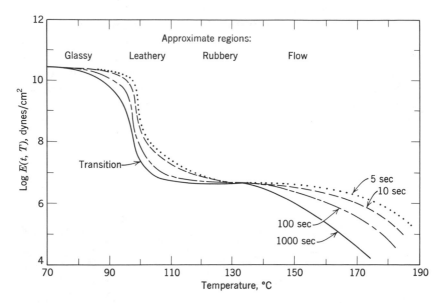

Fig. 12.18. The relaxation moduli of amorphous polystyrene in uniaxial tension at four different times of observation as a function of temperature [76] [85].

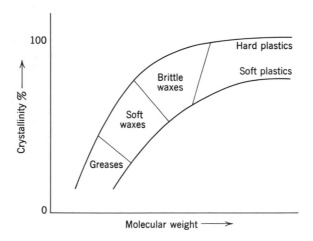

Fig. 12.19. Range of mechanical properties obtainable by variations in molecular weight and branching of polyethylene [77].

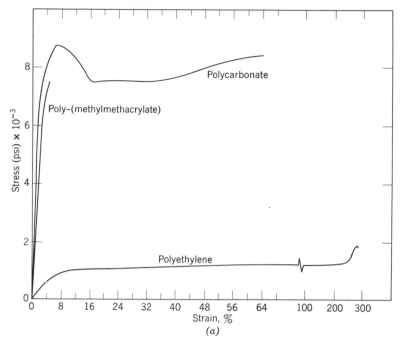

Poly-(methylmethacrylate)

Glass transition temperature, T_g

H   CH$_3$
$\left(\!-\!\!\begin{array}{cc} \text{C} & \text{C} \\ | & | \\ | & | \end{array}\!-\!\right)_n$
H   C—O—CH$_3$
       ‖
       O

+105 °C

Polycarbonate

$\left(\!-\text{O}-\!\!\bigcirc\!\!-\!\!\begin{array}{c} \text{CH}_3 \\ | \\ \text{C} \\ | \\ \text{CH}_3 \end{array}\!\!-\!\!\bigcirc\!\!-\text{O}-\!\!\begin{array}{c} \text{O} \\ ‖ \\ \text{C} \end{array}\!-\!\right)_n$

+150 °C

Polyethylene

$\left(\!-\!\!\begin{array}{cc} \text{H} & \text{H} \\ | & | \\ \text{C} & \text{C} \\ | & | \\ \text{H} & \text{H} \end{array}\!-\!\right)_n$

−120 °C

(b)

Fig. 12.20. (a) Tensile stress-strain curves for three polymeric materials. Note break in strain scale for polyethylene. Strain rate approximately 0.03 per min [80]. (b) Repeat units and glass transition temperatures of polymers used in (a).

Rubberlike elasticity. Local freedom of movement associated with small-scale movements of chain segments is retained, but large-scale movement (flow) is prevented by the restraint of a diffuse network structure. The elastic restoring force is derived from entropy rather than internal energy, from the smaller numbers of configurations available to a molecular chain when it is stretched out [82, 84]. In rubber the stress resides in molecular chains of length, say 100 A, so that the effective radius of a crack tip is much larger than in metals, which contributes to the toughness of rubber [82].

Viscoelasticity. Deformation of the polymer specimen is reversible but time-dependent, and is associated (as in rubber elasticity) with the distortion of polymer chains from their equilibrium deformations through activated segment motion involving rotation about chemical bonds.

Hookean elasticity. Motion of chain segments is drastically restricted, and probably involves only bond stretching and bond angle deformation: the material behaves as a completely brittle solid.

The three principal types of polymers are elastomers, fibers, and plastics [83]. Elastomers are amorphous and are used above T_g, where the behavior is viscoelastic. Crystallinity can develop on stretching which is reversible and causes the increase in stiffness with large strains. Crystalline fibers are strong and provide the strengthening elements of reinforced plastics. A comparison of some tensile strengths of some fibers with steel is given below [77]:

Fiber	*Tensile Strength, psi*
Polypropylene (isotactic)	120,000
Fortisan	150,000
Nylon	120,000
Steel wire	450,000

Fibers usually have a high modulus (10^6 psi) and are often oriented along the fiber axis for maximum strength.

Plastics are usually polymeric materials of intermediate cohesive strength and of fair toughness. There is often a moderate, but not too high, degree of crystallinity.

The basic properties associated with molecular structures can be altered in the case of amorphous polymers by the addition of small filler particles known as plasticizers in order to provide an improvement in toughness. However, these additives are not compatible with crystalline polymers. A major form of reinforcement is obtained through the use of strong glass fibers in a resin matrix. Resins differ from the thermoplastic polymers thus far described in that strong cross links which make the material rigid are formed upon exposure to elevated temperature. Hence such materials are called thermosetting.

Brittle fracture [85]

Polymeric materials, in common with materials in general fracture at stress levels much below their theoretical strength unless plastic deformation and effects of stress-raising flaws are eliminated. In a polymer if all of the chains are not properly aligned for maximum strength, the full potential will not be achieved, but an increase of temperature even in the glassy range will reduce the strength [86]. In addition, there are usually flaws present, which range from included dust particles, gels, salts and catalysts, to scratches, bubbles, voids, or other defects [87]. Studies of the point of origin of fracture

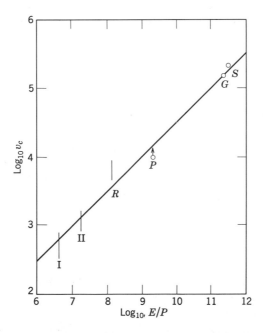

Fig. 12.21. Velocity of crack growth v_c versus modulus E. I and II are silicone rubbers I and II. R, P, G, and S are irradiated polyethylene, plexiglas, glass, and fused silica, respectively [90].

indicate that not only is the surface a critical region as with glass, but that the internal defects can also be important in nucleating fracture. Final failure results from the growth of cracks initiated at these flaws to the critical Griffith size for spontaneous propagation. Super-critical cracks can grow with great rapidity in polymers both above and below the nominal glass-transition temperature, for above a certain crack velocity there is insufficient time for relaxation processes to occur. Measurements of the terminal velocity for a wide range of brittle-behaving materials indicate that this velocity is independent of the volume of the material and can be expressed as a function of the velocity of longitudinal waves [88, 89], as indicated in Fig. 12.21 [90]. However, in similar studies it has been found that the velocity of crack propagation in soft glasses is less than predicted [91], an indication that dissipative processes occur in these substances during failure, as in metals.

The Griffith approach to the analysis of fracture has been applied to polymeric materials [92, 93, 94]. The dependence of the tensile strength on crack size of precracked poly-(methylmethacrylate) (PMMA) specimens is shown in Fig. 12.22 [93]. From the relation

$$\sigma_F = \sqrt{\frac{EG_c}{\pi c(1 - \nu^2)}}$$

and these data a value of G_c can be computed. A comparison of this value with those for steel and glass is as follows:

Fracture Surface Energies at 25°C [94]

Material	$G_c = 2\gamma_s$, calc $\times 10^{-3}$ ergs per cm^2	G_c obs $\times 10^{-3}$ ergs per cm^2
Steel	2.0	2,000
Glass	3.4	1.10
PMMA	1.0	400

It is seen that the observed value of G_c is much larger for PMMA than the surface energy, which is calculated for the rupture of primary, covalent bands. Thus PMMA behaves in a manner more like steel than glass; inelastic processes make a large contribution to the value of G_c. Direct evidence for a viscous flow dissipative process in the case of PMMA is given by the presence of interference colors in a thin layer on the fracture surface, which can arise as the result of an orientation effect of the polymers during fracture. As the length

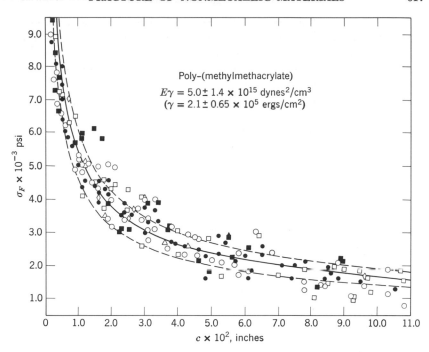

Fig. 12.22. The dependence of tensile strength on crack size in PMMA samples for various sample cross sections and extension rates [93].

of the crack increases (and presumably the velocity) there is a change in color with length which suggests that the dissipative processes are suppressed. As the test temperature is decreased the experimentally evaluated value of G_c increases, Fig. 12.23 [94]. Examination of the fracture surface reveals that the color changes are still present at low temperatures, and the increase in G_c is thought to indicate that the externally supplied energy needed to form the structure must increase as the segmental mobility decreases at lower temperatures. The increase in G_c with decreasing temperature is, of course, opposite to behavior of metals, for the experimental surface energies of metals decrease as the temperature is reduced, and can approach their theoretical values [95].

An increase in molecular weight, or modulus and/or chain orientation can likewise lead to large variations in the value of G_c. Depending on whether cracks run parallel to the chain direction or through the chains, the values of G_c may vary by more than an order of magnitude. As a consequence, the value of G_c for an oriented polymer can be higher or lower than for the oriented case.

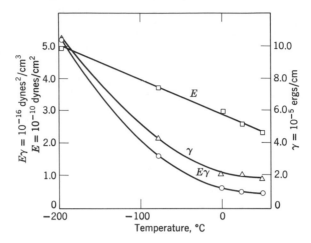

Fig. 12.23. The temperature dependence of the tensile modulus E, the fracture surface energy γ, and the product $E\gamma$ for PMMA [94].

From a knowledge of the tensile strength and EG_c, an inherent flaw size can be calculated. At room temperature a value of 0.1 mm is obtained for PMMA, and about 0.25 mm polystyrene (PS) [94]. Such values are much larger than any flaw present prior to loading, and it therefore *appears that slow crack growth to enlarge the flaws to critical* size must take place during loading, a characteristic not to be expected if the material behaved in a truly elastic manner in the glassy range, but consistent with the large values of G_c determined from tests of precracked specimens. The inherent flaws also give rise to a phenomenon known as crazing in amorphous thermoplastics such as PMMA, PS, and PC below T_g for small stresses. An example of crazing is shown in Fig. 12.24 [96]. These markings can arise either during loading or under sustained stress. They may be nucleated at flaws and also at small groups of molecular chains oriented perpendicular to an applied stress such that they are easily separated internally. In addition, there may be weak regions with a preponderance of chain ends or with a less than average density owing to excess void space. The crazes are not voids but are regions formed by localized deformation in the vicinity of the stress raisers which leads to the formation of a load-bearing oriented structure [87].

In addition to the interference colors visible on the fracture surface in reflected light, which are indicative of a dissipative process in fracture, several other features are usually present on the fracture surface

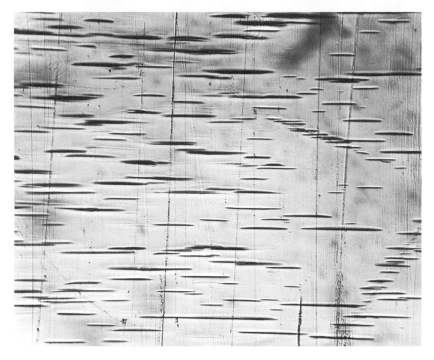

Fig. 12.24. Transmission photomicrograph of a portion of a stress-crazed surface of a PMMA tensile specimen. 100× [96].

of glassy polymers, as indicated in Fig. 12.25 [97]. Near the origin of fracture a mirrorlike surface is generally observed, beyond which the fracture surface becomes more roughened. The boundary between the two is regarded to be the line of demarkation between subcritical and critical crack growth, as in the case of glass. It appears that this critical length is reached when the stress field associated with the crack front is sufficiently high to initiate cracks at flaws ahead of the advancing front, leading to the onset of rapid growth as the multiplicity of cracks so created link up and generate the marked change in fracture appearance. Since these secondary fractures will generally not be coplanar with the main crack, tear lines will develop where they link up. As indicated in Fig. 12.26 [98], these tear lines can develop into characteristic markings, the shape of which will depend on the relative velocity of the two fronts and their radii. In general, the velocity of the main crack will be greater than that of the secondary cracks, and an elongated parabolic type of marking will

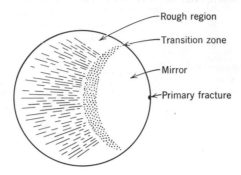

Fig. 12.25. General morphological pattern
of a normal fracture surface [97].

usually develop, as shown in Fig. 12.27 [99]. The particular shape
developed affords a means of determining the relative velocity of the
two crack fronts. The density of the markings generally increases
with distance from the fracture origin, indicating that as the stress
field of the crack front increases more secondary sources are activated
in the region ahead of the front. Wallner lines may also be present,

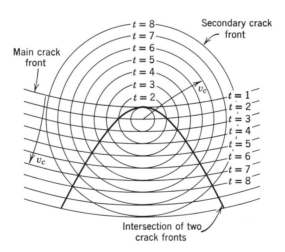

Fig. 12.26. Diagram showing a possible mechanism for formation of parabolas on fracture
surface. Secondary fracture starts at $t = 0$ at
focus of parabola and intersects main crack
front. Both fractures are traveling at the same
velocity v_c. *After Feltner* [98].

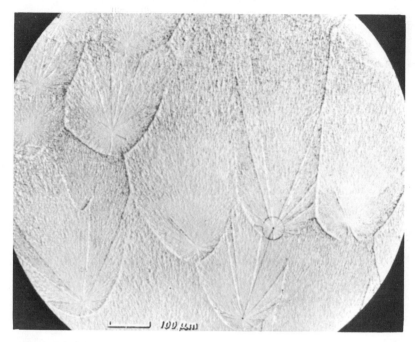

Fig. 12.27. Fracture surface of poly-methylmethacrylate; secondary fracture frontiers [99].

and these can be utilized to determine the absolute crack velocity since the patterns developed will depend upon the relative velocity of stress waves and that of the crack front.

Several factors can modify the appearance of the fracture surface. An increase in strain rate tends to minimize the mirror area, as does an increase in molecular weight. Both of these factors tend to inhibit relaxation processes and hence decrease the critical crack size. A reduction in temperature has the same effect on fracture appearance as an increase in strain rate.

In addition to the mirror region and the parabolic markings, other features characteristic of brittle fracture can be found. For example, radial tear lines of the type shown in Fig. 12.28 [80] may be present, which, in the slow growth region, will contribute to an increase in the effective value of G_c. Circumferential hesitation lines, formed when a fast moving crack momentarily stalls and allows additional relaxation, are also present in Fig. 12.28. Thermosetting materials, that is, thermally cross-linked glasses, are generally more brittle than

Fig. 12.28. Striae on fracture surface of polycarbonate specimen which has been fractured under unidirectional tensile loading. 100×.

thermoplastics; their fracture surfaces are therefore less featured than those of thermoplastics, approaching silica-glasslike in appearance. Epoxy resin fracture surfaces may contain some parabolic markings, but they are not so pronounced as in the case of PMMA [96]. As the temperature of the polmyer is increased above the glassy region, the mode of failure changes from a typical sharp brittlelike separation to ductile separation process involving shear deformation.

Fracture of amorphous rubbery polymers

At test temperatures above T_g the effect on time of the observed properties becomes a major consideration [101]. Molecules above T_g possess enough thermal energy to be able to orient themselves and move with respect to each other under the influence of stress and to adopt equilibrium configurations. This movement is time-dependent, and therefore the properties themselves are much more rate-sensitive than they are below T_g. The influence of frequency and temperature on the dynamic modulus is shown in Fig. 12.29a [100]. At any temperature the modulus increases as the frequency increases. Only for low frequencies and high temperatures can the material achieve an equilibrium condition wherein the deformation and rearrangement of the molecules remain in phase with the stressing system. In this case the material is more compliant and stores less strain energy.

An important aspect of the analysis of the viscoelastic behavior of rubbery polymers is the concept of time-temperature superposition. Data may be shifted to produce a master curve, as shown in Fig. 12.29b [101], by means of a shift factor a_t, which depends on the temperature but not on frequency. Among the most successful of the expressions for this shift factor is that due to Williams, Landel, and Ferry (WLF) [102], which is

$$\log a_t = -\frac{17.44(T - T_g)}{51.6 + (T - T_g)}$$

This equation is applicable from T_g to $T_y + 100°C$. Figure 12.29c [101] is a plot of the data of Fig. 12.29a on this basis. Not only can the effects of temperature on the modulus be correlated by such a shift factor, but the rupture properties may also be so treated. For example, in Fig. 12.30a are shown the rupture points for viton B gum vulcanate. The experimental points were obtained from constant strain-rate tests as a function of rate and temperature [103]. Similarly the strain to rupture can be plotted in forms of a compensated time-to-break, Fig. 12.30b. In analyzing these results we are dealing with a three-dimensional coordinate system which has as its axes α/T, ϵ, and time [101]. The points corresponding to rupture generate a surface in this space. A projection of this rupture surface on the σ/T, ϵ plane generates a curve known as a failure envelope which is independent of time and essentially independent of temperature, Fig. 12.30c [103]. If crystallization should occur during stretching, however, the properties will not superpose in this manner to yield a failure envelope [104].

The physical interpretation placed on the failure envelope is based on the concept that fracture of these rubbery polymers is the result of flaw growth to critical size [105]. At high temperatures and low strain rates equilibrium conditions are attained, and under these conditions a specimen usually breaks before becoming greatly extended. Highly stressed material at the tip of the flaw under these conditions creeps at a much higher rate than does the specimen as a whole; this leads to a large disparity between the tip strain and the nominal strain on the specimen, resulting in rupture with small overall elongation.

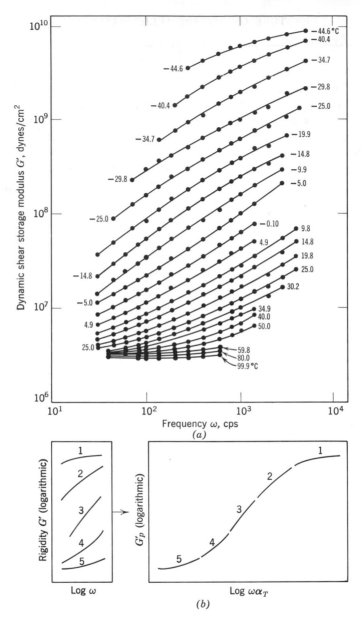

Fig. 12.29. (a) The shear storage modulus (or rigidity) G' as a function of frequency and temperature for the National Bureau of Standards polyisobutylene [100]. (b) Schematic illustration of the superposition of data from Fig. 12.16a. The numbered curves in the figure on the left represent data obtained at different temperatures as a function of the frequency ω. Curve 1 represents data obtained at the lowest temperature; and curve 5, data at the highest temperature. In the figure on the right, the individual isothermal G' curves have been corrected for temperature and density to G'_p, and then shifted horizontally, curve 3 being taken as the reference curve, to effect superposition [101].

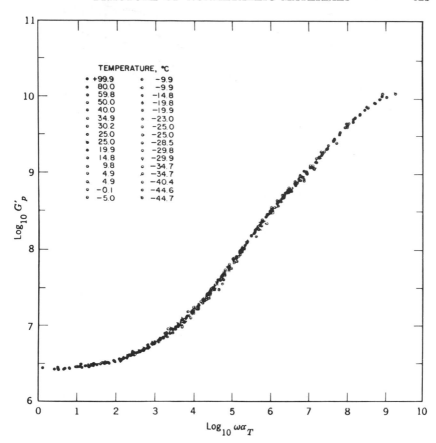

TEMPERATURE, °C

•	+99.9	•	−9.9
•	80.0	•	−9.9
•	59.8	•	−14.8
○	50.0	•	−19.8
•	40.0	•	−19.9
○	34.9	○	−23.0
○	30.2	○	−25.0
○	25.0	○	−25.0
○	25.0	○	−28.5
•	19.9	○	−29.8
○	14.8	○	−29.9
○	9.8	○	−34.7
○	4.9	○	−34.7
○	4.9	○	−40.4
○	−0.1	○	−44.6
○	−5.0	•	−44.7

Fig. 12.29. (c) Superposition of the data of Fig. 12.16b. The individual curves, representing isothermal data, have been corrected for temperature and density. The curve for $T = 25°C$ has been selected as the reference temperature and the remaining curves have been shifted horizontally to effect superposition. The amount of the shift is log a_T [101].

As the temperature is reduced, nonequilibrium conditions exist and the disparity between the rate at which flow occurs at the crack tip and that of the specimen as a whole is reduced, the result is that the overall strain to fracture increases. A maximum in the strain to fracture is observed where the strain rates at the tip of the crack are sufficiently high to cause the tip region to behave in a glassy manner, and thereby minimize the disparity between tip strain and the bulk specimen strain.

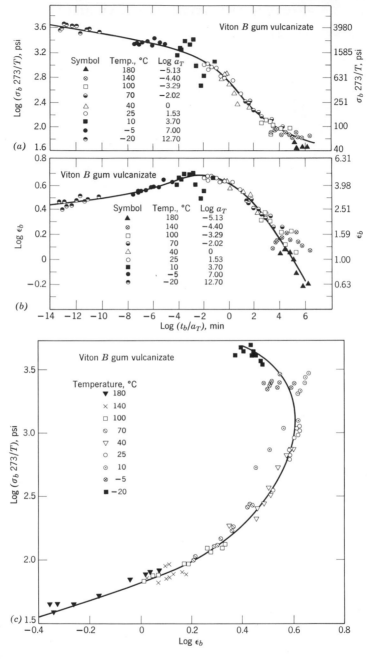

Fig. 12.30. (a) Projection of the rupture points of Viton B to the σ, t plane. Experimental results obtained from constant strain-rate tests at varying rates and temperature. (b) Projection of the rupture points to the ϵ, t plane. (c) Projection to the σ, ϵ plane to generate failure envelope [103].

Time-dependent fracture

Polymeric materials are subject to a variety of *environmental effects* which can lead to a reduction in their resistance to crack growth and impair their structured reliability. In fact, environmental effects pose probably the greatest hindrance to the more extensive use of polymers as structural materials [82]. For example, exposure to ultraviolet light can cause unzipping of long chain molecules by a progressive breakdown starting at the chain end. Chain scission, whereby radiation can cause bond rupture at various places along the chain, can also occur. Chemical environments may also be important; for example, polyethylene is subject to brittle stress and time-dependent fracture quite similar phenomenologically to static fatigue or stress corrosion [106, 107]. The stress-cracking behavior of polyethylene can be much improved by cross linking as by electron irradiation [108]. Fillers such as calcium carbonate may interfere with crack propagation and thereby also improve the resistance to stress cracking. It has been suggested [82] that the dissolution of fibrils between crystalline regions is a critical aspect of environmental crack growth. The *fatigue behavior* of polymers is at times an important consideration, (e.g., in the design of tires). Studies of the behavior of polymeric materials under cyclic load lead to the development of *S-N* curves which are of the same general type as for metals (Chapter 8). Fatigue failure appears to result from growth of cracks nucleated at inherent flaws. The macroscopic mechanism of growth of these cracks is similar to that described for metals, and involves a process of crack-tip blunting and sharpening during each cycle, as shown in Fig. 12.31 [80]. Each load cycle results in the advance of the crack and the creation of a striated surface (Fig. 12.32), as in the case of metals.

Effect of reinforcement

Particle Reinforcement. One of the aims of particle reinforcement of glassy polymers such as polystyrene is to improve their impact toughness. In other cases, improved creep resistance or stiffness can be obtained by the addition of carbon black and by cross linking to improve the high temperature (95°C) strength of polyethylene [108]. Because polymers are sensitive to internal as well as surface flaws, it is necessary to employ modifications which affect the entire volume. One way of improving the bulk toughness is to add rubber particles. To be effective, these particles must be large enough to divert the path of fracture out of its normally straight course [99]. Figure 12.33 indicates the effect of rubber content on both the impact

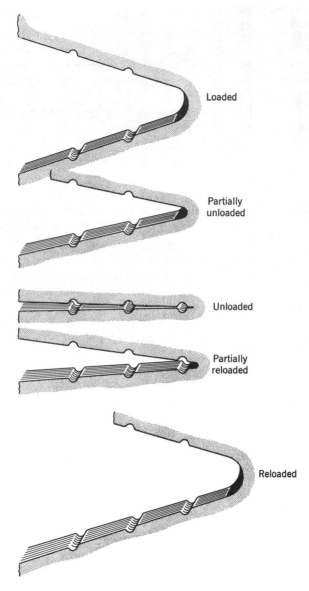

Fig. 12.31. Schematic illustration of deformation at crack tip during one loading cycle in tension-tension loading [80].

Fig. 12.32. Fatigue striae on fracture surface of polycarbonate specimen. 100× [80].

strength and yield strength of polystyrene [108]. In order to offset the drop in yield, part of the polystyrene can be replaced by acrylonitrile, and in this way a new group of reinforced plactics based upon acrylonitrile, butadiene, and styrene (ABS) has been created [108].

Glass Fiber Reinforcement. The high strength to weight ratio of glass fiber reinforced polymers have made them important structural materials, especially for application in pressure vessels. For certain applications limited by compressive strength or consideration of elastic stability, even hollow glass fibers have been considered to achieve greater weight savings [109]. As in the case of fiber-reinforced metals (see Chapter 13), the properties obtained depend on a variety of factors such as the nature of the matrix, the size and type of glass and its distribution, and the thermal expansion and Poisson's ratio characteristics of the two phases [110].

For the case of a brittle resin matrix, it is necessary for the matrix to have lower elastic constants than the fibers, so that for a given strain the stress in the matrix is less than in the fibers [111]. If the shear strength of the resin-matrix interface is relatively weak (1000–

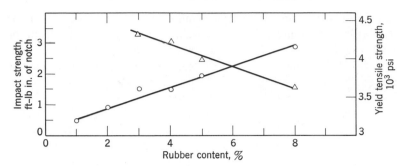

Fig. 12.33. Effect of rubber content on the impact strength and yield tensile strength of experimental high-impact polystyrenes [108].

2000 psi) [112], the method of load transfer is by frictional forces as the resin slides over the surface of the fibers. The normal pressure giving rise to this frictional effect is due to the shrinkage of the resin. If the resin is assumed to be stressed to its fracture strength in tension, owing to shrinkage constraints F, the fiber is subjected to a radial pressure n, given by

$$n = \frac{F\lambda}{d}$$

where d is the diameter of the fiber d and λ is the fiber spacing [110]. The critical transfer length, lc, is then given by

$$l_c = \frac{\sigma_{u(f)}d}{2\mu(\lambda/d)}$$

where $\sigma_{u(f)}$ is the fracture strength of the fiber and μ is the coefficient of friction between the matrix and fiber. For appropriate values of μ (0.4 maximum), F (10,000 psi), it is found that the transfer lengths for resin matrices are an order of magnitude larger than for metal matrices.

In order to observe fracture behavior, simplified model systems are often employed which have single layers of filaments imbedded in an epoxy resin. It has been found that for resin-rich composites, fracture is initiated at defects in the resin which can result in failure of the filaments or in debonding at the interface, whereas in resin-lean composites the fracture is initiated by the rupture of the fibers [113].

If a longitudinal filament breaks within the matrix, or if several short filaments are arranged in a straight line, high interfacial shearing stresses are developed a short distance from the end of the fiber which can lead to fracture of the matrix [114]. The distance from the end of the fiber to the position of maximum shear stress intensity is proportional to the square root of the filament diameter, with the properties of the filament and the matrix entering as a constant factor. When the matrix does not immediately rupture upon the failure of a filament, the tensile fracture of fibrous composites consisting of glass fibers in an epoxy can be treated on the basis of a Weibull distribution of flaws in the fibers. As the stress is increased, failure of the weakest link occurs, throwing more load on the remaining section. Fiber breaks have been observed to occur at stress levels well below the maximum load. Failure results from the accumulation of cracks rather than from the existence of fully ineffective fibers and for this reason the mechanical properties of the matrix material will have a significant effect on composite strength as measured by the efficiency with which the matrix transmits load around a fiber break.

12.4 Summary

1. The tensile strength of glass is extremely dependent on surface condition and the nature of the environment. Under sustained loading reaction with the environment at surface flaws leads to a type of failure known as static fatigue. Recent findings indicate that even glass undergoes some inelastic deformation at high values of stress.

2. The strength of crystalline ceramics also depends on surface condition. Crystals of the NaCl structure are capable of plastic deformation, but the number of independent slip modes is limited, and therefore such crystals are incapable of truly ductile behavior. The presence of grain boundaries and surface imperfections also makes these materials susceptible to failure by thermal shock.

3. Polymeric materials can be brittle when tested below the glass transition temperature. Fracture is nucleated at flaws within the bulk as well as at the surface. Above the glass transition temperature viscoelastic behavior occurs, and fracture behavior therefore becomes more sensitive to time and temperature. The strength of polymers is also influenced by the environment and by cyclic loading. Strength and toughness can be improved by particle or fiber reinforcement.

References

The asterisk indicates that published work is recommended for extensive or broad treatment.

[1] J. E. Gordon, *J. Roy. Aero. Soc.,* **56,** 704 (1952).
*[2] J. J. Gilman, *Mech. Eng.,* **83,** 55 (September 1961).
[3] R. Schmidt, *Metals Progress,* **86,** 133 (1964).
*[4] W. D. Kingery, *Introduction to Ceramics,* Wiley, New York (1960).
*[5] R. J. Charles, *Progress in Ceramic Science,* Pergamon, New York (1961), p. 1.

[6] W. H. Zachariasen and B. E. Warren, *J. Am. Chem. Soc.*, **54**, 3841 (1932);
J. Am. Ceram. Soc. **24**, 256 (1941).

[7] P. W. Bridgman and I. Simon, *J. Appl. Phys.*, **24**, 405 (1953).

[8] W. W. Shaver, *Proc. Roy. Soc.*, **A282**, 52 (1964).

[9] A. A. Griffith, *Phil. Trans. Roy. Soc.*, **A221**, 163 (1921).

[10] W. H. Otto, *J. Am. Ceram. Soc.*, **38**, 122 (1955).

[11] W. F. Thomas, *Phys. of Chem. Glasses*, **1**, 4 (1960).

[12] J. G. Morley, *New Scientist*, **17**, 122 (1963).

*[13] J. J. Gilman, in *Fracture*, B. L. Averbach et al. eds., M.I.T., Wiley, New York (1959).

*[14] H. A. Elliot, *Proc. Phys. Soc.*, **59**, 208 (1947).

*[15] E. B. Shand, *J. Am. Ceram. Soc.*, **37**, 52 (1954).

[16] F. W. Preston, T. C. Baker and J. C. Glahart, *J. Appl. Phys.*, **162**, 17 (1946).

*[17] E. Orowan, *Nature*, London, **154**, 341 (1944).

[18] C. Gurney and S. Pearson, *Proc. Roy. Soc.*, **A192**, 537 (1948).

[19] E. B. Shand, *J. Am. Ceram. Soc.*, **44**, 21 (1961).

[20] R. E. Mould, Glastech. Br., Fifth *Int. Glass Conf. Sonderband*, **32K**, III/18 (1959).

[21] R. E. Mould and R. D. Southwick, *J. Am. Ceram. Soc.*, **42**, 582 (1959).

*[22] W. B. Hillig and R. J. Charles, *High Strength Materials*, Wiley, New York (1965), p. 682.

[23] A. Joffe, M. W. Kirkpitschewa and M. A. Lewitsky, *Z. Phys.*, **22**, 286 (1924).

[24] R. J. Stokes, T. L. Johnston and C. H. Li, *Trans. AIME*, **218**, 655 (1960).

*[25] D. M. Marsh, *Proc. Roy. Soc.*, **A282**, 33 (1964).

[26] D. M. Marsh, *Fracture of Solids*, Interscience, New York (1963), p. 143.

[27] A. J. Dalladay and F. Twyman, *Trans. Opt. Soc.*, **23**, 165 (1921).

[28] E. B. Shand, *J. Am. Ceram. Soc.*, **44**, 451 (1961).

[29] H. Schardin, in *Fracture*, B. L. Averbach et al. eds., M.I.T., Wiley, New York (1959), p. 297.

[30] S. D. Stookey, *High Strength Materials*, Wiley, New York (1965), p. 669.

[31] P. Acloque, Symposium sur la résistance mécanique du verre et les moyens de l'améliorer, Union Scientifique Continentale du Verre, Charleroi, Belgium (1962), p. 851.

[32] F. M. Emsberger, *Proc. Roy. Soc.*, **A257**, 213 (1960).

*[33] E. R. Parker, *Mechanical Behavior of Crystalline Solids*, NBS Monograph 59 (1963), p. 1.

*[34] J. J. Gilman and W. G. Johnston, *Dislocations and Mechanical Properties of Crystals*, Wiley (1957), p. 116.

[35] J. B. Wachtman and L. H. Maxwell, *J. Am. Ceram. Soc.*, **37**, 291 (1954).

[36] M. L. Kronberg, *J. Am. Ceram. Soc.*, **45**, 253 (1962).

[37] R. Chang, *J. Appl. Phys.*, **31**, 484 (1960).

[38] H. Conrad, K. Janowski and E. Stofel, *Aerospace Corp. Report ATN-64-(9236)-16* (1964).

*[39] R. L. Coble, *High Strength Materials*, Wiley, New York (1965), p. 706.

*[40] W. D. Kingery and R. L. Coble, *Mechanical Behavior of Crystalline Solids*, NBS monograph 59 (1963).

[41] R. D. Moeller et al, *Trans. ASM* (1951), p. 39.

[42] R. von Mises, *Z. Angew. Math. Mech.*, **8**, 161 (1928).

*[43] G. W. Groves and A. Kelly, *Phil. Mag.*, **8**, 877 (1963).

[44] R. J. Stokes and C. H. Li, *Fracture of Solids,* Interscience, New York (1963), p. 289.

[45] R. J. Stokes and C. H. Li, *Materials Science Research,* Plenum, New York (1963), p. 133.

[46] J. E. Dorn and J. B. Mitchell, *High Strength Materials,* Wiley, New York (1965), p. 510.

[47] N. S. Stoloff, D. K. Lezius and T. L. Johnston, *J. Appl. Phys.,* **34,** 3315 (1963).

[48] R. J. Stokes and C. H. Li, *Acta Met.,* **10,** 535 (1962).

[49] T. L. Johnston, R. J. Stokes and C. H. Li, *Trans. AIME,* **221,** 792 (1961).

*[50] J. J. Gilman, *Progress in Ceramic Science,* Pergamon, New York (1961), p. 146.

[51] J. J. Gilman, *Trans. AIME,* **209,** 449 (1957).

*[52] T. L. Johnston, *Mech. Behavior of Crystalline Solids,* NBS Monograph 59 (1963), p. 63.

[53] F. J. P. Clarke and R. A. J. Sambell, *Phil. Mag.,* **5,** 697 (1960).

[54] J. Washburn, *Electron Microscopy and Strength of Crystals,* Wiley, New York (1962), Ch. 6.

[55] A. Argon, *Int. Conf. on Fracture,* Sendai, Japan, paper DII-10 (1965).

[56] A. Argon and E. Orowan, *Nature,* **192,** 447 (1961); *Phil. Mag.,* **9,** 1023 (1964).

[57] A. Briggs, F. J. P. Clarke and H. G. Tattersall, *Phil. Mag.,* **9,** 1041 (1964).

[58] A. Briggs and F. J. P. Clarke, *Int. Conf. on Fracture,* Sendai, Japan, paper C-14 (1965).

[59] R. J. Stokes, T. L. Johnston and C. H. Li, *J. Appl. Phys.,* **33,** 62 (1962).

[60] V. L. Indenbom and A. A. Urusovskaya, *Soviet Physics (Crystallography),* **4,** 84 (1960).

[61] R. C. Ku and T. L. Johnston, *Phil. Mag.,* **9,** 231 (1964).

[62] A. J. Forty, *Acta Met.,* **7,** 139 (1959).

[63] M. A. Adams and G. T. Murray, Materials Res. Corp., *Res. Report AT (30-1)-2178* (1960).

*[64] W. D. Kingery, *J. Am. Ceram. Soc.,* **38,** 1 (1955).

[65] E. Glenny and M. G. Rayston, *Trans. Brit. Ceram. Soc.,* **57,** 645 (1958).

*[66] S. S. Manson, *Mechanical Behavior of Materials at Elevated Temperatures,* Wiley, New York (1961), p. 393.

*[67] G. O. Miles and F. J. P. Clarke, *Phil. Mag.,* **6,** 1449 (1961).

[68] A. J. McEvily and E. S. Machlin, in *Fracture,* B. L. Averbach et al. eds., M.I.T., Wiley, New York (1959), p. 450.

[69] A. S. Tetelman, *Fracture of Solids,* Wiley, New York (1963), p. 471.

[70] D. C. Bassett, F. C. Frank and A. Keller, *Phil. Mag.* (1963).

[71] W. P. Schlichter, *Growth and Perfection of Crystals,* Wiley, New York (1958), p. 558.

[72] D. A. Zaukelies, *J. Appl. Phys.,* **33,** 2797 (1962).

[73] C. J. Speerschneider and C. H. Li, *J. Appl. Phys.,* **33,** 1871 (1962).

[74] F. P. Price, *Growth and Perfection of Crystals,* Wiley, New York (1958), p. 466.

[75] A. Keller, *Phil. Mag.,* **3,** 1171 (1957).

*[76] A. V. Tobolsky, *Properties and Structures of Polymers,* Wiley, New York (1960), p. 72.

[77] T. Alfrey, *High Strength Materials,* Wiley, New York (1965), p. 769.

[78] T. W. Campbell and A. C. Haven, *J. Appl. Poly. Sci.,* **1,** 73 (1959).

[79] P. J. Flory et al, *J. Poly. Sci.,* **12,** 97 (1954).
[80] A. J. McEvily, R. C. Boettner and T. L. Johnston, *Fatigue,* Syracuse U. Press (1964), p. 95.
*[81] L. E. Nielsen, *Mechanical Properties of Polymers,* Reinhold, New York (1962).
*[82] F. C. Frank, *Proc. Roy. Soc.,* **A282,** 9 (1964).
*[83] F. W. Billmeyer, *Polymer Science,* Wiley, New York (1962).
[84] T. L. Smith, *Proc. Roy. Soc.,* **282,** 102 (1964).
*[85] B. Rosen, ed., *Fracture Processes in Polymeric Solids,* Interscience, New York (1964).
[86] P. I. Vincent, *Proc. Roy. Soc.,* **A282** (1963), p. 113.
[87] O. K. Spurr and W. D. Niegisch, *J. Appl. Polymer Sci.,* **6,** 585 (1962).
[88] N. F. Mott, *Engineering,* **165,** 16 (1948).
[89] D. K. Roberts and A. A. Wells, *Engineering,* **178,** 820 (1954).
[90] F. Bueche and A. V. White, *J. Appl. Phys.,* **27,** 980 (1956).
[91] H. Schardin, in *Fracture,* B. L. Averbach et al. eds., M.I.T., Wiley, New York (1959), p. 297.
[92] R. S. Rivlin and A. G. Thomas, *J. Poly. Sci.,* **18,** 177 (1955).
*[93] J. P. Berry, *Fracture Processes in Polymeric Solids,* Interscience, New York (1964), p. 157.
*[94] Berry, J. P., *op. cit.* [93], p. 195.
[95] J. J. Gilman, *J. Appl. Phys.,* **31,** 2208 (1960).
*[96] I. Wolock and S. B. Newman, *Fracture Processes in Polymeric Solids,* Interscience, New York (1964), p. 235.
[97] I. J. Leeuwerik, *Rheologica Acta,* **2,** 10 (1962).
[98] C. E. Feltner, Univ. Illinois, *Theoretical and Applied Mechanics Report 224* (August 1964).
[99] A. J. Staverman, *Proc. Roy. Soc.,* **A282** 115 (1964).
[100] E. R. Fitzgerald, L. D. Grandine and J. D. Ferry, *J. Appl. Phys.,* **24,** 650 (1953).
*[101] R. F. Landel and R. F. Fedors, *Fracture Processes in Polymeric Solids,* Interscience, New York (1964), p. 361.
[102] M. L. Williams, R. F. Landel and J. D. Ferry, *J. Am. Chem. Soc.,* **77,** 3701 (1955).
*[103] T. L. Smith, *J. Appl. Phys.,* **35,** 27 (1964).
[104] T. L. Smith, *Proc. Roy. Soc.,* **A282,** 102 (1964).
[105] F. Bueche and J. C. Halpin, *J. Appl. Phys.,* **35,** 36 (1964).
[106] P. Hittmair and R. Ullman, *J. Appl. Poly. Sci.,* **6,** No. 19, 1 (1962).
[107] L. J. Broutman, *SPE J.* (March 1965), p. 283.
[108] R. N. Haward and J. Mann, *Proc. Roy. Soc.,* **A282,** 120 (1964).
[109] B. W. Rosen, General Electric Co., Final Report, Contract No. -63-0674-C (1964).
[110] A. Kelly and W. R. Tyson, *High Strength Materials* Wiley, New York (1965), p. 570.
[111] J. O. Outwafer, *Modern Plastics,* **33,** 156 (1956).
[112] R. D. Mooney and F. J. McGarry, 14th Annual T. and M. Conf. R.P.D. Sect. 12E, Soc. Plastics Ind. (1959).
[113] C. A. Bouc, Univ. Illinois, T. and A. M. Report 234 (November 1962).
[114] Y. M. Malinskii, *Polymer Science U.S.S.R.,* **6,** No. 5, 862 (1964).
[115] B. W. Rosen, *AIAA J.,* **2,** 1985 (1964).

13

Fracture of
Composite Materials

The high modulus, hardness, yield strength and melting point, and low chemical reactivity and density of many nonmetallic solids are extremely desirable properties for certain engineering applications. However, as shown in the preceding chapter, these materials possess minimal tensile ductility and hence minimal notch toughness G_{Ic} at temperatures approaching their melting points. Consequently there are two serious problems associated with their use in bulk form. First, they cannot be easily joined by conventional methods (e.g., brazing) since the smallest fabrication defect could lead to a brittle failure in service. This tends to restrict their usage to simple structures [1] that can be produced by casting and can be carefully machined to final shape. Secondly, failure in these materials almost invariably begins at a surface flaw (e.g., a scratch). Therefore their surfaces must be carefully prepared and a high degree of surface perfection must be maintained during service. This limits the applicability of these materials in bulk form and tends to nullify the benefits gained from their high yield strengths. Even when these structures are loaded in simple compression, the Poisson effect sets up transverse tensile stresses [2] which can lead to unstable fracture when small flaws are present.

Apart from introducing residual compressive stresses at their surfaces (Chapter 12), one means of using these strong brittle solids with a higher degree of reliability is to disperse them discontinuously in a matrix of a softer material. This produces a *composite structure*, two types of which are shown in Fig. 13.1. When the hard phase is aligned in the form of long, thin fibers (e.g., fiberglass and bamboo), the structure is known as a *fiber composite*. The fibers may be continuous (Fig. 13.1a) or discontinuous (Fig. 13.1b) along the fiber axis, but they are usually discontinuous in the plane of maximum

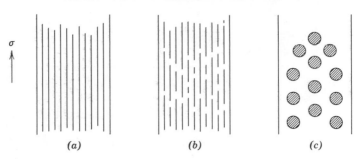

Fig. 13.1. Various types of composite systems. (*a*) Continuous fibers. (*b*) Discontinuous fibers. (*c*) Aggregate containing dispersed phase in a continuous matrix.

resolved tensile stress. When equiaxed particles of a hard phase are distributed in a softer matrix, the structure is known as an *aggregate* (e.g., concrete). When the hard phase is nonmetallic and the matrix is metallic, the aggregate is called a *cermet*. Strictly speaking, hypo-eutectoid steels (Fe_3C in Fe) or dispersion-strengthened aluminum alloys such as 7075 fit these definitions of a composite. However, in these systems the relatively low volume fraction of hard phase produces strengthening by its effect on the plastic properties of the matrix, rather than by carrying a large fraction of the applied load. The term *composite* is generally used to denote systems where the volume fraction of the hard phase is large (i.e., 20–80%) and strengthening is primarily achieved by the load-carrying capacity of the hard phase, rather than by its effect on the matrix.

When the matrix is inherently notch-tough (e.g., low strength FCC metals and epoxy resins), a crack that initiates at a surface flaw in the brittle phase will be blunted and stopped when it reaches the softer matrix. This may result from plastic deformation of the matrix or, if the interface between phases is weak, by splitting of the interface. This splitting deflects the crack out of the plane in which it had been propagating and greatly minimizes the danger of service failure due to a preinduced flaw. In fiber composites the effect is similar to that discussed previously in connection with ausformed steels (see Chapters 7 and 10).

In addition to preventing the formation of a continuous brittle path through the structure, the matrix serves two other important functions [3]. First, it binds the hard particles together and protects their surfaces from flaws which could lead to a loss of strength. Secondly, it transfers stress to the hard phase, both average (nominal) stress and any local stress that was carried by a particle that broke prematurely (e.g., owing to a surface flaw). If the remaining particles can take up this stress without themselves breaking, structural integrity is maintained. This principle is similar to that used on a gross scale in the fail-safe design of large structures.

During the last few years the field of composite materials has received a great deal of attention, and numerous conferences [4, 5, 6, 7] and review articles [3, 8, 9, 10, 11] have been devoted to summarizing and explaining their mechanical behavior. There is no need to reiterate the reviews, and only those aspects dealing primarily with fracture behavior will be discussed here.

13.1 The Fracture of Fiber-Reinforced Metals

Mechanics of fiber strengthening

Continuous Fibers. When a composite containing *continuous*, uniaxially aligned fibers in a metal matrix is stressed along the fiber axis, the deformation behavior is typically *isostrain* (i.e., the strains in the matrix and fiber are equal, hence equal to the composite strain). At low strains [12] both fiber and matrix deform elastically. When strains that are of the order of the yield strain of the matrix are achieved, the matrix deforms plastically but the fibers, having a higher yield strength, remain elastic. When the composite strain is of the order of the yield strain of the fibers, both fibers and matrix can deform plastically, provided the fibers are somewhat ductile. When the fibers do fracture, after small or large tensile strains (depending on test temperature, grain size, and so on, as given in Chapter 6) the composite itself fails (by normal rupture) provided that the fiber content is greater than some minimum value. Consequently the ultimate strength of the composite σ_c is determined primarily by the ultimate tensile strength of the fibers.

For a fiber composite containing a volume fraction V_f of continuous fibers, the composite strength is given by [3]

$$\sigma_c = \sigma_{u(f)}V_f + \sigma'_m(1 - V_f) \qquad (V_{\min} < V_f < V_t) \qquad (13.1)$$

where $\sigma_{u(f)}$ is the UTS of the fibers (which is the same as their fracture strength when they are brittle) and σ'_m is the tensile stress borne by the matrix when the fibers fail (i.e., at a strain corresponding to the failure strain of the fibers) [10]. In general, the fibers are sufficiently separated such that they exert little influence on the strain-hardening characteristics of the matrix, but when V_f is large (e.g., 60%) σ'_m may be greater than predicted from tensile tests carried out on the matrix material alone [3]. Numerous experimental observations [10, 12, 13, 14] indicate that Eq. 13.1 is obeyed over a wide

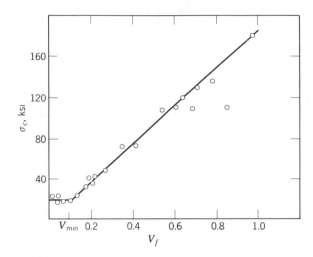

Fig. 13.2. Effect of volume fraction of continuous W fibers. V_f, on the tensile strength σ_c of W-Cu composites. *After A. Kelly and W. Tyson* [10].

range of values of V_f. Figure 13.2 shows this behavior for continuous brittle wires of tungsten in copper.

Equation 13.1 does not apply when $V_f < V_{min}$, where V_{min} is typically of the order of 10% or less [10]. In this range of fiber contents, the matrix is able to absorb the load that had been carried by the fibers before they broke and is then able to strain-harden before its ultimate strength is reached. Similarly, when $V_f > V_t$ (i.e, $V_f > 80\%$ or so) the fibers are in contact with one another and crack prematurely [15] (Fig. 13.3); σ_c then drops below the value predicted by Eq. 13.1, as seen in Fig. 13.2.

Discontinuous Fibers

Most practical composites that are being developed for engineering applications contain discontinuous fibers. The reason is that long, continuous fibers having high tensile strengths are difficult to obtain (especially when they are ceramic or metal whiskers) and some or most of the fibers will break during fabrication. In these cases the initially elastic deformation (and subsequently plastic deformation in a metal matrix) is able to transfer stress to the fibers, provided that the slope of the stress-strain curve of the matrix is less than that of the fibers.† Larger displacements will then be produced in the matrix

---

† This requires that the matrix have a lower modulus than the fiber when it deforms elastically (e.g., the resin in a fiberglass); when the matrix is a metal

than in the fibers. As the matrix starts to flow past the fibers, tangential (shear) stresses are set up at the matrix-fiber interface which load the fibers up to a high level. To a first approximation, the stresses in the fibers σ_{zz} build up linearly from the ends of the fibers [10]. For a fiber of radius r

$$\sigma_{zz} = \frac{2\tau z}{r} \qquad \left(z < \frac{l_c}{2}\right) \tag{13.2}$$

where τ is the flow strength of the matrix in shear or the shear strength of the fiber-matrix interface, whichever is the smaller,† and z is the distance measured along the fiber axis from one of its ends.

---

which can deform plastically, $(d\sigma/d\epsilon)_m \approx E_m/150$ so that this requirement will always be met.

† For metal matrices and good interface bonding, τ is of the order of one-half the ultimate tensile stress of the matrix. Resin matrices do not deform plastically, so that load is transferred by the frictional forces set up when the resin breaks away from the ends of the fibers (producing voids there) and slides over them [16]. τ is then equal to μP where μ is the coefficient of friction between the fiber and the matrix and P is the normal pressure of the resin on the fibers.

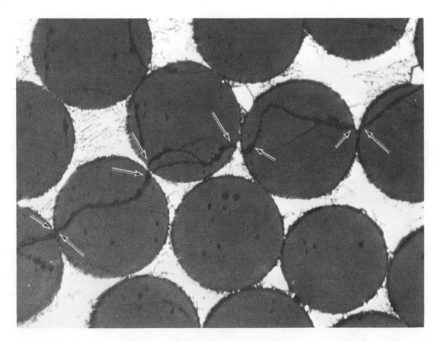

Fig. 13.3. Cracks across contiguous glass fibers in an aluminum matrix. *Courtesy W. Sutton and J. Chorne* [15] *and American Society for Metals.*

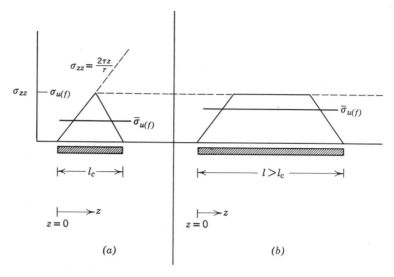

Fig. 13.4. Stress distribution in a fiber; maximum stress permitted at any point along fiber length is $\sigma_{u(f)}$, the fracture strength or ultimate strength of the fiber. (a) $l = l_c$, $\bar{\sigma}_{u(f)}$, the average strength of fiber, equals $\frac{1}{2}\sigma_{u(f)}$. (b) $l > l_c$, $\bar{\sigma}_{u(f)} \to \sigma_{u(f)}$.

The fiber breaks (or necks) when $\sigma_{zz} = \sigma_{u(f)}$ so that

$$\frac{l_c}{2} = \frac{r\sigma_{u(f)}}{2\tau} \tag{13.3}$$

l_c is the critical length of fiber which is required for the stresses in the fiber to build up to the ultimate strength (Fig. 13.4a). When the fiber length $l > l_c$, the stress in the fiber remains constant, as shown in Fig. 13.4b. The critical aspect ratio is defined [10] as

$$\frac{l_c}{d} = \frac{\sigma_{u(f)}}{2\tau} \tag{13.4}$$

where d is the fiber diameter. For tungsten in copper, l_c/d is about 4 at room temperature and 31.2 at 600°C. These variations result primarily from the decrease in τ with increasing temperature, at least in this composite.

When $l = l_c$ (Fig. 13.4a), the *average stress* in the fiber $\bar{\sigma}_{u(f)}$ when it breaks or necks at its midsection will be $\beta\sigma_{u(f)}$ where $\beta \approx \frac{1}{2}$. On the other hand, $\bar{\sigma}_{u(f)}$ will be higher (Fig. 13.4b) when $l \gg l_c$. In this case [10, 15]

$$\bar{\sigma}_{u(f)} = \sigma_{u(f)} \left[1 - (1 - \beta) \frac{l_c}{l} \right] \tag{13.5}$$

so that $\bar{\sigma}_{u(f)} \to \sigma_{u(f)}$ as $l \to \infty$. The strength of the composite is then

$$\sigma_c = \sigma_{u(f)} \left[1 - (1 - \beta) \frac{l_c}{l} \right] V_f + \sigma'_m (1 - V_f) \tag{13.6}$$

$$\sigma_c = \sigma_{u(f)} \left[1 - (1 - \beta) \left(\frac{(l_c/d)}{l/d} \right) \right] V_f + \sigma'_m (1 - V_f) \tag{13.7}$$

and it approaches the strength of the continuous fiber composite as l/l_c or l/d increase. Consequently *maximum tensile strength* is achieved when l/d (or l/l_c) are as large as possible.

On the other hand, when $l < l_c$ the fibers do not break since $\sigma_{zz} < \sigma_{u(f)}$. The composite fails (in tension) when the fibers are pulled out of the matrix, as shown in Fig. 13.5a. In bending, the fiber-matrix interface splits (Fig. 13.5b) and the fracture is similar to that obtained in other anisotropic materials (see Fig. 7.24). Although the tensile strength of flaw-free composites is reduced (e.g., $\bar{\sigma}_{u(f)} = \frac{1}{2}\sigma_{u(f)}$ when $l \approx l_c$), the *fracture toughness* G_{Ic} is increased, as in the case of other materials discussed previously. This is important for practical applications since the strength of real, brittle materials such as these composites is usually governed by their nominal fracture strength σ_F in the presence of notches (i.e., by G_{Ic}) rather than by σ_c.

Assuming that the shear stress τ is maintained during pull out, the work done in drawing a single fiber out of the matrix is $\pi r (l/2)^2 \tau$, using Eq. 13.2. Since l cannot exceed l_c (otherwise the fiber would break) the maximum work that can be done during pull out is $\pi r \tau (l_c/2)^2$. This leads to the result [17] that the maximum work of fracture per unit area is

$$G_{Ic} \cong \frac{V_f}{12} \sigma_{u(f)} l_c \tag{13.8}$$

This relation is analogous to Eq. 2.54 with l_c a measure of the critical displacement for fracture and with $(V_f/12)\sigma_{u(f)}$ the tensile stress level that exists in the plastic zone ahead of the crack.

Maximum toughness is achieved when l is slightly less than l_c and l_c and $\sigma_{u(f)}$ are as large as possible [12]. For W-Cu composites [10] $\sigma_{u(f)} = 180$ ksi, $l_c \cong 0.008$ in., so that $G_{Ic} \cong 5$ ft-lb per in.^{2} (10^7 ergs per cm^2) at ambient temperature with $V_f = 0.5$. Although this is

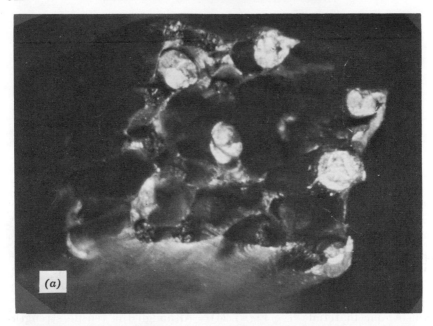

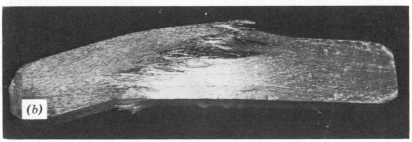

Fig. 13.5. Fracture by fiber pull out when $l < l_c$. (a) Cu-W in tension. (b) Al-SiO$_2$ in bending. *Courtesy A. Kelly and W. Tyson* [10].

about an order of magnitude lower than a desirable value of $G_{Ic} = 50$ ft-lb per in.^{2} for engineering applications (see Chapter 2), it seems likely that higher toughness values will be obtained as research in this field progresses. From Eq. 13.4,

$$G_{Ic} = \frac{V_f}{24} \frac{d}{\tau} \sigma_{u(f)}^2 \qquad (13.9)$$

so that toughness increases as the fibers of a given diameter become stronger and the matrix or its interface becomes weaker.

Unfortunately the parameters in Eq. 13.9 cannot always be varied independently of one another. For example, when V_f is raised beyond 40–60%, the fibers may begin exerting a sufficient constraint on the matrix to raise τ to the point where fibers of a given length are fractured rather than drawn out of the matrix. This has been observed [18] in Ag-steel and Cu-Mo composites. As seen in Fig. 13.6, this effect can actually cause the toughness of a composite to decrease since, in effect, l_c is decreasing faster than V_f is increasing. Similarly, while Eq. 13.9 would indicate that G_{Ic} increases with d, for many fibers (especially nonmetallics) $\sigma_{u(f)}$ is itself a function of d and decreases as d increases [19–23] (Fig. 13.7). Finally, if the matrix or the matrix-fiber interface becomes extremely weak (e.g., at temperatures near its melting point or if there is little wetting between fibers and the matrix), then l_c becomes extremely large and l will, in general, be much less than l_c. Physically this means that the maximum stress that can be built up in a fiber will be less than $\sigma_{u(f)}$, so that G_{Ic} will be small. This effect is analogous to the situation existing in conventionally processed materials near their melting points (Fig. 2.11) where, even though the fracture displacements are large, the stress level in the plastic zone is so small that G_{Ic} is low.

However, as shown in Fig. 13.8, this effect does not become important until the melting point is almost reached, and consequently many fiber composites possess superior properties at elevated temperatures [24, 25] compared with conventionally processed materials. In fact, with few exceptions (e.g., fiberglass) fiber composites will

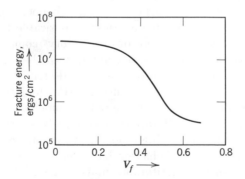

Fig. 13.6. Impact strength of composites of steel fibers in silver, as a function of volume fraction V_f of fibers. *After N. Parikh* [18].

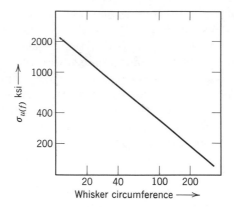

Fig. 13.7. Effect of the size of Al_2O_3 whiskers on their fracture strength at ambient temperature. *After R. Regester* [19].

probably be most beneficial for use in high temperature service applications since relatively inexpensive, tough materials already exist for use at ambient temperatures.

Metallurgical aspects of fiber strengthening

Composites containing *metal wires* have been the subject of numerous fundamental investigations of composite strength because they are readily available, relatively cheap, and exhibit reproducible mechanical properties. From a practical point of veiw, most metals (with the exception of the refractories) exhibit poor high temperature strength properties, and this makes them undesirable for use in creep-resistant composites. Research on refractory composites indicates that there are certain problems associated with their fabrication that may limit their usage. The refractory metals tend to have poor oxidation resistance at high temperatures; if the matrix is not able to prevent oxidation of the fibers, these will become embrittled (Chapter 11) and the composite will fail in a relatively short lifetime. The short-time properties (e.g., high temperature tensile strength) may still be quite good. This behavior has been observed in composites of molybdenum wires in a titanium matrix [28]. A second problem in developing refractory composites is the choice of a compatible matrix for the refractory wire. In some cases (e.g., beryllium in aluminum [29], tungsten in stainless steel [30]) a brittle intermetallic compound forms during fabrication which causes the composite to have a lower strength than the matrix material alone, at least at ambient temperatures.

In composites of tungsten wires in a copper-5% cobalt alloy, re-crystallization of the outer surfaces of the wires occurs during fabrication [31] and this significantly lowers their strength (Fig. 13.9). The sharp decrease in strength that occurs when more than 30% of the fiber is recrystallized results from the fact that cracks in the brittle layer act as miniature notches which weaken the fiber as a whole rather than the recrystallized layer alone. Similar effects take place when other alloying additions which can diffuse rapidly in tungsten

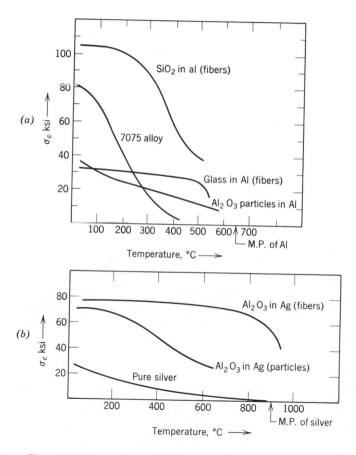

Fig. 13.8. Effect of temperature on the fracture strength of (a) some aluminum alloys and composites; (b) some silver composites. *After W. Sutton and J. Chorne* [15].

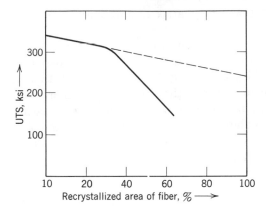

Fig. 13.9. Effect of recrystallization of W fibers on the strength of a W-Cu-Co composite. *After D. Petrasek and J. Weeton* [31].

(e.g., nickel and aluminum) are present in the alloy matrix. In this case the individual fiber strength is $\delta\sigma_{u(f)}$ where δ is the *degradation factor* which measures the reduction of strength of a fiber when it is introduced into a matrix. Equations 13.5 and 13.6 then become

$$\bar{\sigma}_{u(f)} = (\delta\sigma_{u(f)}) \left[1 - (1 - \beta) \frac{l_c}{l} \right] \tag{13.10}$$

$$\sigma_c = (\delta\sigma_{u(f)}) \left[1 - (1 - \beta) \frac{l_c}{l} \right] V_f + \sigma'_m(1 - V_f) \tag{13.11}$$

Since the theoretical (and in some cases measured) [3] strength of *whiskers* is about $E/10$, there is considerable interest in using them as fibers to strengthen metallic matrices. Although metal whiskers are strong, they are drastically weakened if mobile dislocations are introduced into them [22]. As there is a good chance of this occurring during composite fabrication, and since the whiskers can also be weakened if they react with a metal matrix, it seems unlikely that they will be used in practical materials. On the other hand, non-metallic materials such as carbides, borides, silicides, nitrides, some oxides, and graphite have high resistance to dislocation motion at most temperatures. The strength of these materials is primarily limited by surface flaws so that small, perfect whiskers are extremely strong. Means of producing large numbers of whiskers for some of these materials (i.e., Al_2O_3, Si_3N_4, SiC) have already been developed. Al_2O_3 is particularly attractive for high-temperature application since it does not oxidize or recrystallize and shows extremely high strength at elevated temperatures [15, 32] (Fig. 13.10). These whiskers are brittle right up to the melting point so that their strength (when perfect) is determined by their yield stress (i.e, $\sigma_f = \sigma_Y$). The decrease in yield strength with increasing temperature is attributed [32] to a thermally activated process of dislocation nucleation. Because thermally activated processes will occur at a lower stress if this stress is maintained for a longer period of time, delayed cleavage failures occur (Fig. 13.11) under static loading. This may pose a problem if the composites are to be used for long periods of time at elevated temperatures.

A second metallurgical problem associated with the use of ceramic whiskers, particularly oxides, is the lack of wetting between whisker and matrix material. When Al_2O_3 whiskers are dispersed in silver, for example, it is necessary first to coat the whiskers with nickel (e.g., by vapor deposition or drawing them through a molten bath) [15]. This also prevents direct contact of the fibers and reduces the danger of introducing flaws on their surfaces by abrasive action.

Glass [26], silica [13, 25], and silicon nitride [34] have also been used to strengthen metallic matrices. These fibers are attractive because long lengths of them can be produced rather easily. Furthermore, reasonably reproducible strengths can be obtained if they are coated or polished to remove surface damage. Reasonably high composite strengths can be maintained in SiO_2-Al composites up to 400°C (Fig. 13.8a), and delayed failure does not appear to be a serious

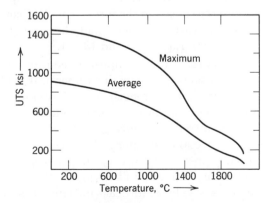

Fig. 13.10. Effect of temperature on the maximum and average strength of Al_2O_3 whiskers. *After S. Brenner* [32].

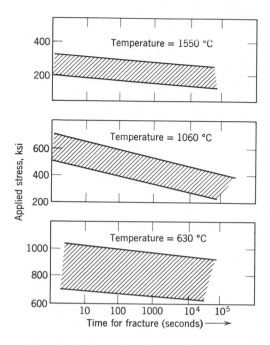

Fig. 13.11. The time for fracture of Al_2O_3 whiskers as a function of applied stress, at the indicated elevated temperatures. *After S. Brenner* [32].

problem in this temperature range [9]. Above this temperature, diffusion of aluminum into the silica and the formation of silicon and alumina cause the fiber surfaces to be damaged and the strength to decrease rapidly (i.e., δ decreases in Eqs. 13.10 and 13.11).

In considering the strength of composites containing nonmetallic fibers (e.g., ceramic whiskers, glass, or silica fibers), it is also necessary to account for variations in the strength of the individual fibers, which scatter about an average (or mean) value. These variations result from surface damage introduced during production of the fibers or from variations in their diameters along their lengths. In Eqs. 13.10 and 13.11 $\sigma_{u(f)}$ then represents an *average fiber strength* as determined from numerous tests on individual fibers [3, 15, 25]. In general, the l/d ratio of these fibers is so large (at least initially) that $\bar{\sigma}_{u(f)} = \delta\sigma_{u(f)}$. Values of δ range from 0.26 to 0.66 for glass or silica in aluminum and from 0.80–0.97 for Al_2O_3 in silver when these composites are tested at room temperature [15].

More exact treatments of composite failure based on a statistical distribution of fracture strengths of the fibers have been performed [35, 36, 37, 38]. These indicate that the number of breaks in the individual fibers increases with increasing applied load, and this has been observed [36] in fiberglass (Fig. 13.12). Composite failure occurs when a sufficient number of breaks have occurred in one cross section such that the remaining fibers cannot support the applied load. When these break, the overloaded matrix fails by combination of shear and interface fracture (pull out), depending on the relative shear strengths of the matrix and the interface.

Because of the difficulties encountered in growing large numbers of whiskers and introducing them into metallic matrices without causing serious damage, some attention has been focused on more direct methods of producing fiber composites. One promising approach is the use of unidirectionally solidified eutectic systems where alternating, parallel plates (or rods) of a matrix and intermetallic compound are produced directly from the melt [39]. Systems investigated thus far include Al-CuAl$_2$ [40], Al-Al$_3$Ni [40], Cu-Cr [41], and Sn-Cu$_6$Sn$_5$ [42]. The strong interface bond in these systems causes tensile

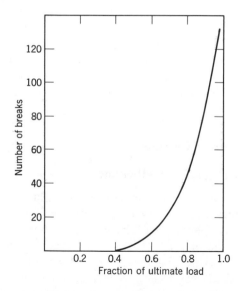

Fig. 13.12. The number of breaks in glass fibers, imbedded in epoxy resin, at indicated fractions of the ultimate load of the composites. *After B. Rosen* [36].

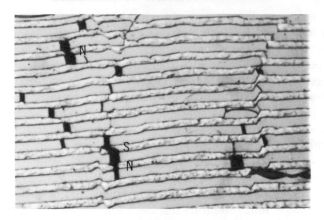

Fig. 13.13. Necking and shear of Al platelets after fracture of CuAl₂ platelets. CuAl₂-Al unidirectionally solidified eutectic system. 1000×. *Courtesy R. Hertzberg et al [40] and AIME.*

failure to occur by matrix shear after sufficient numbers of fiber breaks have occurred (Fig. 13.13). One interesting observation made on these systems is that the mechanical properties are dependent on fiber orientation. When $CuAl_2$ platelets are oriented at 45° to the tensile axis, failure occurs primarily by matrix (Al) shear parallel to the plates rather than platelet cracking. Both the ductility and toughness are increased as compared with the usual case when they are parallel to the tensile axis ($\phi = 0$) but the composite strength is reduced.

13.2   The Fracture of Cermets

Mechanics of aggregate strengthening

Most cermets are prepared by powder metallurgy methods [4, 11] wherein a mixture of powders of a metal matrix (the binder) and ceramic (the hard phase) are sintered at a temperature above the melting point of the matrix. Ideally the liquid completely wets each of the particles so that the resulting microstructure is composed of individual spherical, ceramic particles surrounded by a continuous thin film of matrix (Fig. 13.1c). As in the case of fiber composites, the mechanical strength is determined primarily by the strength of the hard phase. The ductile matrix helps arrest any cracks that form in the particles (i.e., to remove a continuous brittle path through the structure) to protect the particles from surface damage and to transfer stress to them.

The important microstructural parameters that determine the strength and fracture behavior of these composites are the volume fraction of hard phase V_f; the diameter of the particles of hard phase d; and the mean free path through the matrix λ. In general, λ increases as d increases and as V_f decreases. In addition, when λ is very small (i.e., large V_f, small d) there is certain probability that some of the hard particles will come in contact with one another. In this case the degree of particle contiguity C, the average fraction of surface area that one hard particle shares with adjacent particles, also influences the mechanical properties. Techniques for obtaining these parameters by quantitative metallography are well documented in the literature [43, 44, 45] and need not be repeated here.

The tungsten carbide-cobalt system, which is used extensively as a cutting tool material and rock drill bit, has been investigated more thoroughly than any other cermet. Since many of the principles of the fracture behavior of cermets occur in this particular system, it also serves as a good model material for understanding the fracture behavior of cermets in general.

Figure 13.14 summarizes the effect of composition on the mechanical properties of the WC-Co system when the particle size is maintained constant at about 2 μ [46, 47]. It is noted that both the impact strength and tensile elongation decrease and the hardness and elastic modulus increase with increasing volume fraction of hard phase. The elastic modulus does not vary linearly with composition, as predicted for isostrain behavior; the measured values fall in between that predicted for either isostrain or isostress behavior. Of primary interest is the fact that the fracture strength σ_c increases with increasing amounts of WC up to 63% and then decreases with further additions of WC.

As in the case of other alloy systems discussed in preceding chapters, the effect of composition on strength is partially indirect since composition affects the microstructure which, in turn, influences the mechanical properties.† In Fig. 13.15 the variation of σ_c with mean free path λ is plotted [48, 49] for a variety of WC contents V_f. It is apparent that for any value of V_f the strength increases up to a maximum value as λ decreases and then decreases with further decreases in λ. Both the maximum strength attained in an alloy of given composition,

† For example, small carbon (pearlite) additions raise the yield strength of mild steel by refining the ferrite grain size rather than by actually interfering with dislocation motion in the ferrite to any significant degree (Chapter 10).

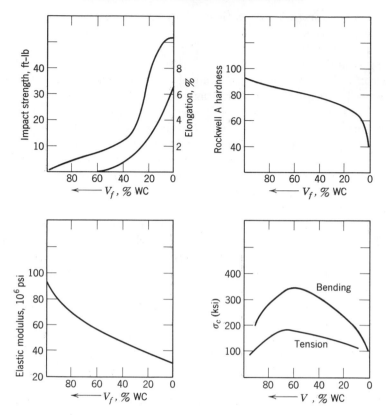

Fig. 13.14. Mechanical properties of WC-Co composites as a function of the volume fraction of WC particles V_f having approximately constant size $d = 2\ \mu$. *After J. Gurland et al* [46, 47].

$\sigma_{c(\max)}$, and the values of σ_c at all values of λ decrease with increasing amounts of hard phase V_f. On this basis the increase in strength observed with increasing V_f (Fig. 13.14) at constant d results from the fact that λ is decreasing with V_f, so that the data of Fig. 13.14 actually lie along the dashed line AB in Fig. 13.15.

The flow strength of the matrix is much lower than that of the hard phase so that plastic deformation begins in the matrix while the hard phase is still elastic [47]. Since the matrix is surrounded by essentially nondeforming particles, transverse tensile stresses are set up in the matrix which, in turn, require that greater nominal stresses be applied to continue plastic deformation [47, 50]. Consequently the potential load-carrying capacity of the matrix is increased by a plastic constraint which is similar to that which occurs in thin brazed joints or in notched cross sections of ductile materials (see Chapter 3) under tensile loading. The magnitude of the constraint increases with (d/λ) and hence with increasing V_f. Recent estimates [50] give the plastic constraint factor at 1.05 when $V_f = 0.30$ and 5.92 when $V_f = 0.90$.

The fracture mode depends on the relative values of carbide strength and on V_f. Three cases are of interest. (1) When V_f is small (e.g., 0.10) the constraint is small and the stress level in the matrix and hence in the particles is too low to cause the particles to crack. Fracture initiates by separation along particle-matrix interfaces [47, 51] and propagates by the standard process of normal rupture (void formation and coalescence) that occurs in low strength, ductile materials (Fig. 13.16a). (2) At larger values of V_f and with λ relatively large, the constraint factor is able to raise the stress level sufficiently to cause some of the carbides to crack after a small amount of matrix deformation. Fracture results from the linking of cracked carbides

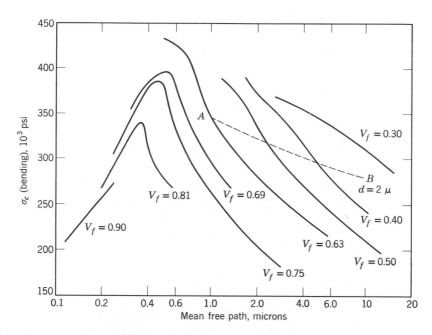

Fig. 13.15. The fracture strength in bending of WC-Co composites, as a function of mean free path through the cobalt matrix λ, at the indicated value fractions V_f of WC. The dashed curve AB is obtained from Fig. 13.14 and is a plot of σ_f versus λ at constant d. *After J. Gurland et al [48, 49].*

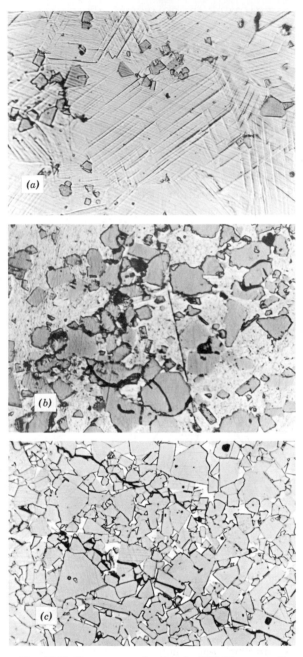

Fig. 13.16. Fracture mechanism in WC-Co composites at various volume fractions of WC. (a) $V_f = 0.10$. (b) $V_f = 0.50$. (c) $V_f = 0.90$. 1050×. *Courtesy C. Nishimatsu and J. Gurland* [47] *and ASM.*

by regions of matrix rupture [47] (Fig. 13.16b).   (3) Finally, at large V_f (i.e., small λ) the large degree of constraint prevents appreciable matrix deformation before the carbides crack.   Fracture takes place almost exclusively by the linking of cracked carbide grains that are contiguous with one another [47].

In general, case 2 describes fracture behavior in the range of mean free paths where σ_c increases with decreasing λ (Fig. 13.15), while case 3 describes fracture behavior when σ_c decreases with λ and contiguity is an important factor.   It is convenient to discuss these cases separately, starting with case 2.   Experimental observations of fracture in the TiC-inconel [52] (Fig. 13.17) and Al-Si systems [53], as well as on WC-Co [51], indicate that the probability of fracturing a particle at a given stress level increases as the size of the particle increases.   This is reminiscent of the behavior noted for ceramic fibers (Fig. 13.7).   If the strength of the composite is determined by the strength of the hard particles, with failure occurring when a sufficient number of particles have broken such that the remainder cannot

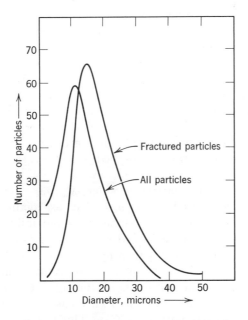

Fig. 13.17. Distribution of TiC particles and cracked TiC particles as a function of particle size in a 50% TiC-inconel composite strained 0.3%. *After J. Low* [52].

support the increased stress without themselves breaking, then the strength of the composite should increase as the particle size decreases. It has been shown [49, 50] that σ_c does vary directly with $d^{-1/2}$ when fracture data (case 2) from all compositions ($V_f = 0.3 - 0.75$ in WC-Co) are included on the plot. It has also been suggested [50] that a Griffith-type relation determines σ_c and that particle size *alone* is important in determining the strength level of the composites.

In view of the fact that the number of cracked particles increases with applied stress [47, 52, 53] and that fracture occurs when these cracks link together by matrix rupture [47, 52], this appears to be an oversimplified view of a statistical problem. Instead, the experimental observations [47, 53] suggest that fracture of a particle is initiated by inhomogeneous slip in the adjacent matrix. Consequently the length of a slip line (which is proportional to λ) will control the number of dislocations n piled up against the particle, hence the stress concentration factor and the probability of crack nucleation [48]. For $d = $ const, an increase in V_f implies a decrease in λ, hence a decrease in n, and thus a decrease in the probability of crack nucleation. This accounts for the behavior in Fig. 13.14, replotted as curve AB in Fig. 13.15. Recent calculations have also shown [54] that when a group of n dislocations are piled up against a large, hard particle, the stress concentration of the pile up decreases as the size of the particle decreases. This partially accounts for the increase in strength that occurs (at $\lambda = $ const) when V_f is decreased (Fig. 13.15), since a decrease in V_f implies a decrease in d at constant λ. Finally the increase in strength that occurs at $V_f = $ const when λ is decreased results from both of these effects: (i) the smaller number of dislocations in pile ups and (ii) the lower strength of a given pile up when d is decreased; the latter, of course, is implied in a decrease in λ at constant V_f.

When λ decreases to a particular value, λ_{max}, $\sigma_c = \sigma_{c(max)}$. Further decreases in λ at constant V_f cause σ_c to decrease again. In this range of mean free paths (case 3) the particles are partially contiguous and fracture occurs when a critical-sized flaw (assumed to be a weak particle-particle boundary) begins to propagate as a Griffith crack [48]. The effect of V_f, λ, and d on σ_c in this case results from the fact that σ_c is a function of the degree of separation $(1 - C)$; this parameter, in turn, increases as λ increases and V_f decreases. Figure 13.18 shows that σ_c increases linearly with $(1 - C)$. This has been explained [48] on the grounds that the initial length of crack is proportional to $d/(1 - C)$ while the effective surface energy is proportional to $d(1 - C)$, so that $\sigma_c \propto \{\gamma/c\}^{1/2} \propto (1 - C)$. Although this hypothesis remains

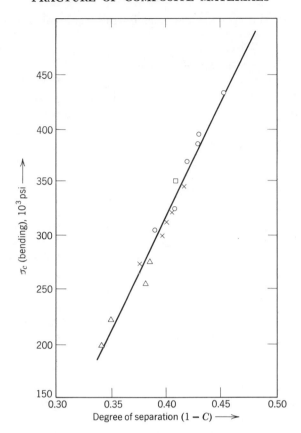

Fig. 13.18. Effect of the degree of separation between WC particles on the facture strength of WC-Co composites at small values of mean free path. *After J. Gurland* [48].

to be proved, there is little doubt that the degree of contiguity is the most important factor in controlling the strength of the composites when V_f is very large and λ is small.

Metallurgical aspects of the fracture of cermets

The preceding discussion has emphasized the fact that high fracture strengths can only be achieved when the particle size, mean free path, and degree of contiguity are all small and the composite is as homogeneous as possible. This requires [55] that the metal matrix completely wet the ceramic particles during liquid phase sintering so that the contact angle θ

$$\theta = \cos^{-1} \frac{\gamma_{SV} - \gamma_{LS}}{\gamma_{LV}}$$

between liquid matrix and solid ceramic be equal to zero. γ_{SV} is the surface energy of the solid, γ_{SL} is the surface tension of the liquid, and γ_{LS} is the solid-liquid interfacial energy. This indicates that wetting is favored when the surface energy of the liquid is as low as possible relative to the surface energy of the solid. Alternatively, when $\theta > 90°$, wetting is not favored (e.g., Ni-MgO system) and the matrix will sweat out of the composite during sintering, producing a porous, brittle structure [55]. When $90° > \theta > 0°$, the systems exhibit partial wettability and particle coalescence, and growth occurs during sintering [51, 55] (e.g., WC-Cu, TiC-Ni, TiC-inconel). In these cases the strengths are relatively low for reasons outlined above and fracture occurs transgranularly through the large ceramic particles or along the boundary between contiguous particles. However, when small amounts of molybdenum are added to the TiC-Ni system, the surface energy of the liquid is decreased and complete wetting occurs [55], as in the WC-Co system. This prevents the growth of TiC grains during sintering (i.e., $d = 0.5 - 1.0\mu$) and causes the particles to be so strong that fracture occurs exclusively through the matrix (case 1) even when $V_f = 0.60$ [51]. Besides the use of alloying additions to promote wetting, changes in sintering conditions can also produce beneficial results. For example, most oxides are not completely wet by liquid metals. However, when the Cr-Al$_2$O$_3$ system is sintered in an oxidizing atmosphere, chromic oxide films are produced which form solid solutions with Al$_2$O$_3$, and wettability and mechanical strength are increased [56].

Finally it must be appreciated that even the strongest cermets (small d, high wettability) exhibit minimal tensile and bend ductility and hence low impact strength and thermal shock resistance, especially when notches are present. This has tended to limit their reliability and use in structural applications, such as for turbine blade materials. It has been suggested [51] that one possible way of improving these properties is by preparing laminated structures of alternating layers of cermet and a ductile material such as nickel. It is interesting to note that the impact strength is increased from 10 to 70 ft-lb when the volume fraction of cermet (TiC-Ni) is decreased from 75 to 37% in these composites. It is very possible that the development of composites, one phase of which is itself a composite, will lead to the more extensive use of cermet materials in structural applications.

13.3. Summary

1. Composite materials are those in which fibers or particles of a hard, brittle material are dispersed in a softer and tougher matrix. The functions of the matrix are: (1) to blunt out any cracks that form in the hard phase, thereby removing a continuous brittle path in the structure, (2) to bind the particles of hard phase together and protect their surfaces from flaws, and (3) to transfer stress to the hard phase, which typically carries most of the load.

2. The ultimate strength of composites containing continuous fibers increases as the strength $\sigma_{u(f)}$ and volume fraction of the fibers V_f increase, for V_f greater than about 0.10.

3. Most practical fiber composite systems contain discontinuous fibers. In these systems the stress carried by the fiber builds up from zero at the end of the fiber to the fracture stress $\sigma_{u(f)}$ when the fiber has a length $l = l_c$. For $l > l_c$, the mean stress carried by a fiber increases from $\sigma_{u(f)}/2$ to $\sigma_{u(f)}$ (continuous fiber). Consequently the strength of a composite increases as l/l_c increases. $l_c = \sigma_{u(f)}d/2\tau$, where τ is the maximum shear stress that can be set up in the matrix or at the fiber-matrix interface.

4. When $l < l_c$, the stress in the fiber cannot build up to the fracture stress and failure occurs by fiber pull out (splitting). The work done in fracture $G_{Ic} = \frac{1}{12} V_f \sigma_{u(f)}l_c$ so that maximum toughness is achieved when l is slightly less than l_c and l_c and $\sigma_{u(f)}$ are as large as possible.

5. Fiber composites appear especially promising for use as high temperature, creep-resistant materials. The biggest difficulty is one of mass producing long lengths of fibers (e.g., whiskers) that have high temperature strength, without reducing fiber strength during manufacture. One potential means of reducing the manufacturing problems is by the use of unidirectionally solidified eutectic systems.

6. Composites containing hard spherical particles dispersed in a softer matrix are known as aggregates. High strength and hardness result from the plastic constraint, set up by the hard particles, which inhibits the deformation of the matrix. Except at very low volume fractions of hard phase, the strength of the composite is determined by strength of the particles or the degree of contiguity between them.

7. Decreasing the size of the particles, and the mean free path between them, increases the particle strength and hence the fracture strength of most aggregates. At very low values of mean free path (i.e., at high volume fractions of hard phase) the degree of contiguity increases, and hence the fracture strength decreases, as the mean free path decreases. Thus there is a particular value of mean free path at which the strength of an aggregate of given composition is a maximum.

8. High strength composites, in which the particle size, mean free path, and degree of contiguity are all small, can only be produced in systems in which the metal matrix completely wets the particles during liquid phase sintering.

References

The asterisk indicates that published work is recommended for extensive or broad treatment.

*[1] J. J. Gilman, *Mech. Eng.* (September 1961), p. 55.

[2] E. Orowan, *Repts. Prog. Phys.,* **12**, 214 (1948).

*[3] A. Kelly and G. J. Davies, *Met. Rev.,* **10**, 2 (1965).

[4] *Cermets,* J. R. Tinklepaugh and W. B. Crandal, eds., Reinhold, New York (1960).

*[5] *Discussion on New Materials, Proc. Roy. Soc.,* **282A** (1964).

*[6] *Fiber Composite Materials,* ASM, Cleveland (1965).

*[7] *High Strength Materials,* V. F. Zackay, ed., Wiley, New York (1965).

[8] A. Kelly, *Proc. Roy. Soc.,* **282A**, 63 (1964).

[9] D. Cratchley, *Met. Rev.,* **10**, 79 (1965).

[10] A. Kelly and W. R. Tyson, *High Strength Materials,* Wiley, New York (1965), p. 578.

[11] A. G. Thomas, J. B. Huffadine and N. C. Moore, *Met. Rev.,* **8**, 461 (1963).

[12] D. L. McDanels, R. W. Jech and J. W. Weeton, *NASA Tech. Note D1881* (1963).

[13] D. Cratchley, *Powder Met.,* **11**, 59 (1963).

[14] H. Piehler, *ASRL Report TR 94–95* (1963).

[15] W. H. Sutton and J. Chorne, *Fiber Composite Materials,* ASM, Cleveland (1965), p. 173.

[16] J. O. Outwater, *Mod. Plastics,* **33**, 156 (1956).

*[17] A. H. Cottrell, *Proc. Roy. Soc.,* **282A**, 2 (1964).

[18] N. M. Parikh, *Fiber Composite Materials,* ASM, Cleveland (1965), p. 115.

[19] R. F. Regester, M. S. thesis, University of Pennsylvania (1963).

[20] W. H. Sutton, J. Chorne and A. Talento, Eighth Prog. Rept., Contract NOw 60-0465-d, U.S. Navy (1962).

[21] G. J. Davies, *Phil. Mag.,* **9**, 953 (1964).

[22] S. S. Brenner, *Growth and Perfection of Crystals,* Wiley, New York (1958), p. 157.

[23] D. M. Marsh, *Fracture of Solids,* Interscience, New York (1963), p. 119.

[24] W. H. Sutton, J. Chorne and A. Talento, 10th Prog. Report, source cited in [20], (1962).

[25] D. Cratchley and A. A. Baker, *Metallurgia,* **69**, 153 (1964).

[26] H. B. Whitehurst and H. B. Ailes, Fifth Quart. Report, Contract NOrd 15764, U.S. Navy (1956).

[27] E. H. Dix, *Symposium on Structures for Thermal Flight,* ASME Conf. 1956 (paper No. 56-AV-8).

[28] R. W. Jech, E. P. Weber and S. D. Schwope, *Reactive Metals,* Interscience, New York (1959), p. 109.

[29] K. Farrell and N. M. Parikh, I.I.T. TR (R1-B241), (1963).

[30] R. H. Baskey Clevite Corp. Final Report, Contract AF 33 (657)-7139.

[31] D. W. Petrasek and J. W. Weeton, *Trans. AIME,* **230,** 977 (1964).

[32] S. S. Brenner, *J. Appl. Phys.,* **33,** 33 (1962).

[33] S. S. Brenner, *Fiber Composite Materials,* ASM, Cleveland (1965), p. 11.

[34] N. J. Parratt, *Powder Met.,* **7,** 152 (1964).

[35] N. J. Parratt, *Rubber and Plastics Age,* **41,** 263 (1960).

[36] B. W. Rosen, *J. AIAA,* **2,** 1985 (1964).

[37] B. W. Rosen, *Fiber Composite Materials,* ASM, Cleveland (1965), p. 37.

[38] J. A. Kies, *Trans. Plastics Inst.,* **33,** 145 (1965).

[39] R. W. Hertzberg, *Fiber Composite Materials,* ASM Cleveland (1965), p. 77.

[40] R. W. Hertzberg, F. D. Lemkey and J. A. Ford, *Trans. AIME,* **233,** 342 (1965).

[41] R. W. Hertzberg, *Trans. ASM,* **57,** 434 (1964).

[42] G. J. Davies, *High Strength Materials,* Wiley, New York (1965), p. 603.

[43] C. S. Smith and L. Guttman, *Trans. AIME,* **197,** 81 (1953).

[44] R. L. Fullman, *Trans. AIME,* **197,** 447 (1953).

[45] J. Gurland, *Trans. AIME,* **212,** 452 (1958).

[46] P. B. Bardzil and J. Gurland, *Trans. AIME,* **203,** 311 (1955).

*[47] C. Nishmutsu and J. Gurland, *Trans. ASM,* **52,** 469 (1960).

[48] J. Gurland, *Trans. AIME,* **227,** 1146 (1963).

[49] K. N. Tu and J. Gurland, TR 12, U.S. A.E.C. Contract A7(30-1)-2394 (1965).

[50] D. C. Drucker, *High Strength Materials,* Wiley, New York (1965), p. 795.

[51] N. M. Parikh, *High Temperature Materials,* Wiley, New York (1959), p. 169.

[52] J. R. Low, Jr., *Trans. AIME, J. Metals* (August 1956).

[53] A. Gangulee and J. Gurland, TR 8, U.S. A.E.C. Contract A7(30-1)-2394 (1964).

[54] D. Barnett and A. S. Tetelman, to be published.

*[55] N. M. Parikh and M. Humenick, Jr., *High Strength Materials,* Wiley, New York (1965), p. 155.

[56] C. A. Hauck, J. C. Donlevy and T. S. Shevlin, WADC TR (54-173), (1956).

[57] T. S. Shevlin, *News in Engineering,* **31,** 17 (1959).

Appendixes

Table of Notation

A.1 Notation Used Throughout Text

Symbol	*Meaning*
a_0	Lattice parameter.
b	Burgers vector of a dislocation.
c	One-half the length of an internal crack or the length of crack exposed to the surface.
c_F	Crack half-length at onset of instability.
C_V	Charpy V notch impact energy.
d	One-half the grain size.
E	Elastic modulus.
G	Shear modulus.
$G_{\mathrm{I}}, G_{\mathrm{II}}, G_{\mathrm{III}}$	Crack extension forces for various modes of crack opening.
G_c	Crack-resistance force, toughness, work done in initiating unstable fracture at the tip of a flaw.
$G_{\mathrm{I}c}$	Plane-strain toughness associated with cleavage or normal (tensile) rupture.
$G_{c(45°)}$	Plane-stress toughness associated with shear rupture under unixial tensile loading.
h	One-half the crack height.
k	Yield strength in shear.
k_y, k'_y	Parameters that determine grain-size dependence of yield strength.
K_σ	Elastic-stress concentration factor of a flaw.
$K_{\sigma(p)}$	Plastic-stress concentration factor of a flaw.
$K_{\epsilon(p)}$	Plastic-strain concentration factor of a flaw.
$K_{\mathrm{I}}, K_{\mathrm{II}}, K_{\mathrm{III}}$	Stress intensity factor for various modes of crack opening.
K_c	Stress intensity factor required to cause brittle (unstable) fracture, fracture toughness.
$K_{\mathrm{I}c}$	Plane-strain fracture toughness.

Symbol	*Meaning*
K_{ISCC}	Lower limit of stress intensity factor required to cause brittle fracture by stress corrosion cracking.
M	Applied bending moment.
N_0	Density of grown-in dislocations.
N_{tot}	Total dislocation density.
P	Applied load.
P_{GY}	Load required to cause general (complete) yielding across a member; fully plastic load.
r	Coordinate of a point near the crack tip; usually a point along the x axis, directly in front of the crack.
R	Plastic zone size.
R^*	Plastic zone size required to initiate unstable fracture near a flaw.
R_β	Value of R at which maximum possible degree of triaxiality is set up ahead of a particular flaw.
t	Thickness.
t	Time.
T	Temperature.
T_B	Propagation-transition temperature in (initially) flaw-free materials.
$T_{B(N)}$	Propagation-transition temperature in structures containing flaw.
T_D	Brittleness-transition temperature in (initially) flaw-free materials. The highest temperature at which $\sigma_f = \sigma_Y$.
$T_{D(N)}$	Brittleness-transition temperature in structure containing flaw $\sigma_F = \sigma_{GY}$.
T_M	Absolute melting temperature (°K or °R).
T_R	Crack-arrest temperature.
T_S	Fracture initiation mode transition temperature in (initially) flaw-free materials.
$T_{S(N)}$	Fracture initiation mode transition temperature in structure containing a flaw.
T_t	Highest temperature at which yielding occurs by twinning.
T^*	Ductility-transition temperature in (initially) flaw-free materials.
$T^*_{(N)}$	Ductility-transition temperature in structure containing a flaw.
v_c	Crack velocity.
v_0	Velocity of sound.
V	Dislocation velocity.

Symbol	*Meaning*
V_f	Volume fraction of dispersed particles or fibers.
$V(c)$	One-half the crack-tip displacement.
$V^*(c)$	One-half the crack-tip displacement at instability.
W	Plate width.
x, y, z	Coordinate axes.
Y	Tensile yield strength.
$\dfrac{1}{\beta}$	Maximum value of plastic stress concentration factor, due to plastic constraint ahead of crack or notch.
ϵ	True tensile strain.
ϵ_E	Engineering strain.
ϵ_F	Fracture strain of large structure containing a flaw.
ϵ_L	Luder's strain.
ϵ_u	Fracture strain, also, cleavage fracture strain.
ϵ'_f	Fracture strain when fracture is specifically initiated by rupture.
$\epsilon(c)$	Tensile strain at crack tip.
$\epsilon_f(c)$	Tensile strain at crack tip required for unstable (brittle) crack propagation.
$\epsilon_{f(R_\beta)}$	Tensile strain required to initiate cleavage ahead of a notch or crack, at $r = R_\beta$.
$\dot{\epsilon}_A$	Applied strain rate.
$\dot{\epsilon}_e$	Elastic strain rate.
$\dot{\epsilon}_p$	Plastic strain rate.
ϵ_p	Plastic strain.
γ_s	True surface energy.
γ_{GB}	Grain boundary energy.
γ_P	Plastic work done near tip of moving micro- or macro-crack. Usually the same as the total work done.
$\gamma_P{}^*$	One-half the work done in initiating unstable fracture at the tip of a flaw. $\gamma_P{}^* = G_c/2$.
γ_m	Work done near the tip of a cleavage crack that is propagating within a single crystal or grain in a polycrystalline aggregate. Usually applies only to microcracks.
γ_B	Work done near crack tip when crack crosses grain boundary
ν	Poisson's ratio.
$\Phi(c)$	Shear displacement at tip of crack loaded in antiplane strain.
$\Phi^*(c)$	Critical shear displacement required to initiate unstable fracture in antiplane strain.
ρ	Root radius.

Symbol	*Meaning*
ρ_{eff}	Limiting effective value of root radius.
σ	Applied (nominal) tensile or compressive stress, based on gross cross-section area.
σ_{net}	Net section stress.
σ_E	Engineering stress.
$\sigma_{xx}, \sigma_{yy}, \sigma_{zz}$	Normal stress components, usually used to describe stresses at or near crack tip.
σ_{yy}	Longitudinal (tensile) stress near crack tip.
σ_{yy}^{max}	Maximum tensile stress level near crack tip.
σ_Y	Yield strength in uniaxial tension or compression.
σ_u	Ultimate tensile strength.
σ_f	Fracture stress of (initially) flaw-free tensile specimen, also, cleavage fracture stress.
σ'_f	Fracture stress when fracture is specifically initiated by rupture.
$\sigma_f{}^*$	Fracture strength of material near tip of macrocrack.
σ_F	Fracture stress of large structure containing a flaw.
σ_M	Stress required to keep moving crack in motion.
σ_L	Concentrated tensile stress at tip of piled-up group of dislocations.
σ_i	Friction stress–tensile stress required to cause yielding of a constrained single crystal.
$\Delta\sigma$	Amount that tensile flow stress is raised by strain hardening.
$\Delta\sigma_f$	Amount of strain hardening required to initiate brittle fracture.
σ_G	Tensile stress level required to cause unstable growth of a microcrack.
σ_B	Tensile stress level required to force microcleavage crack through grain boundary.
σ_p	Fracture strength of a particle or inclusion in a matrix.
τ	Nominal shear stress.
$\tau_{xy}, \tau_{xz}, \tau_{yz}$	Shear stress components.
τ_c	Theoretical shear strength.
τ_Y	Yield strength in shear ($\tau_Y = k$).
τ_f	Shear stress required to cause fracture of (initially) flaw-free specimen.
τ_0	Shear stress required to produce unit dislocation velocity.
τ_L	Concentrated shear stress at tip of piled-up group of dislocations.

Symbol	Meaning
τ'	Shear stress acting on leading dislocation of pile up.
τ^*	Critical stress at which yielding occurs near piled-up group of dislocations.
τ_i	Friction stress in shear.
$\Delta\tau$	Amount that flow stress in shear is raised by strain hardening.
$\Delta\tau_F$	Amount of strain hardening (in shear) required to initiate brittle fracture.
τ_N	Shear stress required to nucleate cleavage microcrack.
ω	Flank angle of notch or crack (radians).

A. Important Symbols Used in Particular Chapters

Symbol	Meaning	Chapter
a	Distance from center of crack $(x = 0)$ to elastic-plastic boundary.	2, 5
a	Depth of notched cross section: $a = W - c$.	7
d	Particle or fiber diameter.	13
D	Crack height.	5
H	Activation energy for creep process.	9
J	A constant for a particular notch and structural geometry.	7
K_E	Kinetic energy of a moving crack.	2
l	Gage length of tensile specimen.	1
l	Gage length of "hypothetical tensile specimen" near crack tip.	2
l	Distance of closest approach between particles.	4
l	Fiber length.	13
l_0	Void spacing.	5, 7
l_c	Fiber transfer length.	13
L	Slip-line length.	4, 6
L	Plastic constraint factor.	7
n	Strain-hardening exponent.	1, 5, 7
n	Dislocation velocity exponent.	4, 5
n	Number of piled-up dislocations.	4, 5, 6
p	Hydrostatic pressure.	1
p	Particle size.	7
P	Pressure inside crack.	5, 9
P_n	Pressure inside microcrack.	9
Q	Activation energy for diffusion.	9
R	Gas constant.	9
S	Relative amount of plane-strain fracture.	3

Symbol	Meaning	Chapter
S	Parameter that characterizes the magnitude of a thermal shock.	12
t	Particle thickness.	6
t_0	Largest thickness at which plate fracture is entirely shear rupture.	3
t_R	Time at which rupture occurs under static loading.	9
T_g	Glass—transition temperature.	12
T_n	Temperature above which necking precedes fracture in tensile test.	6
v	Activation volume.	9
w	Width of gage section of hypothetical tensile specimen near crack tip.	2
W_E	Elastic strain energy of cracked volume.	2
W_S	Surface energy of crack.	2
α	Geometrical factor in equation for stress intensity factor K.	2, 3, 7
α	Parameter that relates dislocation density and strain.	4
α	Measure of mean free path between boundaries that can stop microcleavage crack.	6
α	Parameter that relates strain and time during creep.	9
α	Linear coefficient of thermal expansion.	12
β, β_{Ic}	Toughness parameters that describe effect of plate thickness.	3
β	Parameter that relates dislocation density and strain.	4
β	Angular coordinate of point near crack tip.	5
β	Measure of the ratio of hydrostatic stress to shear stress.	6, 7
β	Parameter that relates strain and time during creep.	9
β	Measure of average fiber strength.	13
δ	Deflection.	7
δ	Stacking fault energy.	9
χ	Angle between tensile axis and pole of slip plane in single crystal.	4
λ	Angle between tensile axis and slip direction in single crystal.	4
λ	Mean free path between particles.	13
K	Thermal conductivity.	12
ϕ	Angle between cleavage planes in adjacent grains.	5
ϕ	Angle between tensile axis and pole of cleavage plane.	6

Symbol	Meaning	Chapter
ϕ	Fast neutron flux (neutrons per cm^2 sec with energy greater than 1 MEV).	9
σ_c	Theoretical cohesive strength.	2
σ_c	Composite strength.	13
σ_m	Maximum thermally induced stress.	12
$\sigma_{u(f)}$	Ultimate strength of fibers.	13
σ'_m	Strength of matrix at ultimate strength of fibers.	13
θ	Angle of misorientation across dislocation wall.	6
θ	Bend angle.	7
ψ	Angular coordinate of point near crack tip.	2

B

Useful Unit and Grain Size Conversion Factors

Load
1 pound (lb) = 454 grams (g)
= 2.2 kilograms (kg)
= 4.5×10^5 dynes.

Length
1 foot (ft) = 12 inches (in.)
= 30.48 centimeters (cm)
= 304.8 millimeters (mm)
= 3.048×10^5 microns (μ)
= 3.048×10^9 angstroms (A).

Stress
1 lb per in.^{2} (psi)
= 6.7×10^{-2} atmospheres (atm)
= 6.88×10^{-2} bar
= 7.03×10^{-4} kg per mm^2
= 6.9×10^4 dyne per cm^2
= 6.9×10^2 dyne per mm^2.

Stress Intensity
1 psi $\sqrt{\text{in.}}$ = 5.04 psi $\sqrt{\text{mm}}$
= 3.54×10^{-3} (kg per mm^2) $\sqrt{\text{mm}}$
= 1.1×10^5 (dynes per cm^2) $\sqrt{\text{cm}}$
= 3.48×10^3 (dynes per mm^2) $\sqrt{\text{mm}}$.

Energy
1 ft-lb = 0.138 kg meter
= 1.38×10^7 erg (dyne cm).

Energy per Unit Area
1 ft-lb per in.^{2} = 2.1×10^6 erg per cm^2
= 2.14×10^{-4} kg m per mm^2.

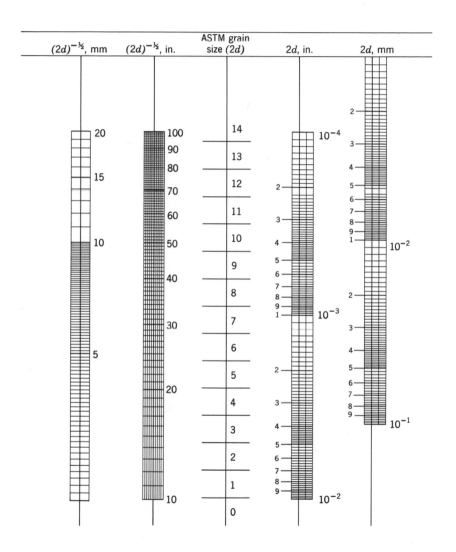

Author Index

The numbers in parentheses indicate the number of times an author's name appears on the indicated page, if more than once.

Subject Index